Lecture Notes in Computer Science 7289

Commenced Publication in 1973
Founding and Former Series Editors:
Gerhard Goos, Juris Hartmanis, and Jan van Leeuwen

Robert Bestak Lukas Kencl Li Erran Li
Joerg Widmer Hao Yin (Eds.)

NETWORKING 2012

11th International IFIP TC 6 Networking Conference
Prague, Czech Republic, May 21-25, 2012
Proceedings, Part I

 Springer

Volume Editors

Robert Bestak
Lukas Kencl
Czech Technical University in Prague
Department of Telecommunication Engineering
Technicka 2, 166 27 Prague 6, Czech Republic
E-mail: {robert.bestak, lukas.kencl}@fel.cvut.cz

Li Erran Li
Bell Labs, Alcatel Lucent
600 Mountain Avenue, Murray Hill, NJ 07974-0636, USA
E-mail: erranlli@research.bell-labs.com

Joerg Widmer
Institute IMDEA Networks
Avenida del Mar Mediterraneo 22, 28918 Leganes (Madrid), Spain
E-mail: joerg.widmer@imdea.org

Hao Yin
Tsinghua University, Tsinghua-ChinaCache Joint Laboratory
FIT 3-429, Haidian District, Beijing 100016, China
E-mail: h-yin@mail.tsinghua.edu.cn

ISSN 0302-9743   e-ISSN 1611-3349
ISBN 978-3-642-30044-8   ISBN 978-3-642-30045-5 (eBook)
DOI 10.1007/978-3-642-30045-5
Springer Heidelberg Dordrecht London New York

Library of Congress Control Number: 2012936974

CR Subject Classification (1998): C.2, H.4, D.2, K.6.5, D.4.6, H.3, E.3

LNCS Sublibrary: SL 5 – Computer Communication Networks and Telecommunications

Typesetting: Camera-ready by author, data conversion by Scientific Publishing Services, Chennai, India

Printed on acid-free paper

Springer is part of Springer Science+Business Media (www.springer.com)

Message from the General Chairs

It is our honor and pleasure to welcome you to the proceedings of the 2012 IFIP Networking Conference. This was the 11th edition of what is considered to be one of the best international conferences in computer communications and networks.

The objective of this edition of IFIP Networking was to attract innovative research works in the areas of network architecture, applications and services, wireless and sensor networks and network science. This goal was more than achieved and we would especially like to thank the Technical Program Committee Co-chairs, Joerg Widmer, Li Erran Li and Hao Yin, who efficiently organized the review of around 230 submissions and composed an outstanding technical program. The selected 64 high-quality papers were arranged into two parallel tracks.

The conference was also greatly honored by hosting the 40th IFIP TC6 Anniversary event. Some of the most prestigious networking scientists on the planet addressed the event: Vint Cerf (over video), Jon Crowcroft and Louis Pouzin, joined by the rising stars of the networking research community: Adam Dunkels, Dina Katabi and Pablo Rodriguez Rodriguez. We would like to express our gratitude to all of them for accepting the invitation; their presence was a privilege to all attendees.

This latest conference edition took place in Prague, Czech Republic, at one of the oldest technical universities in Central Europe, at the Czech Technical University in Prague (CTU). Many of the attendees also took time to enjoy Prague and visit its major cultural monuments.

All this would have not been possible without the hard and enthusiastic work of a number of people who contributed to making Networking 2012 a successful conference. We would like to thank all of them, from the Technical Program Committee Chairs and members, across the Publication and Workshop Chairs and the Organizing Committee to the authors and CTU staff that helped with the local matters. Great thanks also go to the Steering Committee of Networking, to all the members of the IFIP-TC6 for their support and especially to Gunnar Karlsson, Guy Leduc, Peter Radford and Otto Spaniol for helping organize the 40th IFIP TC6 Anniversary event.

Last but not least, we would very much like to encourage current and future authors to continue working in this exciting direction of research and participate in forums similar to this conference to promote explorative engineering and exchange of scientific knowledge and experience.

May 2012

Robert Bestak
Lukas Kencl

Technical Program Chairs' Message

It is a great pleasure to welcome you to the proceedings of Networking 2012. Networking 2012 was the 11th event of the series of International Conferences on Networking, sponsored by the IFIP Technical Committee on Communication System (TC 6), and this year hosted by the Czech Technical University in Prague.

We would like to thank all authors for the high number of submissions the conference received this year. The 225 submitted papers were distributed over the areas of network architecture (37%), network science (21.5%), applications and services (21%), and wireless and sensor networks (20.5%) and came from Europe, Middle East, Africa (58.3%), Asia/Pacific (26.4%), United States and Canada (12%), and Latin America (3.3%). With so many good papers to choose from, the task of the Technical Program Committee (TPC) of selecting the final technical program was not easy. The TPC was formed by 128 researchers from 25 different countries. In addition, 149 reviewers helped to provide further reviews. All papers underwent a thorough review process, with each paper receiving between three and five reviews as well as a meta-review that summed up the online discussion of the paper. After careful consideration, 64 papers were selected for the technical program, resulting in an overall acceptance rate of 28%.

We would like to express our thanks to the members of the TPC and the additional reviewers for all their hard work that made Networking 2012 possible. We would further like to thank the General Chairs, Robert Bestak and Lukas Kencl, for their support throughout the whole review process, and the Steering Committee Chair, Guy Leduc, for his invaluable advice and encouragement. Last not least, we would like to thank all participants for attending the conference.

Li Erran Li
Joerg Widmer
Hao Yin

Organization

Executive Committee

General Chairs

Robert Bestak   Czech Technical University in Prague,
Czech Republic
Lukas Kencl   Czech Technical University in Prague,
Czech Republic

Technical Program Chairs

Li Erran Li   Bell Laboratories, Alcatel-Lucent, USA
Joerg Widmer   Institute IMDEA Networks, Spain
Hao Yin   Tsinghua University, China

Publication Chairs

Josep Domenech   Universitat Politècnica de València, Spain
Jan Rudinsky   University of Iceland, Iceland

Finance Chairs

Hana Vyvodova   Czech Technical University in Prague,
Czech Republic
Petr Hofman   Czech Technical University in Prague,
Czech Republic

Publicity Chair

Michal Ficek   Czech Technical University in Prague,
Czech Republic

Workshop Chair

Zdenek Becvar   Czech Technical University in Prague,
Czech Republic

Local Arrangements Chairs

Mylada Balounova   Czech Technical University in Prague,
Czech Republic
Tomas Hegr   Czech Technical University in Prague,
Czech Republic

Webmaster

Marek Nevosad   Czech Technical University in Prague,
Czech Republic

Steering Committee

George Carle	TU Munich, Germany
Marco Conti	IIT-CNR, Pisa, Italy
Pedro Cuenca	Universidad de Castilla-la-Mancha, Spain
Guy Leduc	University of Liège, Belgium
Henning Schulzrinne	Columbia University, USA

Supporting and Sponsoring Organizations

Faculty of Electrical Engineering, Czech Technical University in Prague
IFIP TC 6
Microsoft Research

Technical Program Committee

Rui Aguiar	University of Aveiro, Portugal
Ozgur Akan	Koc University, Turkey
Khaldoun Al Agha	University of Paris XI, France
Kevin Almeroth	University of California, Santa Barbara, USA
Robert Bestak	Czech Technical University in Prague, Czech Republic
Jun Bi	Tsinghua University, China
Andrea Bianco	Politecnico di Torino, Italy
Chris Blondia	University of Antwerp, Belgium
Fernando Boavida	University of Coimbra, Portugal
Doreen Böhnstedt	Technische Universität Darmstadt, Germany
Olivier Bonaventure	Université Catholique de Louvain, Belgium
Azzedine Boukerche	University of Ottawa, Canada
Raouf Boutaba	University of Waterloo, Canada
Torsten Braun	University of Bern, Switzerland
Wojciech Burakowski	Warsaw University of Technology, Poland
Albert Cabellos-Aparicio	Universitat Politècnica de Catalunya, Spain
Eusebi Calle	University of Girona, Spain
Damiano Carra	University of Verona, Italy
Augusto Casaca	Instituto Superior Técnico in Lisbon, Portugal
Renato Lo Cigno	University of Trento, Italy
Marco Conti	IIT-CNR, Italy
Pedro Cuenca	University of Castilla la Mancha, Spain
Alan Davy	Waterford Institute of Technology, Ireland
Marcelo Dias de Amorim	UPMC Paris Universitas, France
Christian Doerr	Delft University of Technology, The Netherlands
Jordi Domingo-Pascual	Universitat Politècnica de Catalunya, Spain
Wolfgang Effelsberg	University of Mannheim, Germany
Gunes Ercal	University of California, Los Angeles, USA

Laura Feeney	Swedish Institute of Computer Science, Sweden
Wu-chi Feng	Portland State University, USA
Markus Fiedler	Blekinge Institute of Technology, Sweden
Pierre Francois	Institute IMDEA Networks, Spain
Luigi Fratta	Politecnico di Milano, Italy
Laura Galluccio	University of Catania, Italy
Silvia Giordano	University of Applied Science - SUPSI, Switzerland
Vera Goebel	University of Oslo, Norway
Sergey Gorinsky	Institute IMDEA Networks, Spain
Carmen Guerrero	University Carlos III of Madrid, Spain
Andrei Gurtov	Helsinki Institute for Information Technology, Finland
Guenter Haring	Universität Wien, Austria
Volker Hilt	Bell Labs/Alcatel-Lucent, USA
Markus Hofmann	Bell Laboratories, Alcatel-Lucent, USA
David Hutchison	Lancaster University, UK
Baek-Young Choi	University of Missouri, Kansas City, USA
Piotr Cholda	AGH University of Science and Technology, Poland
Xiaowen Chu	Hong Kong Baptist University, Hong Kong
Mohan Iyer	Oracle Corporation, USA
Hongbo Jiang	Huazhong University of Science and Technology, China
Carlos Juiz	Universitat de les Illes Balears, Spain
Martin Karsten	University of Waterloo, Canada
Andreas J. Kassler	Karstads Universitet, Sweden
Lukas Kencl	Czech Technical University in Prague, Czech Republic
Kimon Kontovasilis	NCSR Demokritos, Greece
Georgios Kormentzas	University of the Aegean, Greece
Yevgeni Koucheryavy	Tampere University of Technology, Finland
Ulas Kozat	DoCoMo-Labs, USA
Udo Krieger	Otto Friedrich University Bamberg, Germany
Fernando Kuipers	Delft University of Technology, The Netherlands
Thomas Kunz	Carleton University, Canada
Sung-Ju Lee	HP Labs, USA
Seungjoon Lee	AT&T Research, USA
Kenji Leibnitz	Osaka University, Japan
Qun Li	College of William and Mary, USA
Jorg Liebeherr	University of Toronto, Canada
Benyuan Liu	University of Massachusetts Lowell, USA
Bin Liu	Tsinghua University, China
Yong Liu	Polytechnic Institute of NYU, Italy
John Chi Shing Lui	Chinese University of Hong Kong, Hong Kong
Richard Ma	National University of Singapore, Singapore

Huijuan Wang	Delft University of Technology, The Netherlands
Michael Welzl	University of Oslo, Norway
Cedric Westphal	Docomo Labs, USA
Lars Wolf	Technische Universität Braunschweig, Germany
Tilman Wolf	University of Massachusetts, USA
Adam Wolisz	Technical University of Berlin, Germany
Chuan Wu	The University of Hong Kong, Hong Kong
Haiyong Xie	US Corporate Research, Huawei Technologies, USA
Guoliang Xue	Arizona State University, USA
Fan Ye	IBM T.J. Watson Research Center, USA
Sheng Zhong	University at Buffalo, USA
Martina Zitterbart	Karlsruhe Institute of Technology, Germany

Additional Reviewers

Luca Abeni	Lin Chen
Kaouther Abrougui	Richard Clegg
Reaz Ahmed	Florin Coras
Salvador Alcaraz	Tiago Cruz
Carlos Anastasiades	Felix Cuadrado
Emilio Ancillotti	Davide Cuda
Antonio Fernández Anta	Yong Cui
Spyridon Antonakopoulos	Emrecan Demirors
Sebastien Ardon	Nikos Dimitriou
Baris Atakan	Desislava Dimitrova
Jeroen Avonts	Michael Duelli
João Paulo Barraca	Roman Dunaytsev
Christian Bauer	Philipp Eittenberger
Marco Beccuti	Ozgur Ergul
Youghourta Benfattoum	Israel Martin Escalona
Steven Benno	Pedro M. Fonseca Marques Ferreira
A. Ozan Bicen	Alessandro Finamore
Robert Birke	Adriano Fiorese
Chiara Boldrini	Siddharth Gangadhar
Roksana Boreli	Fernando Pereñiguez García
Diego Borsetti	Andres Garcia-Saavedra
Manel Bourguiba	Katja Gilly
Ruud van de Bovenkamp	Beatriz Gomez
Radovan Bruncak	Zhangyu Guan
Raffaele Bruno	Yang Guo
Sonja Buchegger	Thomas Haenselmann
Valentin Burger	Hans Vatne Hansen
Jin Cao	Marco Happenhofer
Egemen Çetinkaya	Khaled Harfoush

Syed Hasan
Werner Henkel
Matthias Hirth
Michael Hoefling
Chengchen Hu
Philipp Hurni
Esa Hyytiä
Abdul Jabbar
Loránd Jakab
Michael Jarschel
Emmanouil Kafetzakis
Harsha Kalutarage
Merkourios Karaliopoulos
Naeem Khademi
Csaba Kiraly
Dominik Klein
Evangelia Kokolaki
Gerald Kunzmann
Rami Langar
Isaac Lera
Nanfang Li
Noura Limam
Morten Lindeberg
Leonardo Maccari
Anirban Mahanti
Lefteris Mamatas
Jose Marinho
Angelos Marnerides
Fabio Martignon
Alfons Martin
George Mastorakis
Christoph Mayer
Paulo Mendes
Gen Motoyoshi
Hassnaa Moustafa
Mu Mu
Maurizio Munafo'
Alberto Rodríguez Natal
Fawad Nazir
Konstantinos Oikonomou
Antonio de la Oliva
Panagiotis Pantazopoulos
Vasco Pereira
Henrik Petander
Pedro Vale Pinheiro
Maxim Podlesny

Stefan Podlipnig
Marc Portoles-Comeras
Daniele Puccinelli
Scott Pudlewski
Krishna Puttaswamy
Zhengrui Qin
Michal Ries
Justin Rohrer
Andreas Sackl
Pablo Salvador
Lambros Sarakis
Alberto Schaeffer-Filho
Christian Schwartz
Srini Seetharaman
Cigdem Sengul
Ghalib Shah
Gwendal Simon
Kathleen Spaey
Piotr Srebrny
George Stamoulis
Jan Stanek
Rafal Stankiewicz
Rade Stanojevic
Rocco Di Taranto
Nikolaos Thomos
Stefano Traverso
Alicia Triviño
Fani Tsapeli
Matteo Varvello
Francisco Velazquez
Erwin Van de Velde
Ivan Vidal
Michael Voorhaen
Florian Wamser
Wei Wei
Robert Wójcik
Jin Xiao
Xun Xiao
Yufeng Xin
Kaiqi Xiong
Ke Xu
Qi Zhang
Yifan Zhang
Alf Zugenmaier
Patrick Zwickl

Table of Contents – Part I

Content-Centric Networking

Efficient User-Assisted Content Distribution over Information-Centric
Network 1
    HyunYong Lee and Akihiro Nakao

On Inter-Domain Name Resolution for Information-Centric Networks ... 13
    Konstantinos V. Katsaros, Nikos Fotiou, Xenofon Vasilakos,
    Christopher N. Ververidis, Christos Tsilopoulos,
    George Xylomenos, and George C. Polyzos

Cache "Less for More" in Information-Centric Networks 27
    Wei Koong Chai, Diliang He, Ioannis Psaras, and George Pavlou

Collaborative Forwarding and Caching in Content Centric Networks 41
    Shuo Guo, Haiyong Xie, and Guangyu Shi

Social Networks

Crawling and Detecting Community Structure in Online Social
Networks Using Local Information 56
    Norbert Blenn, Christian Doerr, Bas Van Kester, and
    Piet Van Mieghem

Distributed Content Backup and Sharing Using Social Information 68
    Jin Jiang and Claudio E. Casetti

Trans-Social Networks for Distributed Processing 82
    Nuno Apolónia, Paulo Ferreira, and Luís Veiga

Context-Sensitive Sentiment Classification of Short Colloquial Text..... 97
    Norbert Blenn, Kassandra Charalampidou, and Christian Doerr

Reliability and Resilience

Resilience in Computer Network Management 109
    Marcelo F. Vasconcelos and Ronaldo M. Salles

An Experimental Study on the Impact of Network Segmentation to the
Resilience of Physical Processes 121
    Béla Genge and Christos Siaterlis

On the Vulnerability of Hardware Hash Tables to Sophisticated
Attacks 135
   Udi Ben-Porat, Anat Bremler-Barr, Hanoch Levy, and
   Bernhard Plattner

Degree and Principal Eigenvectors in Complex Networks 149
   Cong Li, Huijuan Wang, and Piet Van Mieghem

Virtualization and Cloud Services

Resilient Virtual Network Design for End-to-End Cloud Services 161
   Isil Burcu Barla, Dominic A. Schupke, and Georg Carle

Dynamic Scaling of Call-Stateful SIP Services in the Cloud 175
   Nico Janssens, Xueli An, Koen Daenen, and Claudio Forlivesi

Remedy: Network-Aware Steady State VM Management for Data
Centers 190
   Vijay Mann, Akanksha Gupta, Partha Dutta, Anilkumar Vishnoi,
   Parantapa Bhattacharya, Rishabh Poddar, and Aakash Iyer

Building a Flexible and Scalable Virtual Hardware Data Plane 205
   Junjie Liu, Yingke Xie, Gaogang Xie, Layong Luo, Fuxing Zhang,
   Xiaolong Wu, Qingsong Ning, and Hongtao Guan

IP Routing

Permutation Routing for Increased Robustness in IP Networks 217
   Hung Quoc Vo, Olav Lysne, and Amund Kvalbein

Routing On Demand: Toward the Energy-Aware Traffic Engineering
with OSPF 232
   Meng Shen, Hongying Liu, Ke Xu, Ning Wang, and Yifeng Zhong

Minimization of Network Power Consumption with Redundancy
Elimination 247
   Frédéric Giroire, Joanna Moulierac, Truong Khoa Phan, and
   Frédéric Roudaut

Sign What You Really Care about – Secure BGP AS Paths
Efficiently 259
   Yang Xiang, Zhiliang Wang, Jianping Wu, Xingang Shi, and Xia Yin

Network Measurement

Estimating Network Layer Subnet Characteristics via Statistical
Sampling 274
   M. Engin Tozal and Kamil Sarac

Sparsity without the Complexity: Loss Localisation Using Tree
Measurements 289
    Vijay Arya and Darryl Veitch

Efficient and Secure Decentralized Network Size Estimation 304
    Nathan Evans, Bartlomiej Polot, and Christian Grothoff

A Panoramic View of 3G Data/Control-Plane Traffic: Mobile Device
Perspective... 318
    Xiuqiang He, Patrick P.C. Lee, Lujia Pan, Cheng He, and
    John C.S. Lui

Network Mapping

Towards a Robust Framework of Network Coordinate Systems 331
    Linpeng Tang, Zhiyong Shen, Qunyang Lin, and Junqing Xie

BSense: A Flexible and Open-Source Broadband Mapping
Framework 344
    Giacomo Bernardi, Damon Fenacci, Mahesh K. Marina, and
    Dimitrios P. Pezaros

Validity of Router Responses for IP Aliases Resolution 358
    Santiago Garcia-Jimenez, Eduardo Magaña, Mikel Izal, and
    Daniel Morató

Semantic Exploration of DNS...................................... 370
    Samuel Marchal, Jérôme François, Cynthia Wagner, and
    Thomas Engel

LISP and Multi-domain Routing

On the Dynamics of Locators in LISP 385
    Damien Saucez and Benoit Donnet

A Local Approach to Fast Failure Recovery of LISP Ingress Tunnel
Routers 397
    Damien Saucez, Juhoon Kim, Luigi Iannone,
    Olivier Bonaventure, and Clarence Filsfils

An Analytical Model for the LISP Cache Size 409
    Florin Coras, Albert Cabellos-Aparicio, and Jordi Domingo-Pascual

Path Computation in Multi-layer Multi-domain Networks 421
    Mohamed Lamine Lamali, Hélia Pouyllau, and Dominique Barth

Author Index 435

Table of Contents – Part II

Video Streaming

Quality Adaptation in P2P Video Streaming Based on Objective QoE
Metrics 1
    Julius Rückert, Osama Abboud, Thomas Zinner,
    Ralf Steinmetz, and David Hausheer

Playback Policies for Live and On-Demand P2P Video Streaming 15
    Fabio V. Hecht, Thomas Bocek, Flávio Roberto Santos, and
    Burkhard Stiller

SmoothCache: HTTP-Live Streaming Goes Peer-to-Peer 29
    Roberto Roverso, Sameh El-Ansary, and Seif Haridi

Leveraging Video Viewing Patterns for Optimal Content Placement 44
    K.-W. Hwang, D. Applegate, A. Archer, V. Gopalakrishnan, S. Lee,
    V. Misra, K.K. Ramakrishnan, and D.F. Swayne

Peer to Peer

Enhancing Traffic Locality in BitTorrent via Shared Trackers 59
    Haiyang Wang, Feng Wang, Jiangchuan Liu, and Ke Xu

A Task-Based Model for the Lifespan of Peer-to-Peer Swarms 71
    Yong Zhao, Zhibin Zhang, Ting He, Alex X. Liu, Li Guo, and
    Binxing Fang

Using Centrality Metrics to Predict Peer Cooperation in Live Streaming
Applications 84
    Glauber D. Gonçalves, Anna Guimarães, Alex Borges Vieira,
    Ítalo Cunha, and Jussara M. Almeida

Content Publishing and Downloading Practice in BitTorrent 97
    Seungbae Kim, Jinyoung Han, Taejoong Chung, Hyun-chul Kim,
    Ted "Taekyoung" Kwon, and Yanghee Choi

Interdomain

Towards a Statistical Characterization of the Interdomain Traffic
Matrix 111
    Jakub Mikians, Amogh Dhamdhere, Constantine Dovrolis,
    Pere Barlet-Ros, and Josep Solé-Pareta

Characterizing Inter-domain Rerouting after Japan Earthquake   124
    Yujing Liu, Xiapu Luo, Rocky K.C. Chang, and Jinshu Su

Measuring the Evolution of Internet Peering Agreements   136
    Amogh Dhamdhere, Himalatha Cherukuru,
    Constantine Dovrolis, and Kc Claffy

Obscure Giants: Detecting the Provider-Free ASes   149
    Syed Hasan and Sergey Gorinsky

Security

Detecting Stealthy Backdoors with Association Rule Mining   161
    Stefan Hommes, Radu State, and Thomas Engel

Security Adoption in Heterogeneous Networks: The Influence of
Cyber-Insurance Market   172
    Zichao Yang and John C.S. Lui

Secure Client Puzzles Based on Random Beacons   184
    Yves Igor Jerschow and Martin Mauve

Heterogeneous Secure Multi-Party Computation   198
    Mentari Djatmiko, Mathieu Cunche, Roksana Boreli, and
    Aruna Seneviratne

Cooperation and Collaboration

Competition in Access to Content   211
    Tania Jiménez, Yezekael Hayel, and Eitan Altman

Modelling the Tradeoffs in Overlay-ISP Cooperation   223
    Raul Landa, Eleni Mykoniati, Richard G. Clegg, David Griffin, and
    Miguel Rio

Reducing the History in Decentralized Interaction-Based Reputation
Systems   238
    Dimitra Gkorou, Tamás Vinkó, Nitin Chiluka, Johan Pouwelse, and
    Dick Epema

On the Problem of Revenue Sharing in Multi-domain Federations   252
    Isabel Amigo, Pablo Belzarena, and Sandrine Vaton

DTN and Wireless Sensor Networks

On the Impact of a Road-Side Infrastructure for a DTN Deployed on a
Public Transportation System 265
*Sabrina Gaito, Dario Maggiorini, Christian Quadri, and
Gian Paolo Rossi*

Estimating File-Spread in Delay Tolerant Networks under Two-Hop
Routing 277
*Arshad Ali, Eitan Altman, Tijani Chahed, Dieter Fiems,
Manoj Panda, and Lucile Sassatelli*

A Distributed Smart Application for Solar Powered WSNs 291
*T.V. Prabhakar, S.N. Akshay Uttama Nambi,
R. Venkatesha Prasad, S. Shilpa, K. Prakruthi, and
Ignas Niemegeers*

A Two-Layer Approach for Energy Efficiency in Mobile Location
Sensing Applications 304
Yi-Yin Chang, Cheng-Yu Lin, and Ling-Jyh Chen

Wireless Networks I

Maximizing Lifetime of Connected-Dominating-Set in Cognitive Radio
Networks 316
*Zhiyong Lin, Hai Liu, Xiaowen Chu, Yiu-Wing Leung, and
Ivan Stojmenovic*

Traffic-Aware Channel Assignment for Multi-radio Wireless Networks... 331
Ryan E. Irwin, Allen B. MacKenzie, and Luiz A. DaSilva

QoE Analysis of Media Streaming in Wireless Data Networks 343
*Yuedong Xu, Eitan Altman, Rachid El-Azouzi,
Salah Eddine Elayoubi, and Majed Haddad*

Competition between Wireless Service Providers Sharing a Radio
Resource 355
Patrick Maillé, Bruno Tuffin, and Jean-Marc Vigne

Wireless Networks II

Relay Placement for Two-Connectivity 366
Gruia Calinescu

A 2-Approximation Algorithm for Optimal Deployment of k Base
Stations in WSNs 378
Hui Wu and Sabbir Mahmud

A Semi-dynamic Evolutionary Power Control Game 392
   Majed Haddad, Eitan Altman, Julien Gaillard, and Dieter Fiems

Gossip-Based Counting in Dynamic Networks........................ 404
   Ruud van de Bovenkamp, Fernando Kuipers, and Piet Van Mieghem

Author Index 419

Efficient User-Assisted Content Distribution over Information-Centric Network

HyunYong Lee and Akihiro Nakao

The University of Tokyo, 4-6-1 Komaba, Meguro-ku, Tokyo 153-8904, Japan
ifjesus7@gmail.com, nakao@iii.u-tokyo.ac.jp

Abstract. To solve the fundamental limitations of current Internet in supporting today's content-oriented services, information-centric networking (ICN) concept has been proposed. ICN has attractive features (e.g., name-based routing, in-network caching and multicast) supporting efficient content-oriented services. However, the attractive features may not be fully utilized by all existing contents due to the resources limitation, which means additional technique may be required for improving content-oriented services. In this paper, as one possible way for this, we examine P2P technique exploiting user resources in ICN. We first examine how P2P looks like in ICN. Then, we introduce the contribution-aware ICN and corresponding incentive mechanism to utilize the user resources efficiently. We also show how the contribution-aware ICN can be implemented over the existing ICN architectures. Through simulations, we evaluate an effect of user participation on the content distribution performance in ICN. We also verify the feasibility of the contribution-aware ICN in terms of resources utilization efficiency.

Keywords: Information-centric networking, user-assisted content distribution, P2P, incentive mechanism.

1 Introduction

Current Internet (supporting communications between any pair of machines identified with an IP address) has fundamental limitations in supporting today's content-oriented services (caring about the content itself rather than the IP address of content source). To fundamentally address the mismatch between the current Internet and the today's content-oriented services, ICN [2]-[8] has been proposed as one architecture for future Internet. As a revolutionary approach, ICN replaces addressed machines with named contents in the network level and includes some add-on functions of the current Internet (e.g., multicast and caching) as native in-network functions. For example, in ICN, a user just needs to specify a content name that it wants to download. Then, ICN satisfies the user request with data from any source storing a copy of the content, enabling efficient caching as part of the network service.

Even though ICN supporting attractive features is well suited for efficient content-oriented services, all existing contents may not be able to fully utilize the

R. Bestak et al. (Eds.): NETWORKING 2012, Part I, LNCS 7289, pp. 1–12, 2012.

attractive features of ICN. For example, the in-network caching may be only useful for some popular contents due to the limited size of cache. Then, we may need additional approach for improving the content-oriented services in case of contents that cannot fully utilize the attractive features of ICN. As one possible way for this, in this paper, we focus on P2P technique exploiting user resources (i.e., user-assisted content distribution) with proper incentive mechanism. In particular, we argue that 1) P2P technique exploiting user resources will be still useful for content distribution in ICN as the case of current Internet and 2) existing ICN architecture needs to be extended to utilize the user resources efficiently.

To this end, we introduce the contribution-aware ICN (where the contribution-related information is added to both users and contents) to utilize the user resources efficiently in ICN. A contribution of each user in terms of content distribution is managed by the contribution-aware ICN. In addition, the content is published with a required contribution level that needs to be satisfied by a user who wants to download the content. Therefore, a request of user with enough contribution is only handled so as to encourage the users to contribute their resources. Through simulations, we first show that the user participation has significant impact on the content distribution in ICN, especially when the content cannot fully benefit from the features of ICN (i.e., the in-network caching in our simulation). We also show that the contribution-aware ICN can utilize the user resources efficiently and improve the content distribution performance by using the explicit contribution-related information of users and contents.

This paper is organized as follows. In Section 2, we introduce main features of ICN and discuss its implications on challenging issues of current P2P content distribution. From this, we identify one important research issue in ICN: how to utilize the user resources efficiently. Then, we introduce the contribution-aware ICN in Section 3. We also discuss how the contribution-aware ICN can be implemented over existing ICN architectures. We discuss the incentive mechanism that can be used in the contribution-aware ICN in Section 4. After evaluating the feasibility of the contribution-aware ICN in Section 5, we conclude this paper in Section 6.

2   Contribution-Aware ICN

In this Section, we introduce the contribution-aware ICN architecture to incentivize the user-assisted content distribution in ICN. The user resources will be still useful in ICN, because all existing contents may not be able to fully utilize the attractive features of ICN. For example, the in-network caching is strictly on an opportunistic basis. A cached content can be deleted at any time based on replacement policy such as Least Recently Used (LRU), since the cache size is limited. Due to this reason, only some popular contents can be cached. The native multicast support may be only meaningful for real-time streaming where users are watching same part of content at the same time. Then, the contents that cannot leverage the attractive features of ICN will be provided through a client-server model that is not so efficient in distributing contents unless additional resources (e.g., dedicated servers of CDN and user resources of P2P in the

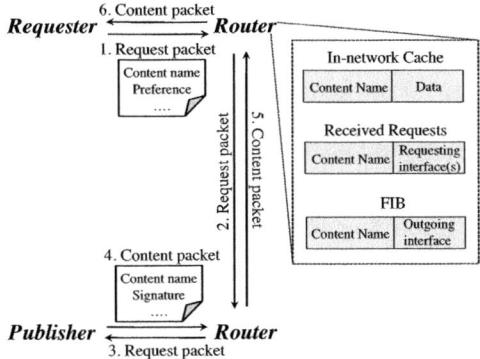

Fig. 1. Conceptual illustration of ICN

current Internet) are utilized. Therefore, we posit that the user resources can be utilized for improving the content distribution in ICN as P2P utilizes the user resources in the current Internet.

2.1 P2P Content Distribution in ICN

Before discussing the contribution-aware ICN trying to utilize the user resources in detail, we first identify one challenging research issue when the P2P technique exploiting the user resources is applied into ICN.

Features of ICN

Even though each existing ICN proposal has different details, there are common features characterizing ICN as follows (Fig. 1)[1]:

– Content has an *unique identifier* decoupled from its location. When a user wants to download a content, the user just needs to issue a content request including the content name instead of its location (e.g., URL of the current Internet). The location-independent content identifier enables efficient caching as the in-network service.
– *Name-based (anycast) routing* forwards the user request specifying the content name to the closest copy of the content with such a name based on the content identifier. In other words, in terms of routing, IP address of the current Internet is substituted by the content name in ICN.
– Router supports *native in-network caching*. The in-network caching may be able to ensure efficient network utilization and improve content availability.
– With *content-based security* [3] (i.e., protection and trust travel with the content, e.g., the content includes a digital signature of its publisher) or *self-certifying name* [6] (i.e., using the cryptographic digest of the content as its name), the user and even the router can examine an integrity of content.

---

[1] For the purpose of illustration, we use CCN [3] as a basic design example.

This feature is a critical enabler for the in-network caching. From now on, we use the content-based security as the representative term.

- *Native multicast* is supported in a network level. Router manages a list of received content requests for efficient forwarding (i.e., pending interest table (PIT) in [3]). When the router receives the requested content, it forwards the content to all requesters, so realizing multicast.

The Benefits of Employing P2P on ICN

Now, we discuss how the P2P looks like over ICN by examining potential implications of ICN features on the challenging issues of current P2P content distribution. *User-driven dynamics of user participation*, or *churn* (i.e., user's unexpected join and leave) [9] is one of the major obstacles for building stable P2P system, since it degrades P2P system performance by making content source unstable. This problem can be mitigated by the name-based routing and the in-network caching. ICN router routes the content request to the copy of content as long as the content exists. The cached contents at ICN router may improve the content availability.

Most P2P applications are network-oblivious and thus build an *underlying topology-unaware overlay network*. Network-oblivious user matching results in extensive inter-domain traffic, causing expensive operational cost to the network operators [10]. Furthermore, to reduce the inter-domain P2P traffic, the network operators often utilize various traffic shaping devices that can degrade P2P networking performance. The issues caused by the network-obliviousness can be solved by the name-based routing that is network-aware.

Open and anonymous nature of P2P (i.e., no control by central authority) attracts large number of users including *malicious users* [11]. Malicious users interfere with normal content distribution (e.g., by uploading inauthentic file). This problem can be mitigated with the name-based routing and the content-based security. A user may satisfy as long as the received content is authentic regardless of content source and transfer channel. Since ICN router can check an integrity of content based on the content-based security, the inauthentic file cannot be easily distributed by the malicious users.

Open and anonymous nature also introduces another type of non-honest user, *free rider* [12]. Free riders degrade the content distribution performance by only consuming the resources without contributing any resources. To encourage the users to contribute their resources, incentive mechanisms (e.g., tit-for-tat of Bit-Torrent [1]) are required. Unlike other challenging issues that can be mitigated by the ICN features, the free riding problem may get worse in ICN if there is no proper incentive mechanism. In the current Internet, each user determines its communication partners based on contribution information of neighboring users. Therefore, the user usually has to upload its content to download a content from other users. On the other hand, in ICN, ICN router matches a content publisher and a content requester based on the name-based routing protocol. In other words, a user can download a content regardless of its contribution by simply specifying the content name. Then, most users may try to exploit the name-based routing while not contributing their resources, which can lead to the tragedy of digital commons. Therefore, in this paper, we propose the contribution-aware ICN to utilize the user resources efficiently in ICN by mitigating the potential free riding problem of P2P over ICN.

2.2 Approach

We first discuss main features of the contribution-aware ICN. Then, we will show how the contribution-aware ICN can be implemented over existing ICN architectures. Unlike current P2P applications where each user usually manages contribution information of its neighboring users and determines whom to upload, the user cannot do the same job in ICN, since a matching between a publisher and a requester is done by ICN router. Because the users cannot manage the information of other users, even existence of other users, the management of user information for incentivizing user contribution needs to be done not by the user itself but by a network side. For this purpose, we propose followings (Fig. 2(b) and Fig. 3(b)).

Management of User Contribution. We introduce one additional entity, called content distribution manger (CDM) to manage contribution history of users in terms of content distribution. The contribution history can include content name, transfer time, publisher ID, and requester ID. Here, we assume that each user has unique ID. ICN router can generate the contribution history after checking the integrity of transferred content and report it to CDM. CDM generates a contribution level (CL) for each user according to corresponding contribution history when it receives a user request for CL. CL can include target content name, CL value, expiration time, and target user ID. CL shows how much each user contributes to the content distribution. For example, CL can be calculated as a ratio of uploaded content to downloaded content. The content publisher can encourage the user contribution by manipulating CL calculation. For example, CL can be generated based on only publisher ID to foster persistent contribution incentives by recognizing and rewarding contributions made by a user across various contents and time. CDM signs CL with its private key to prevent the users from generating false CL if it needs to return CL to users. The content provider can deploy CDM for their content distribution service. Each CDM has an unique identifier so that the user request for CL can be routed to appropriate CDM based on the name-based routing. Here, we assume that the user can download a metadata file including a name of corresponding CDM through a search engine before requesting the content.

Content Publication with Required CL. Basic approach for incentivizing user-assisted content distribution is to allow a user to download a content only when the user uploads a content. For this, the content publisher specifies a required contribution level (RCL) that needs to be satisfied by the requester's CL to download the content when it publishes the content.

Contribution-Aware Routing. ICN router compares the content name and CL of the requester with the name and corresponding RCL of routing table.

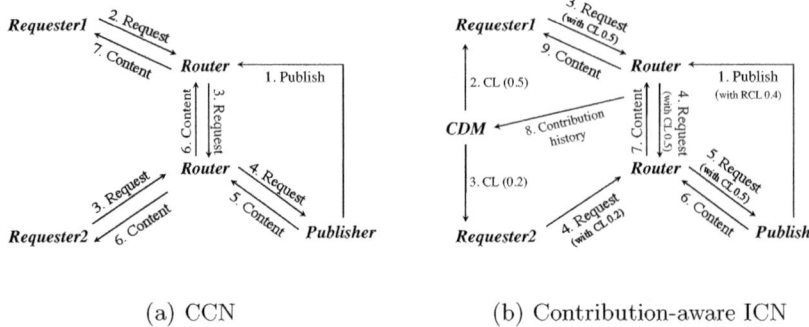

(a) CCN   (b) Contribution-aware ICN

Fig. 2. Contribution-aware ICN over CCN

If there is matched one, it forwards the user request. If there is no matched entry in the routing table (e.g., no entry for the requested content, lower CL than RCL, or an expiration time of CL is over), ICN router drops the user request or returns an error to the requester. For this, the routing table of ICN router needs to be extended to include RCL of published content.

2.3   Implementation of Contribution-Aware ICN

Here, we discuss how the contribution-aware ICN can be implemented over existing ICN architectures. For this purpose, we choose content centric networking (CCN) [3] and publish-subscribe internet routing paradigm (PSIRP) [5] as representative existing ICN architectures.

Contribution-Aware ICN over CCN. The main concept of CCN is to route a content request towards a location where the content has been published (Fig. 2(a)). Once an instance of content is located, it is delivered to the requester along the path the request came from. All the nodes along that path may cache the content in case they have more requests for it. The contribution-aware ICN can be implemented over CCN as follows (Fig. 2(b)). CDM can be deployed as a separate entity. When the requester wants to download a content, it first acquires its CL from CDM. CDM calculates CL of the requester according to a policy given by a content publisher and returns CL to the requester. Then, the requester issues a content request together with its CL. The content request is routed to a copy of content and the content is delivered to the requester if the requester's CL is higher than RCL of the requested content. When ICN router forwards the content to the requester, it generates the contribution history and reports it to CDM.

Contribution-aware ICN over PSIRP. In PSIRP, content is published into the network (Fig. 3(a)). Then, requesters can subscribe to the published content. The publication and the subscription are matched by a rendezvous system. The matching procedure returns a transport ID that can be used for routing of content through a forwarding network. In PSIRP, the contribution-aware ICN can

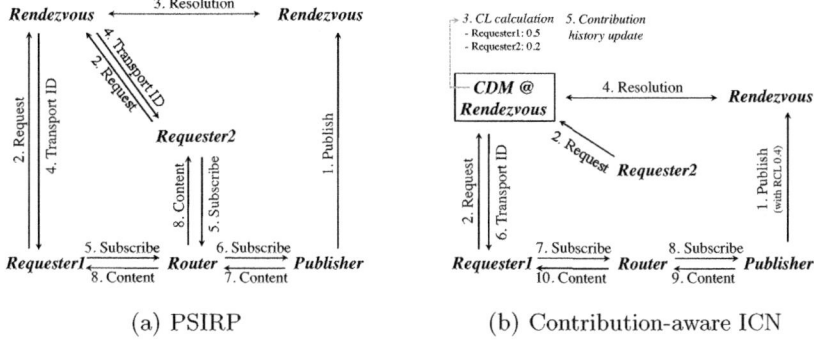

Fig. 3. Contribution-aware ICN over PSIRP

be implemented as follows (Fig. 3(b)). CDM can be deployed at the rendezvous system. When a requester wants to download a content, it sends a content request to the rendezvous system. Then, the rendezvous system first calculates CL of the requester. If CL of the requester is higher than RCL of the requested content, the rendezvous system matches the requester and the publisher. After processing the user request, the rendezvous system returns a transport ID to the requester who has enough CL. The rendezvous system also updates the contribution history of the matched requester and the publisher. A connection for content forwarding is setup based on the returned transport ID and the content is delivered to the requester.

3   Contribution-Aware Content Distribution

Now, we turn our attention to an incentive mechanism to encourage the user-assisted content distribution over the contribution-aware ICN. The incentive mechanism over the contribution-aware ICN can be achieved by CL generation policy and RCL assignment policy. Two policies can be managed by one entity (i.e., a content provider or a user) or different entities (e.g., a user can publish a content with its own RCL assignment policy through public CDM that has pre-defined policy for CL generation set by a content provider). According to a content distribution policy of the publisher, various policies for CL generation and RCL assignment are possible. In this paper, we introduce our choices for a file sharing as a starting point of our research while leaving other possible approaches for future work. In this paper, we assume that RCL of original content publisher is used for re-publication of downloaded content by users. We also assume that a content is divided into segments and each segment is named together with its content name (e.g., ContentName.SegmentID). The original content publisher publishes its content with RCL incrementally increasing. For example, in case of content consisting of 300 segments, first 100 segments have 0 RCL, next 100 segments have 0.1 RCL, and remaining 100 segments have 0.2 RCL. For bootstrapping of newly joining users, some segments are published with zero RCL (i.e., no CL requirement for download). The newly joining peers need some segments to upload, since they need to increase their CL by uploading the content. By downloading and uploading the segments with 0 RCL, the newly joining user may be able to increase its CL enough to download the segments with non-zero RCL. The incrementally increasing RCL may encourage the users to keep uploading the downloaded segments to increase CL enough to satisfy RCL of remaining segments requiring higher RCL than the downloaded segments.

CDM generates CL according to a pre-defined method when it receives a request from a user. As a basic form, in this paper including the evaluation part, CL is calculated as a ratio of the uploaded content to the downloaded content. CL is valid for 10 seconds. Due to some reasons, a user may do not have chance to upload its content even though they willingly want to contribute their resources. In this case, CDM generates *free pass* CL (FPCL) that can be used to download some segments without satisfying RCL. For example, when there is only one requester or when other users already have segments that the requester has, the requester cannot upload to anybody. When only one requester exists, FPCL for all segments is generated. When the requester does not have segments to upload, FPCL for rare segments is generated so that the requester can have a chance to upload the downloaded rare segments. To prevent the free riders from receiving FPCL from CDM, CDM generates FPCL only for users who publish every downloaded segment under an assumption that the user who publishes the content willingly uploads the content when it receives the content request. For this, in CCN case, the router needs to report the content publication by users to CDM. Please note that CDM can know when what user downloads what content based on the contribution history.

4   Evaluation

Using ns-2 [13], we evaluate (1) an effect of user participation on content download performance in ICN and (2) a feasibility of the contribution-aware ICN and the incentive mechanism in terms of resources utilization efficiency. We use PSIRP architecture as the existing ICN architecture for our simulation. We build the simulation topology by using transit-stub model. Our topology includes 1 transit and 10 stub networks connected to the transit domain. In our simulation, a performance bottleneck is user link capacity. For link capacity of users, we set 120KB/s and 40KB/s for downlink and uplink capacity. We use 1,000 users including one initial content publisher. 100MB-sized content is divided into 400 256KB-sized segments. To study an effect of different user types in terms of participation in the content distribution, we divide users into three types: free rider (no content upload), altruistic user (upload the downloaded segments regardless of its CL), and rational user (upload the downloaded segments to increase its CL just enough to download remaining segments). In addition, to study an effect of RCL assignment policy, we assign increasing RCL (0, 0.1, 0.2, and 0.3 for each 100 segments) or flat RCL (0 for 100 segments and 0.1 for remaining 300 segments) for the content. For comparison purpose, we conduct simulations with the existing ICN and the contribution-aware ICN. In the existing ICN, every

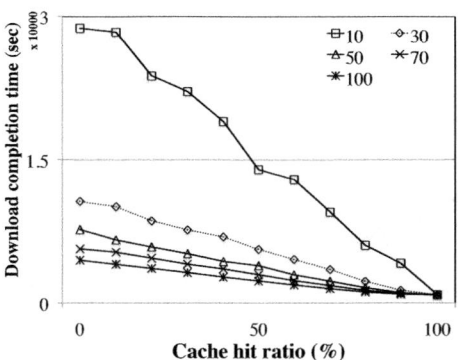

Fig. 4. Existing ICN without RCL

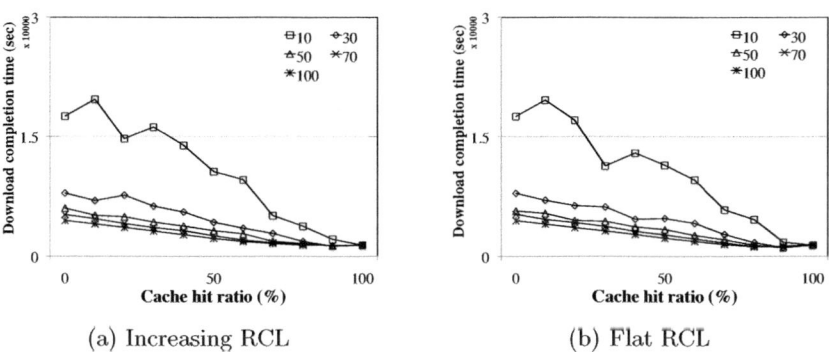

(a) Increasing RCL   (b) Flat RCL

Fig. 5. Contribution-aware ICN with altruistic users

user downloads the content regardless of its contribution, since there is no RCL. On the other hand, in the contribution-aware ICN, only user who satisfies RCL can download a content.

4.1 Effect of User Participation

Through simulations, we find that the user participation has noticeable effect on download performance in ICN (including the existing ICN and the contribution-aware ICN). In the existing ICN (Fig. 4) and the contribution-aware ICN (Fig. 5)[2], the download performance improves linearly as the cache hit ratio increases. On the other hand, the download performance improves significantly as a number of altruistic users increases. Even though the free riders download the content without corresponding contributions in the existing ICN,

---

[2] In the figures, numbers of figure legend indicate % of non-free riders among 1000 users. In the existing ICN, the non-free riders are altruistic users, since there is no RCL.

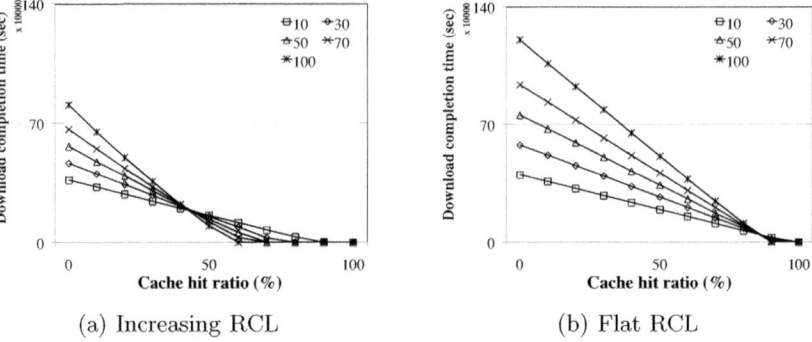

(a) Increasing RCL   (b) Flat RCL

Fig. 6. Contribution-aware ICN with rational users

more number of altruistic users leads to better performance by providing more upload capacity. Due to the same reason (except that the free riders can only download the segments with 0 RCL), in the contribution-aware ICN, the download performance improves as a number of altruistic users increases. This result shows that the user participation has significant impact on the content distribution performance, especially when the cache hit ratio is low. This also means that the download performance can be improved through the user participation even though the content is not popular enough to be cached by ICN router.

We have interesting observations from the contribution-aware ICN with rational users (Fig. 6(a)). When the cache hit ratio is low (less than 40% in our simulation), more number of rational users leads to worse content distribution performance. In case of the free riders, they cannot download segments with non-zeo RCL while freely downloading the segments with 0 RCL. On the other hand, the rational users can continue to download the segments as long as they have enough CL. However, the rational users stop their contributions when they have enough CL to download the segments with non-zero RCL. In other words, amount of upload capacity provided by the rational users is not guaranteed while total amount of required download capacity is fixed. Due to a shortage of upload capacity, some content requests of rational users who have enough CL cannot be handled rapidly and thus this leads to poor performance. However, if the cache hit ratio is high enough to complement the shortage of upload capacity, more number of rational users leads to better performance.

4.2   Effect of RCL

From the comparison between the existing ICN (Fig. 4) and the contribution-aware ICN with the altruistic users (Fig. 5), we find that the free riders degrade the content distribution performance by consuming system resources (i.e., upload capacity of publisher and non-free riders) without contribution. Please note that a comparison between the existing ICN and the contribution-aware ICN with rational users is meaningless, since the non-free riders in the existing ICN

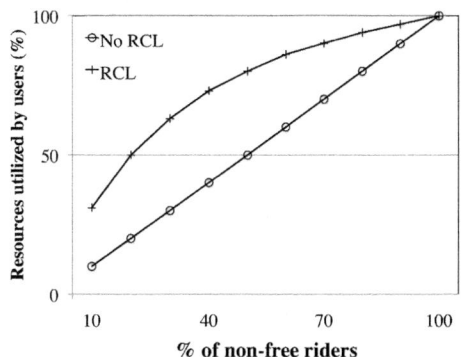

Fig. 7. % of resources utilized by non-free riders

are the altruistic users. In the existing ICN, the free riders and the non-free riders show similar download performance, since they download the content regardless of their upload contributions. This result shows that there should be proper incentive mechanism to prevent the free riders from exploiting the system resources for efficient utilization of system resources in ICN.

In the existing ICN, the portion of system resources utilized by the non-free riders increases linearly as a number of the non-free riders increases, since the free riders and the non-free riders compete with each other to download the content regardless of their contributions. On the other hand, in the contribution-aware ICN (regardless of types of non-free riders), more portion of system resources is utilized by the non free riders (Fig. 7). Remaining portion of the system resources are used by the free riders to download the segments with zero RCL. Thus, that portion decreases as a number of non-free riders increases. Even though the download performance of contribution-aware ICN with rational users is not as good as the case of the contribution-aware ICN with altruistic users due to the unwillingness of user participation in content distribution, most portion of the system resources is utilized by the non-free riders. Above results show that the contribution-aware ICN can utilize the system resources efficiently by preventing the free riders based on the contribution-related information (i.e., RCL, CL, and contribution-aware routing). It also means that the download performance of the contribution-aware ICN can be further improved with a proper incentive mechanism (i.e., RCL assignment and CL calculation policy). We will study this later as future work.

Now, we examine an effect of two RCL assignment policies on the download performance: increasing RCL and flat RCL. The contribution-aware ICN with altruistic users show similar download performance with both the increasing RCL (Fig. 5(a)) and the flat RCL (Fig. 5(b)), since the altruistic users continue to contribute regardless of their CL. On the other hand, in case of the contribution-aware ICN with rational users, the increasing RCL (Fig. 6(a)) is more effective than the flat RCL (Fig. 6(b)) to encourage the rational users to contribute their resources. This result shows that the increasing RCL can encourage the rational users to keep uploading the downloaded segments to increase CL enough to satisfy RCL of remaining segments.

5 Conclusion

In this paper, we introduce the contribution-aware ICN to utilize the user resources for efficient content distribution by mitigating the free-riding issue. The contribution-aware ICN utilizes the centralized entity to manage the user contribution and the content is published with the required contribution level so that the users with enough contribution only can request the content. Through simulation, we show that the contribution-aware ICN encourages the users to contribute their resources. We also show that the content download performance improves significantly as the number of users contributing the resources increases. We are currently focusing on efficient utilization of given user resources and effective incentive mechanism when there are various types of users publishing content with different RCL.

References

1. BitTorrent, http://www.bittorrent.com/
2. Koponen, T., Chawla, M., Chun, B.-G., Ermolinskiy, A., Kim, K.H., Shenker, S., Stoica, I.: A data-oriented (and beyond) network architecture. In: Proc. of ACM SIGCOMM (2007)
3. Jacobson, V., Smetters, D.K., Thornton, J.D., Plass, M.F., Briggs, N.H., Braynard, R.L.: Networking named content. In: Proc. of ACM CoNEXT (2009)
4. Named Data Networking (NDN) project website, http://www.named-data.net/
5. PSIRP project website, http://www.psirp.org/
6. PURSUIT project website, http://www.fp7-pursuit.eu/
7. 4WARD project website, http://www.4ward-project.eu/
8. SAIL project website, http://www.sail-project.eu/
9. Stutzbach, D., Rejaie, R.: Understanding churn in peer-to-peer networks. In: Proc. of ACM IMC (2006)
10. Karagiannis, T., Papagiannaki, D., Faloutsos, M.: Should internet service providers fear peer-assisted content distribution? In: Proc. of ACM IMC (2005)
11. Hoffman, K., Zage, D., Rotaru, C.N.: A survey of attack and defense techniques for reputation systems. ACM Computing Surveys 42(1) (2009)
12. Karakaya, M., Korpeoglu, I., Ulusoy, O.: Free riding in peer-to-peer networks. IEEE Internet Computing 13, 92–98 (2009)
13. The Network Simulator ns-2, http://www.isi.edu/nsnam/ns/

On Inter-Domain Name Resolution
for Information-Centric Networks

Konstantinos V. Katsaros, Nikos Fotiou, Xenofon Vasilakos,
Christopher N. Ververidis, Christos Tsilopoulos,
George Xylomenos, and George C. Polyzos*

Mobile Multimedia Laboratory,
Department of Informatics,
Athens University of Economics and Business,
Patision 76, Athens 104 34, Greece
{ntinos,fotiou,xvas,chris,tsilochr,xgeorge,polyzos}@aueb.gr
http://mm.aueb.gr

Abstract. Information-centric networking (ICN) is a paradigm that
aims to better reflect current Internet usage patterns by focusing on
information, rather than on hosts. One of the most critical ICN functionalities is the efficient resolution/location of information objects i.e.,
name resolution. The vast size of the information object namespace calls
for a highly scalable and efficient name resolution approach. Currently
proposed solutions either rely on a DHT structure, thus ensuring load
balancing and scalability at the cost of inefficient routing, or on hierarchical structures, thus preserving routing efficiency at the cost of limited scalability. In this paper, we study in detail the tradeoff between
state/signaling overhead versus routing efficiency for a generic name
resolution system based on a novel DHT scheme with enhanced routing
properties, and compare it to DONA, an ICN architecture based on hierarchical resolution and routing.

Keywords: content-centric, future internet, named data, rendezvous.

1 Introduction

The Internet has been transformed from an academic curiosity to a global infrastructure for the massive distribution of information, with over 1 billion of
connected devices [6], 1 trillion of indexed web pages [12] and Exabytes of annu-
ally transferred data [6]. Unlike this evolution in usage, at its core the Internet
continues to operate upon a host-centric communication model, even though obtaining and disseminating information is currently the primary interest of users.
This tussle between the core Internet functionalities and its actual usage has
led the networking community to investigate new architectures for the Future
Internet, many of which revolve around information-centrism (e.g. [15],[17]).

---

* Work supported by the ICT PURSUIT project, under contract ICT-2010-257217.

R. Bestak et al. (Eds.): NETWORKING 2012, Part I, LNCS 7289, pp. 13–26, 2012.

Information-centric networking (ICN) places information dissemination at the heart of the architecture. *Information Objects* (IOs) constitute the fundamental entity in ICN. Users in ICN issue requests by naming the desired IOs in order to obtain information; it is the responsibility of the network to locate and deliver the desired IOs to the requester, thus decoupling the communicating parties: end-users need not be aware of each other's network location or, even, identity. Locating IOs is therefore a core function in the context of ICN. In effect, the network operates as a system in which producers of information announce the availability of IOs and requests for IOs are resolved through a name-based anycast scheme, resulting in a *name-resolution system* (NRS).

To face the scalability requirements imposed by the vast number of IOs, several research initiatives have considered the use of *Distributed Hash Table* (DHT) schemes (e.g., [7,16]), which exhibit significant scalability due to their inherent load balancing. However, this advantage comes at the cost of inefficient routing, which may overlook physical network proximity, administrative domain boundaries, routing policies, or a combination thereof. Other approaches, such as DONA [17], adapt to the underlying routing mechanism, requiring however extensive replication of routing information across the inter-domain topology.

In this paper, we examine the emerging tradeoff between state/signaling overhead versus routing efficiency in the context of ICN. To this end, we first design a *DHT-based name-resolution system* (DHT-NRS) based on H-Pastry, a hierarchical version of the Pastry DHT [22]; H-Pastry adapts to the underlying network topology, administrative structure and routing policies. We then present a detailed performance evaluation and comparison against the DONA scheme. Our purpose is to shed light on the expected performance of each approach, paying particular attention to the effect of the underlying inter-domain topology and the traffic workload. Our evaluation is based on a full-fledged simulation environment and considers a wide range of performance aspects, including routing efficiency and signaling overhead, as well as memory and processing load.

In the following, we describe name-resolution in the context of ICN and present our requirements for the envisioned name-resolution system (Section 2). Next, we discuss alternative approaches and position our work in the solution space (Section 3). In Section 4 we first describe the H-Pastry DHT scheme and then proceed with the description of DHT-NRS. We present and discuss the results of our performance evaluation in Section 5 and conclude in Section 6.

2   Name-Resolution in ICN

In ICN, the network focuses on information itself rather than on the endpoints participating in the communication. The network acts as a mediator that decouples content providers from content consumers, radically changing the model of interaction with the network. The network (a) accepts end-user requests for IOs with the purpose of locating the content and enabling its delivery and (b) interacts with content providers which supply information about the availability and location of content. The basic functionality of the NRS in ICN is to match end-user requests with the accumulated information on available content and trigger the delivery of the corresponding data. This matching is based on the IO name, which thus becomes a first class entity in ICN. In this context, we outline below a set of requirements for an NRS, as derived by its central role in an ICN architecture, also identifying the key challenges in this new paradigm.

Flat Identifiers. A critical factor in the design of an NRS is the form of the IO *identifiers* (IDs) employed. Existing systems achieve scalability by applying extensive aggregation on hierarchically structured identifiers (e.g., DNS, IP routing). A critical constraint of hierarchical systems is that the nodes in the root(s) of the tree(s) must be aware of all available prefixes in the name space. The use of flat, semantic-free identifiers has been proposed to decouple location from identity and thus achieve persistence and increased security [11,17,25]. Recent developments in DNS [14] suggest that the increasing demand for naming information cannot be easily handled within the constraints imposed by hierarchical naming. Therefore, our NRS must operate on top of a flat namespace.

Scalability. An NRS must scale as the number of items and corresponding requests grows. Considering that (a) the current amount of unique web pages indexed by Google is greater than 1 trillion [12] and that (b) billions [6] of devices ranging from mobile phones to sensors and home appliances will be joining the network to offer additional content, we should plan for unique IOs in the order of 10^{13}; other studies raise this estimate to 10^{15} [7]. This vast amount of IOs and related requests must be handled through the NRS.

Fault Tolerance and Fault Isolation. The criticality of the NRS calls for a non-central point of failure, which is the major vulnerability of centralized architectures. Furthermore, the architecture should provide fault isolation i.e., the failure of a part of this distributed system should only have a local impact.

Low Signaling Overhead. The expected amount of available information, coupled with the corresponding demand by end-users, is expected to yield an increased signalling load for the NRS, in terms of IO announcements and requests for IOs. Hence, the required information exchange between the nodes of the distributed NRS should be kept to a minimum.

Low Latency. Resolution should complete with minimum delay, yielding low response times for end-users. This calls for efficient routing and load distribution among system nodes.

Routing Policy Compliance. Finally, due to the high volumes of expected resolution traffic, routing should respect the policies of the *Autonomous Systems* (ASes). These policies reflect business relationships established between ASes, with policy violations having an economic impact on their operation.

3   Related Work

The current equivalent of an NRS for the Internet is DNS, therefore we must first explain why DNS is inadequate for our purposes. To begin with, DNS is susceptible to *Denial of Service* (DoS) attacks. This vulnerability stems not only from the limited redundancy in name-servers, but also by the fact that many servers have a single point of attachment to the Internet [21]. In addition, load is not equally balanced between root servers due to the fact that names are not equally distributed among top level domains (e.g. .com names are far more than .edu names). Finally, DNS supports a hierarchical namespace, which constrasts with our requirement for flat identifiers.

The limitations of DNS have led the research community to look into DHT-based approaches for name resolution e.g., [21,25,20]; typical DHTs used in these approaches are Chord [24] and Pastry [22]. The main problem with DHTs is the logarithmic number of hops required (on average) for lookup/resolution. The common workaround for this problem is caching and/or replication, as well as incrementing the average node degree in the DHT. Furthermore, overlay routing in DHT approaches suffers from non-compliance to inter-domain routing policies i.e., the overlay routes followed when resolving a name may violate the routing policies of the underlying physical network [23]. To the best of our knowledge, no solution has been proposed for this problem in any of the available DHT overlays, except for [9], which was developed for the purposes of this work.

In MDHT [7] a multilevel DHT architecture is proposed for name resolution in the ICN context. MDHT aggregates IO registration entries at higher levels of the inter-domain hierarchy, which raises scalability concerns. The proposed solution is an indirection level which allows the NRS to resolve content provider names, with the resolution of the content itself taking place at a lower level. The description of the proposed system lacks details on the operation of the employed DHT, as well as a performance evaluation of the proposed architecture. A similar approach is followed in [16], where an inter-domain name-resolution architecture is presented. Again, the aggregation of routing information at the higher levels of the hierarchy poses significant scalability concerns. In a similar manner to [7], an indirection level is introduced to map *scopes* of information to lower level resolution nodes. The presented performance evaluation is based on significant abstractions e.g., intra-domain routing overhead and caching have been coarsely modeled based on observations made in different contexts which do not reflect the proposed architecture's intrinsic characteristics. In both approaches ([7] and [16]), the routing inefficiency of DHTs is circumvented by the aggregation of information at higher levels of the inter-domain structure.

The *Data-Oriented (and beyond) Network Architecture* (DONA) [17] follows a similar approach in that it also concentrates routing information at the higher levels of the inter-domain structure. DONA builds an information-centric layer over the Internet, based on a network of *Resolution Handlers* (RHs). The deployment of the RHs strictly follows the AS-level structure of the Internet, allowing DONA to directly adapt its structure and operation to the underlying network. Flat, self-certifying names are used to represent information. Content providers issue REGISTER messages to advertise their content to the closest RH which then propagates this information upwards in the hierarchy, also forwarding it across inter-domain peering links. Name resolution is based on a similar propagation of requests (FIND messages) upwards in the hierarchy, until a relevant entry is found. At that point, requests follow the (downhill) reverse path of the corresponding REGISTER message. In this process, shortest path routing is followed and inter-AS routing policies are taken into account.

In the work most similar to ours [5], a performance comparison of content delivery between a DONA-like, hierarchical architecture and a DHT based alternative is presented. The authors study several aspects such as transfer latency and robustness, as well as the impact of in-network caching. However, this investigation does not focus on name resolution and is based on oversimplified network topologies: the DONA-like architecture uses a tree overlay consisting of randomly selected nodes of a two-level, hierarchical inter-domain topology. Our work is far more accurate in that (i) we consider a more realistic topology model that takes into account both multihoming and peering relationships (see Section 5), (ii) we fully implement DONA, allowing the structure of the RH network to mimic the underlying inter-domain topology and (iii) we consider a DHT-scheme that better adapts to the underlying topology (see Section 4.1).

4   An Enhanced DHT-Based NRS

As already discussed, the load balancing and robustness characteristics of DHTs can directly address the scalability concerns posed for an NRS in the context of ICN. However, these features also imply several important disadvantages i.e., name resolution follows longer paths than those offered by the underlying shortest path routing fabric, often unnecessarily crossing administrative domain boundaries and/or violating the established inter-domain routing policies.

In our work we have investigated the extent to which DHT routing can be improved, without sacrificing its scalability advantages. Our purpose is to assess the ability of a DHT-based approach to support an NRS in the context of ICN. To this end, we have designed DHT-NRS, an NRS based on an enhanced DHT design named H-Pastry [9]. In the following, we first provide an overview of the H-Pastry scheme and then we proceed with the description of DHT-NRS. We do not discuss security issues, as they are outside the scope of this paper. However, existing solutions for securing DHTs are also applicable to our work.

4.1   H-Pastry

Based on Pastry [22] and the Canon paradigm [10], H-Pastry is a multi-level DHT scheme that improves routing by taking physical network proximity, administrative domain boundaries and inter-domain routing policies into account. In order to adapt to the multi-level structure of an inter-network, H-Pastry employs a corresponding multi-level structure to partition the information identifier space, and the corresponding routing state. Distinct routing tables are maintained for each level of the inter-domain topology. At each level, H-Pastry nodes maintain routing information about nodes that are numerically closer to certain points in the identifier space (i.e., their own identifier and the identifiers required to fill their Routing table) than any other node at that particular level. In this manner, the participating H-Pastry nodes recursively create H-Pastry rings that adapt to the AS-level topology of the network. Multihoming and peering agreements are also taken into account by appropriately introducing virtual nodes in the inter-domain topology graph [9]. H-Pastry presents good fault resilience properties: as described in [9], H-Pastry routing will fail in the event of a maximum of $H \cdot |L/2|$ concurrent failures of nodes grouped in H sets of $|L/2|$ adjacent IDs each (the corresponding number of failures for Pastry is $|L/2|$) where H is the height of the domain level hierarchy and L the size of the leaf set.

By taking into account the aforementioned routing aspects, H-Pastry considerably improves DHT-based routing. As shown in [9], H-Pastry results in shorter routes, lowering path stretch by 55% and 47% compared to Chord [24] and hierarchical Chord [10], respectively, being comparable to regular Pastry. At the same time, H-Pastry also manages to confine traffic within administrative boundaries resulting in 27% less inter-domain hops and 55% shorter intra-domain paths than regular Pastry. Moreover, by taking routing policies into account, H-Pastry reduces valley-free policy violations per routing path by 56%, 31% and 36% compared to Chord, Pastry and hierarchical Chord, respectively.

4.2 DHT-NRS

Basic Functionality. The name resolution function in DHT-NRS follows the Publish/Subscribe paradigm i.e., publishers (content providers) and subscribers (content consumers) interact with DHT-NRS through a set of *brokers* responsible for matching publications and subscriptions. DHT-NRS is realized by the deployment of these brokers, named *Rendezvous Nodes* (RNs), across the inter-domain topology. Each RN is an H-Pastry node with a unique ID. We envision at least one RN per AS; more RNs will increase system scalability. For each IO a statistically unique ID is created (e.g., with a secure hash function). Publishers and subscribers interact with DHT-NRS via their local domain RNs, designated during network attachment. Name-resolution is then based on the exchange of the following control messages (an example is given in Figure 1):

- `Adv`: An advertisement message sent by the publisher(s) of an IO to register the IO with DHT-NRS (step I in Figure 1). This message contains information that maps the ID of the advertized IO to the network location of the issuing publisher. For each unique ID-publisher mapping, a corresponding *IO entry* (IOE) is maintained by an RN designated by H-Pastry as responsible for that ID (denoted as the *Rendezvous Point* (RVP) for the ID). Advertisement messages are routed towards the RVP through H-Pastry.
- `Sub`: A subscription message is issued by a subscriber to request a specific IO (step II in Figure 1). Subscription messages contain the ID of the requested IO along with information on the subscriber's network location. Subscription messages reach the appropriate RVP through H-Pastry routing. Upon reception, an RVP searches its IOEs for the indicated IO ID. If the subscription message refers to an unknown IO, the RVP discards the subscription.
- `Ntf`: A notification message is sent by an RVP to notify the publisher(s) of an IO to start the delivery of the corresponding data (step IV in Figure 1).

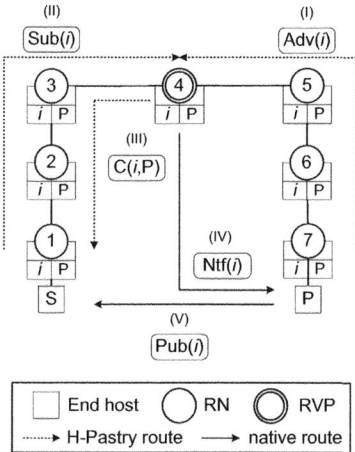

Fig. 1. DHT-NRS Example; the (i, P) records are stored at RNs 4, 5, 6 and 7 during step I, and at RNs 1, 2 and 3 during step III

- Pub: Upon notification from the RVP, a publisher sends the IO to the subscriber(s) (step V in Figure 1) via an underlying forwarding mechanism.

Caching. Even though DHT-NRS is based on a novel DHT scheme that adapts to the underlying network topology, routing is still based on DHT primitives which are characterized by their logarithmic relation to the number of RNs in the inter network. Adopting established practice (see Section 3), we employ caching to shorten the resolution paths. More specifically, we cache the route towards the publisher of an IO. During advertisement of an IO, a cache entry is created in all intermediate RNs encountered by the Adv message (step I in Figure 1). Moreover, when a Sub message reaches an RVP, the RVP actively pushes a copy of the respective IOE to all RNs in the path followed by the subscription (step III in Figure 1). Further subscriptions for an IO trigger Ntf messages directly towards the publisher upon a cache hit, reducing the part of the resolution path that follows H-Pastry's routing. Each cache entry contains a *time to live* (TTL) value that allows for its invalidation and is subject to application level criteria. Cache replacement follows a *Least Recently Used* (LRU) mechanism.

Node Failures. Due to the vital role of the NRS in locating information in ICN, we expect the deployment process to follow a path similar to DNS i.e., rely on highly provisioned RN servers. However, the multitude of DoS attacks against DNS signifies the critical role of system resilience, while the vast size of the expected workload further urges for a design able to cope with node failures. The use of a highly distributed DHT-based solution is a design choice aligned with this desire. Based on the uniform distribution of IO and node IDs in the identifier space, a DHT based solution accomplishes an even distribution of IOEs to nodes, so that a node's failure has a a modest impact to the entire system, proportional to deployment density. The fault resilience properties of H-Pastry add to the DHT-NRS robustness (see Section 4.1). In contrast, hierarchical designs (e.g., DONA) allow even a single node failure to affect the routing of messages to/from the entire sub-tree rooted at the failed node. Obviously, the impact of a node failure is isolated to that part of the inter-domain topology, but the consequences become more severe for nodes at higher levels. Multiple physical incarnations of resolution nodes ameliorate this problem, but only at an additional cost.

5   Performance Evaluation

In order to investigate the extent to which the DHT-NRS proposed fulfills our requirements, we focus on two major aspects: (i) system load, including memory, signaling and processing overheads, and (ii) routing performance. Our investigation offers a detailed comparison with DONA, with the purpose of revealing, as well as quantifying, the inherent (dis)advantages of both architectures. Particular attention is paid to (i) the effect of the underlying inter-network structure on perceived performance and its relation to the structural characteristics of each architecture and (ii) the effect of the popularity characteristics of content on the effectiveness of DHT-NRS's caching scheme.

Topologies. Understanding the Internet topology is a crucial factor in designing new inter-domain protocols. Typically, the AS-level topology is described as a multi-tier hierarchy of interconnected ASes, mostly using transit customer-provider links. However, several measurements and studies argue that the Internet topology is evolving into a mesh dominated by multi-homing and peering relationships [3,18,19]. These studies converge on the number of the ASes, which is estimated to be around 35.000. Evaluating a protocol using a realistic topology of 35.000 ASes (and approximately 200.000 annotated links) is obviously a technical challenge on its own. To overcome this constraint, while also preserving memory and processing resources for the evaluation of non trivial DHT sizes, we used inter-domain topologies generated by the algorithm presented in [8]. The size of these topologies is manageable for evaluation and they maintain the same characteristics (i.e., business relationships) as in the measured graphs.

Workload. Our evaluation considers a detailed model of inter-domain Internet traffic. Based on the measurements presented in [18], we generate a mixture of various traffic types (e.g., Web, Video, P2P)[1]. The resulting traffic mix does not distinguish between user and control plane traffic i.e., the signaling required to locate and start content transmission, except for DNS which only constitutes 0.17% of all traffic. However, our evaluation focuses on the control plane requests made to the NRS to support user plane traffic. Therefore, we translated the user plane traffic mix into a control plane equivalent. To this end, we derived the actual number of IOs for each traffic type by dividing the corresponding data volume with the median IO size of that traffic type as measured in relevant

---

[1] We did not use DNS trace data as they reflect the current Internet architecture. For example, they omit requests sent directly to content owners (e.g., HTTP requests).

studies [2,1,4]. We used the median instead of the mean object size, so as to avoid skewing the results due to the long-tail characteristics of some distributions (e.g., the Pareto tail of Web object size distribution). The set of subscription messages is then shaped based on studies for each traffic type which model its popularity characteristics, including its temporal evolution [2,4,13].

5.1 Results

Our evaluation employs a 400 domain topology, which follows a hierarchical model with six levels, but also contains multi-homing and peering links between the domains. We deploy a population of 4400 RNs/RHs uniformly across the domains. The cache size in DHT-NRS is expressed as a proportion of the median number (m) of registration entries per RH in DONA. We choose the median value, since the distribution of state in DONA is considerably skewed, as we show below. Moreover, we examine scenarios with *infinite cache size* (ICS) i.e., no limitation on the available cache size; this reflects the upper limit for the performance of caching in DHT-NRS. In each scenario, we use a workload corresponding to 25 GBs of traffic, resulting in an average[2] of 2430379 subscription messages for 1032030 IOs. This size limit was imposed by the limitations of the simulation environment. Since end hosts typically reside in access networks i.e., networking domains that have no customer domains and thus do not act as transit domains, all Adv/Sub messages in DHT-NRS (REGISTER/FIND in DONA) are injected into the network from a randomly chosen RN (or RH respectively) residing in an access network. Finally, we considered a single publisher per IO.

Routing. The routing performance of the name-resolution system affects both the latency perceived by the end-users, as well as the traffic load on the network. We express the routing performance of DHT-NRS with the *stretch* metric, defined as the ratio of the number of inter-domain hops required for a Sub message to reach the RVP and the corresponding Ntf message to reach the RN serving the publisher of the desired IO, over the hop count required by an identical FIND message to reach the same target in DONA i.e., the RH that issued the corresponding REGISTRATION.

Figure 2(a) shows the stretch values derived for several scenarios with different cache sizes. Stretch ranges from 2.84 without caching to 1.95 in the ICS scenario. Under current traffic patterns, DONA significantly outperforms DHT-NRS in routing efficiency. This is only possible however because DONA extensively replicates IOEs throughout the hierarchy, thus guarantying the existence of the desired registration on the shortest, policy-compliant path towards the publisher. DONA is superior to DHT-NRS regardless of cache size, since caching is less aggressive than replication, reactively adapting to, and therefore highly depending on, the request patterns. Hence, non popular IOEs are evicted from DHT-NRS caches, or even never requested again, yielding low cache hit ratios. As shown in Figure 2(b), the cache hit ratio is 27.17, 31.75, 37.76 and 53% for

---

[2] We used a different workload instance in each experiment to increase randomness.

the $50\%m$, $100\%m$, $150\%m$ and ICS scenarios respectively. For each scenario, the corresponding savings in the resolution path lengths are 13, 15.39, 18.24 and 31%, respectively. These values are subject to the popularity characteristics of the workload, and show that with current workload patterns route caching in DHT-NRS cannot exceed a cache hit ratio value of 53%, even with unrealistically large cache sizes, resulting in low routing improvements.

In order to provide an insight of the potential benefits of caching in scenarios where item popularity does allow higher cache hit ratios, we also consider the stretch value of resolution paths in which the Sub messages have hit a cache. As shown in Figure 2(a), stretch ranges from 1.34 in the hypothetical, best case scenario (ICS) to 1.67 in the $50\%m$ scenario. These results show that caching achieves a substantial reduction of resolution path lengths in the case of popular IOs, which are still however at least 34% longer compared to DONA. This is due to the indirection mechanism in DHT-NRS, as the RVP may not reside in the shortest path connecting a subscriber and a publisher. However, in cases of multiple publishers per IO (e.g., replication points), we expect the routing properties of H-Pastry and the caching mechanism to further shorten resolution paths by selecting the closest publisher.

State. In our evaluation, we use the term "state" to refer to the IOEs maintained at each node of the name-resolution system. The state size is related to the total number of IOs and determines the amount of resources required to support the operation of the architecture *wrt* memory and lookup processing load. Figure 2(c) shows the cumulative distribution function of the state size for both DHT-NRS and DONA. Note that the x axis is in log scale. The difference between the two architectures is significant, with the size of the state maintained by RN nodes in DHT-NRS being considerably lower than the corresponding load imposed on DONA's RHs. More importantly, we see that DHT-NRS achieves a more uniform distribution of state across the participating nodes compared to DONA. For instance, 95% of the RNs in DHT-NRS maintain less than 550 IOEs ($0\%m$ scenario), while 95% of the RHs in DONA maintain up to almost 33000 IOEs. The highly skewed state distribution in DONA is due to the accumulation of registrations at higher levels of the inter-domain hierarchy, and poses a significant challenge for network operators, who would have to resort to dense deployments of RHs in order to cope with the associated resource requirements.

Moreover, it is important to point out the relation of the observed state distribution skewness to the structure of the inter-domain topology. In the considered topologies, slightly less than 50% of access networks reside at level 2, meaning that a major part of the registrations is stored only at the first two levels of the hierarchy. This is demonstrated in Figure 2(d) which shows the average state size per node at each level of the hierarchy. Note that the y axis is presented in log-scale. This figure suggests that the inter-domain topology structure causes the concentration of excessive state in DONA at the top-most levels of the hierarchy. On the one hand, these top-level domains face disproportionate overhead compared to lower level domains, incurring the corresponding CAPEX/OPEX overheads. On the other hand, however, it also suggests that the scalability

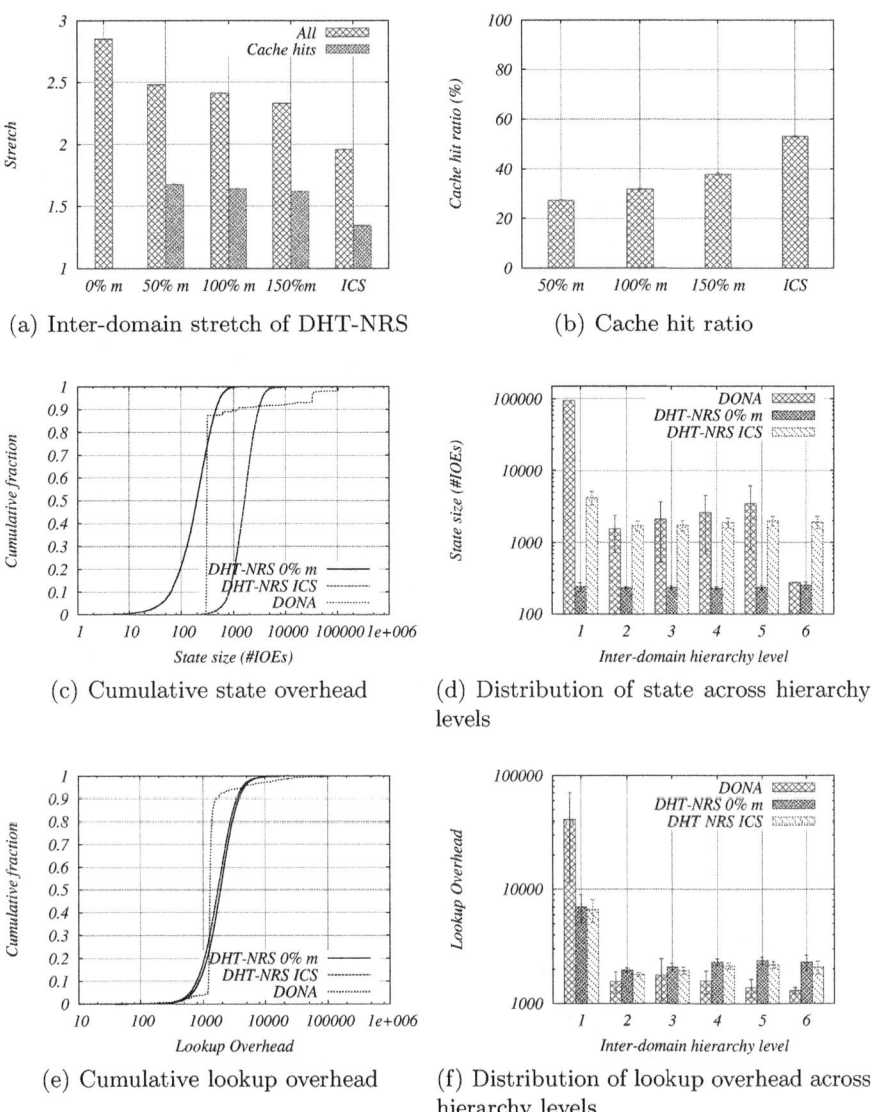

Fig. 2. Performance evaluation results: DHT-NRS vs. DONA

challenge in DONA is mostly concentrated in a limited sub-area of the inter-domain topology, making cloud-based solutions a compelling option.

Figures 2(c) and 2(d) also show the amount and distribution of state required to avoid cache replacement (i.e., the ICS scenario). This hypothetical scenario requires a considerable amount of memory which is highly unlikely to be available in practice (see Section 2). However, even in this scenario, almost 8% of RHs in DONA have higher memory requirements than any RN in DHT-NRS. Moreover, we see that in DHT-NRS state is distributed more evenly across the hierarchy levels, with the exception of level 1 RNs in the ICS scenario where the routing scheme causes a concentration of cached IOEs (see also the next section).

Processing Overhead. The state size per node determines the processing overhead per IOE lookup at the RVP and the overall processing overhead is determined by the actual number of lookups performed by a node. For each architecture, we define the *lookup overhead* (LO) metric as the sum of the total number of Sub/FIND messages forwarded by each RN/RH respectively, and the total number of messages *terminating* at each node i.e., triggering communication with the publisher. Figure 2(e) shows the cumulative distribution function of LO for both architectures. A major fraction of DONA RHs is subject to less LO than RNs in DHT-NRS, due to the considerably worse routing performance of DHT-NRS that results in Sub messages traversing longer networking distances and thus consuming resources at more intermediate RNs. However, a closer look at the LO distribution across the inter-domain hierarchy reveals a disproportionate overhead for the top-level domains in DONA (Figure 2(f)). Evidently, the fact that a major part of the access domains resides at level 2 of the hierarchy results in a corresponding resolution overhead for the top-level domains, which again calls for the use of considerable processing (as well as memory) resources.

Interestingly, Figure 2(f) also shows that RNs at the top-most domains of the hierarchy in DHT-NRS are subject to considerably higher processing overhead compared to lower level RNs. The additional overhead is due to the forwarding of Sub and Ntf messages (ID ownership is evenly distributed across the RNs) and is attributed to the structure of the inter-domain topology and the design of H-Pastry: to adapt to the inter-domain hierarchy structure, H-Pastry causes the top level domains to be included in the first non-local ring of each of the access networks at level 2. Therefore, paths towards non-local RVPs are most likely to pass through these domains. As a substantial fraction of access networks resides at level 2, a proportional number of IOs are served from publishers at these domains, resulting in a significant part of the overall Ntf messages reaching RNs at these domains via the top-level domains. Similarly, H-Pastry causes the Sub messages originating from level 2 access domains to first traverse the top-level domains before taking a downhill direction towards the RVP.

Advertisement/Registration Overhead. In addition to subscriptions, signaling overhead is also generated by IO registrations. We characterize this overhead as the total number of single inter-domain hop transmissions required for the registration of a single IO to complete. Our measurements show that DHT-NRS requires on average 6.34 inter-domain transmissions per IO in the $0\%m$ scenario, while DONA requires on average 35.56 transmissions. These numbers indicate an excessive inter-domain traffic load in the case of DONA, attributed to the (limited) flooding method used to disseminate registration messages to the upper levels of the inter-domain hierarchy (as well as to peering domains). Multihoming plays an important role in this as it results in registration messages being transmitted to multiple domains at higher levels. In the employed topologies 56.75% of all domains are multihomed, with 2.4 providers on average. Unlike DONA, by carefully partitioning the identifier space DHT-NRS forms a structured overlay that only requires the targeted routing of Adv messages.

6   Conclusions

In this paper we investigated name-resolution in the context of ICNs, focusing on the emerging tradeoff between routing state overhead and routing efficiency. Motivated by the significant scalability characteristics of DHTs, but also concerned about their routing inefficiency, we engaged in the design of a name resolution system based on an enhanced DHT scheme, aiming to improve routing performance. Then we engaged in a detailed performance evaluation and comparison against DONA, an alternative that yields efficient routing performance but raises significant scalability concerns. Our findings reveal the effect of the inter-domain topology structure and traffic patterns. With current traffic patterns, route caching in DHT-NRS cannot compete with the extensive replication of routing information in DONA, yielding stretch values ranging from 1.95 to 2.84. However, this comes at the cost of a heavily skewed distribution of memory and processing overhead, as well as significant signaling overhead for the registration of content in DONA. The replication mechanism of DONA along with the structural characteristics of the inter-domain topology result in disproportionate resource requirements for a limited area of the internetwork, at the top-most level domains in the hierarchy, indicating the need for large-scale centralized solutions (e.g. cloud computing).

References

1. Bellissimo, A., Levine, B.N., Shenoy, P.: Exploring the use of BitTorrent as the basis for a large trace repository. Tech. rep., University of Massachusetts Amherst (2004)
2. Busari, M., Williamson, C.: ProWGen: a synthetic workload generation tool for simulation evaluation of web proxy caches. Computer Networks 38(6), 779–794 (2002)
3. CAIDA (2011), http://www.caida.org
4. Cheng, X., Dale, C., Liu, J.: Understanding the Characteristics of Internet Short Video Sharing: YouTube as a Case Study. CoRR abs/0707.3670 (2007)
5. Choi, J., Han, J., Cho, E., Kim, H., Kwon, T., Choi, Y.: Performance comparison of content-oriented networking alternatives: A tree versus a distributed hash table. In: Proc. of the IEEE 34th Conference on Local Computer Networks (LCN), pp. 253–256 (2009)
6. Cisco: Cisco Visual Vetworking Index 2010-2015 (June 2011)
7. D'Ambrosio, M., Dannewitz, C., Karl, H., Vercellone, V.: MDHT: a hierarchical name resolution service for information-centric networks. In: Proc. of the 2011 ACM SIGCOMM Workshop on ICN, pp. 7–12. ACM, New York (2011)
8. Dimitropoulos, X., Krioukov, D., Vahdat, A., Riley, G.: Graph annotations in modeling complex network topologies. ACM Transactions on Modeling and Computer Simulation 19, 17:1–17:29(2009)

9. Fotiou, N., Katsaros, K.V., Vasilakos, X., Tsilopoulos, C., Ververidis, C.N., Xylomenos, G., Polyzos, G.C.: H-Pastry: An adaptive multi-level overlay internetwork. Tech. Rep. 2011-MMLAB-TR-002, Athens University of Economics and Business (2011),
    http://mm.aueb.gr/technicalreports/2011-MMLAB-TR-003.pdf
10. Ganesan, P., Gummadi, K., Garcia-Molina, H.: Canon in G Major: Designing DHTs with Hierarchical Structure. In: Proc. of the 2004 ICDCS, pp. 263–272 (2004)
11. Ghodsi, A., Koponen, T., Rajahalme, J., Sarolahti, P., Shenker, S.: Naming in content-oriented architectures. In: Proc. of the ACM SIGCOMM ICN Workshop, New York, NY, USA, pp. 1–6 (2011)
12. Google: We knew the web was big (July 2008),
    http://googleblog.blogspot.com/2008/07/we-knew-web-was-big.html
13. Guo, L., Chen, S., Xiao, Z., Tan, E., Ding, X., Zhang, X.: A performance study of BitTorrent-like peer-to-peer systems. IEEE Journal on Selected Areas in Communication 25(1), 155–169 (2007)
14. ICANN: ICANN Approves Historic Change to Internet's Domain Name System (2011), http://www.icann.org/en/announcements/announcement-20jun11-en.htm
15. Jacobson, V., Smetters, D.K., Thornton, J.D., Plass, M.F., Briggs, N.H., Braynard, R.L.: Networking named content. In: Proc. of the 2009 ACM CoNEXT, pp. 1–12. ACM, New York (2009)
16. Rajahalme, J., Särelä, M., Visala, K., Riihijärvi, J.: On name-based inter-domain routing. Computer Networks 55, 975–986 (2011)
17. Koponen, T., Chawla, M., Chun, B.-G., Ermolinskiy, A., Kim, K.H., Shenker, S., Stoica, I.: A data-oriented (and beyond) network architecture. In: Proc. of the 2007 ACM SIGCOMM, pp. 181–192. ACM, New York (2007)
18. Labovitz, C., Iekel-Johnson, S., McPherson, D., Oberheide, J., Jahanian, F.: Internet inter-domain traffic. In: Proc. of the 2010 ACM SIGCOMM, pp. 75–86. ACM, New York (2010)
19. Oliveira, R., Pei, D., Willinger, W., Zhang, B., Zhang, L.: The (in)completeness of the observed internet AS-level structure. IEEE/ACM Transactions on Networking 18, 109–122 (2010)
20. Pappas, V., Massey, D., Terzis, A., Zhang, L.: A Comparative Study of the DNS Design with DHT-Based Alternatives. In: Proc. of the 2006 IEEE INFOCOM, pp. 1–13 (2006)
21. Ramasubramanian, V., Sirer, E.G.: The Design and Implementation of a Next Generation Name Service for the Internet. In: Proc. of the 2004 ACM SIGCOMM, pp. 331–342. ACM, New York (2004)
22. Rowstron, A., Druschel, P.: Pastry: Scalable, Decentralized Object Location, and Routing for Large-Scale Peer-to-Peer Systems. In: Guerraoui, R. (ed.) Middleware 2001. LNCS, vol. 2218, pp. 329–350. Springer, Heidelberg (2001)
23. Seetharaman, S., Ammar, M.: Inter-domain policy violations in multi-hop overlay routes: Analysis and mitigation. Computer Networks 53, 60–80 (2009)
24. Stoica, I., Morris, R., Liben-Nowell, D., Karger, D.R., Kaashoek, M.F., Dabek, F., Balakrishnan, H.: Chord: a scalable peer-to-peer lookup protocol for internet applications. IEEE/ACM Transactions on Networking 11(1), 17–32 (2003)
25. Walfish, M., Balakrishnan, H., Shenker, S.: Untangling the web from DNS. In: Proc. of the 2004 USENIX NSDI, p. 17. USENIX Association, Berkeley (2004)

Cache "Less for More" in Information-Centric Networks

Wei Koong Chai, Diliang He, Ioannis Psaras, and George Pavlou

Department of Electronic and Electrical Engineering,
University College London,
WC1E 6BT, Gower Street, London, UK
{w.chai,diliang.he.10,i.psaras,g.pavlou}@ee.ucl.ac.uk

Abstract. Ubiquitous in-network caching is one of the key aspects of information-centric networking (ICN) which has recently received widespread research interest. In one of the key relevant proposals known as Networking Named Content (NNC), the premise is that leveraging in-network caching to store content in every node it traverses along the delivery path can enhance content delivery. We question such indiscriminate universal caching strategy and investigate whether caching *less* can actually achieve *more*. Specifically, we investigate if caching only in a subset of node(s) along the content delivery path can achieve better performance in terms of cache and server hit rates. In this paper, we first study the behavior of NNC's ubiquitous caching and observe that even naïve random caching at one intermediate node within the delivery path can achieve similar and, under certain conditions, even better caching gain. We propose a centrality-based caching algorithm by exploiting the concept of (ego network) betweenness centrality to improve the caching gain and eliminate the uncertainty in the performance of the simplistic random caching strategy. Our results suggest that our solution can consistently achieve better gain across both synthetic and real network topologies that have different structural properties.

Keywords: Information-centric networking, caching, betweenness centrality.

1   Introduction

Information-centric networking (ICN) has recently attracted significant attention, with various research initiatives (e.g., DONA [1], NNC [2], PSIRP/PURSUIT [3][4] and COMET [5]) targetting this emerging research area. The main reasoning for advocating the departure from the current host-to-host communications paradigm to an information/content-centric one is that the Internet is currently mostly used for content access and delivery, with a high volume of digital content (e.g., 3D/HD movies, photos etc.) delivered to users who are only interested in the actual content rather than the source location. As such, we no longer need a natively supported content distribution framework. While the Internet was designed for and still focuses on host-to-host communication, ICN shifts the emphasis to content objects that can be cached and accessed from anywhere within the network rather than from the end hosts only.

R. Bestak et al. (Eds.): NETWORKING 2012, Part I, LNCS 7289, pp. 27–40, 2012.

In ICN, content names are decoupled from host addresses, effectively separating the role of identifier and locator in distinct contrast to current IP addresses which are serving both purposes. Naming content directly enables the exploitation of in-network caching in order to improve delivery of popular content. Each content object can now be uniquely identified and authenticated without being associated to a specific host. This enables application-independent caching of content pieces that can be re-used by other end users requesting the same content. In fact, one of the salient ICN features is in-network caching, with potentially every network element (i.e., router) caching *all* content fragments[1] that traverse it; in this context, if a matching request is received while a fragment is still in its cache store, it will be forwarded to the requester from that element, avoiding going all the way to the hosting server. Out of the current ICN approaches, NNC [2] advocates such indiscriminate content caching.

We argue that such an indiscriminate universal caching strategy is unnecessarily costly and sub-optimal and attempt to study alternative in-network caching strategies for enhancing the overall content delivery performance. We address the central question of whether caching only at a specific sub-set of nodes en route the delivery path can achieve better gain. If yes, which are these nodes to cache and how can we choose them?

Our contribution in this study is three-fold. First, we contribute to the understanding of ubiquitous caching in networked systems by providing insights into its behavior for specific topology types. Second, we demonstrate that selective instead of ubiquitous caching can achieve higher gain even when using simplistic random selection schemes. Third, we propose a centrality-driven caching scheme by exploiting the concept of (ego network) betweenness derived from the area of complex/social network analysis, where only selected nodes in the content delivery path cache the content. The rationale behind such a selective caching strategy is that some nodes have higher probability of getting a cache hit in comparison to others and by strategically caching the content at "better" nodes, we can decrease the cache eviction rate and, therefore, increase the overall cache hit rate.

In the next section, we define the system of interest and layout our arguments and rationale with a motivating example illustrating that caching *less* can be *more*. We then describe our centrality-based caching scheme that can consistently outperform ubiquitous caching. We carry out a systematic simulation study that explores the parameter space of the caching systems, diverging from existing work in networked caches which mostly considers topologies with highly regular structure (e.g., string and tree topologies [6][7][8]), with the content source(s) usually located at the root of the topology forcing a sense of direction on content flows for tractable modeling and approximation. We present results for both regular and non-regular topologies, including scale-free topologies whose properties imitate closely the real Internet topology.

---

[1] In our study, the basic unit of a content can be a packet, a chunk or the entire object itself.

2   Caching in ICN

2.1   Model Description and Problem Statement

As a foundation, we first assume that the network has an ICN publish/subscribe framework in place (e.g., [1][2][3][4][5]). Specifically, we assume that a content request and resolution mechanism is already in place. As pointed out in [9], all the different ICN proposals in the literature invariably have such common functions (although with different primitives). Let $G = (V, E)$ be an undirected network with $V = \{v_1, ..., v_N\}$ nodes and $E = \{e_1, ..., e_M\}$ links. We denote $F = \{f_1, ..., f_R\}$ the content population in the system and $S = \{s_1, ..., s_P\}$ the set of content servers, each associated to a $v \in V$. The content population is randomly hosted in S and we assume that each content object is hosted permanently in only one server.

Content requests are assumed to arrive in the network exogenously and the content request arrival process for content unit r, $1 \leq r \leq R$, follows the Poisson process with mean rate, $\lambda = \sum_{r=1}^{R} \lambda_r$, whereby λ_r is the rate of exogenous content request for f_r. A *cache hit* is recorded for a request finding a matching content along the content delivery path. Otherwise, a *cache miss* is recorded. In the event of a cache miss, the content request traverses the full content delivery path to the content server. Following the convention in the literature, we assume that content units are of the same size and each cache slot in a cache store can accommodate one content unit at any given time. When a cache store is full, the least recently used content will be discarded in the event of an arrival of a new uncached content.

The objectives of this study are: (1) to examine the caching performance of such a system under different caching schemes, (2) to gain insights into the behavior of ubiquitous caching and (3) to develop more sophisticated caching algorithms for achieving better gain.

2.2   Related Work and Motivation

In the networking area, caching has been studied in standalone caches [10][11] focusing on the performance of different cache replacement policies. This isolates the effect of connected caching nodes (i.e., a network of caches). Caching has also been studied in the context of content distribution networks (CDNs) and in the World-Wide Web (web caching), in both cases in a network overlay fashion with some forms of collaborative (e.g., cooperative / selfish caching through game theory [12][13]) or structured (e.g., hierarchical caching [14][15]) caching approaches being considered. In ICN, caching takes place within the network, requiring line-speed operation; in this context, complex algorithms executed by multiple collaborating entities that require information exchanges are simply not feasible. One of the key ICN proposals, networking named content (NNC) [2], defines its ubiquitous caching as follows:

- A router caches every content chunk that traverses it with the assumption that routers are equipped with (large) cache stores.
- A least recently used (LRU) cache eviction policy is used.

This ubiquitous caching strategy ensures a quick diffusion of content copies throughout the network. Hereafter, we refer to this scheme as *NNC+LRU* and treat it as the benchmark for performance comparison.

Such a ubiquitous caching scheme has already raised doubts (e.g., [9]). In the general cache-related literature, some authors have already questioned this aggressive "cache-everything-everywhere" strategy [14][15][16]. The basic reasoning is that since the caching capacity is usually much smaller than the overall population of the items to be cached, it has the property of high cache replacement *error*. We illustrate this property of ubiquitous caching with a motivating example. We define a naïve random caching strategy, *Rdm+LRU*, which simply caches randomly at only *one* intermediate node along the delivery path per request, using LRU cache eviction policy. We compare the two caching schemes in a 7-node string topology where $s_1, P = 1$, is located at v_1 (root) while content requests originate exogenously from other nodes. We observe, in Fig. 1, that even random caching at just a single node along the content delivery path can reduce both the number of hops required to hit the content and the server hits in comparison to ubiquitous caching (*NNC+LRU*).

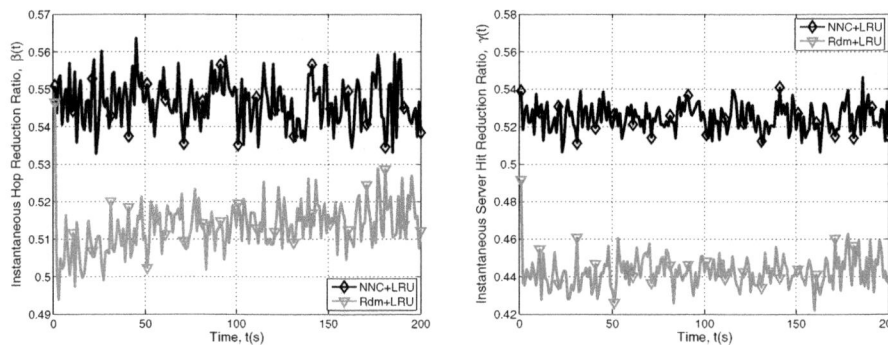

Fig. 1. Simple random caching outperforming ubiquitous caching in the number of hops to hit the content (left) and reduced server hits (right).

3   A Centrality-Based Caching Scheme

3.1   Basic Algorithm

Based on the above observations, we realize that caching indiscriminately does not necessarily guarantee the highest cache hit rate. On the other hand, this result cannot be used as conclusive evidence that caching less is better since the string topology constrains to a large extent the diversity of the content delivery paths (i.e., all delivery paths are fully or partially overlapping), a fact that indirectly increases the probability of a cache hit. Following this argument, we propose a novel caching scheme based on the concept of betweenness centrality [17] which measures the number of times a specific node lies on the content delivery path between all pairs of nodes in a network topology. The basic idea is that if a node lies along a high number of content delivery paths, then it is more likely to get a cache hit. By caching only at those more "*important*" nodes, we reduce the cache replacement rate while still caching content where a cache hit is most probable to happen.

Let's consider the topology in Fig. 2. At time $t=0$, all cache stores are empty and client A requests a content from s_1. The content is being routed via $v_1 \rightarrow v_2 \rightarrow v_3 \rightarrow v_4$ from s_1 to client A. With *NNC+LRU*, all four nodes will retain a copy of the content while under *Rdm+LRU*, only one of them will cache the content. Let's assume now that client B requests the same content. For *NNC+LRU*, the request is satisfied by v_3 but the cached copies at v_1, v_2 and v_4 are redundant. On the other hand, under *Rdm+LRU*, there is ¼ chance to get a cache miss (i.e., content cached at v_4) and ½ chance that the hop count reduction is worse than *NNC+LRU* (i.e., the copy is cached at either v_1 or v_2). However, with a bird's eye view, it is clear that caching the content only at v_3 is sufficient to achieve the best gain without caching redundancy at other nodes. This can be verified by using the betweenness centrality, whereby v_3 has the highest centrality value with most content delivery paths passes through it (i.e., 9 paths).

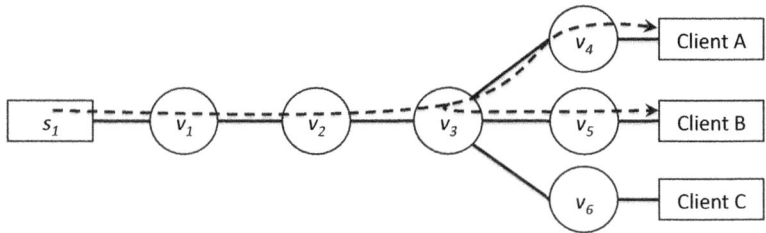

Fig. 2. An example topology with optimal caching location at v_3

We now present our algorithm which we call *Betw+LRU* hereafter. We assume that the betweenness centrality for each node is pre-computed offline (e.g., by the central network management system) as follows.

$$betweenness\ centrality, C_B(v) = \sum_{i \neq v \neq j \in V} \frac{\sigma_{i,j}(v)}{\sigma_{i,j}} \tag{1}$$

where $\sigma_{i,j}$ is the number of content delivery paths from i to j and $\sigma_{i,j}(v)$ is the number of content delivery paths from i to j that pass through node v. Computation of the delivery paths can be done online or dynamically, as our scheme does not require *a priori* path knowledge. Without loss of generality, we use the shortest path as the content delivery path in this paper.

Betw+LRU operates at *per request* level whereby the selected caching node may differ from one delivery path to another. Hence, there is no *fixed* pre-configured caching node in the network (e.g., solutions to *k*-median problems). Specifically, when a content client initiates a content delivery, the request message (e.g., `Find` in [1], `Interest` in [2], `Consume` in [5]) records the highest centrality value among all the intermediate nodes. It may be inserted into the request packet header. This value is copied onto the content messages during the data transmission at the server.

On the way to the requesting user, each router matches its own C_B against the attached one and the content is cached only if the two values match. If more nodes have the same highest centrality value, all of them will cache the content. Note that our solution is highly lightweight as each node independently makes its caching decision solely based on its own C_B, neither requiring information exchange with other nodes nor inference of server location or of traffic patterns as it is the case with collaborative or cooperative caching schemes. In this case, the C_B value is pre-computed offline and configured to every router by the network management system. The pseudo-code for forwarding both the request and the actual content is given below:

Content Request	1. Initialize (C_B=0) 2. **foreach** (v_n from i to j) 3. **if** data in cache 4. **then** *send(data)* 5. **else** 6. Get $C_B(v_n)$ 7. **if** $C_B(v_n)$ > C_B 8. **then** $C_B = C_B(v_n)$ 9. forward request to the next hop towards j
Content Data	1. *Record C_B from corresponding content request* 2. **foreach** (v_n from j to i) 3. Get $C_B(v_n)$ 4. **if** $C_B(v_n)$ == C_B 5. **then** cache(data) 6. forward data packet to the next hop towards i

3.2   Approximation via Distributed Computation

We now sketch a distributed implementation for *Betw+LRU* where the full network topology may not be readily available because of an infrastructure-less network with relatively dynamic topology (e.g., for self-organizing, ad hoc and mobile networks). Since in this case it is not practical for dynamic nodes to efficiently obtain the knowledge of delivery paths between all pairs of nodes in the network, we envision that the nodes themselves can compute an *approximation* of their C_B. This approximation is based on the ego network betweenness concept [18]. The ego network consists of a node together with all of its immediate neighbours and all the links among those nodes. The idea is for each node, v to compute its $C_B(v)$ based on its ego network rather than the entire network topology. From [18], if A is the $N \times N$ symmetric adjacency matrix of G, with $A_{i,j} = 1$ if there exists a link between i and j and 0 otherwise, then the ego network betweenness is $A^2[\mathbf{1} - A]_{i,j}$ where $\mathbf{1}$ is a matrix of 1's.

From an implementation point of view, the construction of the ego network for each node can be done by simply requiring each node to broadcast the list of its one-hop neighbours with message Time-To-Live=1 when it first joins the network and whenever there are changes to its one-hop neighbour set. The overhead is thus limited as the message propagation is limited to one hop only. The ego network can then be built by adding links that connect to itself or its own neighbors based on the received neighbor lists and ignoring the entries to nodes not directly connected to itself. The ego network betweenness is simply the $\sigma_{i,j}(v)$ of v's ego network. The rest of the caching operations remain unchanged (as described in the previous section).

Although the ego network betweenness only reflects the importance of a node within its ego network, it has been found that it is highly correlated with its betweenness centrality counterpart in real-world Internet service provider (ISP) topologies [19]. Coupled with its low computation complexity (reduced from $\mathcal{O}(NM)^2$ to $\mathcal{O}(d_{max}^2)$ where d_{max}^2 is the highest node degree in the network), it presents itself as a good alternative for large / dynamic networks. This caching algorithm using ego network betweenness centrality along with the LRU cache eviction policy is referred to as *EgoBetw+LRU* hereafter. Referring back to Fig. 2, the outcome of *Betw+LRU* and *EgoBetw+LRU* is the same since v_3 remains the node having the highest centrality value.

4   Performance Evaluation

4.1   Performance Metrics and Simulation Scenarios

Caching in networks aims to: (1) lower the content delivery latency whereby a cached content near the client can be fetched faster than from the server, (2) reduce traffic and congestion since content traverses fewer links when there is a cache hit and (3) alleviate server load as every cache hit means serving one less request. We use the *hop reduction ratio, β* as the metric to assess the effect of the different caching schemes on (1) and (2) above while we use the *server hit reduction ratio, γ* on (3).

$$Hop\ reduction\ ratio, \beta(t) = \frac{\sum_{r=1}^{R} h_r(t)}{\sum_{r=1}^{R} H_r(t)} \quad (2)$$

where $H_r(t)$ is the path length (in hop count) from client(s) to server(s) requesting f_r from time t-1 to t and $h_r(t)$ is the hop count from the content client to the first node where a cache hit occurs for f_r from t-1 to t. If no matching cache is found along the path to the server, then $h_r = H_r$. In other words, the hop reduction ratio counts the percentage of the path length to the server used to hit the content given caching in intermediate nodes. In a non-caching system, $\beta = 1.0$.

$$Server\ hit\ reduction\ ratio, \gamma(t) = \frac{\sum_{r=1}^{R} w_r(t)}{\sum_{r=1}^{R} W_r(t)} \quad (3)$$

where $W_r(t)$ is the number of request for f_r from t-1 to t and $w_r(t)$ is the number of server hits for f_r from t-1 to t. Note that high hop reduction does not directly translate to high server hit reduction.

---

[2] Based on the best known betweenness computation algorithm in U. Brandes, "A faster algorithm for betweenness centrality", Journal of Mathematical Sociology 25(2):163-177.

We seek to draw insights from the inspection of network topologies with very different structural properties – (1) k-ary trees which have almost strict regular structure (i.e., all nodes besides the root and leaves have the same $k + 1$ valence) and (2) scale-free topologies following the Barabasi-Albert (B-A) power law model [20] which accounts for the preferential attachment property of the Internet topology and results in graphs with highly skewed degree distribution. It is interesting to note that the betweenness distribution of B-A graphs also follows the power law model [21].

Content requests for different content are generated based on Zipf-distribution with $\sum_{r=1}^{R}(C/r^{\alpha}) = 1$ where the probability for a request for the rth popular content is C/r^{α} with α being the popularity factor. We use $\alpha = 1.0^3$ and requests originate randomly from all nodes. Each simulation run begins with all cache stores being empty (i.e., cold start). Unless otherwise specified, the simulations are run with the following parameters: total simulation time = 200 s, $\lambda = 5,000$ request/s, content population = 1,000 and uniform cache store size = 10% of total content population.

4.2   Experiments with k-ary Trees

(a) Instantaneous Behavior

A k-ary tree is defined via two parameters, namely k, the spread factor, denoting the number of children each node has and D is the depth of the tree from root. We first show in Fig. 3 the instantaneous behavior of the different caching schemes for both β and γ in a 5-level binary tree ($k=2$, $D=4$). All caching schemes reach a stationary performance after a few seconds. We point out that since all simulations go through a *warm-up* phase, *NNC+LRU* always reaches the stable performance level first. This is due to its *always cache* policy.

We observe that both *Betw+LRU* and *Rdm+LRU* perform better than *NNC+LRU* for both metrics. Tracking the evolution of the cache stores over time revealed that this is due to the high cache replacement rate in *NNC+LRU*. Replacing cached content rapidly causes content often being evicted before the next matching request is received. We have shown this in [6]. The effect is magnified considering that the whole chain of caches on the delivery path is affected. This is the fundamental basis on why the counter-intuitive caching *"less for more"* can be true. We further observe that the argument that caching selectively may increase cache miss is untrue in k-ary trees. We do find that there are more cache misses if the caching node is randomly selected rather than caching at nodes with high betweenness. Finally, an interesting observation is that instead of approximating the performance of the *Betw+LRU* scheme as it was meant to be, *EgoBetw+LRU* actually performs at the same level as *NNC+LRU*. This is the due to the regularity of the topology whereby nodes between the root and the leaves have the same ego network and thus, have the same C_B. Since the algorithm specifies that all nodes with equal highest C_B along the delivery path should cache, in this case *EgoBetw+LRU* is simply reduced to a similar behavior with *NNC+LRU*.

---

[3] From our results, we note that the order of performance amongst the caching schemes remains unchanged for $0.6 \leq \alpha \leq 1.5$. So, the results presented here are valid for these values of α.

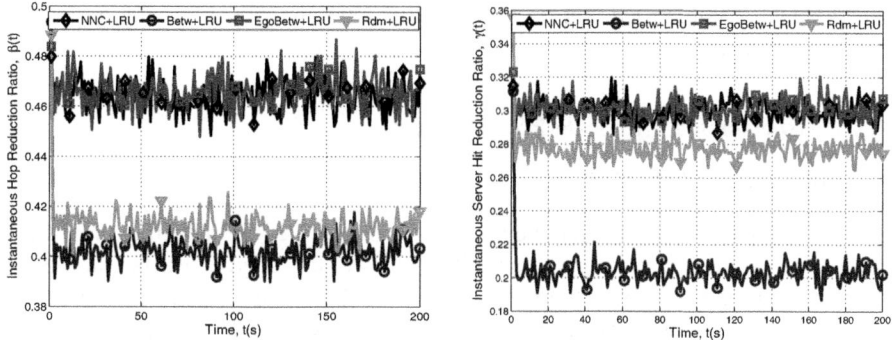

Fig. 3. Instantaneous behavior of the caching schemes for a binary tree; (left) β, (right) γ

(b) Effect of Topology Features on Performance

In k-ary trees, D affects the expected path lengths and k impacts the path diversity. We now study the validity of the previous observations in different configurations of k-ary trees by obtaining the β at 95% confidence interval for a range of depths and spread factors. Our results in Fig. 4 suggest that the caching schemes exhibit consistent behavior for different k-ary trees.

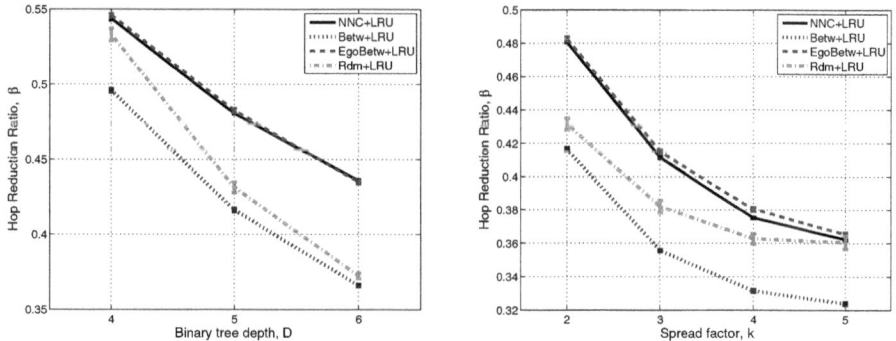

Fig. 4. Betw+LRU consistently outperforms the rest over different D (left) and k (right)

We find that while the performance distance between *NNC+LRU* and *Betw+LRU* remains approximately constant, *Rdm+LRU* does not exhibit such consistency. *Rdm+LRU* performs increasingly better in terms of hops saved when D is increased and k is decreased. This is due to the fact that each node has equal probability to cache content and in effect, distributes cache replacement operation uniformly across different nodes. In turn, this results in content being cached longer when compared to *NNC+LRU*. It increases the cache hit probability especially in topologies with very low number of content delivery paths. This, however, is counter-balanced by the increased number of branches in the topology, whereby a greater number of cache misses will occur. Our *Betw+LRU* scheme does not suffer from such a drawback since the caching node always has the highest probability of getting a cache hit and thus maintains stable cache hit (reducing server hits) and gain (reducing the content delivery hop count).

4.3   Experiments with Scale-Free Topologies

(a) Instantaneous Behavior

Although regular graphs lend themselves to tractability in modeling, real-world Internet topologies are not regular but follow a power law degree distribution [20]. As such, we consider scale-free topologies following the construction method described in [20] (referred to as B-A graphs hereafter). We show in Fig. 5 the performance of the different caching schemes in a B-A graph with $N = 100$ over time. First and foremost, we see that the performance of both our centrality-based caching schemes (*Betw+LRU* and *EgoBetw+LRU*) perform better than *NNC+LRU* for both metrics and *EgoBetw+LRU* now approximates closely *Betw+LRU*. This is because, without the regular structure, the ego networks of the nodes within the B-A graphs reflect correctly their actual betweenness. This result, thus, suggests that the more scalable and distributed *EgoBetw+LRU* algorithm can be used for irregular graphs.

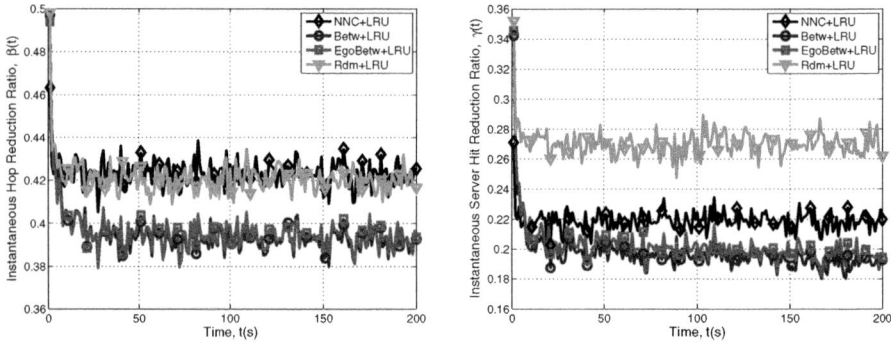

Fig. 5. Instantaneous behavior of the caching schemes in a B-A graph; (left) β, (right) γ

Secondly, we observe that *Rdm+LRU* no longer outperforms *NNC+LRU*. In fact, it performs at the same level as *NNC+LRU* with respect to hop reduction and due to the highly skewed degree distribution in the topology, it fails to alleviate load from the server (i.e., it has the highest number of cache misses).

(b) Effect of Topology Features on Performance

Unlike k-ary trees which are fully described via the tuple (k, D), each generation of a B-A graph with the same parameters results in a different topology since the links are created based on the probability proportional to the attractiveness of existing nodes (i.e., preferential attachment). We evaluate the caching schemes over ten B-A graphs with $N = 100$ and mean valence = 2. From Fig. 6 (left), both centrality-based caching schemes perform better than the rest. However, *Rdm+LRU* is worse than *NNC+LRU* in most cases even with the topology having the same properties. This is due to the skewed node degree distribution of the graph that increases the probability of the scheme caching at nodes having low cache hit probability. Fig. 6 (right) shows how ego network betweenness approximates betweenness in a B-A graph.

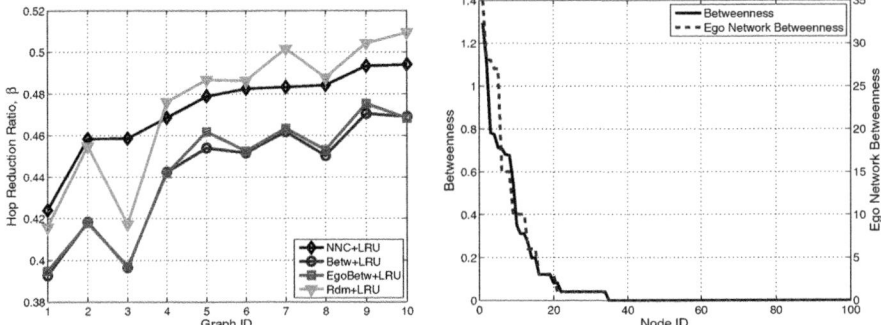

Fig. 6. Caching performance with different 100-node B-A graphs (left) and a sample ego network betweenness and betweenness values of the nodes in a B-A graph (right)

From Fig. 7 (left), we observe again that centrality-based caching schemes provide the best hop reduction ratio while *Rdm+LRU* exhibits inconsistent gain across B-A graphs with different sizes. We observe that as the size of the topology increases, *Rdm+LRU* gradually performs worse than *NNC+LRU*. The power-law distribution of betweenness in B-A graphs plays a vital role in this phenomenon as it results in high number of nodes having low probability of getting a cache hit. Since *Rdm+LRU* does not differentiate the centrality of the nodes, there is higher probability of *Rdm+LRU* caching at these "unimportant" nodes. Note that this observation is untrue for k-ary trees (the case when D is increased) due to the high number of overlapping shortest paths (an obvious example being the string topology).

From Fig. 7 (right), we see that different request intensities do not affect the order of performance amongst the caching schemes. This is due to the fact that all caching schemes converge to a stable performance level (cf., Fig. 1, 3 and 5).

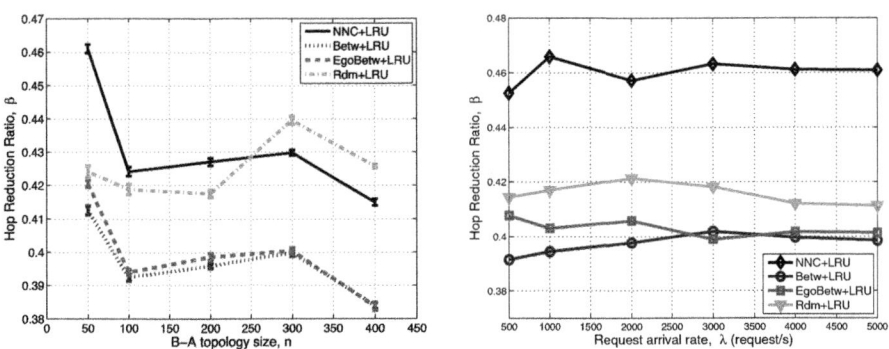

Fig. 7. Hop reduction ratio for different B-A graph sizes (left) and request rates, λ (right)

In Table 1, we provide representative results of the different caching schemes across the different topologies in terms of number of hops and server hits saved. It is clear that *Betw+LRU* reliably achieves better gains (both in terms of hop and server hit reduction) in comparison to *NNC+LRU*. For instance, it reduces server hits over 30% and hop count over 17% in comparison to *NNC+LRU* in the string topology.

Table 1. Sample performance achieved after 200s in different types of topology

Caching Scheme	String (D = 10, k=1)		k-ary Tree (D = 4; k = 2)		B-A (N = 100)	
	$\sum h$	$\sum w$	$\sum h$	$\sum w$	$\sum h$	$\sum w$
NNC+LRU	2,6839,45	498,603	2,684,325	299,657	2,137,015	211,852
Betw+LRU	2,211,248	337,362	2,331,061	203,673	2,045,852	204,479
EgoBetw+LRU	2,680,614	497,146	2,698,153	301,797	2,074,089	207,628
Rdm+LRU	2,206,002	377,289	2,386,569	277,575	2,195,303	291,560

4.4   Experiments with the Real Internet Topologies

To further verify our findings, we proceed to assess the caching performance of the different caching schemes in a real-world Internet topology. We focus on a large domain-level topology, extracting a sub-topology from the CAIDA dataset [22]. The topology is rooted at a tier-1 ISP (AS7018) and contains 6804 domains and 10205 links. We do not aggregate stub domains while sibling domains/links are not considered. In a similar manner to the previous simulation setup, all content servers and clients are randomly distributed across the topology. Fig 8 shows both the hop reduction and server hit reduction ratios achieved in this setup.

The results show that the different caching schemes behave in a similar fashion to the B-A graphs but not to k-ary trees, reinforcing the notion that B-A graphs reflect better real network topologies. These results further confirm the validity of our centrality-based caching scheme even in large real network topologies.

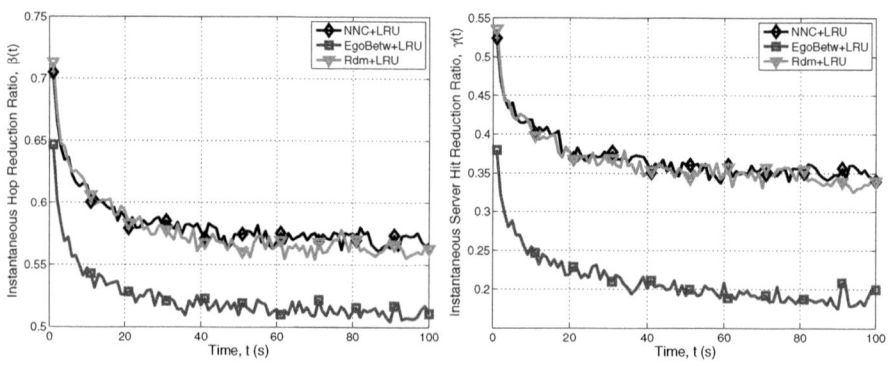

Fig. 8. Instantaneous behavior of the caching schemes in a large-scale real Internet topology; (left) β, (right) γ

5   Summary and Conclusions

We argue against the necessity of a ubiquitous caching strategy in ICN and investigate the possibility of caching less in order to achieve higher performance gain. We first demonstrated that a simple random caching strategy (*Rdm+LRU*) can outperform (though inconsistently) the current pervasive caching paradigm under the conditions that the network topology has low number of distinct content delivery paths and high average delivery path length. We, then, proposed a caching strategy based on the concept of betweenness centrality (*Betw+LRU*) such that content is only cached at the nodes having the highest probability of getting a cache hit along the content delivery path. We also proposed an approximation of it (*EgoBetw+LRU*) for scalable and distributed realization in dynamic network environments where the full topology cannot be known a priori. We compared the performance of our proposals against the ubiquitous caching of the NNC proposal [2] (*NNC+LRU*). Based on our extensive simulations, we observed that *Betw+LRU* consistently achieves the best hop and server reduction ratios across topologies having different structural properties without being restricted by the operating conditions required by *Rdm+LRU*. Our results further suggest that *EgoBetw+LRU* approximates closely *Betw+LRU* in non-regular topologies (e.g., B-A graphs) and thus presents itself as a practical candidate for the deployment of this approach. Besides synthetic topologies (i.e., *k*-ary trees and B-A graphs), the observations are further verified with a large-scale real Internet topology. Thus, we conclude that indeed caching *less* can achieve *more* and our proposed *(Ego)Betw+LRU* is a candidate for realizing this promise.

Acknowledgements. This work was undertaken under the Information Society Technologies (IST) COMET project, which is partially funded by the Commission of the European Union. We would also like to thank our project partners who have implicitly contributed to the ideas presented here.

References

1. Koponen, T., et al.: A Data-oriented (and Beyond) Network Architecture. In: Proc. ACM SIGCOMM 2007, Kyoto, Japan (August 2007)
2. Jacobson, V., Smetters, D.K., Thornton, J.D., Plass, M., Briggs, N., Braynard, R.L.: Networking Named Content. In: Proc. ACM CoNEXT, pp. 1–12 (2009)
3. Trossen, D., et al.: Conceptual Architecture: Principles, Patterns and Sub-components Descriptions (May 2011), http://www.fp7-pursuit.eu/PursuitWeb/
4. Jokela, P., Zahemszky, A., Rothenberg, C., Arianfar, S., Nikander, P.: LIPSIN: Line Speed Publish/Subscribe Inter-networking. In: Proc. ACM SIGCOMM, Barcelona, Spain (2009)
5. Chai, W.K., et al.: CURLING: Content-ubiquitous resolution and delivery infrastructure for next-generation services. IEEE Commun. Mag. 49(3), 112–120 (2011)
6. Psaras, I., Clegg, R.G., Landa, R., Chai, W.K., Pavlou, G.: Modelling and Evaluation of CCN-Caching Trees. In: Domingo-Pascual, J., Manzoni, P., Palazzo, S., Pont, A., Scoglio, C. (eds.) NETWORKING 2011, Part I. LNCS, vol. 6640, pp. 78–91. Springer, Heidelberg (2011)

7. Carofiglio, G., Gallo, M., Muscariello, L., Perrino, D.: Modelling data transfer in content centric networking. In: Proc. International Teletraffic Congress, ITC (2011)
8. Arianfar, S., Nikander, P., Ott, J.: Packet-level caching for information-centric networking. Finnish ICT-SHOK Future Internet Project, Tech. Rep. (2010)
9. Ghodsi, A., et al.: Information-centric Networking: Seeing the forest for the trees. In: ACM Workshop on Hot Topics in Networks (HotNets-X), Cambridge, MA (November 2011)
10. Dan, A., Towsley, D.: An approximate analysis of the lru and fifo buffer replacement schemes. In: ACM SIGMETRICS, pp. 143–152 (1990)
11. Jelenkovic, P., Radovanovic, A., Squillante, M.S.: Critical sizing of lru caches with dependent requests. Journal of Applied Probability 43(4), 1013–1027 (2006)
12. Laoutaris, N., Smaragdakis, G., Bestavros, A., Matta, I., Stavrakakis, I.: Distributed selfish caching. IEEE Trans. on Parallel and Distributed Systems 18(10) (2007)
13. Dán, G.: Cache-to-Cache: Could ISPs cooperate to decrease peer-to-peer content distribution costs? IEEE Trans. on Parallel and Distributed Systems 22(9) (2011)
14. Che, H., Tung, Y., Wang, Z.: Hierarchical web caching systems: modelling, design and experimental results. IEEE Journ. on Selected Areas of Communications 20(7) (2002)
15. Laoutaris, N., Che, H., Stavrakakis, I.: The LCD interconnection of LRU caches and its analysis. Performance Evaluation 63(7), 609–634 (2006)
16. Wong, T.M., Wilkes, J.: My cache or yours? Making storage more exclusive. In: Proc. USENIX Annual Technical Conference, Monterey, CA, pp. 161–175 (2002)
17. Izquierdo, L.R., Hanneman, R.A.: Introduction to the Formal Analysis of Social Networks Using Mathematica. University of California, Riverside
18. Everett, M., Borgatti, S.: Ego network betweenness. Social Networks 27, 31–38 (2005)
19. Pantazopoulos, P., Karaliopoulos, M., Stavrakakis, I.: Centrality-driven scalable service migration. In: Proc. International Teletraffic Congress, ITC (2011)
20. Barabasi, A.L., Albert, R.: Emergence of scaling in random networks. Science 286(5439), 509–512 (1999)
21. Wang, H., Hernandez, J.M., Van Mieghem, P.: Betweenness centrality in a weighted network. Physical Review E 77, 046105 (2008)
22. CAIDA dataset, http://www.caida.org/research/topology/#Datasets

Collaborative Forwarding and Caching in Content Centric Networks

Shuo Guo[1], Haiyong Xie[2,3,*], and Guangyu Shi[3]

[1] University of Minnesota, Minneapolis, Minnesota 55455
sguo@umn.edu
[2] University of Science and Technology of China, Hefei, China
haiyong.xie@ustc.edu
[3] Central Research Institute, Huawei Technologies, Shenzhen, China
shiguangyu@huawei.com

Abstract. Content caching plays an important role in content-centric networks. The current design of content-centric networks adopts a limited, en-route hierarchical caching mechanism, and caching and forwarding are largely uncoordinated. In this paper, we propose a novel collaborative caching and forwarding design. In this design, collaboration is guided by content popularity ranking, based on which we introduce a collaborative forwarding table to allow coordination between caching and forwarding. We also propose a self-adaptive dual-segment cache division algorithm to deal with dynamic inconsistent content popularity. We evaluate our design via extensive simulations and demonstrate that our design improves content access cost and cache miss rate by at least 30% in a diverse network settings.

Keywords: content-centric network, name-based routing, collaborative forwarding and caching.

1 Introduction

Content caching plays an important role in content-centric networks (CCN) [25] or named-data networks (NDN) [38]. With routers being able to cache contents in such networks, it is likely that not only the content distribution costs incurred to the network but also the quality of service experienced by end users are significantly improved. Internet content caching, especially collaborative caching, has drawn much attention (*e.g.*, [8,14,16,20,21,23,26,29,32,35,37]) and some have become commercially successful since more than a decade ago. This leads us to believe that collaborative caching in CCN is a key to success in that the network performance could be significantly improved by letting routers collaborate with each other to optimize overall caching performance. Furthermore, forwarding, if coordinated with caching, is likely to further optimize the network performance.

The current design of CCN adopts a hierarchical caching mechanism allowing only *limited* collaboration in content caching. More specifically, for a given

---

* Corresponding author.

R. Bestak et al. (Eds.): NETWORKING 2012, Part I, LNCS 7289, pp. 41–55, 2012.

content, caching in CCN takes place only at en-route routers (*i.e.*, routers on the paths between a requesting host and one or multiple content origins), and thus forms a hierarchical caching mechanism. An en-route router that has the requested content will directly respond with the content from its local content store and then suppress further forwarding the request (*i.e.*, Interest) to the next router in the routing hierarchy. With its unique name-based routing architecture and Interest forwarding, CCN advocates a *"host-to-content"* communication model differing from the *"host-to-host"* model in Internet. In CCN, where content comes from is no longer important to the requesting host[1]. Additionally, not only en-route routers but also routers in the same administrative domain (particularly those nearby en-route routers) could have possibly cached a requested content. These observations suggest that collaborative caching beyond the current limited hierarchical mechanism is feasible and could be beneficial.

However, collaborative caching in CCN, if not well designed, could significantly increase the communication overhead. For instance, control messages exchanged among routers, as an example of such overhead, are necessary to enable collaborations. Such messages normally contain information about what contents are stored in a particular router; due to the enormously large number of distinct contents, such messages could consume a significant portion of the network bandwidth. Additionally, the extra latency of exchanging such messages may further slow down the collaborative decision making process and thus reduce the effectiveness. A naive approach to collaborative caching is to adopt a broadcast mechanism, *i.e.*, each request is forwarded to all routers and only those with the requested content respond with the data. However, such an approach is too costly and inefficient. A key challenge to collaborative caching in CCN is how to make routers know what contents are available from other collaborative routers in an economic and efficient manner. Furthermore, since routers have knowledge about such availability information, routers should leverage it when making forwarding decisions; namely, forwarding and caching should be coordinated and collaborative.

In this paper, we go beyond the en-route caching mechanism and propose a novel name-based distributed collaborative forwarding and caching design (referred to as *CFC* for short) for content-centric networks. More specifically, collaboration is guided by content popularity, and content popularity is measured distributively by content routers. Each router maintains an Availability Information Base (*AIB* for short), estimating which content could be available from which router. Each router also generates a *popularity ranking sequence* periodically through local measurements and propagated such sequences; after aggregating sequences announced by other routers, each router is able to update its AIB and leverage it to optimize forwarding decisions. However, in practice content popularity is likely different when measured from different routers. In order to deal with such inconsistency seen by different routers, we are inspired by the PodNet Project [24] and propose a self-adaptive dual-segment cache division design, using an additional cache space

---

[1] CCN architecture has measures to ensure content security, which is beyond the scope of this paper.

to handle inconsistent content requests and dynamically adjusting cache division based on different levels of inconsistency.

We summarize our contributions as follows. Firstly, to the best of our knowledge, our popularity-ranking based collaborative forwarding and caching scheme for content-centric networks is the first to coordinate forwarding and caching decisions through the availability information base, allowing us to utilize the information of content popularity ranking to reduce the network cost for cache collaboration. When assuming consistent popularity, we theoretically prove the optimality of our design. Secondly, We propose a novel self-adaptive dual-segment cache design to deal with popularity inconsistency. Thirdly, we evaluate the performance of our design via extensive simulations and demonstrate that our design outperforms the hierarchical caching design significantly.

The rest of the paper is organized as follows: Section 2 summarizes the related work. Section 3 introduces the network model. Sections 4 and 5 present our collaborative caching design, followed by the evaluations in Section 6. Section 7 concludes the paper with future work.

2   Related Work

In recent years there has been a line of work on emerging future Internet architectures (see, *e.g.*, [5,6,15,25,38]). In such architectures, content caching becomes an inherent capability of network elements such as routers. Without specifying the details of content caching, these architectures are designed to allow flexible design and implementation of new caching schemes. However, they also pose new challenges to caching schemes; in particular, it remains unclear how content caching should be provisioned (independently or collaboratively), and how it should be implemented efficiently. To the best of our knowledge, our work is among the first attempts to investigate these issues and provide new insights through comparative evaluations.

There is also a large body of literature on content caching in traditional network architectures (see, *e.g.*, [8,14,16,17,20,21,23,26,29,32,33,35,37]). Content caching has been an integral component of Internet-based services for many years, and this has been reflected by the proliferation of content delivery networks (*e.g.*, [3,4,11,27,28,30]). Collaborative caching (or cooperative caching) has been a long-lasting research topic. Researchers have not only investigated the effectiveness of collaborative caching (see, *e.g.*, [21,35]), but also proposed numerous collaborative caching schemes for both general networks and networks with specific structures (see, *e.g.*, [8,14,19,20,23,26,29,32,33,37]); for instance, general Internet-based content distribution (see, *e.g.*, [20,26,34]), delivering content of special types (*e.g.*, [29]), content caching in networks with special topological structures (*e.g.*, [8]), and content caching in special networks such as ad hoc networks (*e.g.*, [24]), content-centric networks (*e.g.*, [10,31,36], and peer-to-peer networks (*e.g.*, [23]).

Our work clearly differs from the above work in that we take advantage of the unique properties of content-centric networks, propose the Availability Information Base guided by the popularity ranking sequence to achieve coordinated forwarding and caching for content routers, and propose the self-adaptive dual-segment division algorithm to deal with inconsistent popularity.

3   Network Model

We consider an autonomous system managed by an administrative domain. The network consists of N content routers. Each router i has a local Content Store (CS) that can cache up to C_i content objects ("contents" for short). The size of each content is u at the largest. We assume that content can be chunked into pieces[2], and each piece fits one cache unit (*i.e.*, the size of each piece is no greater than u). Then, the entire network can cache at most Cu of data, where $C = \sum_{i=1}^{N} C_i$. Users send requesting packets of "Interest" to their nearest routers (see, *e.g.*, [25]).

We use a ranking sequence $\{r_1, r_2, ..., r_C\}$ to denote the most popular C contents, sorted in descending order of popularity. This ranking sequence can be measured in real time as routers receive Interests. All routers may not see the same distribution of content popularity; however, we assume that the ranking sequence $\{r_1, r_2, ..., r_C\}$ measured by different routers have a certain percentage of mismatches or shifts, refer to as the *popularity inconsistency*.

In intradomain, topological information, *e.g.*, link status information and link costs between any pair of adjacent routers i and j (denoted by d_{ij}), is typically distributed by intradomain routing protocols such as OSPF and ISIS. Link costs can correspond to IGP link weights, or other metrics that the content-oriented network cares (*e.g.*, distance, latency). When other metrics are used, they can be distributed across the whole domain via OSPF TLV messages. Other topological information including sizes of Content Stores (C_i's) and the average rate of arriving Interests (*Received Interest Rate* for short), denoted by I_i for router i, can also be distributed in the same way as link costs.

Note that an Interest contributes to a router's Received Interest Rate only if it is sent to this router by a client, rather than another router. We assume that a router can easily distinguish (by, *e.g.*, adding an additional flag bit in the Interest packets) if an Interest comes from a user, or instead from a collaborative router.

4   Ranking-Based Collaborative Forwarding and Caching

In this section we present our design for the collaborative forwarding and caching scheme.

4.1   Overview

We introduce a new component, the *Availability Info Base* (*AIB* for short), to allow us coordinate forwarding and caching in content routers, as shown in

---

[2] Content objects are segmented into pieces in many content-centric networks (*e.g.*, [25, 38]) as well as in many content-oriented overlay networks (*e.g.*, [7, 18]).

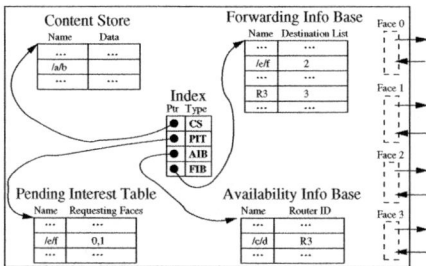

Fig. 1. Content router with collaborative forwarding and caching

Fig. 1. AIB keeps track of content availability information. More specifically, AIB can be thought of as a table, where each entry has two columns, *Name* and *RouterID*, suggesting that a given named content is available from a router. Note that we assume that each router in the network has a name and routers' names are propagated through the network via intradomain routing protocol such as OSPF. As a result, routers' names are treated in the same way as content names and put in FIB. For instance, the outbound face to reach Router R3 is face 3, and content /c/d is available from R3, as shown in Fig. 1.

Each content router periodically announces the pairwise link cost and collaborative forwarding/caching related metrics via OSPF or ISIS intradomain routing protocols. Each router also measures the ranking of incoming Interests, namely, examine the received Interests from the users, and generates its local ranking sequence of the most popular C contents. Each router implements the collaborative forwarding and caching mechanism, namely, a distributed mechanism to make joint decisions for forwarding and caching. Additionally, each router measures the miss rate of the interests in order to further improve the caching/forwarding efficiency.

Upon receiving an Interest, a router first checks whether the content is available and fresh in its local CS. If yes, the router responds with the locally cached content. Otherwise, it looks up the Pending Interest Table (PIT), and either this Interest should not be forwarded if it is already pending in PIT, or it should be forwarded and PIT be updated accordingly. In the latter case, the router looks up AIB to check whether the content is available from other collaborative routers. If not, the Interest should be forwarded using the default policy in content-centric networks (*e.g.*, look up the outbound face in FIB and forward to the designated face; if FIB lookup fails, use a broadcast-like approach to forward the Interest). Otherwise, the Interest should be forwarded to the designated collaborative router. In order to do so, the router needs to look up FIB to determine the outbound face to reach the designated router. Note that most likely retrieving contents from collaborative routers within an autonomous system saves a noticeable time than getting it from the origin, as the latter typically requires traversing multiple autonomous systems and multiple interdomain links.

4.2   Collaborative Forwarding and Caching Mechanism

We next describe the collaborative forwarding and caching mechanism. In our design, each content router keeps track of the most popular C contents. We now assume a consistent popularity model where the most popular C contents at each router results in the same ranking sequence $\{r_1, r_2, ..., r_C\}$; under this assumption, we need to understand how to optimally distribute these C contents in the N caches in the corresponding N routers, whose sizes are $C_1, C_2, ..., C_N$, so that the average content access cost can be minimized in the network.

Recall that I_i denotes the average number of interests received by router i and d_{ij} denotes the link cost of nodes i and j (d_{ij} becomes the cost of accessing content from the local Content Store when $i = j$). Such costs can correspond to either intradomain routing weights or other performance-related metrics such as distance and latency. Then, for any cache unit in router i, the average cost of accessing this content requested by users is

$$cost_i = \frac{\sum_{k=1}^{N} I_k d_{ki}}{\sum_{k=1}^{N} I_k}, \tag{1}$$

where $cost_i$ is the weighted sum of pair-wise access costs from all N routers.

The following theorem states how contents should be optimally distributed to minimize the average content access cost:

Theorem 1. *Suppose the average cost for accessing the content in a cache unit of router i is $cost_i$. Without loss of generality, assume that after sorting, $cost_1 \leq cost_2 \leq ... \leq cost_N$. Also, suppose the ranking sequence is $r_1, r_2, ...r_C$ in descending order of popularity. Then, the solution that minimizes the average cost of accessing the most popular C contents in the network is to let the more popular content be cached at a place that has a lower cost, i.e., router 1 caches $r_1, ..., r_{C_1}$, router 2 caches $r_{C_1+1}, ..., r_{C_1+C_2}$, and so on.*

Proof. The proof is straightforward using contradiction. Suppose in the optimal solution that minimizes the average cost of accessing the C most popular contents, there exists two contents $r_i < r_j$ (r_i is more popular than r_j) stored at routers p and q, respectively. Routers p and q follows $cost_p > cost_q$. Then, by swapping content r_i and r_j, we get a smaller average content access cost because the frequency of accessing r_i is higher than r_j.

Theorem 1 sheds light on how we should design the distributed collaborative mechanism. More specifically, with the help of intradomain routing protocol, topological information is generally available to each router; as a result, each router can calculate $\{cost_i | i \in [1, N]\}$ for all collaborative routers in the network and sort all routers using these values. Following Theorem 1, the most popular contents should be stored by the router with the least cost, and the less popular contents by the router with a larger cost. Therefore, for any top-C popular content, each router knows not only which contents it should keep in its local Content Store, but also which collaborative router it can request this content from, if not locally available. Such availability information for the top-C contents is stored in AIB.

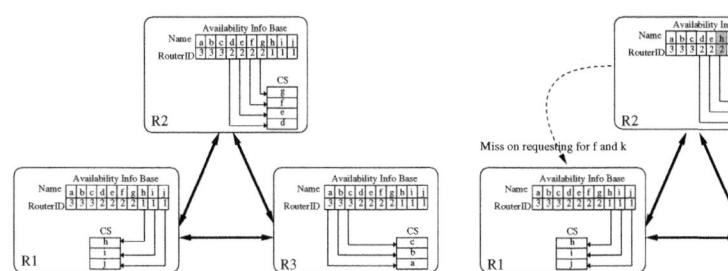

Fig. 2. A design for collaborative forwarding and caching

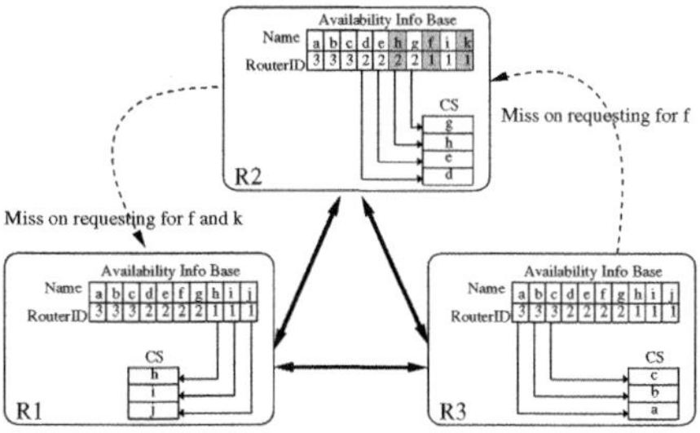

Fig. 3. An example of inconsistent ranking sequences

Example 1. We illustrate our design using a simple example shown in Fig. 2. In the example, there are three collaborative routers R1, R2 and R3. These routers can cache 10 contents in total (the size of these three caches are 3, 4, and 3, respectively). The most popular 10 contents measured by these routers are consistent. Suppose $cost_1 \geq cost_2 \geq cost_3$. Based on Theorem 1, the most popular 3 contents should be cached in router R3, the next 4 contents should be cached in R2, and the next 3 contents should be cached in R1. As shown in the figure, for any incoming Interest requesting for the most popular C contents, AIB tells where it should be forwarded to (the content is either available from the router's local CS or other collaborative routers).

4.3   Dealing with Inconsistent Popularity Ranking

In practice, popularity rankings seen by different routers are more likely inconsistent. In such cases, the efficiency of our CFC scheme will be degraded.

Example 2. We illustrate inconsistency in popularity ranking through an example shown in Fig. 3. In this example, router R2's ranking sequence is slightly different from R1's and R3's. In R2's ranking sequence, the positions of content h and f are swapped and content k replaces j. As a result, router R2 caches contents $\{d, e, h, g\}$; however, routers R1 and R3 expect that R2 caches $\{d, e, f, g\}$ instead. Whenever R1 and R3 forward Interests for content f to R2, such Interests have to be further forwarded towards the origin by R2 (not shown in the figure). Similarly, R2 always forwards Interests for content f and k to R1, resulting in cache misses and further forwarding.

To address the above problem resulted by inconsistent popularity, we adopt a dual-segment cache design, namely, divide a router's Content Store into two segments: (1) the *Advertised Content Store* (*ACS* for short), denoted by C_i', is the regular collaborative cache that is operated the same way as described in the preceding subsection assuming consistent popularity; and (2) the *Complementary Content Store* (*CCS* for short), denoted by C_i'', is the cache space used for adapting to the inconsistency of popularity distribution. The rationale behind this dual-segment design is to leverage CCS to absorb contents that are

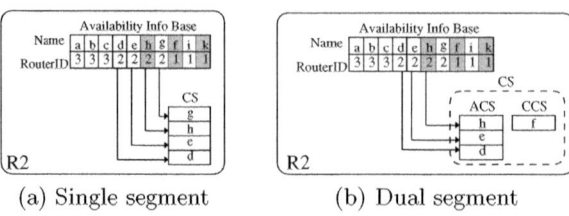

(a) Single segment   (b) Dual segment

Fig. 4. An example of cache space division

supposed to be store at a router but are missing in its ACS due to inconsistent popularity.

Upon receiving an Interest, if the requested content is available locally, the router directly responds with the data. Otherwise, if the Interest comes from another collaborative router, the router forwards the Interest towards the content origin and stores the returned data into its Complementary Content Store when the data comes back; if the Interest comes from a requesting host directly, the router applies its knowledge of popularity-ranking sequence and checks whether the ranking of the requesting content is less than C; If yes, it forwards the Interest to the collaborative router designated by the sequence; otherwise, it forwards the Interest towards the origin.

Example 3. Fig. 4 shows an example of this design for router R2 in the previous example. With the single-segment design shown in Fig. 4(a), content f, which is supposed to be cached in R2, is always missing due to the inconsistency of R2's ranking statistics. However, with the dual-segment design shown in Fig. 4(b), R2 only advertise 3 as its cache size. As a result, an extra cache unit can be used to store the missing content f. Therefore, future Interests requesting for f forwarded from routers R1 and R3 can be fulfilled by R2.

5   Adaptive Content Store Division

The impact of the preceding dual-segment design on the performance of collaborative forwarding and caching in content-centric networks could be subtle. On the one hand, a sufficiently large CCS is more favorable to adapt to the popularity inconsistency; and on the other hand, when the total size of the Content Store is fixed, a smaller CCS is more favorable, as the ACS could be larger to store more frequently requested contents in the network. Clearly there exist trade-offs when determining their sizes.

A straightforward solution is fixed division of ACS and CCS, *e.g.*, 90% dedicated to ACS and the remaining to CCS. However, the problem of fixed division is one size does not fit all, namely, routers may experience different levels of popularity inconsistency, and a fixed size may either over-estimate or under-estimate the inconsistency level, thus resulting an inefficient use of the cache space. This can be observed in the examples in Fig. 3 and Fig. 4. Router R1 and

R3 experience less popularity inconsistency than R2 does. In fact, they have no cache misses after R2 switches to the dual-segment cache as shown in Fig. 4. As a result, there is no need to allocate any space for CCS in router R1 and R3.

We note that the cache miss rate plays an important role in the efficiency of dual-segment collaborative caching. On the one hand, when the miss rate is low, it implies a potentially oversized CCS and thus a waste of cache space. On the other hand, when the miss rate is high, then contents supposed to be stored in a designated collaborative router are actually not stored in it, resulting in additional costs to forward Interests to the designated router and then towards the origin.

We design a distributed, self-adaptive algorithm to address the above division problem; specifically, we adjust the division of ACS and CCS based on the dynamics experienced by the content routers. We define the *Locking Miss Rate* (LMR for short) to characterize the maximum miss rate (corresponding to a maximum level of popularity inconsistency) that a router would like to tolerate. Every router distributively adjusts its size of CCS to make its miss rate closely approach to LMR.

More specifically, a router starts with a pre-configured initial cache division, *e.g.*, 90% for the Advertised and 10% for CCS. It then begins measuring the cache miss rate for all Interests received from other collaborative routers. Recall that the size of ACS and CCS are denoted by C_i' and C_i'' respectively.

We denote the measured cache miss rate by MR. If $MR \leq LMR$, it is likely that the router experiences less popularity inconsistency than expected, we may have an oversized CCS, so the size of CCS C_i'' is halved so that we increase the size of ACS C_i' to cache more contents in the network. If $MR > LMR$, the popularity inconsistency is likely under-estimated; therefore we should reduce the size of ACS to have a larger CCS in order for accommodating the inconsistency. However, instead of reducing the size of ACS C_i' aggressively, we linearly reduce it and allocate more space to CCS C_i'' based on the following equation:

$$C_i' \leftarrow (C_i' - \Delta C_i'), \qquad (2)$$

where $\Delta C_i'$ is the size reduced by ACS C_i' and increased by CCS C_i''. In this equation, the right hand side is a rough estimation of the number of contents expected to be missed (with the current size of CCS C_i'' unchanged) after the size of ACS C_i' is reduced to $(C_i' - \Delta C_i')$. Ideally, these missed contents are expected to store at the extra space allocated to CCS C_i''. With this equation, the percentage the size of ACS C_i' is reduced can be calculated by

$$\frac{\Delta C_i'}{C_i'} = \frac{MR}{1 + MR}, \qquad (3)$$

And as a result, the new ACS size should be $\frac{C_i'}{1+MR}$. This algorithm is shown in Algorithm 1.

Algorithm 1. Self-Adaptive Division Algorithm

1: Initialize C_i' and C_i'' such that $C_i' + C_i'' = C_i$.
2: τ is a pre-determined threshold, *e.g.*, $\tau \in [0, 0.5]$.
3: **while** TRUE **do**
4:     Measure MR for interests from other collaborative routers
5:     **if** $(\text{MR} < (1 - \tau) \cdot \text{LMR})$ **then**
6:         $C_i'' \leftarrow 0.5 C_i''$.
7:         $C_i' \leftarrow C_i - C_i''$.
8:     **else if** $(\text{MR} > (1 + \tau) \cdot \text{LMR})$ **then**
9:         $C_i' \leftarrow C_i'/(1 + MR)$.
10:        $C_i'' = C_i - C_i'$.
11:    **end if**
12: **end while**

---

6 Evaluations

In this section we systematically evaluate the collaborative forwarding and caching design via simulations.

6.1 Simulation Setup

Network Topologies. We use the pop-level network topology of Abilene [2] in simulations. Link costs are approximated by the measured end-to-end average latencies using PlanetLab and are in the range of $[10, 40]$. Note that we also run simulations on other network topologies (*e.g.*, CERNET [12]); however, the results are consistent and thus we omit them here due to space limit. We include two additional nodes to represent content origins in Europe and Asia. For each of these additional nodes, we choose the geographically closest node in each network and connect them. Thus we have a trans-Atlantic link and a Trans-Pacific link. We refer to the topology with these additional nodes and links as the *extended topology*. In this extended topology, the Trans-Atlantic link cost is 120 and the Trans-Pacific link cost is set to 200. Unless otherwise specified, the cache size of each node is 100 units; we assume contents are chunked into unit-sized pieces, and the total number of content pieces varies from 2000 to 20,000. We assume a 5% *LMR* by default, and the initial splitting ratio is 90%-10%, *i.e.*, 10% of the cache space is used as the Complementary Cache in the beginning.

Content Popularity and Requests. To generate inconsistent content popularity distribution, we use a single ranking sequence following the Zipf distribution [9, 13, 22] and randomly inject noises to 10% of the contents by shifting them to new positions up to a distance of 100 from their original positions in the sequence. We also evaluate different levels of inconsistency by shifting a different percentage of contents in the original sequence in Section 6.4.

Interests are generated following the Zipf distribution. The number of interests each router receives directly from users per unit time is randomly chosen from 100 to 500. The adaptive cache division algorithm is run every 50 unit time. The simulation duration in total is 1000 unit time.

Algorithms. We evaluate three schemes using simulations. The first is our collaborative forwarding and caching scheme, referred to as the *global collaborative forwarding and caching (GCC)*. The second is a hierarchical caching approach [8], denoted as *HIE*, where nodes are divided into two clusters based on their locations and there are two additional nodes serving as the upper-level caches whose size is three times of the size of lower-level caches. The third is a locally clustered variant of GCC, referred to as the *local collaborative forwarding and caching (LCC)*, where nodes are divided into two clusters (in the same way as HIE) and each cluster run its own GCC independently.

Note that when evaluating these algorithms, all nodes have an identical caching size in order to make the comparison fair. However, due to the extra two upper-level caching nodes, the total cache size of HIE is larger than that of GCC and LCC.

Performance Metrics. We quantify the performance using two metrics. The first metric is the average cumulative link costs of accessing content by a user, referred to as the *Average Access Cost*. The second metric is the percentage of interests received by each cache that are not fulfilled locally, referred to as the *Cache Miss Ratio*.

6.2   Impacts of Number of Contents

We first quantify the performance by varying the number of delivered contents. Fig. 5 summarizes the results of average content access cost in the Abilene networks. We make the following observations. Firstly, as the number of delivered contents increases, the average access cost increases monotonically. This is because with more contents, the percentage of the contents that can be accessed from caches decreases and more contents have to been accessed directly from content sources, leading to higher costs.

Secondly, our design, GCC, outperforms HIE and cuts the cost by more than 30%. The reason is that GCC can take full advantage of peer caches, while HIE allows only very limited collaboration between the low-tier and high-tier nodes.

Thirdly, the localized collaborative scheme, LCC, enforces a clustered structure, therefore fewer nodes are collaborating with each other, resulting approximately 10% higher cost on average than the global collaborative scheme.

Fig. 6 summarizes the results of average cache miss rate. We observe that with HIE, the cache miss rate increases as the total number of contents increases, and that the miss rates are almost an order of magnitude higher than GCC and LCC. HIE does not use popularity information to guide forwarding and caching; instead, it searches local cache and upper-level cache only. As the number of contents increases, the miss rate increases since the percentage of contents can be stored in cache is limited by the size of cache storage. This explains why we see an increasing curve for HIE. For GCC and LCC, we observe that the miss rates largely remain the same, due to LMR being fixed to 5%.

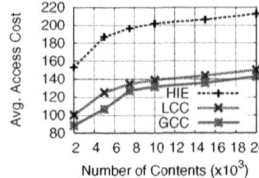
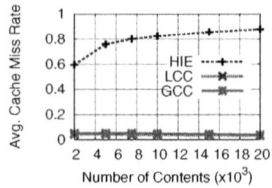
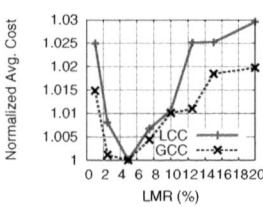

Fig. 5. Average content ac- **Fig. 6.** Average cache miss **Fig. 7.** Normalized average
cess cost   rate   access cost vs. LMR

6.3   Impact of Adaptive Cache Division

We next evaluate the impact of the adaptive cache division algorithm. More specifically, we quantify the performance of global and local collaborative caching by varying LMR.

Fig. 7 plots the impact of cache adaptation on average access cost. Note that the cost is normalized and the total number of contents is 10000 in this figure. We observe that the best performance can be achieved when LMR is approximately 5%. Note that when LMR is in the range of $[0.025, 0.1]$, the average access cost is within 4% of the best performance, suggesting that the performance of GCC and LCC are not sensitive to LMR.

Fig. 8 plots the results of cache miss rate. We observe an increasing miss rate as LMR increases, however, it increases slower when LMR is larger than approximately 10%. With a larger LMR, the cache division algorithm tries to keep the cache miss rate around LMR, in order to make the most efficient use of the cache space. When LMR increases from 10% to 20%, the cache miss rate only slightly increases because the miss rate is mainly contributed by the inconsistency of content popularity, which is generated by shifting 10% of contents from a given ranking sequence.

6.4   Impact of Popularity Inconsistency

We next evaluate the impact on the average access cost when the inconsistency in popularity changes. More specifically, we change the percentage of popularity inconsistency from 1% to 30%, while setting the total number of contents to 10000 and keeping all the other network settings the same as in Fig. 5. We summarize the results in Fig.9.

We make the following observations. Firstly, the higher the percentage of inconsistency, the higher the average access cost for GCC and LCC. We note that when the percentage of inconsistency is higher, each node allocates larger space to the Complementary Content Store in order to handle requests for contents that are not stored in its Advertised Content Store. The reason these contents are not stored in the Advertised Content Store is that different routers see different sets of contents for a given ranking range. With an increasing percentage of contents shifting from their original positions in the ranking sequence, the union

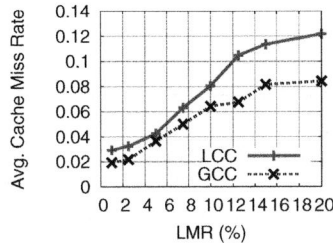
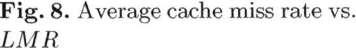
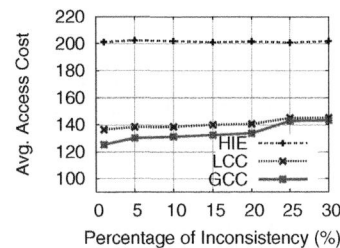

Fig. 8. Average cache miss rate vs. *LMR*

Fig. 9. Average access cost vs. popularity inconsistency

of these different sets becomes larger, leading to a larger Complementary Content Store and a smaller Advertised Content Store, which eventually increases the average content access cost due to the reduced number of unique contents stored in the network.

Secondly, the average access cost of HIE remains almost the same. The reason is that HIE does not rely on consistency of popularity seen by different nodes, and thus is not affected by popularity inconsistency. However, even when the percentage of inconsistency is as high as 30%, GCC and LCC still improve the average access cost by approximately 30%, due to the reason that HIE does not take full advantage of cache available from other routers nearby, leading to an inefficient use of caching.

7 Conclusion

In this paper we propose a novel distributed, popularity-guided collaborative forwarding and caching design for content-centric networks, where we introduce an Availability Information Base to allow coordination between forwarding and caching in content routers. In order to deal with popularity inconsistency in realistic networks, we also propose a self-adaptive dual-segment cache division algorithm. We evaluate our design via extensive simulations and demonstrate that our design improves content access cost and cache miss rate by at least 30% in a diverse network settings.

There are many avenues to future work. We plan to implement a prototype based on CCNx [1] and conduct medium- to large-scale experiments. This also gives us opportunities to investigate the complexity and feasibility of the proposed framework. We also plan to theoretically model and analyze the impacts of popularity inconsistency on the effectiveness of the proposed design.

Acknowledgements. This work was partially supported by the National Natural Science Foundation of China under Grant No. 61073192, by the Grand Fundamental Research Program of China (973 Program) under Grant No. 2011CB302905, by the New Century Excellent Talents Program, and by the Fundamental Research Funds for Central Universities under Grant No. WK0110000014.

References

1. CCNx, http://www.ccnx.org
2. Abilene: Internet2 IP IGP Metrics,
   http://noc.net.internet2.edu/i2network/maps--documentation/maps.html
3. Akamai Technologies, http://www.akamai.com
4. Amazon CloudFront Express, http://www.amazon.com/cloudfront
5. Anand, A., Dogar, F., Han, D., Li, B., Lim, H., Machadoy, M., Wu, W., Akella, A., Andersen, D., Byersy, J., Seshan, S., Steenkiste, P.: XIA: An architecture for an evolvable and trustworthy internet. Tech. Rep. CMU-CS-11-100, Carnegie Mellon University (February 2011)
6. Anderson, T., Birman, K., Broberg, R., Caesar, M., Comer, D., Cotton, C., Freedman, M., Haeberlen, A., Ives, Z., Krishnamurthy, A., Lehr, W., Loo, B.T., Mazires, D., Nicolosi, A., Smith, J., Stoica, I., van Renesse, R., Walfish, M., Weatherspoon, H., Yoo, C.: NEBULA - a future internet that supports trustworthy cloud computing. White Paper (2010)
7. BitTorrent, http://www.bittorrent.com
8. Borst, S., Gupta, V., Walid, A.: Distributed caching algorithms for content distribution networks. In: IEEE INFOCOM 2010, San Diego, CA (March 2010)
9. Breslau, L., Cao, P., Fan, L., Phillips, G., Shenker, S.: Web caching and zipf-like distributions: evidence and implications. In: Proc. IEEE INFOCOM 1999, vol. 1 (March 1999)
10. Carofiglio, G., Gallo, M., Muscariello, L., Perino, D.: Modeling data transfer in content-centric networking. In: ITC 2011, pp. 111–118 (2011)
11. CDNetworks, http://www.cdnetworks.com
12. Cernet: Cernet Topology, http://www.edu.cn/20060111/3170194.shtml
13. Cheng, X., Dale, C., Liu, J.: Statistics and social network of youtube videos. In: Proc. IWQoS 2008, pp. 229–238 (June 2008)
14. Chow, C.Y., Leong, H.V., Chan, A.T.S.: Distributed group-based cooperative caching in a mobile broadcast environment. In: Proc. of Mobile Data Management 2005, pp. 97–106 (2005)
15. Dannewitz, C.: NetInf: An information-centric design for the future internet. In: Proc. 3rd GI/ITG KuVS Workshop on the Future Internet (May 2009)
16. Dilley, J., Maggs, B., Parikh, J., Prokop, H., Sitaraman, R., Weihl, B.: Globally distributed content delivery. IEEE Internet Computing, 50–58 (September/October 2002)
17. Dykes, S., Robbins, K.: A vaibility analysis of cooperative proxy caching. In: IEEE INFOCOM 2001, pp. 1205–1214. Anchorage, AK (2001)
18. eMule-Project.net, http://www.emule-project.net
19. Erman, J., Gerber, A., Hajiaghayi, M.T., Pei, D., Sen, S., Spatscheck, O.: To cache or not to cache: The 3g case. IEEE Internet Computing 15, 27–34 (2011)
20. Fan, L., Cao, P., Almeida, J., Broder, A.Z.: Summary cache: a scalable wide-area web cache sharing protocol. IEEE/ACM Trans. Netw. 8 (June 2000)
21. Gadde, S., Chase, J., Rabinovich, M.: Web caching and content distribution: a view from the interior. Computer Communications 24(2), 222–231 (2001)
22. Gill, P., Arlitt, M., Li, Z., Mahanti, A.: Youtube traffic characterization: a view from the edge. In: Proc. ACM IMC 2007. ACM, New York (2007)
23. Hefeeda, M., Noorizadeh, B.: On the benefits of cooperative proxy caching for peer-to-peer traffic. IEEE Transactions on Parallel and Distributed Systems 21, 998–1010 (2010)

24. Helgason, O., Karlsson, G.: Podnet: A system architecture for opportunistic content distribution. Tech. rep., Royal Institute of Technology (KTH) (February 2010)
25. Jacobson, V., Smetters, D.K., Thornton, J.D., Plass, M.F., Briggs, N.H., Braynard, R.L.: Networking named content. In: ACM CoNEXT 2009, Rome, Italy (December 2009)
26. Korupolu, M.R., Dahlin, M.: Coordinated placement and replacement for large-scale distributed caches. IEEE Transactions on Knowledge and Data Engineering 14, 1317–1329 (2002)
27. Level3 Communications, http://www.level3.com
28. Limelight Networks, http://www.limelight.com
29. Ni, J., Tsang, D.: Large-scale cooperative caching and application-level multicast in multimedia content delivery networks. IEEE Communications Magazine, 43(5) (May 2005)
30. Pathan, M., Buyya, R., Vakali, A.: Content delivery networks: State of the art, insights, and imperatives. In: Buyya, R., Pathan, M., Vakali, A. (eds.) Content Delivery Networks. LNEE, vol. 9, pp. 3–32. Springer, Heidelberg (2008)
31. Psaras, I., Clegg, R., Landa, R., Chai, W., Pavlou, G.: Modelling and Evaluation of CCN-Caching Trees. In: Domingo-Pascual, J., Manzoni, P., Palazzo, S., Pont, A., Scoglio, C. (eds.) NETWORKING 2011, Part I. LNCS, vol. 6640, pp. 78–91. Springer, Heidelberg (2011)
32. Sailhan, F., Issarny, V.: Cooperative Caching in Ad Hoc Networks. In: Chen, M.-S., Chrysanthis, P.K., Sloman, M., Zaslavsky, A. (eds.) MDM 2003. LNCS, vol. 2574, pp. 13–28. Springer, Heidelberg (2003)
33. Sarkar, P., Hartman, J.H.: Hint-based cooperative caching. ACM Trans. Comput. Syst. 18 (November 2000)
34. Wessels, D., Claffy, K.: Internet cache protocol (icp), version 2 (1997)
35. Wolman, A., Voelker, M., Sharma, N., Cardwell, N., Karlin, A., Levy, H.M.: On the scale and performance of cooperative web proxy caching. In: Proc. ACM SOSP 1999, pp. 16–31 (1999)
36. Xie, H., Shi, G., Wang, P.: TECC: Towards collaborative in-network caching guided by traffic engineering. In: IEEE INFOCOM 2012, Orlando, FL (March 2012)
37. Yin, L., Cao, G.: Supporting cooperative caching in ad hoc networks. IEEE Transactions on Mobile Computing 5, 77–89 (2006)
38. Zhang, L., Estrin, D., Burke, J., Jacobson, V., Thornton, J.D., Smetters, D.K., Zhang, B., Tsudik, G., Claffy, K.C., Krioukov, D., Massey, D., Papadopoulos, C., Abdelzaher, T., Wang, L., Crowley, P., Yeh, E.: Named data networking (NDN) project. Tech. Rep. NDN-0001, Palo Alto Research Center (PARC) (October 2010)

Crawling and Detecting Community Structure in Online Social Networks Using Local Information

Norbert Blenn, Christian Doerr, Bas Van Kester, and Piet Van Mieghem

Department of Telecommunication,
TU Delft, Mekelweg 4, 2628CD Delft, The Netherlands
{N.Blenn,C.Doerr,S.vanKester@student.,P.F.A.VanMieghem}@tudelft.nl

Abstract. As Online Social Networks (OSNs) become an intensive subject of research for example in computer science, networking, social sciences etc., a growing need for valid and useful datasets is present. The time taken to crawl the network is however introducing a bias which should be minimized. Usual ways of addressing this problem are sampling based on the nodes (users) ids in the network or crawling the network until one "feels" a sufficient amount of data has been obtained.

In this paper we introduce a new way of directing the crawling procedure to selectively obtain communities of the network. Thus, a researcher is able to obtain those users belonging to the same community and rapidly begin with the evaluation. As all users involved in the same community are crawled first, the bias introduced by the time taken to crawl the network and the evolution of the network itself is less.

Our presented technique is also detecting communities during runtime. We compare our method called *Mutual Friend Crawling (MFC)* to the standard methods Breadth First Search (BFS) and Depth First Search (DFS) and different community detection algorithms. The presented results are very promising as our method takes only linear runtime but is detecting equal structures as modularity based community detection algorithms.

Keywords: Social Networks, Community Detection, Crawling.

1 Introduction

Analyzing human social behavior depends on observations of large scale networks. Online Social Networks (OSNs) proved to be good sources of information facilitating this kind of research, as the most popular OSNs such as Twitter, Facebook or LinkedIn consists of hundreds of millions of user accounts, thereby allowing an analysis at a sufficiently large scale.

However, it is usually not possible to obtain datasets from the operators of OSNs. Therefore, a common approach is to crawl the network per user. In this way, a user is randomly chosen and a list of his friends is downloaded. Out of the list of friends, again one user is selected and a list of friends retrieved.

R. Bestak et al. (Eds.): NETWORKING 2012, Part I, LNCS 7289, pp. 56–67, 2012.

This method repeats in principle until every user in the network has been visited once. This method varies in the selection of the next friend, are ranging from Breadth First Search towards Depth First Search.

This automated process of downloading users typically requires a robot (a software program) to look at the profile page of a user and store the names of all friends. Such an operation usually takes between 0.1 to 2 seconds as it includes multiple HTTP requests to a server in order to iterate through the whole list of friends. An optimistic[1] calculation shows that with one crawling computer, obtaining LinkedIn's database of 120 million users (as of Nov. 2011) would take approximately half a year. The same calculation for Facebook's dataset of 650 million users leads to a crawling time of ca. 2 years. By using massively parallel crawling techniques those times can be decreased. Clearly, by the time the last records have been obtained, the much of the retrieved information will be outdated.

A lot of work on social network analysis has been done using communities of users as a level of abstraction. A natural question is therefore whether it is possible to direct the crawling procedure in such a way that it is crawling the network community-wise. This would enable researchers to analyze useful subgraphs of the whole network while still obtaining data. In contrast, using the standard crawling methods like Breadth First Search or Depth First Search, one literally has to wait until the whole network is crawled before starting to analyze the data because there might be a few users critical to a particular single community still missing from the dataset, and their existence and criticality cannot be determined until all data is in.

In this paper we present a simple approach to crawl a network community-wise and detect communities at the same time. Our algorithm called *Mutual Friend Crawling* is compared to existing community detection methods.

The remainder of this paper is structured as follows. Section 2 summarizes the related work, in Section 3 our method of crawling is described, evaluated and the community detection is explained. Section 4 will compare the community detection with well known community detection algorithms and Section 5 will summarize our findings and give an outlook.

2   Related Work

Communities are defined in terms of the fraction of nodes of a network, that share more connections with each other than with the rest of the network. When analyzing a social network, some relevant questions are:

1. How many communities are there in the network?
2. Which nodes are in the same community?
3. How well are users in communities connected?

---

[1] In this context, optimistic means that no mechanisms against crawling or screen scraping are enforced.

A well known metric to capture the community structure of a network is modularity. Modularity m as defined in Clauset, Newman and Moores's work [1] is "the number of edges falling within groups minus the expected number in an equivalent network with edges placed at random." The definition of modularity is given in equation 1.

$$m = \frac{1}{2L} \sum_i^N \sum_j^N (a_{ij} - \frac{d_i d_j}{2L}) 1_{\{i \text{ and } j \text{ belong to the same community}\}} \qquad (1)$$

For a given graph G with N nodes, L links and a given partition, the modularity denotes how well the community structure is expressed. The element a_{ij} denotes the element corresponding to the ith row and jth column of the adjacency matrix of G and d_i is the degree of node i. $1_{\{i \text{ and } j \text{ belong to the same community}\}}$ is the indicator function returning 1 if i and j are in the same community otherwise 0.

A modularity value of 0 defines that the number of links belonging to the same community is equal to the number a random graph would have. The higher the modularity the more pronounced the community structure, except for the trivial case of modularity $= 1$, in which all links of G are in the same community. Conversely, this means that negative values are a definition of something like an "anti-community" structure. A nice overview over modular graphs and how to achieve high modularity is given in Trajanovski [2].

As we will present an algorithm to crawl community structure in real world OSNs having linear complexity, we compared our results to crawling techniques like Breadth First Search (BFS) and Depth First Search (DFS) as described in Cormen et al. [3]. In both techniques the graph is crawled node per node adding all discovered nodes to a list of nodes to visit. The difference between BFS and DFS is based on the procedure how the next node to visit is selected. In BFS the first node of this list is selected to be visited next and removed from the list whereas in DFS the last node in the list is selected and marked as visited. Both techniques are leading the crawling procedure towards the inner core of the network due to the friendship paradaxon. This paradaxon, first observed by Field [4] stating originally that your friends have more friends than you, will force a crawl towards nodes having a high centrality in the network. A related effect, noted by Kurant et al. [5] describes that BFS and DFS is introducing an bias towards high degree nodes for an incomplete traversal of the network.

To our knowledge, BFS and DFS are the most used techniques to traverse a graph. We will show in section 3 that, in order to reveal the community structure of a graph BFS and DFS are not the best choice. One possible technique of crawling a network and detecting communities at the same time is to facilitate random walks. Random walks are known to stay inside communities as described in Pons and Latapy [6] and Lai and Lu [7]. The main idea behind random walk community detection is that a community has more links between nodes of the community than between communities. Because of this definition, a random walk would traverse nodes of the same community more often than the ones of different communities. However, a random walk allows steps backwards to already visited nodes which is increasing the time taken to crawl the network.

Different community detection algorithms like fast and greedy community detection by Clauset et al. [1], Spinglass by Reichold and Bornholdt [8], edge betweenness clustering by Girvan and Newman [9] or label propagation by Raghavan et al. [11] cannot be used to detect communities during crawling as those algorithms are meant to be applied onto the full topology of a network.

A related approach to detect community structure while crawling is presented by Nguyen et al. [12]. Their algorithm Quick Community Adaptation (QCA) assumes that the community structure is already known for a complete network and manages to calculate community structure in dynamic networks. QCA tries to maximize modularity by assigning a "force" which attracts a node towards a community. However, this method also needs the whole network including assignments of nodes into communities. In their approach the used algorithm to estimate the initial community memberships is presented by Blondel et al. [13] called the Louvain(-la-Neuve) method. This algorithm calculates a modularity maximizing partition of a given graph by using the change in modularity when discovering a new node and adding it to an existing community. If the difference is not positive the node stays in its initially assigned community.

To compare the result of different clustering algorithms we decided to use the Jaccard similarity coefficient next to the already defined modularity. The Jaccard similarity coefficient defines the similarity of sample sets by measuring the quotient of the intersection and the union of both sets. To compare the result of different cluster assignments a definition of Fortunato and Castellano [14] given in equation 2 is used.

$$I_J(s_1, s_2) = \frac{n_{11}}{n_{01} + n_{11} + n_{10}} \tag{2}$$

In equation 2, n_{11} denotes the number of node pairs found in the same community whereas n_{01} and n_{10} are the number of pairs of nodes assigned to the same community by algorithm s_1 but not s_2 and vice versa.

3   Crawling and Detecting Communities

We introduce *Mutual Friend Crawling* which crawls nodes of a network in such a way that communities are visited one after another. First we will show that our approach is crawling all nodes belonging to one community before continuing with nodes of a connected community. Afterwards we will show how to detect communities using this approach.

In contrast to BFS and DFS, our algorithm assumes the knowledge about the degree of neighboring nodes. This assumption is reasonable in OSNs as the number of friends is very easy to obtain, whereas the process of receiving the actual links towards them needs more effort. For example in the OSN Twitter, each message contains a field containing the number of followers and friends the originating author has. Also, OSNs are most commonly crawled using a technique called screen scraping. Here, the OSN is accessed the same way a user does, by using HTTP requests to analyze web pages for relevant data.

The number of friends is usually listed at the profile page of a user but several clicks (requests) on the list of friends are needed to obtain all node ids (friends) having a relationship with this user. If one is interested in more details of a user, for example the real name or group affiliations, this profile information will need to be obtained in any case and at the same time a friend count is also available without any additional overhead. In this way the needed crawling effort has not increased but only the order in which the data is gathered has changed.

Mutual Friend Crawling is based on the "reference score" (S_R) defined in equation 3. This score denotes the fraction of the number of already discovered links pointing to node f (references) so far in the crawling process and the total degree, i.e., total number of friends, of node f.

$$S_R = \frac{\text{number of found references to f}}{\text{degree of node f}} \tag{3}$$

During the crawl, the next node to process is chosen from the list of the already discovered nodes having the largest S_R. The full algorithm is specified in pseudo code in algorithm 1.

Algorithm 1. MUTUAL FRIEND CRAWLING

1: create a queue Q
2: create a map R
3: add starting node to Q
4: store starting node and 0 as number of found references in R
5: **while** Q is not empty **do**
6:     **for** all elements in R **do**
7:         reference_score $\leftarrow \dfrac{\text{value in R}}{\text{degree of the node}}$
8:         max_score $\leftarrow max$(max_score, reference_score)
9:     **end for**
10:     next_node $\leftarrow$ dequeue element having max_score from Q
11:     delete next_node from R
12:     **if** next_node has not been visited yet **then**
13:         **for** all neighbors of next_node: **do**
14:             add neighbor to Q
15:             increment number of found references to neighbor by 1 and store it in R
16:         **end for**
17:         remember that node (next_node) was visited
18:     **end if**
19: **end while**

Mutual Friend Crawling is based on a BFS algorithm having two major differences. The first one is a map used to store the number of found references as indicated in line 2, 4 and 15. The second difference is based on the way the next node to visit is chosen: instead of choosing simply the next one from the list as BFS does, MFC calculates the reference score (lines 6-9) and chooses the next node based on the maximum of the reference score (line 10 & 11).

The algorithm will therefore first visit nodes where to which the largest number of the overall links have already discovered. If a network has a community structure based on the definition of having more links in the community than links connecting communities our algorithm will crawl communities one after another.

In order to apply the algorithm on weighted graphs, a simple definition of the strength of a node as the sum of the weights of adjacent edges is sufficient. In this case the reference_score S_R is defined as the fraction of the sum of weights of already discovered links to the strength of the node as defined in 4.

$$S_R = \frac{\sum (\text{weights of found references to f})}{\text{strength of node f}} \quad (4)$$

3.1 Community Crawling

Figure 1 illustrates our crawling approach intuitively using a small example. Using the definition of communities as stated above (more links "within" the group than ending "outside"), simple visual inspection already shows that there are six clusters. If we were to explore each cluster one after the other, the algorithm should first explore all nodes belonging to one color before starting with the next group of nodes. The nodes labels denote one possible order in which the graph could be crawled in order to visit communities one after another, thus leading to the intended and perfect exploration order.

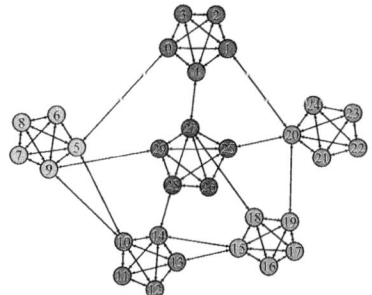

Fig. 1. A simple example graph. Nodes are labeled by the order of traversal during the crawling process. Different colors denote different communities.

In order to test the proposed algorithm on multiple graphs a metric defining how strong the community structure in a given graph is expressed was needed. A simple metric is given by the ratio of links inside communities to the total number of links. We define this value as P_{in} reflecting the probability an arbitrary chosen link is an intra community link.

We created a total of 100,000 artificial networks with different P_{in} values using a graph generator, described in van Kester [15], each with 10,000 nodes, 100,000 links and 100 equally sized communities have been generated. Two general types of graphs commonly found in network science were evaluated:

The degree distribution of the first type follows a uniform distribution and the degree distribution of the second type is approximating a power-law function. All graphs have been crawled from all possible starting nodes. During the crawl we kept track of the order when nodes are visited, which allows us to analyze if a complete community has been crawled before going to the next one.

Figure 2 shows the crawling trajectories of those crawls. The figure is expressing how many nodes have to be crawled in order to visit all nodes of a community. In order to crawl the network community-wise the optimal traversal would need to visit all nodes of one community first before visiting the next node belonging to the next community. Because all communities are equally sized, the optimal traversal of the graph would lead to a diagonal line in figure 2. A bended line is expressing the fact that nodes from different communities were visited before all nodes of the previous visited community have been finished. The order in which multiple communities are crawled is not reflected in figure 2, as we are primarily interested in obtained one completely crawled community after another and not which one is obtained first. The used colors express trajectories of different crawling methods whereas green lines belong to DFS, blue to BFS and red to our *Mutual Friend Crawling*.

For high P_{in} values, figure 2 illustrates that our algorithm performs as expected and leads the walk on the graph to all nodes contained in one community before crawling the next. For BFS and DFS a larger fraction of the network has to be crawled to finish one community. Interestingly, BFS perform "closer" to the optimum than DFS. This is because BFS explores the local neighborhood whereas DFS explores the nodes furthest away from the starting node. Thus, the "chance" of BFS to visit all nodes of one community earlier than in DFS is higher.

The proposed crawling method performs reasonably better than BFS and DFS in terms of crawling along the community structure. For P_{in} values larger 0.3,

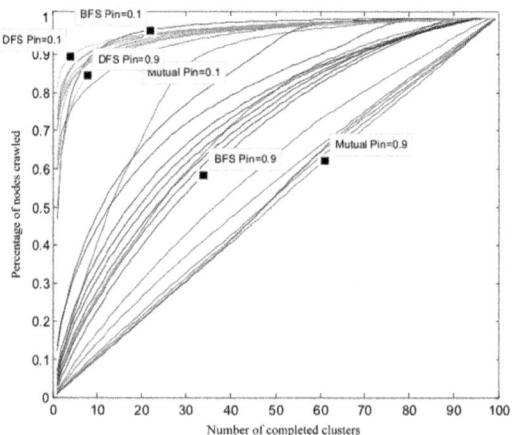

Fig. 2. Crawling communities. Depicts the percentage of nodes that have to be visited in order to crawl a full community.

the order in which the nodes are traversed fulfills our requirements. However, P_{in} values smaller than 0.5 somehow define negative communities in terms of the definition and therefore a BFS approach by chance performs better.

3.2 Community Detection

We already demonstrated empirically that our method crawls communities of a graph one after another. In order to detect communities while crawling the graph, traces of the reference_score of visited nodes can be analyzed. Figure 3 shows the trace of reference_scores performed on the example graph (figure 1) starting from node 0. As mentioned earlier the proposed method always selects the next node to visit having the highest reference_score (line 8 in algorithm 1). Hence the reference_scores inside communities should always increase or stay roughly the same while traversing the graph. When detecting a node that is interconnecting communities, a large number of links of this node are ended in the previously unknown community. Therefore, its reference_score will be smaller than the ones of nodes connected to this in the current community. As this node is selected as the last one, a drop can be observed in the trajectory of reference_scores of visited nodes. Figure 3 shows five major drops of the score. Those drops in the reference_score of the chosen nodes reflect the creation of new communities. We define the difference between the next reference_score to the previous one as $\Delta refscore$. During traversing the graph, all nodes are added to the same community as long as $\Delta refscore$ is large enough. If the score decreases a new community can be created. To prevent the creation of single node communities, we found that the drop in the reference_score should be at least half the difference between the maximum S_{max} and the minimum S_{min} of the reference_score in this community. By using this method, the *Mutual Friend Crawling* is creating six communities on the sample graph having the community assignments as indicated by different colors in figure 1.

One problem however still remains: a possibly incorrect classification of certain nodes during the first visit of a new community. In case neighbors of the starting

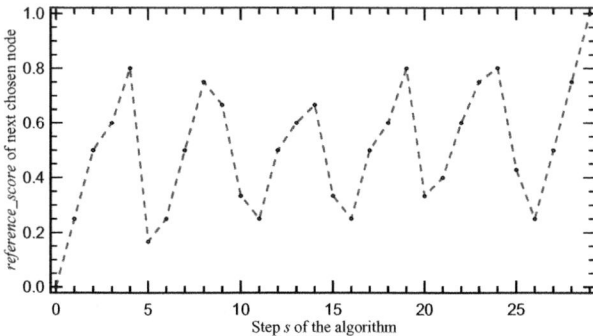

Fig. 3. Plot of reference_scores versus the number of visited nodes

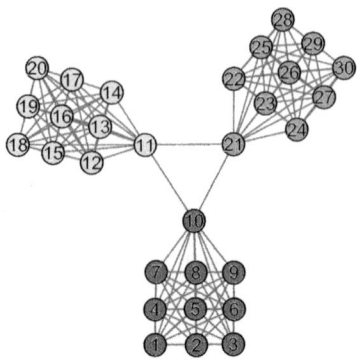

Fig. 4. Example graph where a misclassification may occur when visiting node 11

node as well as neighbors of the first node of a community have the same degree as the visited node, a node of a different community could be assigned incorrectly. Such misclassification is exemplified for the case of three equally-sized, fully-connected communities which are pairwise connected through one node as shown in figure 4. In this case when starting in one clique, all nodes of this clique are added to the first community. When reaching one of the 3 nodes connecting communities (10, 11 or 21 in figure 4) the score drops, a new community is generated and all following nodes are added to this community. When reaching for example node 10, the reference_score for the 2 peers (11 and 21) is the same. One of them is chosen to be next node to visit. But now, (e.g., by visiting node 11) 2 links towards the third node are discovered doubling its score. Having a higher score than all other neighbors of 11, 21 will be added to the same community as 11. Afterwards one of 11's or 21's neighbors are visited where the reference_score drops again leaving 11 and 21 in one community. all other nodes afterwards are correctly classified. Our solution is to check for this kind of misclassification by iterating through all nodes in a already discovered community checking if a node has more connections with another community then inside the "own" community. If so this node is merged to the connected community. As this procedure is raising the density in the community the node is merged to, the modularity value is increasing as stated in Trajanovski [2].

4   Verification and Performance Evaluation

To prove the correctness of *Mutual Friend Crawling* in terms of community detection, we compare the results of community assignments to existing approaches. However, as the variety of community detection algorithms is too large to compare against, we chose algorithms which (with slight modifications) can be used to identify communities while crawling. This means we are comparing against algorithms which iteratively assign nodes to communities without having the knowledge about the whole graph.

The chosen methods are:

1. Newman and Clauset's fast and greedy modularity maximizing method [1]
2. Pons & Latapy's random walk method [6]
3. the Louvain(-la-Neuve) method by Blondel et al. [13]

As all mentioned approaches are not directly providing a map of community ids to node ids we chose the partition resulting from the merges of nodes leading to a maximum of modularity. This partition is then compared to the output of our algorithm. For the given example graph in figure 1, all community detection algorithms found the same result as indicated by the colors in figure 1.

A comparison of the mentioned methods on some selected datasets is given in the following table 1. The datasets we compared against contain "Zachary's karate club" [16], Girvan and Newman's "American College football games" [9] and the network of all Digg users as described in Tang et. al. [17].

Table 1. Comparison of Mutual Friend Crawling to well known community detection procedures on different datasets

Dataset	Method	Number of communities	P_{in}	Modularity
Karate club	original partition	2	0.86	0.36
	Louvain method	4	0.74	0.42
	Fast and greedy method	3	0.74	0.38
	Random walk method	5	0.63	0.35
	Mutual friend crawling	2	0.86	0.36
Football	original partition	12	0.64	0.554
	Louvain method	10	0.708	0.604
	Fast and greedy method	5	0.746	0.544
	Random walk method	9	0.726	0.603
	Mutual friend crawling	9	0.736	0.57
Digg	Louvain method	26646	0.94	0.478
	Fast and greedy method	37591	0.92	0.393
	Mutual friend crawling	78308	0.83	0.142

As given in table 1, our method is comparable to existing and well known procedures when compared in terms of the P_{in} value and modularity. Except for the last dataset, a large scale directed network of all users of Digg.com where the number of detected communities is higher than given by the Louvain(-la-Neuve) or fast and greedy method.

However, this could be based on the resolution limit of modularity, as described in Fortunato and Barthélemy [18]. A partition having a high modularity could lead to a relatively small number of large communities which is not reflecting the real community structure. The communities found by *Mutual Friend Crawling* are smaller than the ones found by the other methods still having the same properties like a power law shaped community size distribution. Also the number of users in a group given by our method is reasonable. The largest community found by Mutual Friend Crawling has a size of 9443 users whereas the largest one found by the Louvain method contains 186271 users. Without further investigation one may arguments for both numbers to be better than the other one. Therefore we leave this question to be solved by further research.

Table 2. Comparison of the partitions discovered by our method to other community detection algorithms on the college football dataset [9] using the Jaccard similarity index

method	original	Louvain	Fast and greedy	Random walk	Mutual friend crawling
original	1	0.719	0.354	0.615	0.468
Louvain		1	0.424	0.721	0.483
Fast and greedy			1	0.422	0.324
Random walk				1	0.487
Mutual friend crawling					1

As the P_{in} value and the modularity are global values which cannot be used to compare two partitions directly the Jaccard similarity index may be used. As mentioned earlier, if pairs of nodes are assigned to the same community this similarity will have a high value.

Table 2 shows the similarity of node assignment into communities between the different community detection algorithms. While this metric is not very sensitive to the number of communities it shows that our approach is equally good as well known methods. A more complete analysis of the Jaccard similarity index is given in van Kester [15].

5   Conclusion and Outlook

In this paper we presented *Mutual Friend Crawling*, an algorithm to crawl a large scale OSN in such a way that the community structure of the network is detected and communities are crawled one after another. To our knowledge this is the first analysis directing a crawling process towards community structure. We showed that our method crawls communities one after another. Especially when obtaining large scale networks, researchers could begin to analyze datasets based on communities while the crawling process is still running.

Further work is needed if the community consists of overlapping groups as the partition of a large scale network into clearly separated communities does not make sense in most Online Social Networks. Also the application of standard metrics (like modularity or the used P_{in} value) to compare partitions to real world community structure should be the focus of additional, future research.

Acknowledgements: We thank Wynand Winterbach and Stojan Trajanowski for useful discussions. This project was partially supported by the Transsector Research Academy for Complex Networks and Services (TRANS).

References

1. Clauset, A., Newman, M.E.J., Moore, C.: Finding community structure in very large networks. Physical Review E 70 (December 2004)
2. Trajanovski, S., Wang, H., Van Mieghem, P.: Maximum Modular Graphs. European Physics Journal B (2011) (submitted)

3. Cormen, T.: Introduction to algorithms. MIT electrical engineering and computer science series. MIT Press (2001)
4. Feld, S.L.: Why Your Friends Have More Friends Than You Do. American Journal of Sociology 96(6), 1464–1477 (1991)
5. Kurant, M., Markopoulou, A., Thiran, P.: On the bias of BFS (Breadth First Search). In: 22nd International Teletraffic Congress (ITC), pp. 1–8. IEEE (2010)
6. Pons, P., Latapy, M.: Computing Communities in Large Networks Using Random Walks. In: Yolum, p., Güngör, T., Gürgen, F., Özturan, C. (eds.) ISCIS 2005. LNCS, vol. 3733, pp. 284–293. Springer, Heidelberg (2005)
7. Lai, D., Lu, H., Nardini, C.: Enhanced modularity-based community detection by random walk network preprocessing. Phys. Rev. E 81, 066118 (2010)
8. Reichardt, J., Bornholdt, S.: Statistical mechanics of community detection. Phys. Rev. E Stat. Nonlin. Soft. Matter Phys. 74 (July 2006)
9. Girvan, M., Newman, M.E.J.: Community structure in social and biological networks. Proceedings of the National Academy of Sciences 99, 7821–7826 (2002)
10. Newman, M.E.J., Girvan, M.: Finding and evaluating community structure in networks. Phys. Rev. E 69, 026113 (2004)
11. Raghavan, U.N., Albert, R., Kumara, S.: Near linear time algorithm to detect community structures in large-scale networks. Physical Review E 76, 036106+ (2007)
12. Nguyen, N., Dinh, T., Xuan, Y., Thai, M.: Adaptive algorithms for detecting community structure in dynamic social networks. In: 2011 Proceedings IEEE INFOCOM, pp. 2282–2290 (April 2011)
13. Blondel, V.D., Guillaume, J.-L., Lambiotte, R., Lefebvre, E.: Fast unfolding of community hierarchies in large networks. CoRR, abs/0803.0476 (2008)
14. Fortunato, S., Castellano, C.: Community structure in graphs (2007)
15. Van Kester, S.: Efficient Crawling of Community Structures in Online Social Networks. PVM 2011-071, Tu Delft (September 2011)
16. Zachary, W.W.: An Information Flow Model for Conflict and Fission in Small Groups. Journal of Anthropological Research 33(4) (1977)
17. Tang, S., Blenn, N., Doerr, C., Van Mieghem, P.: Digging in the Digg Social News Website. IEEE Transactions on Multimedia 13, 1163–1175 (2011)
18. Fortunato, S., Barthélemy, M.: Resolution limit in community detection. Proceedings of the National Academy of Sciences 104, 36–41 (2007)

Distributed Content Backup and Sharing Using Social Information

Jin Jiang and Claudio E. Casetti

Dipartimento di Elettronica e Telecomunicazioni, Politecnico di Torino, Italy

Abstract. This paper addresses the need for content sharing and backup in household equipped with a home gateway that stores, tags and manages the data collected by the home users. Our solution leverages the interaction between remote gateways in a social way, i.e., by exploiting the users' social networking information, so that backup recipients are those gateways whose users are most likely to be interested in accessing the shared content. We formulate this problem as a Budgeted Maximum Coverage (BMC) problem and we numerically compute the optimal content backup solution. We then propose a low-complexity, distributed heuristic algorithm and use simulation in a synthetic social network scenario to show that the final content placement among "friendly" gateways well approximates the optimal solution under different network settings.

Keywords: Content sharing, social networks, federated homes.

1 Introduction and Motivation

The wealth of digital devices and appliances in everyday's life has brought about dramatic changes in our habits. Perhaps one of the most remarkable is the reliance on digital storage for whatever information content we own or produce. Of course, no savvy user would rely solely on storing precious, irreplaceble data in a single device and backup systems are now common in most households. More recently, the availability of "cloud" storage services, aimed at consumers and companies alike, such as Dropbox, Box.net, Ctera to name a few, has introduced a new opportunity. In the latter case, a wideband Internet connection can be exploited during idle periods to run background backups onto cloud storage.

One of the drawbacks of personal or cloud backup approaches is the fact that data of potential interest of other users sit unused in a storage device. Let us consider the following example. George has a set of pictures of the latest family vacations and he wants to show them to his friend John, while, at the same time backing them up. George remotely uploads the pictures to John's NAS, where a storage quota is reserved for such purpose; John is then notified that a copy of the pictures now exists in his NAS and that he is welcome to have a look, while keeping it in its NAS as a backup. For fairness, a similar quota for John's backups should be set aside at George's. The example could be extended to a close group of friends, as defined within social networks, and the potential of

R. Bestak et al. (Eds.): NETWORKING 2012, Part I, LNCS 7289, pp. 68–81, 2012.
© IFIP International Federation for Information Processing 2012

such a scheme instantly become apparent. By leveraging typical social networks indicators, such as interests, hobbies and preferences, and by having all personal digital data appropriately tagged, the matching of remote users and content to backup would allow to catch two birds with a stone: safe, redundant online backup and social content sharing.

In this paper we will outline an architecture to realize this vision. In line with recent "federated homes" networking architectures [1, 2], we assume that home is equipped with a gateway and a large number of interconnected devices within the household. The gateway allows any content to be downloaded from outside the household, stored on it, and accessed by satellite devices in communication range of the home gateway. We also assume that, in keeping with the federated home vision, multiple neighboring or remote home gateways can be connected in a collaborative fashion, and can exchange various information. Although not explored in this paper, let us assume that a home gateway can collect its users' social networking data, e.g., a list of user's friends and interests, friends' locations and whether they are in the federated home network or not.

The outlook is not as simple as the description implies, though. Firstly, there are gateway selections issues. Choosing a friend's gateway to back up data only because mutual user interests match is not a sound policy from a networking point of view. The remote gateway could have poor connectivity or it could be overloaded. Even though friends in social networks are more likely to be in nearby areas [3], the gateway could be located in a far away country. The remote gateway should enforce a rigid quota management to avoid being swamped by friends' uploads. Additionally, there are management details to address: the user must rely on the backed-up content to be readily available on the remote gateway and it should be notified when the content is about to be deleted. If the content is deleted, a second-best choice should be identified, based on the same criteria that guided the former selection.

Our work falls into the same category as several recent research efforts tacked the problem of multiple backups across different resources [4–6]. None of these works, though, leverages the potential of social networking. Related to our problem are also the works on content placement exploiting information from social networking. [7] proposes ContentPlace, which is a social-oriented framework for data dissemination taking into consideration user interest with respect to content. A similar approach is taken in [8] where it is shown that mobility and cooperative content replication strategies can help bridge social groups. Another relevant work on an efficient social-aware content placement in opportunistic networks is [9] in which the authors model the content placement as a facility location problem.

In our paper, we devise an efficient content placement scheme to determine where to back up the content from a user's gateway to remote gateways belonging to his/her social friends. As remarked above, placing content replicas "outside" the home (i) consumes transmission bandwidth for uploading the content and (ii) incurs a storage cost on the remote friends' home gateways. So we aim at a strategy *that maximizes the friends' benefit by trying to match content type and friends' interests while taking into account both the bandwidth constraints between gateways and the storage space at the remote gateway.* Thus, we model this optimization problem as a Budgeted Maximum Coverage (BMC) problem, as preliminary introduced in [10], and numerically obtain the optimal content placement solutions under a synthetic social networking scenario. Next, we propose and evaluate some heuristic distributed algorithms that federated gateways can implement to realize a social backup strategy. We evaluate the heuristics and discuss the conditions under which they can approximate the optimal solution.

2   Model Description

We consider N federated home gateways, GW_i $(i = 1, 2, \ldots, N)$, each reachable through an Internet connection. A home gateway stores content for all users in the corresponding household, setting aside a "friend quota", Q_i, defined as the available storage capacity for the content uploaded by friends of its users. We will generically indicate by C_{ih} the bandwidth from gateway GW_i to gateway GW_h, and by C_{hi} the bandwidth in the opposite direction.

We then assume that totally there are M users in the network, and each one is registered on a single home gateway through which the user can access/store the content. For the purpose of identifying which users are registered on which gateway, we define an $N \times M$ matrix $\mathbf{P}$ whose generic element is given by:

$$P_{ij} = \begin{cases} 1 & \text{if } U_j \text{ registered on } GW_i \\ 0 & \text{otherwise.} \end{cases} \tag{1}$$

where i indicates the gateways, $i = 1, 2, \ldots, N$, while j is the user index, $j = 1, 2, \ldots, M$.

As explained earlier, we assume that home gateways somehow collect the social information of their users. In particular, we are interested in collecting *user's friend list* and *user's interests*. Also, gateways can collect friends' registration information (i.e., which friend of U_j is registered on which gateway).

We define E_j as friend list of user U_j:

$$E_j = \{U_f : F(j, f) = 1\} \tag{2}$$

where a friendship function $F(j, f)$ tracks whether user U_j and user U_f are friends:

$$F(j, f) = \begin{cases} 1 & \text{if } U_j \text{ and } U_f \text{ are friends, } j \neq f; \\ 0 & \text{otherwise.} \end{cases} \tag{3}$$

As we have seen, a user and its friends cooperate to back up content, which we generically represent as *items*. Items can be videos, photos or any other digital content. In order to handle the mathematics, we assume there is a maximum of K different items. A generic item k, $k = 1, 2, \ldots, K$, is characterized by its size $D^{(k)}$ and classified into a content type l (e.g., movies, books, outdoor photography ...). Content types too are finite, , i.e. $l = 1, 2, \ldots, L$, where L is the interest area size, i.e., the total number of content types considered in our system.

The association between an item and its type is assigned according to a uniformly random distribution.

The user's interests, i.e., the distribution of content preferences of the user, is captured by an *interest vector*, defined as follows. Given an item of type l let I_{jl} denote the interest factor that user U_j has for it, with $0 \leqslant I_{jl} \leqslant 1$. Thus, we can outline user U_j's profile through its interest vector:

$$\bar{I}_j = (I_{j1}, I_{j2}, \ldots, I_{jl}, \ldots, I_{jL}) \tag{4}$$

where $\sum_{l=1}^{L} I_{jl} \triangleq 1 - r^j$. r^j is the probability that user U_j is interested in a content type out of the interest area L. Without loss of generality, we will just assume $\sum_{l=1}^{L} I_{jl} = 1$ or $r^j = 0$, i.e., users do not have interests outside the area L.

For the sake of notation simplicity, we also assume that every user has the same average number of items to back up.

Following the definitions above, each user has the objective of finding a selection of friends from the friend list, on whose gateways to back up its items. The matching of item and remote gateway should benefit both the hosting users, i.e., by closely matching their interests, and, it should transfer data effectively, i.e., by maximizing the utilization of available bandwidth between the respective gateways. Clearly, these objectives are not the only possible choices, and they lead to somewhat arbitrary weight formulations in the optimization problem. Through far from unique, such formulations will attempt to enhance the benefits we have outlined above.

As previously observed, we cast the optimization problem as a BMC problem. In BMC problems, a collection of sets $\sigma = \{\sigma_1, \sigma_2, \ldots, \sigma_m\}$ have associated costs $\{c_i\}_{i=1}^m$. The σ sets are defined to comprise elements $X = \{x_1, x_2, \ldots, x_n\}$ whose associated weights are $\{w_j\}_{j=1}^n$. The goal is to find a collection of subsets $\sigma' \subseteq \sigma$, such that the total weight of elements in σ' is maximized and their total cost is bounded by a budget L. The BMC problem is known to be NP-hard [11].

Gateway GW_i is assumed to have already collected user U_j's friend list E_j. We define $\sigma_j = \{\sigma_{j1}, \sigma_{j2}, \ldots, \sigma_{jh}, \ldots, \sigma_{jN}\}$ the collection of subsets σ_j for user U_j; here, subset $\sigma_{jh} = \{U_f \in E_j : P_{hf} = 1\}$ denotes the set of friends of user U_j who are registered on gateway GW_h, $h = 1, 2, \ldots, N$. We recall that $P_{hf} = 1$ means that user U_f is registered on gateway GW_h, and that $U_f \in E_j$ means that user U_f is in the friend list of user U_j, so $\sigma_{jh} \subseteq E_j$. The cost $c_{(\sigma_{jh})}^{(k)}$ of selecting the subset σ_{jh} is defined as the cost of uploading the content item k of size $D^{(k)}$ onto the gateway GW_h, which can be defined as: $c_{(\sigma_{jh})}^{(k)} = D^{(k)}$.

The *element* set in our problem obviously is the user U_j's friend list E_j. For each element/user $U_f \in E_j$ (user U_f is a friend of user U_j), we can define the weight as the benefit $w_{(U_f)}^{(k)}$ that element U_f can obtain when item k is uploaded onto gateway GW_h where U_f is registered. Such benefit will chiefly depend on the interest I_{fl} that U_f has in the uploaded content. The friend's interest also has a subtle implication for the content owner U_j: the more U_f is interested in the backed-up items, the longer it is likely to store them. Additionally, we factor in the ease of accessibily of the content when its owner U_j wants to retrieve it, i.e., the bandwidth between the hosting gateway and the uploader gateway. The ease of accessibility is also a plus for U_f, because shorter retrievals are less penalizing for its uplink capacity. We can thus define $w_{(U_f)}^{(k)}$ as:

$$w_{(U_f)}^{(k)} = I_{fl} \cdot C_{hi} \tag{5}$$

where I_{fl} denotes the interest of U_f in items of type l which content k belongs to and C_{hi} is the bandwidth of the link from the hosting gateway GW_h to the uploader gateway GW_i. Although bandwidth and user interests are, by definition, quantities that may vary over time, we can safely assume that the time scale of their variations is larger than the time scale of the algorithm execution.

Our constraint is the gateway friend quota, Q_i, which we recall is the available storage capacity for data uploaded by friends.

Finally, our problem can be formulated as follows:

$$
\begin{aligned}
maximize \quad & \sum_{k=1}^{K} \sum_{U_f \in E_j} w_{(U_f)}^{(k)} \cdot y_f^{(k)} \\
subject\ to \quad & \sum_{k=1}^{K} D^{(k)} \cdot x_h^{(k)} \leqslant Q_h \\
& \sum_{P_{hf}=1} x_h^{(k)} \geqslant y_f^{(k)} \\
& x_h^{(k)}, y_f^{(k)} \in \{0,1\}
\end{aligned}
\tag{6}
$$

where $x_h^{(k)} = 1$ indicates that gateway GW_h is selected to host a backup of content item k, while $y_f^{(k)} = 1$ means that user U_f's associated home gateway is chosen as backup for the content item k. The first constraint limits the limitation of available friend quota on each home gateway; and the second one applies to the case of one gateway with multiple associated users: if one user is chosen as backup for one item k, its associated gateway must be selected too.

The number of Boolean decision variables ($x_h^{(k)}$ and $y_f^{(k)}$) is $O(K\langle N\rangle)$, where $\langle N\rangle$ denotes the average number of friends per user. The number of constraints is $O(K\langle N\rangle + M)$. The solution time required by the Gurobi solver for an instance with approximately $1,000$ gateways, $3,000$ users and an average of 5 content items for each user to share or backup, is about 30 minutes using a 4-core 2.3 GHz system and a 4 GB RAM.

3   Distributed Heuristics

The greatest hurdles in translating the optimization model into a working implementation are (i) that the model paints a *static* picture, where all users take instantaneous decisions and (ii) that decisions are taken by a centralized, knowledgeable entity.

In this section we propose a set of distributed heuristic algorithms that strive to achieve the same goal as the model outlined in the previous section. The algorithms take two different viewpoints: that of participating content owners who have data to share or back up and that of remote gateways who provide their own storage space for their social friends. In both cases, we follow the same arguments used in the optimization problem definition.

From the viewpoint of content owners, not only do they wish to back up or restore the data as fast as possible, but, in the long run, they also wish that the remote gateway keeps the content for as long as possible. Therefore, content owners are naturally disposed to choose friends from whose gateways they can retrieve the backed up items more quickly. Also, they would like friends to be interested in the content they upload, because such friends are more inclined to store it for a long time. If one wants to back up their kid's pictures, what better place than the grandparent's gateway?

At the receiving end, the remote gateway can display two types of behavior that are arguably worth investigating. One is a selfish behavior: regardless of the backup requests received by friends of its users, the remote gateway will devote its "friend quota" only to maximize the interests of its associated users, i.e., based only on the first factor in the benefit $w_{(U_f)}^{(k)}$ of eq. (5). The other one is a cooperative behavior: the remote gateway fills up its friend quota while trying to maximize the whole benefit of eq. (5), hence accounting for both its users' interest and the bandwidth toward the content owner's gateway.

We will address either viewpoint through a specific distributed algorithm: a Greedy Placement Algorithm (GPA) run by content owner gateways in order to identify the most suitable places where to back up their items, and the RePlacement Algorithm (RPA), run by each remote gateway upon receiving a backup request.

3.1   The Greedy Placement Algorithm

We assume that a user has available all items it wants to back up when the GPA procedure is started. Further, we assume time to be slotted in intervals of fixed length and that the starting time slot of GPA procedure on a gateway is random. On each gateway, the sequence in which users start GPA is also randomly determined. To achieve the fairness among all the users in the system, each user can run GPA *only once* per time slot.

When starting the GPA, a gateway GW_i will have already collected the following information *from each of its associated users U_j*:

- the friend list E_j;
- the remote friend gateway list RG_j which includes the remote gateways on which user U_j's friends are associated;
- list K_j of items to back up;
- for each item $k \in K_j$, the benefit $w_{(U_f)}^{(k)}$ of each friend $U_f \in E_j$ as defined in Section 2;

– for each remote friend gateway $GW_h \in RG_j$, a quantity referred to as *gateway aggregate benefit* $w_h^{(k)} = \sum_{U_f \in E_j} w_{(U_f)}^{(k)} \cdot P_{hf}$ (recall that $P_{hf} = 1$ indicates the friend U_f is associated to gateway GW_h);
– a query list Z_j where each element is a pair (k, GW_h) representing an item and the IDs of a remote friend gateways, sorted by their gateway aggregate benefit $w_h^{(k)}$.

The main idea behind GPA, detailed in Algorithm 1, is the following: every time the algorithm is scheduled, user U_j sends a backup request to the remote friend gateway whose ID is in the element that tops the query list Z_j. Such element is then removed from the list if the request is accepted; otherwise, it is pushed back to the bottom of Z_j. After sending backup requests on behalf of a user U_j for a total item size of S bytes, GPA stops and it is rescheduled randomly in the next time slot.

Algorithm 1. Greedy Placement $GPA(U_j, Z_j)$

Require: RETRY counters for all elements of Z_j
 $size \leftarrow 0$
 loop
   pop_front element (k, GW_h) from Z_j
   if $\text{RETRY}(k, GW_h) > \text{MAX\_RETRY}$ **then**
     continue
   end if
   if $size + D^{(k)} > S$ **then**
     insert_head element (k, GW_h) into Z_j
     break loop
   else
     $size = size + D^{(k)}$
   end if
   send Backup_REQ to GW_h for k
   if Backup rejected **then**
     push_to_back element (k, GW_h) into Z_j
     $\text{RETRY}(k, GW_h) = \text{RETRY}(k, GW_h) + 1$
   end if
 end loop
 if $Z_j! = \emptyset$ **then**
   schedule $GPA(U_j, Z_j)$ next time slot
 end if
 return RETRY counters for all elements of Z_j

Since the query list Z_j of user U_j is sorted by the gateway aggregate benefit for the corresponding remote gateway, the pop_front operation corresponds to extracting from the list the item k and the ID of the best candidate gateway where it can be backed up in the current time slot. A *Backup_REQ* message is then sent to such gateway.

Once a gateway receives the *Backup_REQ*, it will first check whether it has already cached this item. If not, and there is enough free space in its friend quota[1], it will set aside the corresponding size for this item in its storage space. A *Backup_REP* message is returned to the item owner notifying it whether the backup request was accepted. If the request is accepted, the content owner will start the upload. If the request is denied, or no reply is received, the corresponding list element is pushed at the bootom of the query list, for a later retry, up to a limit of MAX_RETRY times.

Upon reaching the S bytes backup limit, and if the query list is not empty, the gateway schedules the next run of GPA, and the next batch of backup requests for user U_j, at the next time slot. In order to achieve fairness across the federated network, all users should use the same upper backup limit S.

We finally remark that GPA can easily be modified so that the gateway attemps to back up a single item k only onto a limited set of friends' gateways. In this paper, we have only considered the most general (and most challenging) case in which the gateway tries to back up all items on all the friends' gateways. Using the above notation, when GPA evantually stops, an always successful gateway will have dislocated $|RG_j| \cdot |K_j|$ items across the federated network, for each of its users.

3.2   The Replacement Algorithm

If remote gateways "passively" accepted all backup requests until their quota is filled up, their collection of backed up items would not match the optimum allocation, being strongly dependent on which users start the GPA procedure first, hence which greedy requests they receive first. After all, the friends of the users associated to a gateway share the quota on this gateway by competing with each other. In order to alleviate such unbalancement, we introduce a second algorithm, called RPA, RePlacement Algorithm, to be run by every gateway upon receiving a backup request and discovering that the quota is already filled up. We assume that a gateway GW_i holds a list B_i of items backed up in its storage space, sorted by benefit, while we indicate by q_i the free space still available for backups out of the friend quota.

As explained above, when a gateway GW_i receives a *Backup_REQ* message for a new item k of size $D^{(k)}$, it will check whether it has the enough storage space for it; if the space is not sufficient, the gateway will compute the aggregate benefit $w_i^{(k)}$ of the item according to eq. (5) and start the RePlacement Algorithm, described in Algorithm 2.

The gist of the RPA procedure is the following. In order to maximize the benefit of the users associated to the receiving gateway, the replacement strategy considers the removal of backed up items with lower benefit than the incoming item. The B_i list is sorted by benefit, and RPA checks if there are enough items with lower benefit than $w_i^{(k)}$ that can be dropped to leave room for the incoming item. A second check verifies if the product of the total benefit and total size of

---

[1] We will discuss the case of no free space in the next subsection.

Algorithm 2. The RePlacement Algorithm $RPA(GW_i, k)$

Require: $B_i, q_i, w_i^{(k)}, D^{(k)}$
  $replace \leftarrow$ **false**, $FreeSpace \leftarrow q_i$
  $DropWeight \leftarrow 0, DropSize \leftarrow 0$
  while $B_i \neq \emptyset$ **do**
    select $k' \in B_i$ with the lowest benefit
    $DropWeight \leftarrow DropWeight + w_i^{(k')}$
    $DropSize \leftarrow Dropsize + D^{(k')}$
    $FreeSpace \leftarrow FreeSpace + D^{(k')}$
    if $w_i^{(k)} < DropWeight$ **then**
      $replace \leftarrow$ **false**
      break
    else if $w_i^{(k)} == DropWeight$ **then**
      if $DropSize \leqslant D^{(k)}$ **then**
        $replace \leftarrow$ **false**
        break
      else
        $replace \leftarrow$ **true**
        break
      end if
    else
      if $D^{(k)} > FreeSpace$ **then**
        $B_i \leftarrow B_i \setminus k'$
        continue
      else
        $replace \leftarrow$ **true**
        break
      end if
    end if
    if replace **and** $D^{(k)} \cdot w_i^{(k)} \leqslant DropSize \cdot DropWeight$ **then**
      $replace \leftarrow$ **false**
    end if
  end while
  return $replace$

the items selected for dropping is smaller than the benefit/size product of the incoming item. If so, the latter replaces the dropped items in the storage space of GW_i. Ideally, the second check is aimed at preserving the network efficiency, so that a large backed up item is not easily dislodged by much smaller item with a marginally higher benefit.

If the remote gateway GW_i replaces content item k', a $Backup\_DEL$ message must be sent to inform the content owner that it needs to find a new gateway where to store k'. The content owner will thus place a corresponding element in the Z_i queue of algorithm GPA (or start a new instance of GPA if Z_i has been emptied in the meanwhile).

The running time cost of GPA and RPA is minimal. For each user for which GPA is run, the length of the query list Z_j is $O(\hat{K}\hat{E})$, where $\hat{K}$ is the average number of items per user and $\hat{E}$ is the average number of remote friend gateways. So for the individual gateway the running time is $O(\hat{K}\hat{E}m/n)$ with m/n denoting the average number of users per gateway. As for the running time of RPA, the algorithm searches the whole the backed up item list B_i to check what can be replaced, while the maximum number of iterations depends on the number of remote friends and their own items to back up. So the complexity of RPA for one gateway also is $O(\hat{K}\hat{E}m/n)$.

4   Evaluating Optimal Model and Distributed Heuristics

We will now investigate the validity of our approach by following two main directions. Firstly, we numerically solve the model and derive the maximal benefit according to eq. (6). The results will be benchmarked against two other, simpler backup (re)placement strategies, to evaluate the gain that comes at the expense of extra complexity in the backup strategy. Secondly, we compare the content allocation resulting from the optimal solution with what is achieved through the distributed heuristics. In this case too, we will explore variants of GPA and RPA.

Our evaluation will necessarily target a synthetic scenario. Recreating all the conditions and variables of an actual online social network would be a daunting task. We thus extrapolate its essential features and create a scaled-down version for our simulation following the procedure we outlined and validated in [10].

4.1   Optimization, Heuristics and Variants Thereof

In order to extract meaningful comparisons between the optimization approach and the distributed heuristics, sharing the same scenario is not enough. On the one hand, we used Gurobi [12], which runs a variant of the branch-and-cut algorithm, to numerically solve the BMC problem in eq. (6). The solution yielded an optimal joint content item placement, i.e., the set of candidate home gateways to be selected for each content item, as well as the optimal benefit value that each user can obtain by being selected. On the other hand, we simulated the heuristic approach on an ad-hoc simulator. The GPA and RPA algorithms were stripped to their bare bones (i.e., not considering the actual content file transfer while working on GPA timeslot granularity) so that we could focus on the resulting allocation after requests, allocations and replacements have settled. Finally, we compared the steady-state outcome to what the optimization had predicted.

We have tried to gauge the effectiveness of optimization and heuristics not only by comparing one against the other, but also by running some variants of either approach with the aim of catching a glimpse of what we would stand to lose or gain, if we chose a simpler (or a more convolute) strategy than the ones outlined in Sec. 2 and 3.

As far as optimization was concerned we evaluated three different content placement strategies. The first strategy is the *joint* optimization method, described in Section 2, in which the friends who have the largest interest in the corresponding content item and the highest uplink bandwidth will be selected first (assuming their quota is not used up). The second strategy is a *bandwidth-based* method, in which friends reachable through gateways with the highest bandwidth will be selected, regardless of their interests in the uploaded items. The optimal bandwidth-based placement is still obtained from eq. (6) by changing the benefit definition into $w_{(U_f)}^{(k)} = C_{hi}$. The last one is a *random* method, where users randomly choose what friends to share the content item with, as long as enough quota is available, not considering any other factors.

Concerning the heuristics, we considered three versions of GPA:

- GPA-j, which corresponds to the definitions outlined for Algorithm 1;
- GPA-b, where the benefit of a friend U_f on GW_h just depends on the uplink bandwidth from GW_h to GW_i (where U_j is located) , i.e., $w_{U_f}^{(k)} = C_{hi}$;
- GPS-r, where elements in Z_j are randomly sorted.

Likewise, we evaluated two versions of RPA, RPA-ns and RPA-s differing by the sorting of B_i. The former corresponds to the definitions outlined for Algorithm 2. In the latter, the benefit is defined just as the sum of interests of the associated users who are also friends of the content owner, disregarding the uplink bandwidth to the owner gateway; This behavior is "selfish", hence RPA-s, because the receiving gateway only tries to maximize the interest of its own users.

One final variant, which affects the GPA procedure, concerns the size of items to back up. At first, we assumed that all items have the same size. While not realistic per se, it could be meaningful if the implementation of the backup system were limited to a single class of items. In this case, tagged as "fixed size" in our results, we assumed all items to have a 10 MB size. Then, we considered items of any possible size within a bound. Such "variable size" case features random item sizes following a truncated exponential distribution with expected value of 10 MB and maximum size of 50 MB. While studying the latter case, though, we soon found out that allocation results were dangerously biased toward smaller items, so we introduced a *fair item size balancing* mechanism in GPA. Items were divided into size groups of 10 MB each and GPA was modified so that a backup request for one item was sent only if the total amount of data already backed up in the item group did not exceed the total amount already backed up for items of the first size group bigger than it. For instance, if a 15 MB item in the 10-20 MB size group increased the total amount of data already backed up for that size group to 95 MB, and the total amount of the 20-30 MB size group were still 90MB, the request would be put on hold.

4.2   Performance Evaluation

Our first set of plots aims at comparing the optimization results, in their three variants, with the corresponding three variants of the distributed heuristics in which the GPA algorithm alone in employed. The rationale of such comparison is to show the importance of the replacement management introduced by RPA (which is not used in these first results). For reason of space, we cannot show the whole possible parameter space, so we will just focus on a few representative cases to prove our point. Throughout the section, the number of gateways is $N = 1000$, the number of users is $M = 3000$ and the number of item types is $L = 10$. Results with $L = 50$ and $L = 100$ were qualitatively similar.

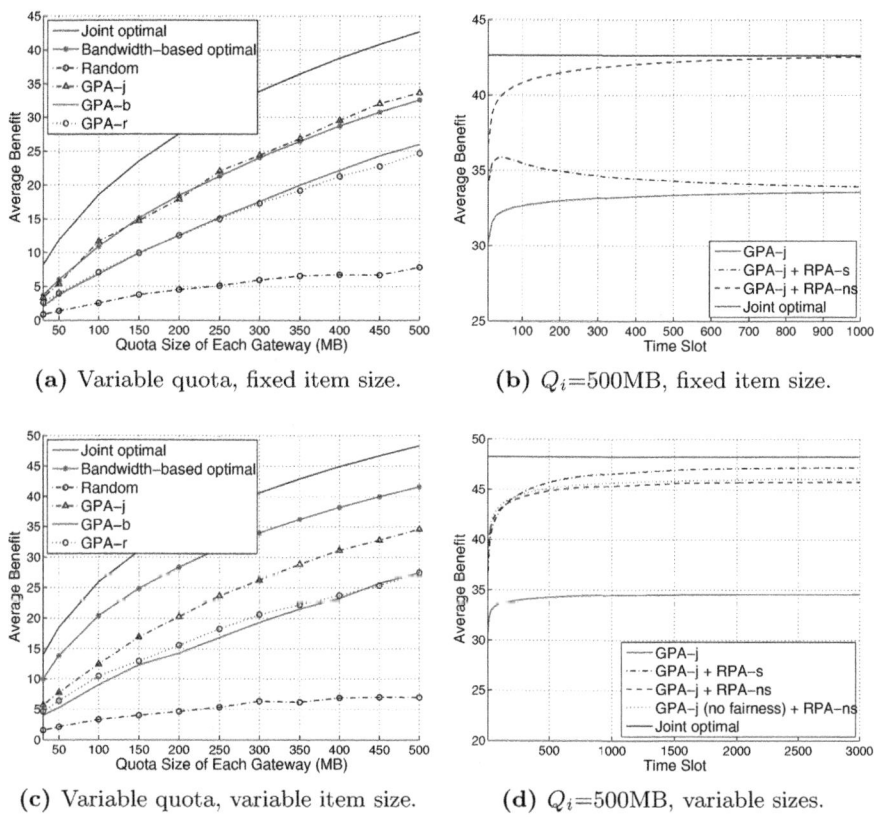

(a) Variable quota, fixed item size.

(b) Q_i=500MB, fixed item size.

(c) Variable quota, variable item size.

(d) Q_i=500MB, variable sizes.

Fig. 1. Avg. user benefit obtained by different methods under different cases

The plot in Fig. 1a clearly ranks the optimization and heuristics variants in terms of average user benefit as defined in eq. (5), for various quota constraints and fixed item size. It shows that jointly optimizing bandwidth and benefit is a clear winner. Even though the average benefit is low (due to the an average of three users per gateway sharing the same quota), the advantage of finding an optimal allocation for backed-up content depending on interest and bandwidth is clearly visible. Additionally, GPA heuristics alone do not match the joint optimization results, as expected. Similar conclusions can be drawn for the variable item size case in Fig. 1c, where GPA-j fares even worse.

In the next set of results, we let receiving gateways run the replacement algorithm, RPA, in its two (selfish and non-selfish) versions. Fig. 1b plays out the $Q = 500MB$ quota case over time, in the same scenarios just examined. It shows that, given some time to converge, GPA-j and RPA-ns together yield an allocation that progressively corrects the initial uneven backup distribution provided by the distributed implementation of GPA-j alone. The selfish version of RPA, RPA-s, instead shows that if the owner and the storing gateway are not on the same page, as it were, when deciding what items are preferable to backup, the performance reverts to that of GPA-j alone. The selfishness of RPA, however, seems to have a lesser impact in the variable item size case, shown in Fig. 1d, where we also plotted a run of GPA-j without the fair item size balancing (tagged "no fairness" in the legend). The use of size balancing, though of marginal impact on the average user benefit, comes in handy in order to control the size of backed up items, as will be shown below.

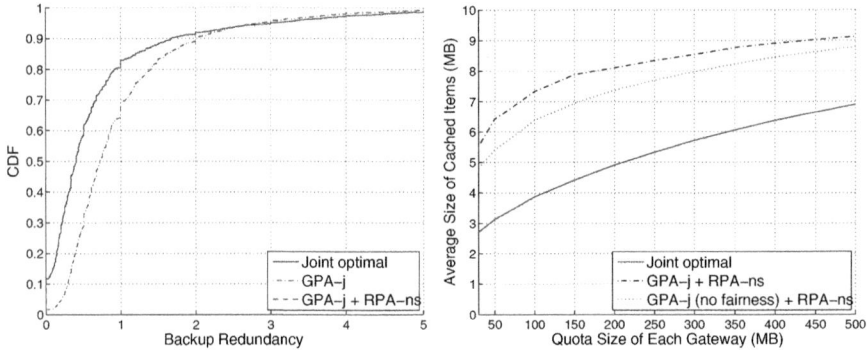

Fig. 2. Backup redundancy CDF; fixed item size and $Q_i = 500$MB (left). Avg. backed-up item size as a function of gateway quota; variable item size (right).

We next plot the *backup redundancy*, i.e. the average number of a user's own items that have been replicated onto its friend gateways, upon reaching convergence of the distributed heuristics. The CDF of the redundancy is useful to understand the thoroughness of the backup process. In the left plot of Fig. 2, RPA-ns allows the remote gateway to replace the items (all of the same size) with smaller benefit and improve the storage thoroughness as much as the optimal method (the two curves indeed overlap). Finally, the right plot of Fig. 2 reports the average size of items backed up in the "friend" storage quota at remote gateways, when convergence is reached. Recalling that the average item size in the variable item case is 10MB, we can conclude that the loss of average user benefit with respect to the optimal case shown in Fig. 1d is offset by a fairer distribution of item sizes in the gateway storage across the federated network (i.e., the average size of backed up item achieved by GPA-j and RPA-ns is just 10% smaller than the average item size in the system, as opposed to 30% smaller in the joint optimization case). Also, the use of fair item size balancing with GPA proves of some consequence to achieve this result.

5   Conclusions

In this work we addressed a novel approach to content sharing and backup in home networks. We envisioned a system where content is placed "outside the home" in a cloud formed by federated home networks and its location is selected exploiting the user's social networking information. We studied its performance in terms of the benefits that remote gateways and their users can enjoy when friends choose them to back up (and share) their content. We defined an optimization problem and compared its numerical solution with several flavors of distributed heuristics.

Acknowledgments. This work was partly funded by the European Union FP7, under grant agreement 258378 - FIGARO project and by the Italian Ministry of Education and Research under the PRIN project GATECOM.

References

1. The FIGARO FP7 ICT project, http://www.ict-figaro.eu/
2. Defrance, S., Gendrot, R., Le Roux, J., Straub, G., Tapie, T.: Home Networking as a Distributed File System View. In: HomeNets (August 2011)
3. Backstrom, L., Sun, E., Marlow, C.: Find me if you can: improving geographical prediction with social and spatial proximity. In: 19th ACM WWW (April 2010)
4. Belaramani, N., Dahlin, M., Gao, L., Nayate, A., Venkataramani, A., Yalagandula, P., Zheng, J.: PRACTI replication. In: USENIX Symposium on Networked Systems Design and Implementation (NSDI) (May 2006)
5. Ramasubramanian, V., Rodeheffer, T., Terry, D., Walraed-Sullivan, M., Wobber, T., Marshall, C., Vahdat, A.: Cimbiosys: A platform for content-based partial replication. In: NSDI (August 2009)
6. Post, A., Kuznetsov, P., Druschel, P.: PodBase: Transparent storage management for personal devices. In: USENIX International Workshop on Peer-to-peer Systems (February 2008)
7. Boldrini, C., Conti, M., Passarella, A.: ContentPlace: social-aware data dissemination in opportunistic networks. In: ACM MSWiM 2008. ACM (October 2008)
8. Jaho, E., Stavrakakis, I.: Joint interest-and locality-aware content dissemination in social networks. In: IEEE/IFIP WONS 2009 (February 2009)
9. Pantazopoulos, P., Stavrakakis, I., Passarella, A., Conti, M.: Efficient Social-aware Content Placement in Opportunistic Networks. In: IEEE/IFIP WONS 2010 (February 2010)
10. Jiang, J., Casetti, C.: Socially-aware Gateway-based Content Sharing and Backup. In: HomeNets (August 2011)
11. Khuller, S., Moss, A., Naor, J.: The budgeted maximum coverage problem. Information Processing Letters 70(1), 39–45 (1999)
12. Gurobi optimizer, http://www.gurobi.com/html/products.html

Trans-Social Networks
for Distributed Processing

Nuno Apolónia, Paulo Ferreira, and Luís Veiga

INESC ID Lisboa / Technical University of Lisbon,
Rua Alves Redol 9, 1000-029 Lisboa, Portugal
nuno.apolonia@ist.utl.pt, {paulo.ferreira,luis.veiga}@inesc-id.pt

Abstract. A natural succeeding process for the Internet was to create Social Networks (e.g. Facebook, among others), where anyone in
the World can share their experiences, knowledge and information, using personal computers or mobile devices. In fact, Social Networks can
be regarded as enabling information sharing in a peer-to-peer fashion.
Given the enormous number of users, sharing could also be applied to
the untapped potential of computing resources in users' computers.

By mining the user friendship graphs, we can perform people
(and resource) discovery for distributed computing. Actually, employing
Social Networks for distributed processing can have significant impact
in global distributed computing, by letting users willingly share their
idle computing resources publicly with other trusted users, or groups;
this sharing extends to activities and causes that users naturally tend to
adhere to.

We describe the design, development and resulting evaluation of a
web-enabled platform, called Trans-SocialDP: Trans-Social Networks for
Distributed Processing. This platform can leverage Social Networks to
perform resource discovery, mining friendship relationships for computing resources, and giving the possibility of resource (not only information)
sharing among users, enabling cycle-sharing (such as in SETI@home)
over these networks.

Keywords: social networks, distributed processing, cycle-sharing,
resource discovery.

1 Introduction

In the past few years the computing power has been increasing. However, many
computational problems requiring enormous amount of computer resources still
exist; e.g. applications for scientific research, multimedia video or image encoding.

One of the earliest initiatives identifying idle computing cycles suitable for
distributed computing, regards the WORM computing project at Xerox PARC
[Sho98], proposed in 1978. Years after, the scientific community realized the
potential benefits, mainly in physics, since the supercomputers employed for
heavy-duty calculations at each institution, would often be under utilized.

R. Bestak et al. (Eds.): NETWORKING 2012, Part I, LNCS 7289, pp. 82–96, 2012.

Thus, by enabling the harvesting and allowing the sharing of such idle processing times, grid computing emerged.

Projects such as SETI@Home [ACK+02], Folding@Home,[1] Distributed.net[2] gather a gigantic pool of resources on the Internet, by using desktop computers from any house hold (also known as global distributed computing), allowing them to process data much quicker than in traditional supercomputers.

Nowadays, Social Networks allow individual distributed computing projects to be easily proposed, and promoted across friendship relationships, as well as making more effective the publicity of their motivation and results.

Motivation. The Internet has made it possible to exchange information more rapidly in a global scale. With the creation of Social Networks, anyone in the world can share their experiences and information using only his Internet enabled personal computer or mobile device.[3] Under this scope, there are many Social Networks such as Facebook, Orkut, and others still being actively created (such as Google+), each one exporting its own API to interact with their users' and groups' databases as well as allowing to gather idle resources scattered across the World; examples of such APIs are Facebook API[4] and OpenSocial.[5]

Shortcomings of Current Systems. The public-resource sharing and cycle-sharing systems that are widely used today, are not concerned with the common users' needs. They are mostly used for intensive computational projects (and proprietary) such as Folding@Home, PluraProcessing.

Some systems are beginning to use technologies previously unavailable to other projects, in order to cover more Internet users. However, they use the users' browsers to do cycle-stealing not addressing the needs of the common users. Moreover, these systems use remote code embedded on Web sites and games (i.e. Adobe Flash based games) to gain access to potential idle resources.

Contribution. The main contribution of this project is the Trans-SocialDP platform with its architecture, messaging protocol and client application. This platform can perform resource and service discovery on top of Social Networks for third-party applications. Furthermore, Trans-SocialDP is able to gather idle cycles from users' computers and communities that are willing and capable of processing tasks, in order to achieve cycle-sharing on Social Networks. It also allows common users to make use of this paradigm to speedup their own (or common) tasks' execution without having to create their own networks.

Trans-Social Networks for Distributed Processing (Trans-SocialDP) is a Web-enabled platform, which was developed and evaluated to interact with Social Networks, and thus being able to mine the Social Networks for users' information which includes their idle resources. It leverages an existing middleware, Ginger [VRF07, SFV10] for task (called Gridlets) creation and aggregation.

---

[1] Folding@Home: `folding.stanford.edu` accessed on 16/02/2012.
[2] Distributed.net: `www.distributed.net` accessed on 16/02/2012.
[3] Facebook Mobile: `facebook.com/mobile` accessed on 16/02/2012.
[4] Facebook Developers: `developers.facebook.com` accessed on 16/02/2012.
[5] OpenSocial Web site: `code.google.com/apis/opensocial` accessed on 14/02/2012.

Our main concerns on the development of Trans-SocialDP were with the resource discovery and the manner with which a user could submit his own Job, processed on others' computers, while also being able to mine any Social Network for user's information.

Document Road-Map. The rest of the paper is organized as follows. In the next section, we address the relevant literature related to our work. Section 3 describes the architecture of Trans-SocialDP and its implementation detailed in Section 4. In Section 5, we offer an extensive evaluation of the platform in the context of two (interconnected) Social Networks. Section 6 finishes the paper with some conclusions and lines for future work.

2   Related Work

This section offers a review on relevant works and technologies more related to our work, addressing: i) Social Networks and mining on Social Networks, and ii) peer-to-peer networks, Grids and Distributed Computing.

Social Networks. Social Networks are popular infrastructures for communication, interaction and information sharing on the Internet. Anyone with a desktop computer and a Browser can access such Web sites, like Facebook, MySpace, Orkut, Hi5, YouTube, LinkedIn and many more.[6] They are used to interact with other people for personal or business purposes, sending messages, posting them on the Web site, receiving links to other Web sites or even sharing files between people, among other uses.

In Social Networks, the basic (real life) behaviors or interaction patterns still apply [Sco88]. By grouping people in the same areas or topics, it is easier to exploit those interactions, because people understand better what the distributed tasks will accomplish and are willing to participate. Social Networks have already began to sprout new ideas to exploit them for uses other than human interactions, such as using it for enhancing Internet search [MGD06] and leveraging infrastructures to enable ad-hoc VPNs [FBJW08].

We focus on Facebook and OpenSocial, because of their size and possibility of access to users' databases by means of the APIs provided. Furthermore, Facebook claims to have reached 845.000.000 users (as of January of 2012) and MySpace (one Web site that uses the OpenSocial API) claims to have more than 130.000.000 registered users. The potential of these networks for global distributed computing is best compared to other networks.

The Facebook and OpenSocial APIs enable Web applications to interact with the Social Networks servers using a *REST*-like interface[7] or, in case of Facebook, also a *Graph* interface.[8] This means that the calls from third-party applications are made over the Internet by sending HTTP GET and POST requests.

---

[6] List of Social Networks on `Wikipedia.org`.

[7] Representational State Transfer: `tinyurl.com/6x9ya` accessed on 14/02/2012.

[8] OpenGraph Protocol: `ogp.me` accessed on 14/02/2012.

Social Cloud. [CCRB10] is described as being a model that integrates Social Networking, cloud computing [AFG+10] and volunteer computing. In this model, users can acquire the resources (the only resource considered is disk space) by exchanging virtual credits, making a virtual economy over the social cloud computing. Users can gather resources from their friends (either by virtual compensation, payment, or with a reciprocal credit model [MBAS06]), allowing this project to approach the objectives for public-resource sharing. Furthermore, there are a number of advantages gained by leveraging Social Networking platforms, such as gaining access to a huge user community, exploiting existent user management functionality, and rely on pre-established trust formed through user relationships. However, the trusting relationship of friends, may not be always the case[9] in Social Networks such as Facebook.

Peer-to-Peer Networks, Grids and Distributed Computing. Peer-to-Peer (P2P) networks and Grids are the most common types of sharing support systems. They evolved from different communities to serve different purposes [TTP+07].

Grid systems interconnect clusters of supercomputers and storage systems. Normally they are centralized and hierarchically administrated, each with its own set of rules regarding resource availability. Resources can be dynamic and thus may vary in amount and availability during time, and have to be known beforehand among the network. Grid systems were created by the scientific community to run computation intensive applications that would take too much time in normal desktops (without being distributed) or on a single cluster, e.g. large scale simulations or data analysis.

P2P networks are typically made from house hold desktop computers or common mobile devices, being extremely dynamic in terms of resource types and whose membership can vary in time with more intensity than with Grids. These networks are normally used for sharing files, although there are a number of projects using those kinds of networks for other purposes, such as sharing information and streaming (e.g. Massive Multi-player Online games using P2P [KLXH04] to alleviate server load, distributing tasks as SETI@Home [ACK+02], data streaming for watching TV[10]). The nodes (or peers) are composed by anonymous or unknown users unlike in Grids, which raises its own problems with security or even with forged results [TTP+07].

SETI@Home. [ACK+02] aims at using globally distributed resources to analyze radio wave signals that come from outer space, hoping to find radio signals originated from other planets on our galaxy. For this project, having more computing power means it is possible to cover a greater range of frequencies, instead of using supercomputers [ACK+02]. Thus, the authors found a way that lets them use computers around the world to analyze such wave signals.

---

[9] How Facebook could make cloud computing better: `tinyurl.com/237ddem` accessed on 14/02/2012.

[10] PPStream: `ppstream.com` accessed on 14/02/2012.

The wave signals are divided in small units of fixed size and distributed among the clients (that would be located in any user computer operating as a screen saver when there are idle cycles). Then, each client computes the results in its spare time and sends it to the central server asking for more work to do. The most important lesson of SETI@Home project was that to attract and keep users, such projects should explain and justify their goals, research subject and its impact.

BOINC. (Berkeley Open Infrastructure for Network Computing) [And04] is a platform for distributed computing through volunteer computers; it emerged from the SETI@Home project and became useful to other projects.[11] Although each project has its own topic and therefore their own computational differences, the BOINC system used for each project (client application) has to be unique.

There are many other projects for distributed computing.[12] However, all of them have only one topic of research (for each project), meaning that each system does not have the flexibility of changing its own research topic. With **BOINC Extensions for Community Cycle Sharing (nuBOINC** [SVF08]), users without programming expertise may address the frequent difficulties in setting up the required infrastructures for BOINC systems and subsequently gather enough computer cycles for their own project. The nuBOINC extension is a customization of the BOINC system, that allows users to create and submit tasks for distributed computing using available commodity applications. They try to bring global distributed computing to home users, using a public resource sharing approach.

The main concept of **Ginger** (Grid Infrastructure for Non-Grid Environments) [VRF07], from which our proposal is derived, is that any home user may take advantage of idle cycles from other computers, much like SETI@Home, given the right incentives [RRV10]. However, by donating idle cycles to other users to speedup their applications, they would also take advantage of idle cycles from other computers, to speedup the execution for their own applications, with arbitration based on users classes [SFV10], reputation and possibly subject to a virtual currency economic model [Oli11].

To leverage the process of sharing, Ginger introduces a novel application and programming model that is based on the Gridlet concept. Gridlets are work units containing chunks of data and the operations to be performed on that data. Moreover, every Gridlet has an estimated cost (CPU and bandwidth) so that they can try to be fair for every user that executes these Gridlets.

Discussion. While there have been works done to approach volunteer computing using Social Networks, as the communication overlay, they either do not have the same objectives, as public cycle-sharing, or they do not give the users the possibility of using others' resources (idle cycles) for their own work. Comparing these projects based on the communication's latencies, can be volatile and misleading because of the unstable conditions, either by the servers' latencies, network latencies or even the type of computers used.

---

[11] BOINC projects: `boinc.berkeley.edu/projects.php` accessed on 14/02/2012.
[12] List of Distributed Computing projects on `Wikipedia.org`.

Furthermore, for our project we assume that the user may not want to give or have idle cycles to spare which would add a small communication time to the overall process.

3   Architecture

Our work makes use of Social Networks, such as Facebook and MySpace to mine these networks locating users' information and their resources in order to execute tasks (*Gridlets* [VRF07]) on their computers. Note that, this information may include user's profile, such as friends and groups.

Trans-SocialDP uses the SIGAR library[13] to acquire local information about resources (i.e. processor information, memory available). Such information can be sent over the Social Network when requested, while also using it to decide whether it should accept a new Job (from someone else).

The main approach for Trans-SocialDP is to have the platform split into two parts: one that interacts with the Social Networks; and another to interact with the users, the local resources and the Ginger Middleware (which is out of the scope of this work [VRF07]).

Design Requirements. The client application interacts with the Social Networks (Facebook, MySpace) through Web Protocols named Graph and REST (which are an added layer to the HTTP protocol). As Social Networks are still developing their own systems, the operations available within the client application may change over time. We use libraries such as *RestFB* library,[14] *myspace-id* library[15] and *OpenSocial-java* library[16] to ease the communication to and from the Social Networks.

Moreover, in order for Trans-SocialDP not to interfere with the users' normal usage of their computers, the client application can schedule processing to another time, while also preventing its overuse by stopping its activities, i.e. the processing of requests and Gridlets only happens when there are idle cycles to spare.

Trans-SocialDP Architecture. The Trans-SocialDP architecture relies on: i) the interaction with the Social Networks through the Social Network's API (Graph or REST protocols), for the purpose of searching and successfully executing jobs; ii) the Ginger Middleware for Gridlet creation and aggregation; and iii) the user's operating system to acquire the information and hardware states that are needed.

Jobs are considered to be tasks initiated by the users, and containing more than one Gridlet to be processed in someone else's computer; all Jobs state what they require in order to execute those Gridlets, so that the client application

---

[13] SIGAR library: `hyperic.com/products/sigar` accessed on 14/02/2012.

[14] RestFb Web site: `restfb.com` accessed on 14/02/2012.

[15] MySpaceID: `developer.myspace.com/MySpaceID` access on 14/02/2012.

[16] OpenSocial Java: `code.google.com/p/opensocial-java-client` access on 14/02/2012.

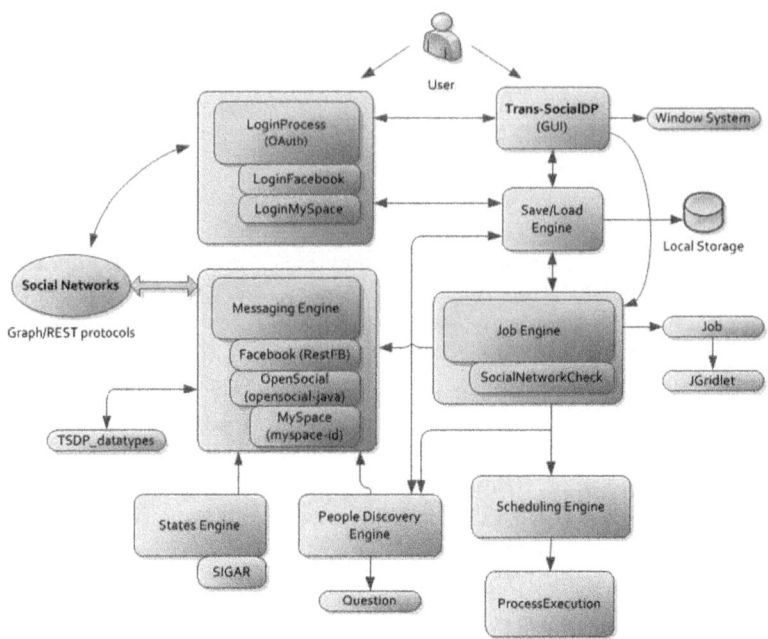

Fig. 1. Trans-Social Networks for Distributed Processing module view

can search for users or groups (computer information) that could help on their processing.

A Gridlet contains the information necessary to process it, meaning that it has the data file(s) to be transferred to another user and the arguments to be given to the executable program. The process of creating and aggregating the Gridlets is managed by the Ginger Middleware and is outside the scope of this work [VRF07].

The architecture for Trans-SocialDP is comprised of a set of modules depicted in Fig. 1, and described as follows.

Trans-SocialDP (GUI): this is the main module, which contains the graphical user interface for the user to interact with the client application. It is responsible to initiate the interaction with the Social Networks and local computer.

Login Process: this module includes the OAuth[17] call definition for each Social Network. It provides to the user the Log-in web site for each Social Network in order for the client application to interact with each one; it uses an embedded web browser (using the JDIC library) for that effect. Note that each Social Network is responsible for implementing its own OAuth system (or other authentication systems); for such the client application needs to implement submodules with each specification for each connection (LoginFacebook, LoginMySpace).

---

[17] OAuth Web site: oauth.net access on 27/02/2012.

Messaging Engine: this is a main module to the client application; it is required for communication with all the Social Networks, containing all the necessary functions to do so. The communication between the client application and the Social Networks chosen are made by each submodule, according to the used protocol of the Social Network. In particular, for the Facebook submodule it uses the Graph protocol to send/retrieve Posts/Comments to/from this Social Network.

MySpace submodule sends/receives key-value pairs to/from a global map used by third-party applications, which is provided by MySpace servers. Furthermore, MySpace is being constructed as if it was an OpenSocial network, thus we make use of the OpenSocial-java library to allow for OpenSocial communication protocol.

The Messaging Engine also contains its own data types, which are converted to the message schemas applied to the messages sent/retrieved for each Social Network, because they may block some types of messages.

Job Engine: this module is divided in two parts, where the first is responsible to start the chain of events for processing a Job (submitted by the user), such as sending search messages, sending Gridlets to other users.

The second part (SocialNetwork Check) is responsible for searching for new Jobs in each Social Network, in order to answer them according to the Job's requirements and the local resource's availability. In the specific case of Facebook the redirected messages (FoF method) are also taken into consideration, when starting or searching for Jobs. Thus, each Messaging Engine submodules need to be aware of the communication protocol previously established for each Social Network, so that the Job Engine only requires to know a generic way of send/retrieving information for any Social Network.

People Discovery Engine: this module mines the Social Networks to retrieve users, friends, groups and other types of information from them. In addition, it is also used to communicate with other client applications in order to establish relations between different users' accounts on different Social Networks (users that are connected to several Social Networks).

As an implementation issue, the client application needs to distribute the tasks (Gridlets) as fair as possible. Thus, the module uses the users' information to assess if the incoming messages (sent by users that accepted the Job), come from different users. In order to send a Gridlet for each user, even if they accepted the Job on more than one Social Network (may not happen when the number of users is less than the number of Gridlets).

Furthermore, this module contains a data structure for the questions that are sent to other client applications, i.e. in order to ask other users for a specific person from a Social Network.

Scheduling Engine: this module gives priorities to the Gridlet's processing, according to a specified criteria (e.g. users' friends tasks may have higher priorities than other users).

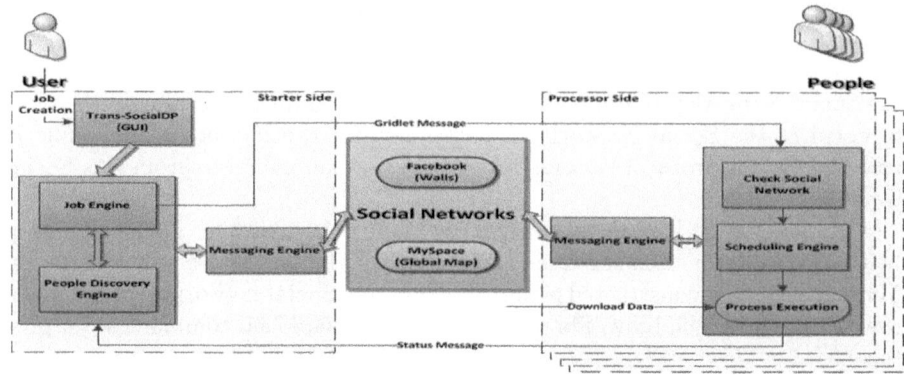

Fig. 2. Trans-SocialDP Prototypical example

***States Engine*:** this module determines the state of the user's computer, taking in consideration the processor's idle times, Internet connectivity (essential to all engines), the user's Social Networks states and the local state (i.e. when the client application has been halted by the user). It also uses the SIGAR library, which reports the system information needed to determine the availability of the resources.

Prototypical Example. The prototypical example as depicted in Fig. 2 gives an idea of the platform's communication flow, from the creation of a new Job (by the Starter user), to using the Messaging Engine to send and retrieve messages from the Social Networks the application is connected to. In this example, the Gridlet message is sent by the Starter's Messaging Engine to a Social Network (Facebook), retrieved by the Messaging Engine on the Processor side (an application that accepted the Job), scheduling it to be processed when it has idle cycles to spare and sending the results (status) back to the Starter in the same manner.

4   Implemention Details

The implementation of Trans-SocialDP aims for a simple use by the end-users. Also, the different types of operating systems lead us to favor portability; therefore, we used Java as the main language. We primarily chose Facebook over other alternatives, because it has a higher number of registered users than any other Social Network, and MySpace because it is well-known by Internet users, while also utilizing some of the OpenSocial concepts.

This section gives an insight on how the technologies were used, such as Graph and REST protocols. It also explains the schemas used for the messages sent/received to/from the Social Networks chosen.

Technology Employed. For the purpose of interacting with the Graph and REST servers, the client application makes use of the RestFb library for Facebook, the OpenSocial-java and MySpaceID libraries for MySpace, which gives a simple and flexible way of connecting to them and conceal the use of XML or JSON objects.[18] However, the functions (using REST) or connections (using Graph) have to be known, in order to use these libraries, e.g. to read the Posts on a user's Wall on Facebook using the Graph protocol, we need a user's ID or Name for the library to access Facebook and retrieve that user's Wall.

As Trans-SocialDP also needs to gather the information about the local resources of the users' computers, we make use of the *SIGAR* library. This allows us to easily access a list of local resources each time it is called, such as processors, cores, memory. Also, it gives us the ability to know the current states of those resources, i.e. it can give us the available memory at the requesting time, or even the current idle time for each of the available cores. This library is also useful for the fact that it can work in multiple environments, such as Windows, Linux, among others, making it possible the portability of Trans-SocialDP to other systems.

Message Schemas. Trans-SocialDP uses Social Networks to send and retrieve messages via their external interfaces. In Facebook, it reads Posts (messages that are contained in the users' Walls, groups' Walls) and Comments (messages contained within the Posts), and writes other messages on users' Walls (which is a space that contains messages) either as Posts or Comments.

As for MySpace, Trans-SocialDP uses a global map of key-value pairs to send and retrieve messages to and from other applications. These keys, are generated by the client application depending on the type of message plus a random number in order for the keys to be different (e.g. for the Job search request message the key becomes transSocialDP.JobSearch.Random Number). The values for each key contains the message to be sent to another user, which is the same as we use on Facebook.

These Schemas are very simple and human readable, in order for Facebook (or other Social Networks) to allow them on the Web site, and not consider them as Spam or other type of blocked messages. They are also human readable to assure the users what information is being sent to other users.

5   Evaluation

The evaluation of Trans-SocialDP is comprised of a scenario that assesses each Social Network and ultimately the combination of interacting with several Social Networks, in order to know the effects each can carry to the processing times, while also evaluating our works' goals.

Scenario Mixed. This scenario was designed to evaluate the performance of Trans-SocialDP on both Social Networks (Facebook, Myspace) with a complex system of connections. As depicted in Fig. 3 the starter has a connection to Facebook (FB) and to MySpace (MS) where it can send/retrieve messages from other client applications, adding that some users may have accounts on both Social Networks and are identified by their UIDs of each Social Network. The starter needs

---

[18] JSON: `json.org` accessed on 14/02/2012.

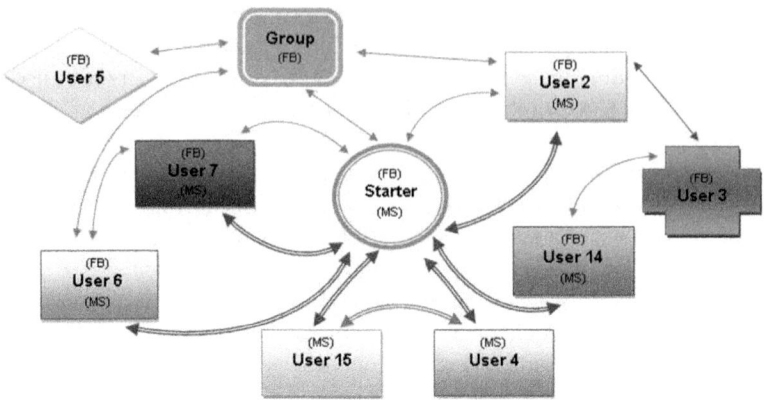

Fig. 3. Trans-SocialDP Scenario Mixed View

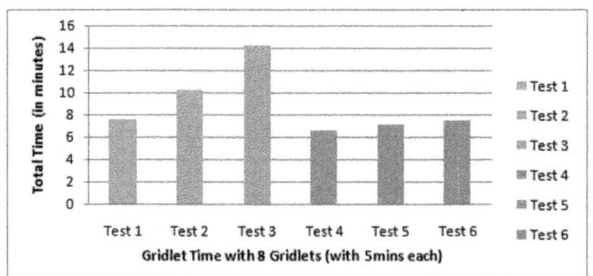

Fig. 4. Rendering Test Times for Scenario Mixed

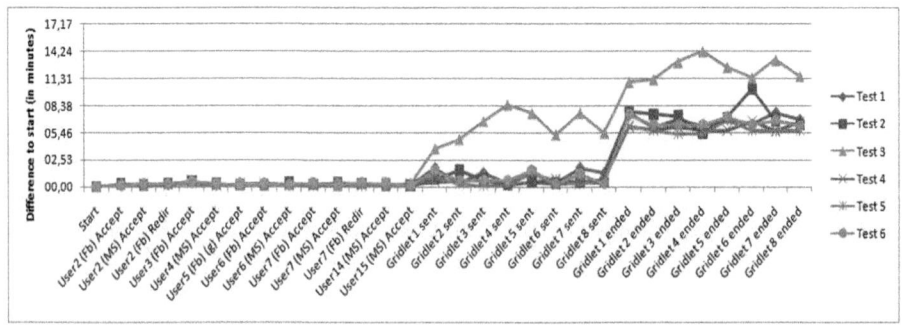

Fig. 5. Communication Times for Scenario Mixed

to know if a user is the same on both networks, even though the UIDs are different. This network is composed by 2 Friends in Facebook, each one having a Friend (FoF), a group (Facebook) with 4 users including the starter, and 6 users connected to MySpace.

The number of *Gridlets* to be processed are 8 in total, and the processing time is 5 minutes each, in order for all users in the network to take at least one Gridlet. The first three tests were made without a local database of users' information, and the later three with it.

For this scenario we assume that i) all client applications are running, and connected to their respective Social Networks; ii) the client applications can retrieve a *Gridlet* message from any place that they have accepted, e.g. User 7 can accept a Job from Facebook (Friend connection) or MySpace and that User 14 can only accept a Job from MySpace, although it has a connection to User 3 (on Facebook), it goes beyond the 2nd degree of friendship we imposed.

The results for scenario Mixed, as depicted in Fig. 4, describes the total time for a Job for each test. Some situations, as in test 3, are caused by latency related problems with the communication between client applications. Also, for test 2 the creation of the local database might not have been completely accurate, thus the starter sent more than one *Gridlet* to a user (the same user on different Social Networks).

This case can happen when the starter asks for information about a user to the network (or friends/other users) and does not get an answer within a pre-defined time period; it then assumes that the user may not be the same in both Social Networks, where in fact they were. However, besides all the latency caused by the communication, the total process times in all tests still gain speedups against local execution (where it should have been about 40 minutes).

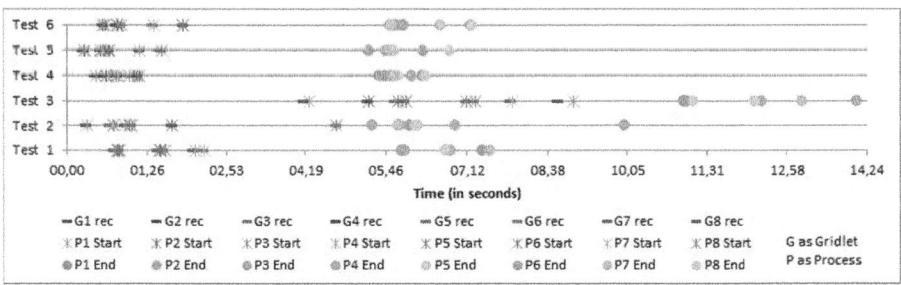

Fig. 6. Wait times for Gridlet process on Scenario Mixed

In Fig. 5 we can see the time each task takes to be completed, and the spikes the connection to the Social Network can cause on the overall process; e.g. the third test took much more time to send the *Gridlets* than in the other tests, because of the added times for mining information about other users and some latency in the servers.

In Fig. 6 we can notice some delay to receive a *Gridlet* message in test 3, which can occur when the network communication is having more latency than usual. We also see in test 2 that a *Gridlet* message was received and its processing was delayed in relation to the other *Gridlets*, when a user received it from another Social Network.

Discussion. We can state that the overhead that Trans-SocialDP imposes on the overall process is minimal compared to the time it takes to process a *Gridlet*, which in realistic terms can be more than 1 hour. However, these times can be hindered by other tasks such as searching for resources that do not return positive results, or even that the total resources available are lesser than the number of *Gridlets* that have to be processed.

Furthermore, we can state that the search for users' information may hinder the total process, because the information on both Social Networks may not be compatible, and thus situations like sending more than one *Gridlet* for the same user via different Social Networks can occur.

When comparing with local execution, Trans-SocialDP decreased the total processing time, compared to what it would have consumed in the users' computers. Trans-SocialDP achieves overall speedups on Jobs, and the added support of other Social Networks may benefit this speedup. By finding capable users, that either the search did not reached on one of the Social Networks, or that the users are only connected to the other Social Network.

Moreover, we can confirm that the users can donate their resources (processors' time) for other users' consumption. While also, taking advantage of other users' resources, that have the same interests (or in the same groups), to further speedup their own tasks.

6   Conclusion

Our project exceeds the boundaries of a common use of Social Networks, allowing any user to consider donating their idle processing cycles to others (while also making use of others) that may need to finish their tasks faster than they would in their own computers. In this project we presented a new method of resource and service discovery through the use of Social Networks and the interaction between users.

The main approach for Trans-SocialDP is to have a client application split in two parts. One that interacts with the Social Network using REST or Graph protocols; and another to interact with the users' computers for local resource discovery, and the Ginger Middleware for creation and aggregation of Gridlets.

We evaluated Trans-SocialDP with several scenarios to determine the viability and the changes it would take by contacting different Social Networks. Thus, we created a scenario that interacts with both Social Networks to regard Trans-SocialDP performance, stability and viability on such networks.

In addition, the scenario described was created with several users' roles in mind, meaning the users could be regarded as Friends, Friends of Friends, group members, or other roles, in order to fully test the communication system between the users' client applications.

With the obtained results, we can conclude that while the total times for processing a Job gained speedups against local execution in the users' computers, this can be hindered by some variables: latency of Social Network servers, the fact that searching for resources among Social Networks users may not return positive results, that the total number of available resources is less than the number of Gridlets that comprises a Job, while also the distribution of tasks among the available users may not completely parallelize them.

In summary, we find all our goals to have been met, in the sense that our proposed platform can leverage Social Networks, by mining users' friendships, relationships, and affiliation to groups and communities, in order to gather computational resources. Such resources can be employed to speedup jobs in a global distributed computing platform. By bringing in the concept of resource sharing to global infrastructures such as Social Networks, we allow virtually any common user to make use of idle resources scattered across the Internet, within a framework they are familiar with. Within, Trans-SocialDP, global cycle sharing can become widely employed as many other features supported by Social Networks.

Future Work. For the future, we believe that Jobs completion and the search for resources would benefit with requirements' semantics, increasing the chance to direct Gridlets to peoples' computers that would satisfy those. In addition, a reputation method would assure users that the distributed tasks are given to those that can actually undertake the responsability of processing the tasks.

Acknowlegments. This work was partially funded by FCT projects PTDC/ EIA-EIA/102250/2008, PTDC/EIA-EIA/113993/2009 and by PIDDAC Program funds (INESC- ID multiannual funding).

References

[ACK+02] Anderson, D.P., Cobb, J., Korpela, E., Lebofsky, M., Werthimer, D.: SETI@ home: an experiment in public-resource computing. Communications of the ACM 45(11), 56–61 (2002)

[AFG+10] Armbrust, M., Fox, A., Griffith, R., Joseph, A.D., Katz, R.H., Konwinski, A., Lee, G., Patterson, D.A., Rabkin, A., Stoica, I., et al.: A view of cloud computing. Commun. ACM 53(4), 50–58 (2010)

[And04]   Anderson, D.P.: BOINC: A system for public-resource computing and storage. In: Proceedings of the 5th IEEE/ACM International Workshop on Grid Computing, pp. 4–10. IEEE Computer Society (2004)

[CCRB10] Chard, K., Caton, S., Rana, O., Bubendorfer, K.: Social Cloud: Cloud Computing in Social Networks. In: 2010 IEEE 3rd International Conference on Cloud Computing, pp. 99–106. IEEE (2010)

[FBJW08] Figueiredo, R.J., Boykin, P., Juste, P., Wolinsky, D.: Integrating overlay and social networks for seamless p2p networking. In: WETICE, pp. 93–98 (2008)

[KLXH04] Knutsson, B., Lu, H., Xu, W., Hopkins, B.: Peer-to-peer support for massively multiplayer games. In: IEEE INFOCOM, vol. 1, pp. 96–107. Citeseer (2004)

[MBAS06] Mowbray, M., Brasileiro, F., Andrade, N., Santana, J.: A reciprocation-based economy for multiple services in peer-to-peer grids. In: 6th IEEE International Conference on Peer-to-Peer Computing, P2P 2006, pp. 193–202. IEEE (2006)

[MGD06]   Mislove, A., Gummadi, K.P., Druschel, P.: Exploiting social networks for internet search. BURNING, 79 (2006)

[Oli11]   Oliveira, P., Ferreira, P., Veiga, L.: Gridlet Economics: Resource Management Models and Policies for Cycle-Sharing Systems. In: Riekki, J., Ylianttila, M., Guo, M. (eds.) GPC 2011. LNCS, vol. 6646, pp. 72–83. Springer, Heidelberg (2011)

[RRV10]   Rodrigues, P.D., Ribeiro, C., Veiga, L.: Incentive mechanisms in peer-to-peer networks. In: 2010 IEEE International Symposium on Parallel & Distributed Processing, Workshops and Phd Forum (IPDPSW), pp. 1–8. IEEE (2010)

[Sco88]   Scott, J.: Social network analysis. Sociology 22(1), 109 (1988)

[SFV10]   Silva, J.N., Ferreira, P., Veiga, L.: Service and resource discovery in cycle-sharing environments with a utility algebra. In: 2010 IEEE International Symposium on Parallel & Distributed Processing (IPDPS), pp. 1–11. IEEE (2010)

[Sho98]   Shostak, S.: Sharing the universe- Perspectives on extraterrestrial life. Berkeley Hills Books (1998)

[SVF08]   Silva, J., Veiga, L., Ferreira, P.: nuboinc: Boinc extensions for community cycle sharing. In: SASO Workshops, pp. 248–253 (2008)

[TTP^{+}07]   Trunfio, P., Talia, D., Papadakis, H., Fragopoulou, P., Mordacchini, M., Pennanen, M., Popov, K., Vlassov, V., Haridi, S.: Peer-to-Peer resource discovery in Grids: Models and systems. Future Generation Computer Systems 23(7), 864–878 (2007)

[VRF07]   Veiga, L., Rodrigues, R., Ferreira, P.: GiGi: An Ocean of Gridlets on a "Grid-for-the-Masses". In: Proceedings of the 7th IEEE International Symposium on Cluster Computing and the Grid, pp. 783–788. IEEE Computer Society (2007)

Context-Sensitive Sentiment Classification
of Short Colloquial Text

Norbert Blenn, Kassandra Charalampidou, and Christian Doerr

Department of Telecommunication
TU Delft, Mekelweg 4, 2628CD Delft, The Netherlands
{N.Blenn,C.Doerr}@tudelft.nl, K.Charalampidou@student.tudelft.nl

Abstract. The wide-spread popularity of online social networks and the resulting availability of data to researchers has enabled the investigation of new research questions, such as the analysis of information diffusion and how individuals are influencing opinion formation in groups. Many of these new questions however require an automatic assessment of the sentiment of user statements, a challenging task further aggravated by the unique communication style used in online social networks.

This paper compares the sentiment classification performance of current analyzers against a human-tagged reference corpus, identifies the major challenges for sentiment classification in online social applications and describes a novel hybrid system that achieves higher accuracy in this type of environment.

Keywords: Online Social Networks, Sentiment Analysis, Text Classification.

1 Introduction

The amble availability of data from online social networks in machine-readable format has made it possible to investigate and evaluate a whole new set of research questions at a large scale, such as "how do trends form?", "what determines how influential a person is?" or "how do our friends and contacts shape our opinion?". Many research questions posed in online social network analysis thus require to be able to assess the context and meaning of a user's statements, identifying in the simplest case whether a sentence is a neutral, objective comment or a subjective opinion, or in more advanced scenarios tracing and quantifying the development and flow of positive/negative thoughts across a user's various communication threads.

In recent years, a number of methods for sentiment analysis have been proposed; these techniques have however been developed for and are consequently geared towards the extraction of meaning in large text corpora, such as product reviews, letters or articles. User communication in online social networks such as Facebook or Twitter has however very specific and challenging features: 1) Messages are short and highly abbreviated and therefore only a small angle of

R. Bestak et al. (Eds.): NETWORKING 2012, Part I, LNCS 7289, pp. 97–108, 2012.

attack for an automatic classifier, 2) Text is very colloquial and typically deprived of any context information, thus making it difficult to infer the subject and reference of the sentiment.

Due to these special characteristics, traditional sentiment analysis methods do not provide sufficiently accurate results when applied towards online social network communications. This paper deviates from these established sentiment classification approaches and describes an alternative method utilizing additionally both grammatical and contextual information for increased detection accuracy. The contributions of this paper are two-fold: First, we evaluate a set of available sentiment classifiers against a dataset obtained from the social microblogging platform Twitter, and measure the detection performance of these automatic tools against a human classification. Second, we identify common problems in sentiment analysis and demonstrate how an alternative approach can provide a higher detection accuracy than previous context-less classifiers, and beside pure identification of sentiment polarization, can additionally provide a magnitude quantification of sentiments.

The remainder of this paper is structured as follows: Section 2 overviews previous work in sentiment classification, with a specific focus on online social network sentiment classification. Section 3 describes the evaluation corpus, compares the detection performance of existing approaches and discusses common problems with short colloquial text analysis. Section 4 introduces our hybrid approach and outlines additional application use cases. Section 5 summarizes our findings.

2   Related Work

Sentiment analysis, i.e., the extraction of an opinion's overall polarization and strength towards a particular subject matter, is a recent research direction [1,2], and typically approached from a statistical, or machine-learning angle. Attention has been given particularly in the domain of movies [3,4], by analysis of social media data, as reflection of common opinion. It is found that prices of the movies industry have a strong correlation with observed outcome frequencies, and therefore they are considered as good indicator of future outcomes. Most recently published work either perform unsupervised learning on a provided corpus of perceived positive and negative texts such as product reviews [5,6], or use a set of curated keywords with positive or negative connotations to classify input [7,1].

Another common approach [5,6] is measuring sentence similarity between given data input and texts of specific polarity, which explores the hypothesis that opinion sentences will be more similar to other opinion sentences that to factual ones. Additionally, previous work [8,9] was focused on learning extraction patterns associated with objectivity (and subjectivity) in order to be used as features of objective/subjective classifiers. It is shown that this approach achieves higher recall and comparable precision than previous techniques. Apart from that, recent publications [10,11], introduced the use of Natural Language Processing modules in order to extract concepts from the processed text and eventually derive sentiment out of them. In the very recent past, several of these general approaches have been specifically extended towards the mining of sentiments from online social media sources, in particular the microblogging platform Twitter [12]. For our analysis, we focus on those approaches for which the original authors made a reference implementation available to us, specifically we compare the classification accuracy with the following classifiers:

Twitter Sentiment. We used the bulk classification service available on the Twitter Sentiment website [13] in order to classify our test-set. This tool attaches to each tweet a polarity value: 0 for negative, 4 for positive and 2 for neutral- therefore we consider the first to describe subjective tweets, while neutral is for objective tweets. The main idea behind Twitter Sentiments approach is the use of emoticons as noisy labels for the training data which is shown that it increases the accuracy of different machine learning algorithms (Naive Bayes, Maximum Entropy, and SVM). It is noted that the web service of Twitter Sentiment uses a Maximum Entropy classifier.

Tweet Sentiments. Our test-set was also tested through the API of Tweet-Sentiments [14], a well known tool for analysing Twitter data and provide sentiment analysis on tweets. TweetSentiments is based on Support Vector Machines (SVM) and is using the LIBSVM library developed at Taiwan National University. It classifies tweets as positive, negative or neutral and these values are treated as stated previously.

Lingpipe. We also used the Sentiment Analysis tool of the LingPipe [15] package which focuses on the subjective/objective (as well as positive/negative) sentence categorisation especially on the movie-review domain. This approach uses the usual machine learning algorithms (Naive Bayes, Maximum Entropy, SVM) and a Java API of the classifier is available online. Even though it comes with its own training set, we used the half of our hand-classified set to train the classifier, in order to have better results. The other half was used as test-set and results were compared to the corresponding hand-classified tweets.

3   Methodology, Test Corpus and Performance Evaluation

To evaluate the performance of established sentiment classifiers and create a benchmark for our developed solution, we randomly sampled a set of some 1,000 publicly readably messages from the microblogging platform Twitter. Prime use cases for sentiment analysis are for example research questions revolving around the spread of information, opinion formation and identification of influential relationships in social networks, and such processes are typically believed to be present in discussions around product and media such as music, book or movie reviews.

Data Acquisition and Processing. For our evaluation, we therefore collected a data-set of 1,073 randomly chosen tweets related to the five most popular films of the 83rd Academy Awards. We used the language detection library of

Cybozu Labs [16], in order to eliminate the tweets written in any language other than English, while we also tried to remove advertising tweets out of the set. Multiple retweets of the same text were also removed to prevent performance over- or underestimation, as well as unnecessary tokens like link urls, "@" tags for mentioning a user, 'RT' tags etc. Each tweet of this test-set was classified by hand before the begin of the evaluation into an objective or subjective statement; this corpus is used throughout this work as a reference benchmark.

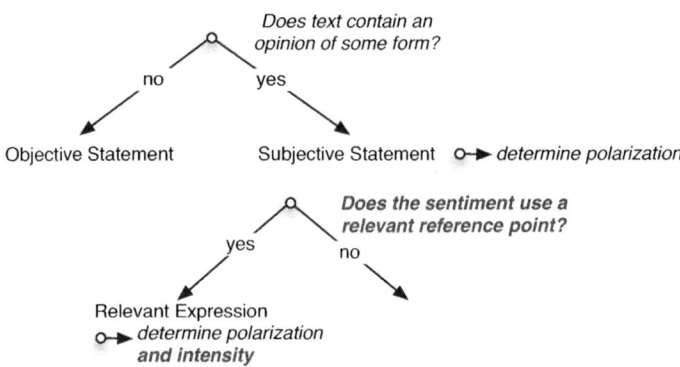

Fig. 1. Sentiment classification involves a multi-stage process, in which not only the existence of a sentiment should be checked, but also the reference point and strength should be assessed. (Contributions of this work are indicated in green.)

Components in Sentiment Analysis. Starting point for any sentiment analysis (as shown in figure 1) is the detection of any form of opinion in written text. If the author expresses some form of judgement, the input can be considered a subjective statement, otherwise the data is classified as an objective claim. Example cases distinguished by such test are for example "I liked *The King's Speech*" versus, "*The King's Speech* was a really long movie.", respectively. Another case is given by subjective messages where the sentiment is not based on the relevant reference point (the title of the movie). For example the tweet "I like you, even when watching *The King's Speech*" is a positive tweet, but its not the opinion about the movie that is positive. Once the existence of a sentiment has been established, typically a classification step is performed to determine whether the speaker is expressing a positive or negative opinion over a particular subject matter.

Challenges for Classification in Short Colloquial Text. In many types of inputs, and specifically in micro-texts such as tweets or chats, however a problem arises: conversations are highly abbreviated. As tweets offer only 140 characters of payload, messages are reduced to a bare minimum and several different thoughts - for example reactions to previous incoming messages - frequently are abbreviated and intertwined: "Watched King's Speech today in class. I love the end of the term." This results in a very small footprint on which sentiment analysis can be conducted, at least compared to the essay- and article-type classification previously used for polarization analysis. Previously established approaches, which for example operate using a statistical word-frequency analysis, are therefore less suited, as the low quantities of text and the high concept compression ratios are resulting in very high statistical fluctuations and noise during the detection. For this reason, we pursue a different approach in this work and analyze the grammatical structure of messages. By detecting which concepts a particular sentiment is referencing to, we can make more fine-grained decisions and consequently achieve a higher prediction accuracy especially in dense, intertwined texts. In the example above this concept is the end of the term (and therefore potentially a looser schedule) rather than the movie itself.

Automatic Detection and Tuning of Polarization Intensity. Finally, many applications and research hypotheses can be better served if not only the existence of a sentiment and its general polarization is known, but an absolute notion of "how positive" or negative a particular opinion can be derived. This would allow both a better assessment of how opinions are propagated and adopted, as a person with a strong negative attitude towards a particular concept is first expected to become less and less negative before developing a positive sentiment if at all. Without a quantification of polarization such trends would go unnoticed. Additionally, a quantitative measurement of attitude would allow of differentiation between alternatives, which in sum are all considered positively, but in a pairwise comparison are not equal.

If considered in previous work, this aspect is typically approached using manually curated word lists, as for example in [2]. When following this strategy in the application context of micro-messages, we however discovered two fundamental difficulties: 1) Users utilize a rich set of vocabulary to describe their opinions about concepts. Capturing and maintaining an accurate ranking of evaluative comments would require a significant effort in a practical setting. 2) Expressions indicating positive and negative sentiments and their relative differences are neither stable over time nor between different people, thus a method to re-adjust and "normalize" sentiment baselines over time or between say generally very positively oriented, neutral or pessimistic speakers will provide an advantage.

Evaluation of the State-of-the-Art. To evaluate the performance of existing sentiment classifiers, we let the set of available classifiers described in section 2 analyze and distinguish a reference body of tweets. As most methods do not allow for a sentiment quantification, we limited this evaluation to only a general polarization detection which is supported by all systems. Comparing the output against the previous human classification, we measured the overall accuracy of the automatic classifiers in distinguishing subjective from objective statements as shown in figure 2(a). Figure 2(b) shows the overall performance in correctly and incorrectly classified statements.

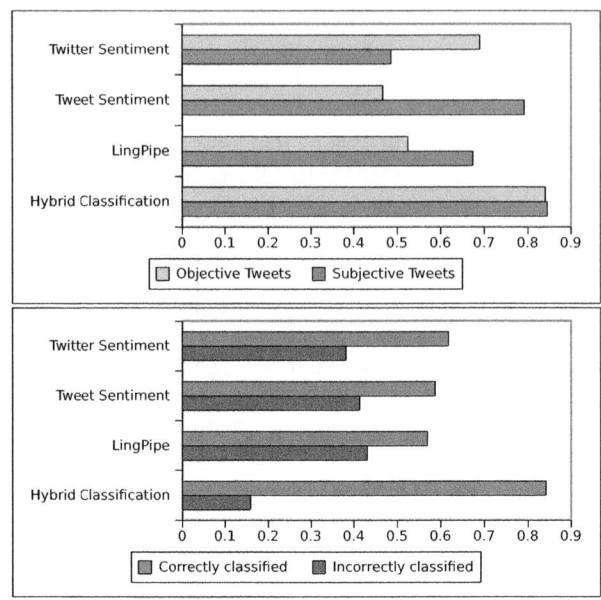

Fig. 2. Classification accuracy of different sentiment analyzation methods

As can be seen in the figure, the classification accuracy of all statistical sentiment analyzers is between 55 and 60%, whereas the proposed statistical-
grammatical hybrid approach yields a correct classification accuracy of about 85%,
a 40% gain over previous work. Note also that the accuracy of existing system also
varies significantly between the type of input data: Twitter Sentiment [13] for example is much stronger identifying objective statements compared to subjective
ones, while Tweet Sentiment [14] shows exactly the opposite behavior. The proposed hybrid solution on the other hand does not show any significant bias.

4   Sentiment Classification and Polarity Estimation

This section describes in detail the underlying concept of the proposed hybrid
sentiment classifier. As shown in figure 1, the general task of sentiment analysis
can be conducted in two general phases, first the detection of opinions in general
(yielding to a categorization in objective and subjective text), after which a
quantification of the polarity can be attempted. The following discussion mirrors
these steps.

4.1   Grammatical Sentiment Classification

In order to classify a given message into subjective and objective we analyze the
grammatical structure of a tweet. Subjectivity is mostly based on adjectives or
verbs expressing the polarity related to the subject of the message. This means

if directly expressing an emotion one usually uses a verb like "like/love/hate" whereas expressing the mood about something usually contains an adjective. Consider the examples: *I'm feeling sick today, I liked the movie.*

To determine the existence of a sentiment, we therefore invert the grammatical structure of a given text to detect the presence of verbs carrying an emotional meaning or to find the adjectives associated with the keywords we are performing the sentiment analysis on, in our case the titles of movies. Grammatical structure analysis is a mature research area and we use Klein and Manning's lexicographical parser [17] to determine the structure of the English texts of tweets after removing special characters as described in section 3. For a given text, this tool is estimating the grammatical structure. An example for the tweet "I liked the movie" is given in the following:

```
[I, liked, the, movie]
(ROOT
  (S
    (NP (PRP I))
    (VP (VBD liked)
      (NP (DT the) (NN movie)))))

nsubj(liked-2, I-1)
det(movie-4, the-3)
dobj(liked-2, movie-4)
```

Here, the parser is reasoning that "I" is the nominal subject of "liked", and "movie" is the direct object of the verb "liked". As mentioned, most tweets expressing a sentiment have this kind of structure. If an adjective is referring to a subject the likelihood is quite high that this tweet expresses the mood about something. Note however that it is in general possible that the speaker was using sarcasm or irony, which could not be detected from a grammatical viewpoint.

In a second step, we cross-check whether the word referring to the subject of a tweet is an adjective. This can be done using either a lexical database such as WordNet [18] or a part-of-speech Tagger [19], which will be used further in this discussion. Through such a tool, every word in a given sentence can be annotated with a tag identifying its purpose in the sentence [20], so the example "I liked the movie" is marked as:

```
I/PRP liked/VBD the/DT movie/NN
```

The part of speech tagger tells us that "I" is a personal pronoun, "liked" a verb in past tense, "the" a determiner and "movie" a noun. By connecting the so gathered information of a message we build simple rules to detect if a message is a subjective statement whereas all the others by inversion have to be objective:

1. if an adjective is referring to the subject
2. if an verb out of a list is referring to the subject
3. adjective + [movie, film]
4. [movie, film] is/looks [adjective]
5. love/hate + [movie, film]

4.2   Automatic Polarity Estimation

After the existence of a subjective component has been established, it is necessary to determine the overall polarity of the sentiment and if possible also the magnitude of the sentiment. In order to estimate the general polarity direction of words in the corpus, we used an unsupervised approach based on word correlations. This approach is inspired by the way a person is learning to judge which words have a positive or negative meaning, which is essentially a result of a lot of exposure to speech and written text, from which the learner infers which words appear in a positive or negative context.

The same basic principle, inferring which words appear together in a positive or negative context, can however be easily mirrored in a machine as well. Here, a computer would simply need to count how often a particular adjective has been encountered with a positive meaning compared to the frequency it has been observed with a negative connotation. To begin such an automatic classification, some notion of what is deemed positive or negative will be necessary. In our work, we have explored two general options, first by manually specifying a set of keywords one would associate with positive expressions such as "fantastic", "incredible", "amazing", which can read from existing databases such as [18], and second by looking at the most basic positive/negative expression commonly used in online messages such as chats, emails or microblogs: a positive smiley :-) and a negative smiley :-(. In the following, we will discuss the results for this second alternative.

From a list of one million tweets, we search for all tweets containing a positive smiley :-), :) or =) – in the following referred to as *positive keywords* – and created a list of texts containing at least one of those symbols. Similarly, a list of all tweets containing a negative smiley such as :-(, :(or =(– deemed *negative keywords* – was prepared. Using the techniques discussed above, we dissect all individual statements and count the number of co-occurrences between every detected word and the positive or negative keywords. To correct for differences in length of those two lists – as users typically write more positive than negative statements –, the two values are then normalized by the number of words in the list of messages containing positive words and the list of messages containing negative words. This results in a relative assessment of a particular word to appear in a positive or a negative context, where context is defined by the positive and negative keywords, respectively.

To arrive at a relative polarity of a particular word between the two extremes "positive" and "negative", it is now simply enough to subtract the relative frequencies. This number is positive if the word is typically used within a positive context and negative if the word typically occurring with a negative meaning. As this number is biased by the number how often a word is used in general, a final correction step is executed in which each rating is multiplied by the term frequency measured in all tweets: the emphasis of frequently observed words is therefore reduced, and the value of unusual ones is lifted.

Fig. 3 shows the output of this simple procedure conducted over a body of one million tweets, and using just positive and negative smileys as corresponding positive and negative keywords. As can be seen, this unsupervised process leads to a clear and meaningful ranking of adjectives. We have repeated this analysis over a selection of datasets of different duration and verified the automatically generated polarity estimation against those done by a human. Typically, less than 7% of the words were considered wrongly placed in the overall order; out of a list of 30 adjectives between one and two placement were deemed higher or lower in the ranking by a human observer than by a machine. As this method can be executed without manual intervention, this new methodology can be continuously conducted to detect the general development of sentiments in entire online communities, as well as to identify whether the polarity of certain words shift in strength over time.

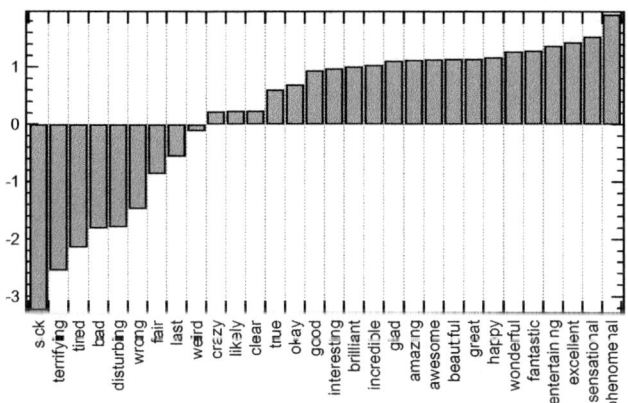

Fig. 3. Automatic polarity ranking of adjectives based on general Twitter messages

A more comprehensive analysis is needed to see if the sentiment of adjectives change over time. However taking twitter data of durations of one week, one and two months the polarity seems to stabilize the longer the duration of observation.

4.3   Detecting Networks of Concepts

This general method is however not limited to determine the polarity strength of words in general, but can be used to detect and identify common concepts and their associated sentiments in general. To do so, it is simply necessary to swap out the two sets of keywords (which in the last section were :-), :), =) and :-(, :(, =(, respectively), and replace them with those terms and synonyms relevant to a particular study.

Consider for example a situation where one would to determine the associations made with the brands and products of two hypothetical tea manufacturers: *McArrow's orange-peppermint tea* and *DrBrew's strawberry-melon tea*. Here, one would populate *keyword group 1* with words from the first area, i.e., McArrows, orange-peppermint, etc. and analogously *keyword group 2* with words such as DrBrew, strawberry-melon, etc., and by the same means described above, this method would derive the set of words frequently used in combination with any of those keyword terms as well as the strength of their typical common appearance as shown in figure 4.

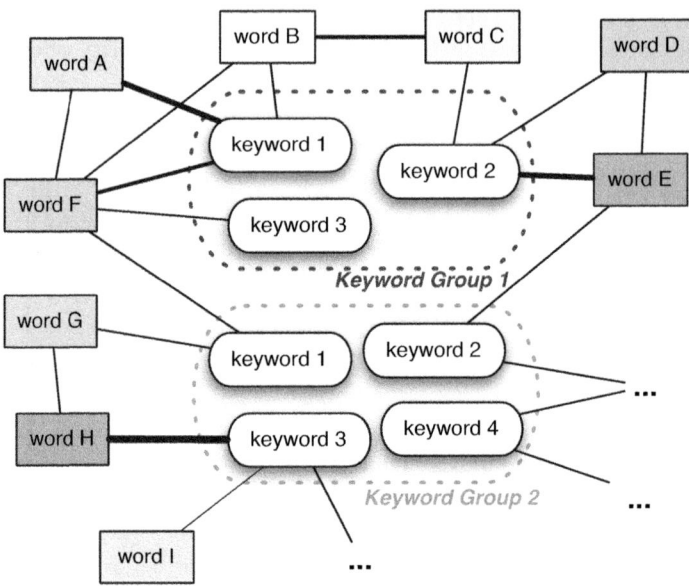

Fig. 4. The technique can be broadened to determine the concepts commonly associated with any keyword as well as the particular strength of the association

The words $A-I$ discovered to be co-located can however be themselves further interpreted, for example, 1) depending on how positively/negatively they are (as discussed above), or 2) which general topic areas or word field the concepts come from. Imagine for example words A and F in figure 4 to be "taste" and "flavor", while words D and E are "packaging" and "price". Clearly, such combined word co-localization, polarization and word-field analysis will provide a significant insight to our hypothetical tea manufacturer, which can also be easily repeated over time to track its overall development, but also to the researcher interested in how particular opinions form, are spread and change over time.

As a general example, we have applied this techniques towards the keywords "coffee" and "tea", and let the system arrange the resulting associated words

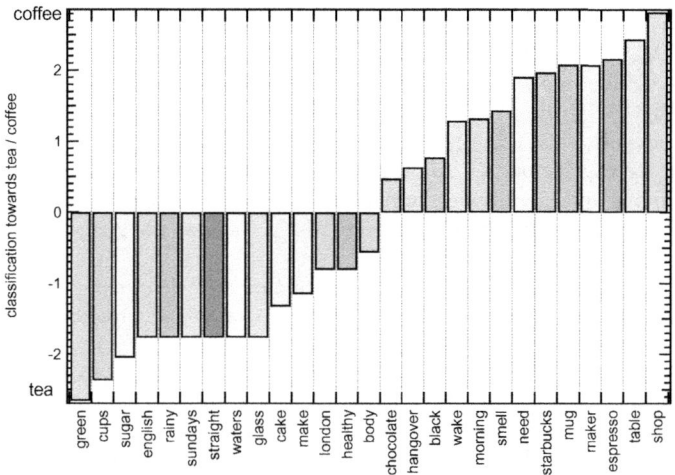

Fig. 5. Polarization analysis of commonly used words with coffee and tea based on general Twitter messages

according to their overall polarity strength. The overall abstract network can then be drawn in a similar manner as figure 3, and figure 5 shows the 27 strongest connections commonly made the terms coffee and tea. All bars indicating their affiliation towards the two beverages are colored by their polarity value. Red indicates a negative, green a positive and white a neutral polarity.

5 Conclusion

In this paper, we have presented an alternative method to determine the existence and strength of subjective opinions in short colloquial text, an application domain where existing approaches do not yield a high detection accuracy. The proposed system works through a combination of grammatical analysis with traditional word frequency analysis, does not need supervised training and improves the accuracy of previous work by about 40%.

Acknowledgements. This project was partially supported by the Transsector Research Academy for Complex Networks and Services (TRANS).

References

1. Pang, B., Lee, L., Vaithyanathan, S.: Thumbs up? Sentiment classification using machine learning techniques. In: Proceedings of the 2002 Conference on Empirical Methods in Natural Language Processing, EMNLP (2002)
2. Turney, P.D.: Thumbs up or thumbs down? semantic orientation applied to unsupervised classification of reviews. In: Proceedings of the 40th Annual Meeting of the Association for Computational Linguistics (2002)

3. Mishne, G., Glance, N.: Predicting movie sales from blogger sentiment. In: AAAI 2006 Spring Symposium on Computational Approaches to Analysing Weblogs, AAAI-CAAW 2006 (2006)
4. Asur, S., Huberman, B.A.: Predicting the future with social media. In: 2010 IEEE/WIC/ACM International Conference on Web Intelligence and Intelligent Agent Technology (WI-IAT), vol. 1, pp. 492–499. IEEE (2010)
5. Yu, H., Hatzivassiloglou, V.: Towards answering opinion questions: Separating facts from opinions and identifying the polarity of opinion sentences. In: Proceedings of the 2003 Conference on Empirical Methods in Natural Language Processing, pp. 129–136 (2003)
6. Barbosa, L., Feng, J.: Robust sentiment detection on twitter from biased and noisy data. In: Proceedings of the 23rd International Conference on Computational Linguistics: Posters, pp. 36–44 (2010)
7. Cilibrasi, R., Vitanyi, P.: Automatic meaning discovery using google. CWI (2004) (manuscript)
8. Riloff, E., Wiebe, J.: Learning extraction patterns for subjective expressions. In: Proceedings of the 2003 Conference on Empirical Methods in Natural Language Processing, pp. 105–112. Association for Computational Linguistics (2003)
9. Wiebe, J., Riloff, E.: Creating subjective and objective sentence classifiers from unannotated texts. In: Computational Linguistics and Intelligent Text Processing, pp. 486–497 (2005)
10. Jiang, L., Yu, M., Zhou, M., Liu, X., Zhao, T.: Target-dependent twitter sentiment classification. In: Proc. 49th ACL: HLT, vol. 1, pp. 151–160 (2011)
11. Cambria, E., Hussain, A., Havasi, C., Eckl, C.: Sentic Computing: Exploitation of Common Sense for the Development of Emotion-Sensitive Systems. In: Esposito, A., Campbell, N., Vogel, C., Hussain, A., Nijholt, A. (eds.) COST 2102 International Training School 2009. LNCS, vol. 5967, pp. 148–156. Springer, Heidelberg (2010)
12. Twitter, http://www.twitter.com
13. Go, A., Bhayani, R., Huang, L.: Twitter sentiment classification using distant supervision. tech. rep., Stanford University (2009)
14. Tweet sentiments, http://tweetsentiments.com
15. Lingpipe, http://alias-i.com/lingpipe
16. Shuyo, N.: Language Detection Library (2010), http://code.google.com/p/language-detection/
17. Klein, D., Manning, C.D.: Accurate unlexicalized parsing. In: Proceedings of the 41st Meeting of the Association for Computational Linguistics (2003)
18. Wordnet. a lexical database for english, http://wordnet.princeton.edu/
19. Kristina Toutanova, D.K.C.M., Singer, Y.: Feature-rich part-of-speech tagging with a cyclic dependency network. In: Proceedings of HLT-NAACL (2003)
20. The university of pennsylvania (penn) treebank tag-set, http://www.comp.leeds.ac.uk/ccalas/tagsets/upenn.html

Resilience in Computer Network Management

Marcelo F. Vasconcelos and Ronaldo M. Salles

Instituto Militar de Engenharia,
Praça General Tiburcio, 80. Rio de Janeiro, Brasil
marcelo.vasconcelos@globo.com, salles@ieee.org

Abstract. The protection of network infrastructure and services is one of the main issues network managers face today. This paper investigates this matter through the study of network resilience. A resilience factor (RF) is proposed to take into account topological aspects of the network and also the amount of traffic losses under stress conditions. The proposed factor is evaluated using a real network backbone (RNP). Results obtained showed that the resilience factor is consistent and can be employed to support network managers decisions about network expansions, link additions and removals. Such decisions improve network robustness making it less vulnerable to attacks and failures.

Keywords: Resilience, Network Management, Network Link Failures, Traffic Loss.

1 Introduction

Among the various tasks a network manager has to perform, the study on how to improve robustness to better support faults and attacks to the network infrastructure and services is a critical issue. Such study usually involves the analysis of redundancies, alterations and expansion of the network resources.

In this sense, decisions about where to efficiently invest at a reasonable cost often depend on the experience of each network manager. Despite taking into account information concerning the operation of the network, the process is very subjective and may lead to poor or not effective choices.

The main goal of this work is to propose a measure of resilience in order to better quantify this important property, and thus provide a methodology for the use of resilience in the management of computer networks. The metric can be employed to assist network design and also to support decision-making regarding expansions or changes in the topology of an operational network. It is thus possible to decide between one or another change in the topology based on the impact of each one to the resilience of the network.

The first step toward this goal is to propose a resilience factor (RF) in order to quantify the network level of resilience. The proposed resilience factor takes into account both the topological aspects as well as the traffic demands posed to the network. After being able to measure network resilience through RF, one can search for alterations in topology in order to obtain a better outcome in terms of RF.

R. Bestak et al. (Eds.): NETWORKING 2012, Part I, LNCS 7289, pp. 109–120, 2012.

Finally, the proposed methodology is evaluated using a real network scenario where results obtained conforms to expectations of network managers.

This paper is organized as follows. Section 2 reviews the related work on the topic. Section 3 presents our proposal: the resilient factor, methodology and algorithm. Section 4 discusses the results obtained using a real network scenario, and finally Section 5 presents the conclusions and suggestions for future work.

2   Related Works

The quest for increasing robustness of the network led to the study of resilience. Network resilience is a broad area, this work considers the following definition: *"Resilience in networks is the ability of an entity to tolerate, endure and automatically recover from challenges under the conditions of the network, coordinated attacks and traffic anomalies"* [1].

From the above definition, several works in the literature studied resilience using different techniques and methods. They can be roughly divided into two main categories: reactive and proactive approaches. Reactive strategies are mostly devoted to routing and other network configuration mechanisms that are trigged to react to a given fault or anomaly condition observed in the network, whereas proactive approaches try to condition the network beforehand to better resist to faults, anomalies and attacks. The focus of this work is on proactive approaches.

A quantitative and statistical analysis of the frequency and duration of faults are carried out in [2], [3] and [4], the results are equivalent and were obtained in real topologies. It was shown that failures are part of network operation and most of them last for less than 10 minutes. Regarding interruptions of network services, scheduled interruptions correspond to 20 % and unplanned interruptions correspond to 80 % of the total. Within the 80 % of unscheduled interruptions, 70 % correspond to a single link failures and 30 % are attributed to multiple failures of equipment and fiber optic networks. This information is important and could be taken into account in the development of resilience metrics.

The work [6] brings a comprehensive analysis of network resilience, addressing topics such as failures in networks, mechanisms of resilience and analysis. The purpose of that paper is to present a methodology based on the calculation of autonomous systems edge connections availability using distribution functions of link occupation. Such functions depend on network failures and traffic variations. Failures can be single, double, triple and so on. The traffic considered is proportional to the size of the population in a given area. Results show that link additions to the network provide a greater network availability.

The work in [7] was one of the first to introduce a method to measure the fault-tolerance of a network. The measure was defined as the number of faults a network may suffer before being disconnected. The authors computed an analytical approximation of the probability of the network to become disconnected and validated their proposal using Monte Carlo simulation results.

The simulation scenario employed three particular classes of graphs to represent network topologies: cube connected-cycles, torus and n-binary cubes – all of them are symmetric and with same node degrees.

The authors in [8] were interested in evaluating the robustness of a network against attacks and node failures. They applied the concepts of node connectivity and topology symmetry. The evaluation of their method was carried out using different groups of networks: symmetric topologies, scale-free, and random topologies. In [9] a similar method was presented, but evaluation was performed considering only network topologies that are scale free to better adhere to real Internet characteristics. A scale free topology [10] is such that the degree of its nodes follows the power law distribution [11].

The method proposed in [12] used the k-connectivity property of a graph to construct a resilience factor for network topologies. The authors showed that the method was consistent and more efficient than previous approaches. However, the computational cost to compute the factor is quite high and the method does not take into account the capacity of links and network traffic.

On the other hand, both network traffic and link capacity were considered by the authors in [13] and [14]. The parameters were related by the concept of *delivered value: the ratio of traffic routed through a network after a capacity decrease or loss and the value sent before the reduced capacity.* According to the previous definition:

$$DV = \frac{(DeV - LOSS)}{DeV} \qquad (1)$$

where DV is the delivered value, DeV is the amount of traffic that should pass through the link and $LOSS$ is the amount of traffic that was not delivered due to network changes (capacity decrease or loss). In their work a complete disconnection was not considered but the capacity of a given link could be reduced due to failures, which might impact DV.

It can be observed from previous works that in order to provide a more complete view about network resilience it is important to consider not only connectivity aspects of the network topology as in [12], but also to compute the impact on traffic caused by network changes as in [13] and [14].

3 Proposed Method

This paper proposes a method to quantify resilience of a given network through the resilience factor, RF. As mentioned before, the proposed method does not focus on a single network aspect but can be viewed as a composite metrics of delivered traffic and topology characteristics as follows:

$$RF = (ADV, R) \qquad (2)$$

where ADV is the average delivered value and R accounts for the number of redundancies the network topology enjoys. In this work, R is also defined as the *network order*. Details about the computation of RF will be presented next.

3.1 Resilience Factor

The first aspect to analyze is how network topology is considered in the term R that composes RF.

From the definition and all previous works discussed in Section 2, it is possible to realize that resilience is associated to redundancy in the following way: a network enjoys a high level of resilience if it has a large number of redundant resources to cover possible losses during attacks and/or failures. We try to quantify this notion using the concept of network order.

In this work, *network order* (or simply R) is computed by the number of redundant links the network has to connect its nodes. In other words, R is defined as the maximum number of links that can be removed and still preserves network connectivity – the resulting topology becomes a spanning tree.

For instance, suppose a network where nodes are connected only by the spanning tree (ST) itself, in this case any failure causes a disconnection and traffic losses since there is no spare resources to keep nodes connected. Hence, a ST network is of zero order: $R = 0$.

On the other hand, a full mesh network has $\frac{n(n-1)}{2}$ links, where n is the number of nodes in the topology. For this network the spanning tree (ST) is composed by $n-1$ links, and so there are $\frac{n(n-1)}{2} - (n-1)$ extra links connecting nodes, yielding $R = \frac{(n-2)(n-1)}{2}$.

Therefore, the parameter *network order* is limited by these two practical situations that can be found in real network topologies:

$$0 \leq R \leq \frac{(n-2)(n-1)}{2} \tag{3}$$

We would like to study resilience taking into account redundant links only and how their disconnection impacts the delivered traffic, or in another way, how the network depends on them to deliver its traffic. For example, a topology that has no redundancy but that can route all traffic, will have $RF = (1,0)$ indicating that $ADV = 1$ (no losses) and $R = 0$ (no redundancy). If in another network we can draw five links out of it and all traffic continues to be delivered then $RF = (1,5)$, and so on.

ADV is based on Eq. 1, in fact it is computed as the average of DV in each x different network conditions given by links removals/failures, leading to $DV(x)$. Note that differently from [13] and [14] Eq. 1 is applied to the network as a whole and not to a specific link only. The objective is to evaluate how topology changes due to link failures and reduce its capacity to deliver traffic.

Let us illustrate the computation of $RF = (ADV, R)$ using the example in Fig.1. The topology in Fig.1a has 4 nodes and 5 links, where the capacity of each link is given in Mbps. The computation of R is directly obtained from the topology: $R = 5 - 3 = 2$, i.e. the topology has two redundant links in addition to the links of the spanning tree. The *network order* (R) is used to evaluate all failure conditions (link removals) regarding network redundancies only, and how they impact the delivered traffic.

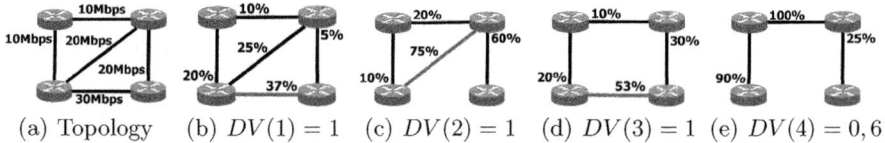

(a) Topology   (b) $DV(1) = 1$ (c) $DV(2) = 1$ (d) $DV(3) = 1$ (e) $DV(4) = 0,6$

Fig. 1. Exemple: obtaining ADV

Hence, according to our approach there are $2^R = 4$ failure conditions to be tested: no failure (Fig.1b), one failure (Figs.1c and 1d) and two failures (Fig.1e). In fact, if more than two failures occur the network will be disconnected and all traffic from and to the disconnected node will be lost. This is an extreme situation that is less likely to happen than the other cases (see Section 2) and will not be considered in this work. Such extreme cases are studied in [12]. It is important to notice that although extreme fault scenarios (nodes disconnection) are not taken into account in the RF, the computation of ADV considers worst case analysis.

First, in Fig.1b it can be seen that the highest network link load is 37%, and so all traffic is delivered since there is no loss in the network, for this first scenario $DV(1) = 1$. The link of highest occupancy is selected to be removed from the topology (worst case) yielding in the scenario of Fig.1c. All traffic crossing that link has to be rerouted to other links, loads on the remaining links increase, however the network has enough spare resources to support such failure and all network demands are still delivered: $DV(2) = 1$. Now the diagonal link is removed to generate the situation in Fig.1d, where still $DV(3) = 1$. Finally, both spare links are removed and the network is connected only by its spanning tree (limiting situation before node disconnection). In Fig.1e there is a link operating at 100%, in fact demands exceed link capacity and there is loss of traffic in the network: $DV(4) = 0.6$.

The parameter ADV is then computed as the mean DV for all failure situations considered:

$$ADV = \frac{DV(1) + DV(2) + DV(3) + DV(4)}{4} = 0.9 \qquad (4)$$

and $RF = (0.9, 2)$ for the network in the example. This can be read as: the network has two redundant links and can deliver almost all demanded traffic even in situations involving failures related to the spare capacity (redundancies).

A generalization of Eq. 4 is possible if it is known beforehand the probability of each one of the failure scenarios illustrated in Fig.1 occurs. Thus, the corresponding DV could be weighted, α_i, according to the probability of the specific failure occurs:

$$ADV = \sum_{i=1}^{2^R} \frac{\alpha_i DV(i)}{\beta} \qquad \beta = \sum_{i=1}^{2^R} \alpha_i \qquad (5)$$

Since such information is not always available and to simplify the approach, we consider in this work all $\alpha_i = 1/\beta$.

The algorithm to calculate the resilience factor is presented below. It takes as inputs the information regarding the network topology (CapM, n and m) and traffic demands (TM), and outputs the resilience factor (ADV and R).

Algorithm 1. RF(In: CapM, TM, n, m; Out: ADV, R)

1: $ADV \leftarrow 0$
2: $R \leftarrow m - (n - 1)$
3: $DeV \leftarrow$ sum all elements in TM
4: compute Q
5: **for all** elements $q \in Q$ **do**
6:    compute $CapM(q)$ from CapM
7:    compute $WM(q)$ from WM
8:    $R \leftarrow$ routes calculated using $WM(q)$
9:    $OM \leftarrow$ occupation obtained from TM and R
10:   $LOSS \leftarrow OM - CapM(q)$
11:   $DV \leftarrow \frac{(DeV - LOSS)}{DeV}$
12:   $ADV \leftarrow ADV + \frac{DV}{2R}$
13: **end for**
14: **return** (ADV, R)

Legend:

- n: number of nodes in the topology
- m: number of links in the topology
- $CapM_{n \times n}$: adjacency matrix with link capacities (topology)
- Q: set of all failure scenarios, as illustrated in Fig.1
- q: element of Q, one particular scenario as in Fig.1d
- $TM_{n \times n}$: traffic demand matrix, origin-destination pair demands
- $WM_{n \times n}$: weight matrix for shortest-path routing (e.g. RIP or OSPF [15])
- R: routing matrix, all path from origin to destination obtained using WM
- OM: link occupation matrix
- LOSS: traffic loss matrix

The first step of the algorithm is to initialize ADV to zero. Then in step 2, the *network order* is computed using the number of links and nodes in the topology. Step 3 computes the network demanded value DeV summing up all elements of the TM matrix. Step 4 computes the set Q of all topology combinations obtained from the removal of redundant links as previously illustrated in Fig.1.

In the next step in line 5, the algorithm enters in a loop to check the behavior of each one of the elements in Q (as seen in Fig.1) and accumulates DV for each of them in the average ADV (line 12). After that, the algorithm ends returning the resilience factor.

Inside the main loop the first step in line 6 is to update CapM for the particular element in Q, obtaining $CapM(q)$. Then in line 7 the weight matrix is also updated to $WM(q)$. Routes are compute and stored in R since they will be used to return OM. In line 10, the loss matrix is calculated by subtracting elements in

OM and $CapM(q)$. DV is computed in line 11 and ADV in line 12. After that, a new element in Q (other topology combination) is processed and the loop goes on until the last element in Q.

The occupation matrix OM is obtained using the values in TM, as follows: for each demand in TM the corresponding route is determined between origin to destination, then the amount of demanded traffic is added to each element (link) of OM that takes part of the route. Later in the following step, CapM is applied to compute possible losses: any element in OM above the corresponding element of CapM results in loss of traffic since demands exceed capacity. The matrix LOSS store such values.

The delivered value DV is the difference between DeV and traffic losses due to the lack of capacity on links composing the network to absorb demands. DV equals one whenever all network demands are delivered, i.e. no loss is experienced in the network.

With the use of the resilience factor proposed, it is possible for instance to verify the effect of additions and removals of links connecting two nodes. For each particular situation the network manager may use RF to assess the impact such changes cause on the network. This method provides support for several management decisions the network manager has to take during the operation of the network under his responsibility.

4   Results

In this section we apply the proposed method to evaluate the resilience of a real network topology: the Brazilian National Research Network (RNP) in Fig. (2). The idea is to employ RF and analyze how changes in topology affect the network and its traffic. RF is also used as a tool to provide important information to support decisions about network expansion and protection against failures and attacks.

The Brazilian National Research Network has 27 nodes, 29 links and 7 redundancies that covers all states of Brazil. This network was selected since there is available information on the web (http://www.rnp.br) to conduct our experiment, for instance it was possible to obtain source destination peak demands of network traffic. However, information about all links occupation is missing and had to be inferred.

We considered two main traffic patterns to obtain the missing information about links occupation: max-min and proportional fair share of network resources [16]. While max-min fair is widely adopted as an ideal form of network sharing, proportional fair is shown to better resemble "TCP-like" sharing of network links.

We also studied different load conditions to better assess our method under a greater number of scenarios: peak (rnp historical peak demands), 100%, 95%, 90%, 50% and 30% of the maximum link capacities in the topology.

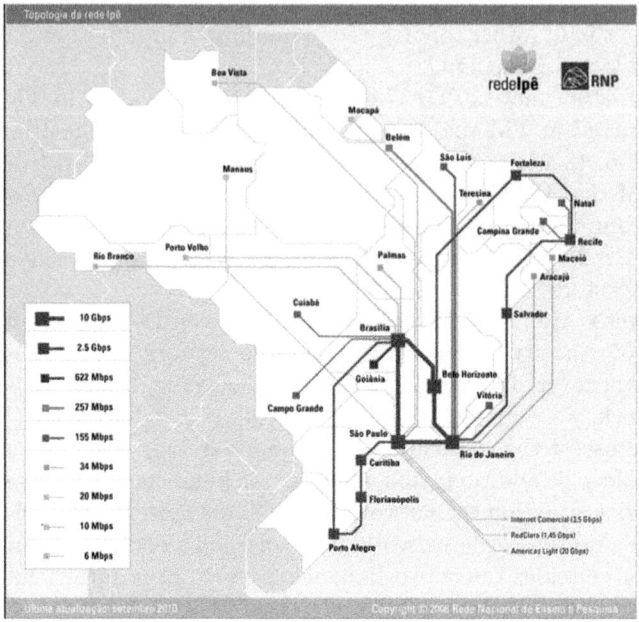

Fig. 2. Brazilian National Research Network

Regarding routing schemes, we considered the two most common forms of routing in our tests: minimum hop count (same link weights as used in RIP) and shortest paths according to link capacity (weights given by 100/capacity (Mbps) as used in OSPF [15]).

The objective of the test is to assess how the network (RNP) behaves when subjected to changes in its topology. How some network expansions (link additions) and removals affect the resilience of the network. This can be of great value to support network manager decisions.

To perform this test, we selected five additions and withdrawals of links, always seeking the best results: increase in network robustness for the case of additions and verification of worst cases of loss when links are removed.

We consider the following perception the network manager may have when analyzing the impacts to the network of a link addition or removal.

a. (link addition): a reasonable increase in robustness may be achieved whenever a link is added to the network core – i.e. when it has the potential to serve a large number of origin-destination demands.

b. (link removal): a network does not enjoy high levels of robustness whenever it heavily depends on certain core links – i.e. when a problem occur on those links it affects a large number of demands

Considering the principles described before, tests were taken according to Table 1, where the addition of a link is represented by the sign + and removal is

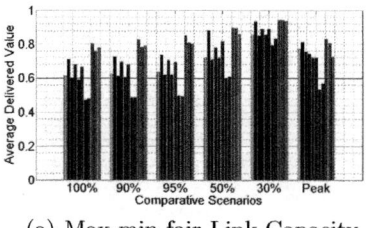

(a) Max-min fair Link Capacity

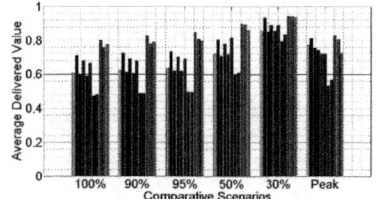

(b) Proportional fair Link Capacity

Fig. 3. Occupation method proportional fair

represented by -. The link connecting nodes A and B is represented by (A, B). In the case of additions, the value of the new link is presented. The first line represents the baseline before any change in the topology. In Figs. 3(a), 3(b), 4(a) and 4(b) the first green bar represents the baseline with no change in network, the blue bars represent the additions and the red bars represent the removals corresponding to Table 1, respectively.

Table 1. Tests in RNP

Change	Gbps	Change	Gbps
Initial	no change	Initial	no change
+(RJO,BSB)	10	-(RJO,SPO)	X
+(BSB,CWB)	10	-(BSB,SPO)	X
+(FZA,SDR)	2.5	-(SPO,CWB)	X
+(BSB,REC)	2.5	-(POA,BSB)	X
+(BHZ,REC)	2.5	-(FZA,REC)	X

The bar graphs presented from Figs. 3(a), 3(b), 4(a) and 4(b) correspond to the ADV obtained for each one of the eleven situations described in the table, respectively. Each figure represents a certain network scenario (bandwidth sharing and routing scheme used) under different load conditions (seven groups of bars). For instance, it was considered in Fig.4 a max-min fair share of link resources and a routing scheme following OSPF. The first bar in the figure indicates that $ADV = 0.6$ for the topology without changes (first line of the table) when links can be fully occupied (100%).

Some general conclusions can be made from the figures. First, bandwidth link sharing methods did not have a great impact on the results. For the same routing strategy and network loads, max-min and proportional fair results were close to each other.

The same does not apply to different routing methods, ADV varies significantly according to the routing scheme adopted. This can be explained given that routing has a direct effect on the distribution of traffic loads among links and then the delivered traffic also suffers from modifications on routing schemes.

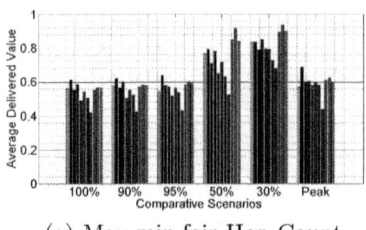
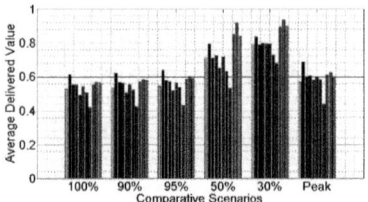

(a) Max-min fair Hop Count   (b) Proportional fair Hop Count

Fig. 4. Occupation method maxmin fair

However, the general behavior was preserved for most of the cases, a change in topology that provides an increase in resilience under a certain network setting was also observed as positive in all other situations. For instance, the second bar in the graphs, which involves an addition of a 10Gbps link between Rio and Brasília (second line in the table). This addition conforms to the perception of a network manager as discussed before.

The first insertion between Rio de Janeiro (RJO) and Brasília (BSB) has the best results in terms of increased robustness to all situations of traffic profile and routing methods, noting this addition as most indicated. This result was confirmed in testing all of combinatorial additions, as shown in Fig. 5(a). The other additions have increased the RF below the insertion between Rio de Janeiro and Brasília, and in some cases, such as adding a link between Brasília and Recife, the value obtained by the ADV is less than the original. Regarding removals, we can see that removal of the link between Rio de Janeiro and São Paulo (SPO) is the most critical and leads to a more significant decrease in network robustness. In possession of this information, the network manager can take actions to ensure the connection between these two cities is preserved.

A not intuitive result regarding removals can be seen when the link between Porto Alegre (POA) and Brasília, or Recife (REC) and Fortaleza (FZA) is removed and an increase in ADV is obtained. This situation does not conform with the normal perception as discussed before but it indeed has an explanation. Such removals provide other ways to route traffic between those cities and this new configuration is preferred in terms of resilience. Those links were heavily used and the removal of them cause the traffic to be better distributed over other network paths yielding in a favorable resilience condition.

Combinational tests of additions and removals are shown in Figs. 5(a) and 5(b). For additions, the best result is the addition between cities of Rio de Janeiro and Brasília, with $RF = (0.7223, 7)$. For removals, the worst result occurs when the link between Rio de Janeiro and São Paulo is lost, with $RF = (0.4588, 6)$. The best result among all removals was between the cities of São Paulo and Curitiba (CWB), with $RF = (0.8229, 6)$. Also it is shown in Figs. 5(a) and 5(b) the results of other possible additions and removals, indicating a complete scenario to support the decision of network managers.

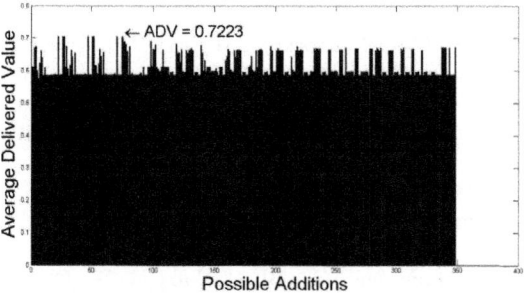

(a) Resilience Factor for all possible additions

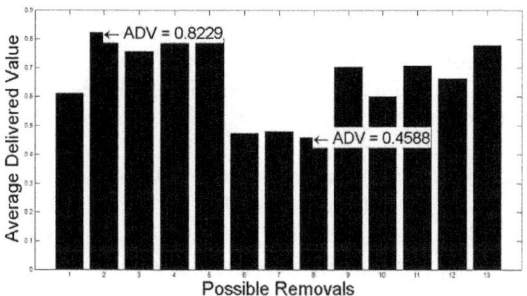

(b) Resilience Factor for all possible removals

Fig. 5. Possible additions and removals

5   Conclusions

This paper investigated the problem of resilience in computer networks under a network manager perspective. A resilience factor (RF) was proposed to evaluate resilience according to the number of redundancies and also the traffic delivered by the network.

The method was tested using a real network backbone scenario and was shown to be consistent and useful to support decisions performed by network managers about expansions, additions and removals of links. General results conform with the general perception of managers, however the methodology highlighted other important aspects that usually need a deeper understanding of networking operations. The use of RF also indicates which are the most advantageous insertions and removals, the ones that provide greater impact to network operation.

Finally, it is important to mention that the proposed factor and methodology can be used in other scenarios where different types of networks operate: electric power transmission, oil flow, gas distribution, water, sewage, etc.

References

1. Aggelou, G.: Wireless Mesh Networking. McGraw-Hill Professional (2008)
2. Iannaccone, G., Chuah, C.-N., Mortier, R., Bhattacharyya, S., Diot, C.: Analysis of link failures in an ip backbone. In: Proceedings of the 2nd ACM SIGCOMM Workshop on Internet Measurment, IMW 2002, pp. 237–242. ACM, New York (2002)
3. Nucci, A., Schroeder, B., Bhattacharyya, S., Taft, N., Diot, C.: Igp link weight assignment for transient link failures. In: Proc. International Teletraffic Congress (2003)
4. Markopoulou, A., Iannaccone, G., Bhattacharyya, S., Chuah, C.-N., Diot, C.: Characterization of failures in an ip backbone. In: INFOCOM 2004. Twenty-third Annual Joint Conference of the IEEE Computer and Communications Societies, vol. 4, pp. 2307–2317 (2004)
5. Li, V., Silvester, J.: Performance analysis of networks with unreliable components. IEEE Transactions on Communications 32(10), 1105–1110 (1984)
6. Menth, M., Duelli, M., Martin, R., Milbrandt, J.: Resilience analysis of packet-switched communication networks. IEEE/ACM Trans. Netw. 17(6), 1950–1963 (2009)
7. Najjar, W., Gaudiot, J.-L.: Network resilience: a measure of network fault tolerance. IEEE Transactions on Computers 39(2), 174–181 (1990) ISSN 0018-9340
8. Centre, C.C., Dekker, A.H., Colbert, B.: Scale-free networks and robustness of critical infrastructure networks. In: Proceedings of the 7th Asia-Pacific Conference on Complex Systems (2004)
9. Dekker, A.H., Colbert, B.D.: Network robustness and graph topology. In: Proceedings of the 27th Australasian Conference on Computer Science, ACSC 2004, pp. 359–368. Australian Computer Society, Inc., Darlinghurst (2004)
10. Barabasi, A.-L., Albert, R., Jeong, H.: Scale-free characteristics of random networks: The topology of the world-wide web. Physica A: Statistical Mechanics and its Applications 281(1-4), 69–77 (2000)
11. Faloutsos, M., Faloutsos, P., Faloutsos, C.: On power-law relationships of the internet topology. SIGCOMM Comput. Commun. Rev. 29, 251–262 (1999)
12. Salles, R.M., Marino, D.A.: Strategies and metric for resilience in computer networks. The Computer Journal (2011)
13. Omer, M., Nilchiani, R., Mostashari, A.: Measuring the resilience of the global internet infrastructure system. In: Proc. 3rd Annual IEEE Systems Conf., pp. 156–162 (2009)
14. Omer, M., Nilchiani, R., Mostashari, A.: Measuring the resilience of the trans-oceanic telecommunication cable system. IEEE Systems Journal 3(3), 295–303 (2009)
15. Moy, J.T.: OSPF Complete Implementation. Addison-Wesley Professional (2000)
16. Kelly, F., Maulloo, A., Tan, D.: Rate control in communication networks: shadow prices, proportional fairness and stability. Journal of the Operational Research Society 49 (1998)

An Experimental Study on the Impact of Network Segmentation to the Resilience of Physical Processes

Béla Genge and Christos Siaterlis

Institute for the Protection and Security of the Citizen,
Joint Research Centre, Via E. Fermi, 21027, Ispra, Italy
{bela.genge,christos.siaterlis}@jrc.ec.europa.eu

Abstract. The fact that modern Networked Industrial Control Systems (NICS) depend on Information and Communication Technologies (ICT) is well known. Although many studies have focused on the security of NICS, today we still lack a proper understanding of the impact that network design choices have on the resilience of NICS, e.g., a network architecture using VLAN segmentation. In this paper we investigate the impact of process control network segmentation on the resilience of physical processes. We consider an adversary capable of reprogramming the logic of control hardware in order to disrupt the normal operation of the physical process. Our analysis that is based on the Tennessee-Eastman chemical process proves that network design decisions significantly increase the resilience of the process using as resilience metric the time that the process is able to run after the attack is started, before shutting down. Therefore a resilience-aware network design can provide a tolerance period of several hours that would give operators more time to intervene, e.g., switch OFF devices or disconnect equipment in order to reduce damages.

Keywords: network segmentation, cyber-physical, resilience, security.

1 Introduction

Modern Critical Infrastructures (CI), e.g., power plants, water plants and smart grids, rely on Information and Communication Technologies (ICT) for their operation since ICT can lead to cost optimization as well as greater efficiency, flexibility and interoperability between components. In the past CIs were isolated environments and used proprietary hardware and protocols, limiting thus the threats that could affect them. Nowadays CIs, or more specifically Networked Industrial Control Systems (NICS), are exposed to significant cyber-threats; a fact that has been highlighted by many studies on the security of Supervisory Control And Data Acquisition (SCADA) systems [7,11]. The recently reported Stuxnet worm [8] is the first malware specifically designed to attack NICS. Its ability to reprogram the logic of control hardware in order to alter physical processes demonstrated how powerful such threats can be. Stuxnet was a concrete

R. Bestak et al. (Eds.): NETWORKING 2012, Part I, LNCS 7289, pp. 121–134, 2012.

proof of a successful cyber-physical attack but by no means a trivial attack. It required a thorough knowledge of the physical system, software and OS vulnerabilities.

The size of physical processes led plant designers to structure SCADA system components into multiple network segments, i.e., Virtual Lans (VLANs), [2] interconnected with wireless devices. One of the main advantages of this approach is that malware infections do not propagate to other VLANs unless the attacker is capable to compromise the protection mechanism, e.g., firewalls, of other VLANs as well. Nevertheless, the compromise of one network segment could cause the physical process to shut down, e.g., physical damage, unless designers take appropriate measures to limit the effects of a single compromised control network segment.

Based on these facts, in this paper we investigate the relationship between control network segmentation and the resilience of physical processes. We consider an adversary with a level of sophistication similar to the case of Stuxnet [8] that is able to take over an entire control network segment, i.e., VLAN. The goal of the attacker is to disrupt the normal operation of the physical process by reprogramming the logic of control hardware, as in the case of Stuxnet. The attack scenario was implemented with our previously developed framework [9] that uses real-time simulation for the physical components and an emulation testbed based on Emulab [17] to recreate the cyber part of NICS, e.g., SCADA servers, corporate network. In the implemented scenario we used the Tennessee-Eastman chemical process [5].

The rest of the paper is structured as follows. After an overview of related work in Section 2, we provide a discussion on the segmentation problem and implemented attack scenarios in Section 3. We continue with the presentation of experimental results in Section 4 and we conclude in Section 5.

2   Related Work

According to Wei and Ji [16], a resilient control system is one that is able to: (i) minimize the incidence of undesirable incidents; (ii) mitigate the undesirable incidents; and (iii) recover to normal operation in a short time. In this context our analysis points out an important factor to increase the resilience of industrial systems: the segmentation of process control networks into VLANs. However, this is only one factor that could be considered. Several others were identified with solutions proposed by other authors as well. This section provides a brief presentation of those approaches that mostly relate to ours.

The work of Cárdenas, et al. [3] clearly pointed out that intrusion detection systems combined with a reaction mechanism that closes the system monitoring loop are able to effectively increase the resilience of the system. Their work showed that control loops implemented in control hardware, i.e., Programmable Logic Controllers (PLCs), can be adjusted in order to counteract the effects of Denial of Service attacks. In the field of Smart Grids, the work of Zhu, et al. [18] showed that routing is a major concern and proposed a secure routing protocol to increase the resilience of Smart Grids. The work of Chen, *et al.* [4] addressed the importance of hierarchical control solutions for increasing the resilience of Power Grids. The proposed solution uses well-established control theory to guarantee accuracy and system stability. Finally, we mention the recent work of Ji and Wei [10] that proposed a method to quantify the resilience of NICS in terms of quality of control. The authors also proposed a control algorithm for wireless NICS that is able to keep the process in a normal operating state while it is confronted with attacks such as Radio Frequency jamming and signal blocking.

Compared to the previously mentioned techniques, the proposed segmentation-based approach addresses more sophisticated attacks, similar to Stuxnet, that might involve the reprogramming of PLCs. Such attacks are not addressed by existing approaches. Moreover, even in the case of techniques that add counter-measures to the process control network, such as the work of Cárdenas, *et al.* [3], more sophisticated attacks are not targeted. Such approaches rely on PLCs running legitimate control code with incorporated countermeasures, that could be rewritten by malware. The proposed segmentation methodology could also be combined with techniques that ensure the security of industrial systems [1,12], leading to a system that is both secure and resilient against cyber threats.

3 Problem Statement and Attack Scenario

The Stuxnet malware was a concrete proof that nowadays attackers are capable not only to infiltrate into the process and control networks, but are also capable to reprogram PLCs. Such attack scenarios have an important impact on the physical process as the code that keeps the process in its operating limits is replaced by malicious code. Therefore, new techniques that also address more sophisticated attacks, i.e., similar to Stuxnet, must be developed. In this section we discuss the applicability of network segmentations to counteract such powerful attacks. We begin with an overview of typical process control architectures and we continue with a discussion on the proposed control network segmentation. Finally, we provide a brief presentation on the implemented adversary model and attack scenario.

3.1 Process Control Architecture Overview

Modern SCADA architectures have two different control layers: (i) the physical layer, which comprises actuators, sensors and hardware devices that physically perform the actions on the system, e.g., open a valve, measure the voltage; and (ii) the cyber layer, which comprises all the information and communications devices and software that acquire data, elaborate low-level process strategies and deliver the commands to the physical layer. The cyber layer typically uses SCADA protocols to control and manage an industrial installation. The entire architecture can be viewed as a "distributed control system" spread among two networks: the control network and the process network. The process network usually hosts the SCADA servers (also known as SCADA masters), human-machine

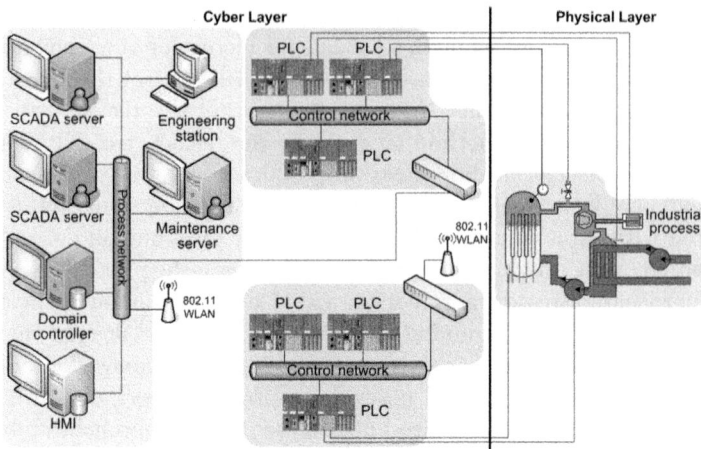

Fig. 1. Process control architecture

interfaces (HMIs), domain controllers and other installation-specific nodes, e.g., engineering stations, maintenance servers. The control network hosts all the devices that on one side control the actuators and sensors of the physical layer and on the other side provide the control interface to the process network. A typical control network is composed of a mesh of PLCs (Programmable Logic Controllers), as shown in Fig. 1.

From an operational point of view, PLCs receive data from the physical layer, elaborate a local actuation strategy, and send commands to the actuators. When requested, PLCs also provide the data received from the physical layer to the SCADA servers (masters) in the process network and eventually execute the commands that they receive. In modern SCADA architectures, communications between a master and PLCs is usually implemented in two ways: (i) through an OPC (Object Linking and Embedding (OLE) for Process Control) layer that helps map the PLC devices; and/or (ii) through a direct memory mapping notation making use of SCADA communication protocols such as Modbus, DNP3 and Profibus.

3.2 Control Network Segmentation

The main goal of the segmentation procedure is to increase the resilience of physical processes. In practice engineers might use network segmentation for a number of reasons such as physical constraints, e.g., location of devices, or protection of mission-critical services. In typical implementations the segmentation is most of the time forced by physical constraints [14] where each individual segment is isolated from the rest and includes network security protection mechanisms, e.g., firewalls. These segments are interconnected by VPNs and are remotely accessible by engineers.

Instead of applying typical segmentation rules such as the ones mentioned previously, in this paper we propose a segmentation that focuses on the physical process. The goal of the procedure is to maximize the resilience of the physical process in case of the full compromise of one or more network segments. The procedure relies on the ability of regular control code to counterbalance the disturbance generated by malicious code running in compromised segments. More specifically, in the proposed approach we separate PLCs controlling input valves (FeedPLCs) from PLCs controlling output valves (FreePLCs), associated to the same unit. This way, the effect of compromised FeedPLCs is balanced by legitimate FreePLCs and vice-versa.

For a better understanding of the impact of the proposed approach, let us assume a simple scenario involving a pipe and 3 valves controlled by 3 PLCs. In this scenario one of the control valves is feeding products into the pipe while the other two are freeing products from the pipe. If designers would place all 3 PLCs on the same network segment (see Fig 2 (a)), in case of an attack that compromises the entire segment the adversary would be able to OPEN the input valve and CLOSE the output valves. This would lead to a sudden increase of the pressure that could cause severe damages to the physical process. On the other hand, by placing FeedPLCs and FreePLCs on separate network segments (see Fig 2 (b)), in case one of the segments is compromised, regular PLCs could balance the generated disturbance and avoid catastrophic consequences.

In the present study we compared the full network compromise setting to the segmentation with the proximity and product flow criteria. Although the analysis is limited to these settings, our main goal was not to be exhaustive, but to show that control network segmentation plays an important role in the resilience of physical processes.

3.3   Adversary Model and Attack Scenario

The employed adversary model reprograms PLCs with malicious code in order to shut down the physical process. Identifying the attack vector that could compromise the system to enable such a scenario is not the main focus of this study. However, the Stuxnet worm together with other studies such as the one performed by Nai Fovino, *et al.* [11] showed that such scenarios are possible in real settings. For instance, corporate firewalls could be compromised by infected user stations within the corporate network. A similar scenario was recently reported by Google [6], the official report stating that errors in Web browser implementations enabled the installation of a malware on a user's machine within the corporate network. From there the malware spread and infected other stations as well. Another example is the Stuxnet worm that included several attack vectors such as USB drives and vulnerabilities in the Operating System, but also vulnerabilities in the Siemens WinCC/Step 7 software. WinCC/Step 7 is the software used to communicate with a variety of PLCs produced by Siemens. By exploiting vulnerabilities in this software, the designers of Stuxnet were able not only to reprogram PLCs but to also hide the changes from human operators.

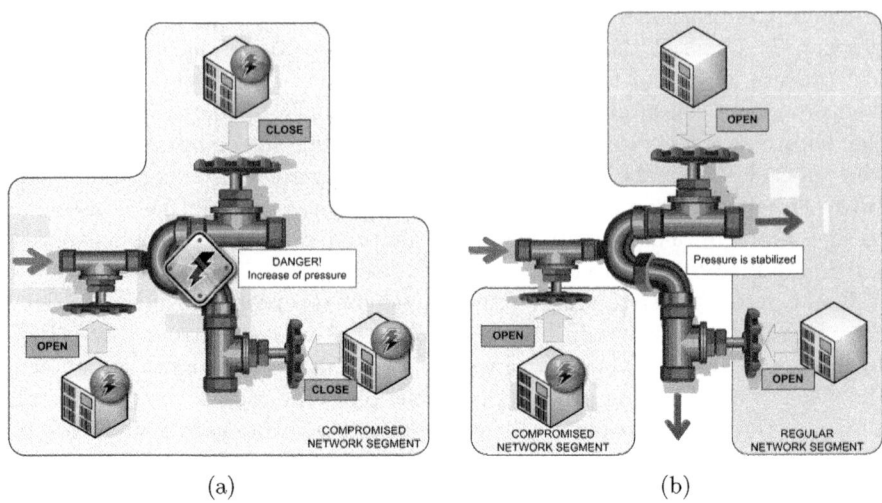

(a)   (b)

Fig. 2. Effect of compromised PLCs on the physical process: (a) proximity-based segmentation, and (b) product flow-based segmentation

As pointed out by Cárdenas, *et al.* [3] attacks that target the minimum/ maximum value of parameters/control variables are the ones that can damage the process in relatively short time periods. Such attacks cause the accumulation of products, e.g., steam or water, by completely opening valves that feed products into units and completely closing valves that free products from units. The attack model employed in this study follows the same procedure to force the physical process to shut down. More specifically, based on the documentation of the physical process, the malicious code completely opens input valves and completely closes output valves.

4   Experimental Setting and Results

The results presented in this section prove that network segmentation can be an effective approach to increase the resilience of physical processes confronted with sophisticated attacks. For this purpose we use as a resilience metric the time that the process is able to run after the attack is started, before shutting down, i.e., *shut down time* (SDT). First, the SDT is measured for each compromised VLAN, as generated by the segmentation procedure mentioned in the previous sections. Then, the SDT is compared to the SDT of the full network compromise setting to show the benefits of product flow-based segmentation over proximity/ad-hoc segmentation.

We start the presentation with an overview of the experimentation framework and of the Tennessee-Eastman chemical plant used as the physical process model. We continue with an overview of the experimental setup and finally we present the experimental results.

4.1   Overview of the Experimentation Framework

In the context of the experimental scenario described in the previous sections we simulated the physical process and we emulated the cyber layer using the experimentation framework developed in our previous work [9]. There are several reasons why we have chosen this approach for our study. First, by testing the resilience of a real system there could be concerns about the potential side effects of the experiment. Second, software based simulation has always been considered an efficient approach to study physical systems, mainly because it can offer low-cost, fast and accurate analysis. Nevertheless, it has limited applicability in the context of cyber security due to the diversity and complexity of computer networks. Software simulators can effectively model normal operations, but fail to capture the way computer systems fail.

The experimentation framework developed in our previous work [9] follows a hybrid approach, where the Emulab-based testbed recreates the control and process network of NICS, including PLCs and SCADA servers, and a software simulation reproduces the physical processes. The architecture, as shown in Fig. 3, clearly distinguishes 3 layers: the cyber layer, the physical layer and a link layer in between. The cyber layer includes regular ICT components used in SCADA systems, while the physical layer provides the simulation of physical devices. The link layer, i.e., cyber-physical layer, provides the "glue" between the two layers through the use of a shared memory region.

The physical layer is recreated through a soft real-time simulator that runs within the SC (Simulation Core) unit and executes a model of the physical process. The cyber layer is recreated by an emulation testbed that uses the Emulab architecture and software [17] to automatically and dynamically map physical components, e.g., servers, switches, to a virtual topology. Besides the process network, the cyber layer also includes the control logic code that in the real world is run by PLCs. The control code can be run sequentially or in parallel to the physical model. In the sequential case, a *tightly coupled* code (TCC) is used, i.e., code that is running in the same memory space with the model, within the SC unit. In the parallel case a *loosely coupled* code (LCC) is used, i.e., code that is running in another address space, possibly on another host, within the R-PLC unit (Remote PLC). The main advantage of TCCs is that these do not miss values generated by the model between executions. On the other hand, LCCs allow running PLC code remotely, to inject (malicious) code without stopping the execution of the model, and to run more complex PLC emulators. The unit that implements global decision algorithms based on the sensor values received from the R-PLC units is also present in the experimentation framework as the *Master* unit. The cyber-physical layer incorporates the PLC memory, seen as a set of registers typical of PLCs, and the communication interfaces that "glue" together the other two layers. Memory registers provide the link to the inputs, e.g., valve position, and outputs, e.g., sensor values, of the physical model.

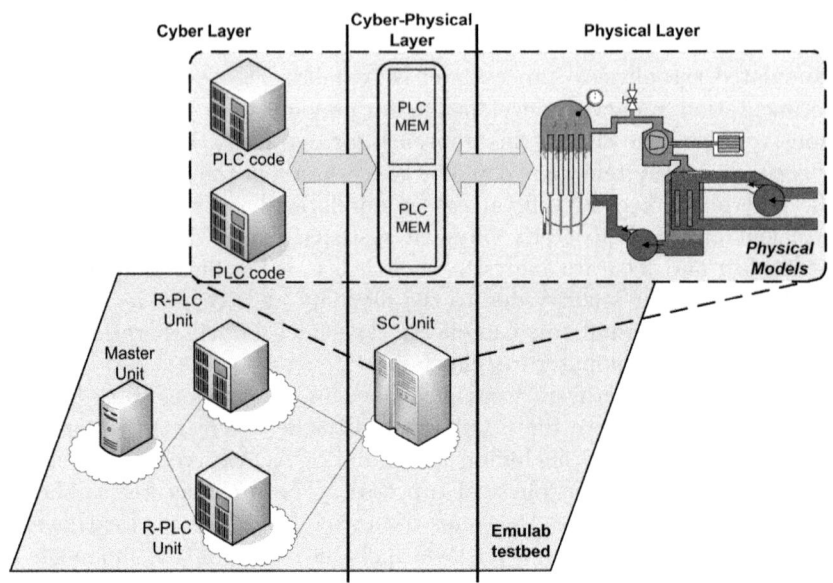

Fig. 3. Experimentation framework overview

Prototypes of SC, R-PLC and Master Units have been developed in C# (Windows) and have been ported and tested on Unix-based systems (FreeBSD, Fedora and Ubuntu) with the help of the *Mono* platform. Matlab Simulink was used as the physical process simulator (physical layer). From Simulink models the corresponding 'C' code is generated using Matlab Real Time Workshop. The communication between SC and R-PLC units is handled by .NET's binary implementation of RPC (called *remoting*) over TCP. For the communication between the R-PLC and Master units, we used the Modbus over TCP protocol.

4.2   Tennessee-Eastman Chemical Process

The TE process is a well-known problem in the automation and process control community mainly because it represents a hypothetical chemical plant that is very similar to an actual plant. The model has been provided by the Tennessee Eastman company [5]. The schematic for the TE process is presented in Fig. 5 where we also show the associated PLCs.

The process is fairly complex: it produces two products from four reactants and the plant has a total of seven operating modes that include a base operating condition. The plant simulation provides a total of 41 measurements and 12 manipulated variables. In this use case we assume that the plant is controlled by the Programmable Logic Controllers (PLCs), i.e., TCCs, that implement the base control strategy proposed by Sozio [15].

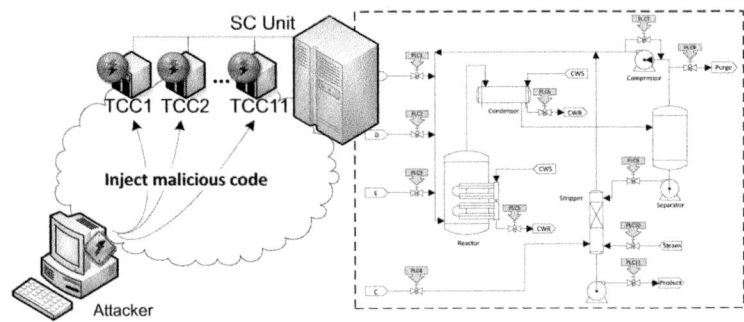

Fig. 4. Experimental setup

4.3   Experiment Setup

The attack scenario described in the previous sections was implemented in the
Joint Research Centre's (JRC) Experimental Platform for Internet Contingen-
cies (EPIC) laboratory. The Emulab testbed included nodes with the following
configuration: FreeBSD OS 8, AMD Athlon Dual Core CPU at 2.3GHz and 4GB
of RAM. In our experiments we used the TE model implementation given as a
Matlab 'C'-based MEX S-Function, developed by Ricker [13], from which the
stand-alone 'C' code was generated using the Matlab Real Time Workshop. The
generated code was integrated into the experimentation framework in order to
interact with the real components of the emulation testbed. Regular and malicious control code were implemented as TCCs. The experimental setup is shown
in Fig. 4.

The results of the segmentation procedure, based on the segmentation criteria
discussed in the previous sections, are given in Fig. 5. For the proximity criteria
(see Fig. 5 (a)) - setting A, we defined 3 segments based on the proximity to the
3 main units (*Reactor*, *Separator* and *Stripper*), each implemented as a separate
VLAN. By using the same 3 main units we also defined 3 segments based on the
product flow criteria - setting B, as shown in Fig. 5 (b). In both figures we used
a white color for PLCs on VLAN 1 and VLAN 1', light gray for PLCs on VLAN
2 and VLAN 2' and dark gray for PLCs on VLAN 3 and VLAN 3'.

4.4   Experimental Results

As a result of the previously described segmentation procedure, 6 independent
VLANs were identified for both settings, i.e., settings A and B. For each VLAN
we implemented the attack scenario described in the previous sub-sections and
we measured the shut down time (SDT).

The operation of the TE process for 40h without any disturbances is shown
in Fig. 6, where the target setpoints are illustrated with a dashed line. With
the implemented control loops the process is able to run in a steady-state, as
shown by the two sub-figures depicting the behavior of two parameters that
could trigger a shut down of the process. Without these control loops, process
parameters would reach their shut down limits after approximately 3.6h [15].

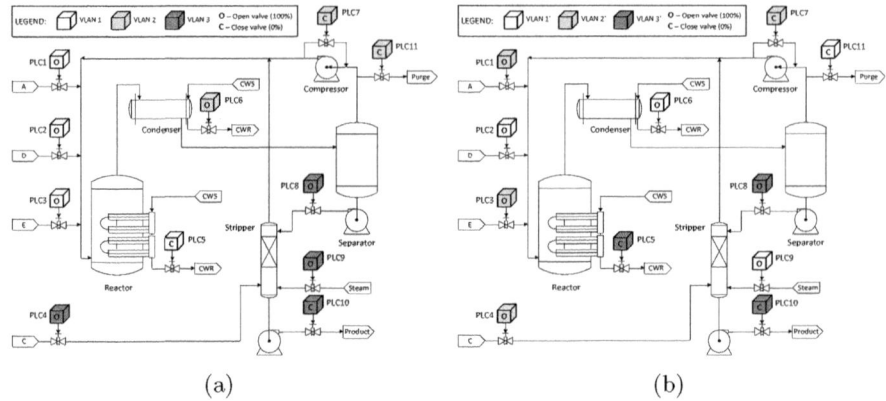

(a)   (b)

Fig. 5. Tennessee-Eastman process and associated PLCs: (a) control network segmentation in setting A, and (b) control network segmentation in setting B

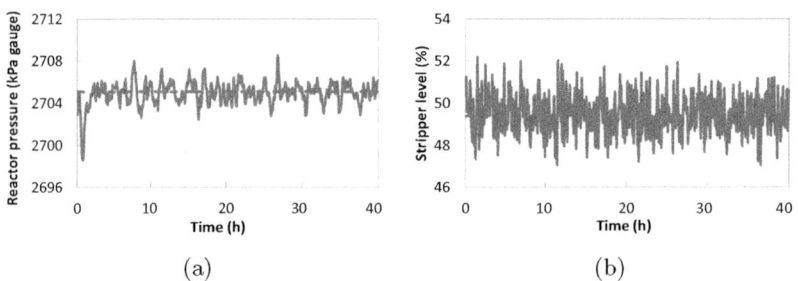

(a)   (b)

Fig. 6. Normal operation of the Tennessee-Eastman process for 40h without any disturbances: (a) Reactor pressure, and (b) Stripper level

After running the TE process for 10h, in the next step we launched the attack scenario described in the previous sections. First, the attack was launched against the full control network and then against each VLAN identified in the segmentation procedure. Because of space considerations we only illustrate the behavior of the process for the maximum SDT for settings A and B. A summary of the results is given in Fig. 7.

In the full network compromise setting, all PLCs were running malicious code. This lead to the shut down of the TE process in 0.05h (3min), caused by an increase in the *Reactor* pressure above the 3000kPa shut down limit. The *Reactor* pressure for this setting was illustrated in Fig. 8 (a). In the remaining of this section we use the measured SDT from this setting to show that the SDT can be increased with control network segmentation.

In setting A, the maximum SDT was measured in case VLAN 2 was compromised, while the minimum SDT was measured in case of VLAN 3. As shown in

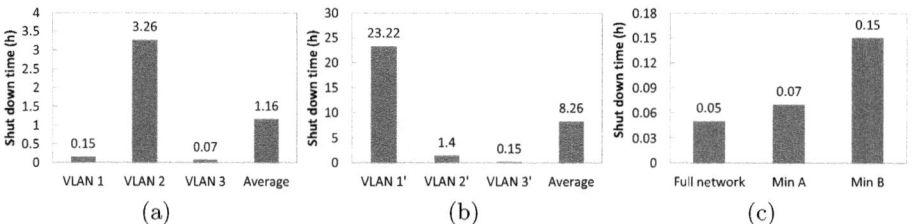

Fig. 7. Shut down time: (a) setting A, (b) setting B, and (c) compared results

Table 1. Reason for shut down of the Tennessee-Eastman process

Setting	Compromised VLAN	Shut down reason
Full network	–	high reactor pressure
A	1	high reactor pressure
	2	high reactor pressure
	3	high stripper liquid level
B	1'	high reactor pressure
	2'	high reactor pressure
	3'	high stripper liquid level

Fig. 8 (b), the attack on VLAN 2 increased the *Reactor* pressure above the shut down limit of 3000kPa in 3.26h. In case of the compromise of the remaining two VLANs we measured a smaller SDT. Thus, for VLAN 1 the measured SDT was 0.15h, while for VLAN 3 it was 0.07h. For VLAN 1, the shut down of the TE process was caused by an excessive increase of the *Reactor* pressure, while for VLAn 3 it was caused by high liquid levels in the *Stripper* unit. We summarized the results for setting A in Fig. 7 (a) and Table 1.

By comparing the results from setting A with the SDT from the full control network setting, we see an increase of 0.02h for the minimum SDT and of 3.21h for the maximal SDT. In the field of Information Security it is a well known fact that the security strength of a system is given by its weakest component. Therefore, in our context it is more important to increase the smallest SDT than the largest or the average value for a specific setting. As in setting A the smallest measured value was of 0.07h (4.2min), this corresponds to an increase of 40% in the value of the minimal SDT. As shown by the results from setting B, the minimal SDT can be further increased by employing process-specific information in the segmentation procedure.

For setting B the segmentation procedure also generated 3 VLANs, but with a different configuration, as shown in Fig. 5 (b). By applying the same experimental strategy for setting B, the maximum SDT was measured for the compromise of

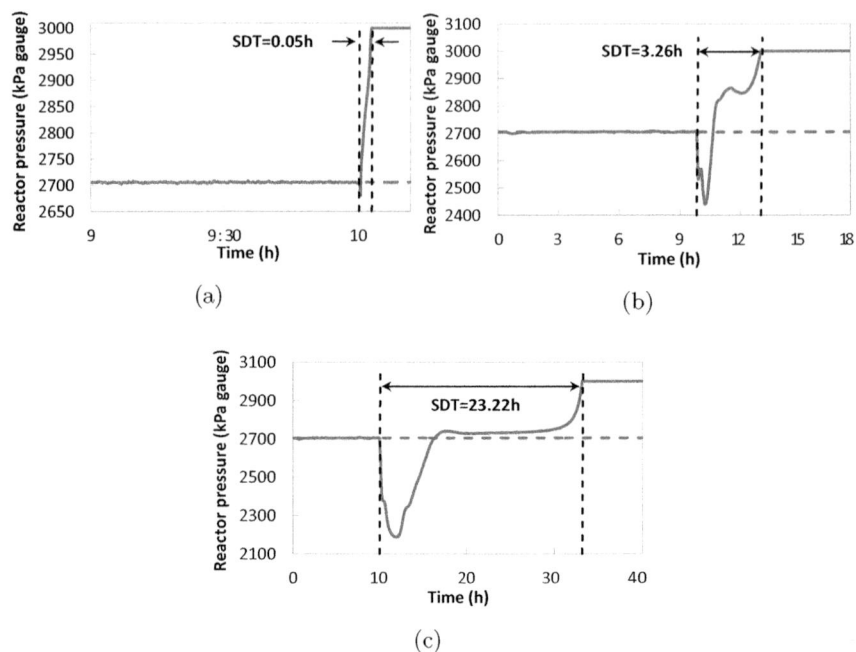

Fig. 8. Disturbed operation of the Tennessee-Eastman process: (a) full control network compromise, (b) compromise of VLAN 2 in setting A, and (c) compromise of VLAN 1' in setting B

VLAN 1', with the effects shown in Fig. 8 (c). In this case the maximum SDT increased to 23.22h, that is more than 7 times the value of the maximum SDT from setting A. The monitored parameter illustrated in Fig. 8 (c) shows that initially the attack causes large deviations on the process parameters. Nevertheless, after 5h legitimate PLCs from non-compromised VLANs bring the process back into the steady-state. The TE process remains in this state for approximately 15h. After this period the accumulated disturbances exceed to capabilities of legitimate PLCs, causing the pressure within the *Reactor* unit to increase until the shut down limit. We also inspected the SDT for the remaining two VLANs. For VLAN 2' the shut down was caused by an excessive increase in the *Reactor* pressure, while for VLAN 3' the shut down was caused by a high liquid level in the *Stripper* unit. We summarized the results for setting B in Fig. 7 (b) and Table 1.

A significant aspect that we should note for setting B is that the minimum SDT increased to 0.15h (9min), that is more than twice the minimum SDT recorded for setting A. This shows that a careful examination of the physical process can lead to a segmentation that increases the resilience of the physical process by more than 100%, compared to a segmentation based on the proximity criteria. Furthermore, if we compare the increase in the minimum SDT to the full network setting, the increase is above 200%. We summarized these results in Fig. 7 (c).

Based on the results from this section we can conclude that a resilience-aware network design can provide a tolerance period of several additional minutes or even hours. This would give operators more time to intervene, e.g., switch OFF devices, and reduce the damages caused to the physical process.

5   Concluding Remarks

In this paper we have shown that network design choices and specifically network segmentation in VLANs can have an important impact to the resilience of physical processes. Compared to existing approaches, the proposed method has several advantages: (i) it can be applied to a wide variety of industrial systems; (ii) it also targets more sophisticated attacks similar to Stuxnet; and (iii) it does not require new error-prone software/hardware to be installed, for each segment existing security techniques can be replicated. Our proposal can also be viewed as complementary to existing approaches and can be implemented together with other techniques that also address the resilience of industrial systems [3,4,18], but do not target more sophisticated attacks. Finally, we also mention that the proposed segmentation methodology can be combined with techniques that ensure the security of industrial systems [1,12], leading to installations that are both secure and resilient against cyber threats. The study reported in this paper is a first step in our work towards the development of a method that maximizes the resilience of physical processes with network segmentation. As part of our future work, we also intend to study the applicability of our proposal in the context of more complex physical processes such as an entire Power Grid.

References

1. dos Anjos, I., Brito, A., Pires, P.M.: A model for security management of SCADA systems. In: Proceedings of IEEE International Conference on Emerging Technologies and Factory Automation, pp. 448–451 (2008)
2. Boyer, S.: Supervisory Control And Data Acquisition. International Society of Automation, USA (2010)
3. Cárdenas, A., Amin, S., Lin, Z.S., Huang, Y.L., Huang, C.-Y., Sastry, S.: Attacks against process control systems: Risk assessment, detection, and response. In: Proceedings of the 6th ACM Symposium on Information, Computer and Communications Security, pp. 355–366 (2011)
4. Chen, M., Nolan, C., Wang, X., Adhikari, S., Li, F., Qi, H.: Hierarchical utilization control for real-time and resilient power grid. In: Proceedings of the 21st Euromicro Conference on Real-Time Systems, pp. 66–75 (2009)
5. Downs, J.J., Vogel, E.F.: A plant-wide industrial process control problem. Computers & Chemical Engineering 17(3), 245–255 (1993)
6. Drummond, D.: A new approach to China (2010),
http://googleblog.blogspot.com/2010/01/new-approach-to-china.html
7. East, S., Butts, J., Papa, M., Shenoi, S.: A taxonomy of attacks on the DNP3 protocol. IFIP AICT, vol. 311, pp. 67–81 (2009)
8. Falliere, N., Murchu, L.O., Chien, E.: W32.stuxnet dossier (2010), http://www.wired.com/images_blogs/threatlevel/2010/11/w32_stuxnet_dossier.pdf

9. Genge, B., Fovino, I.N., Siaterlis, C., Masera, M.: Analyzing cyber-physical attacks on networked industrial control systems. In: Critical Infrastructure Protection, pp. 167–183 (2011)
10. Ji, K., Wei, D.: Resilient control for wireless networked control systems. Journal of Control, Automation, and Systems 9(2), 285–293 (2011)
11. Fovino, I.N., Carcano, A., Masera, M., Trombetta, A.: An experimental investigation of malware attacks on SCADA systems. Journal of Critical Infrastructure Protection 2(4), 139–145 (2009)
12. Pal, O., Saiwan, S., Jain, P., Saquib, Z., Patel, D.: Cryptographic key management for SCADA system: An architectural framework. In: Proceedings of International Conference on Advances in Computing, Control, & Telecommunication Technologies, pp. 169–174 (2009)
13. Ricker, N.: Tennessee Eastman challenge archive (2002), http://depts.washington.edu/control/LARRY/TE/download.html
14. Siemens: Security concept pcs 7 and wincc - basic document (2008), http://support.automation.siemens.com/WW/llisapi.dll?func=cslib.csinfo&lang=en&objid=26462131&caller=view
15. Sozio, J.: Intelligent parameter adaptation for chemical processes. Master's thesis, Virginia Polytechnic Institute and State University, USA (1999)
16. Wei, D., Ji, K.: Resilient industrial control system (RICS): Concepts, formulation, metrics, and insights. In: Proceedings of the 3rd International Symposium on Resilient Control Systems, pp. 15–22 (2010)
17. White, B., Lepreau, J., Stoller, L., Ricci, R., Guruprasad, S., Newbold, M., Hibler, M., Barb, C., Joglekar, A.: An integrated experimental environment for distributed systems and networks. In: Proceedings of the 5th Symposium on Operating Systems Design and Implementation, pp. 255–270 (2002)
18. Zhu, Q., Wei, D., Başar, T.: Secure routing in smart grids. In: Workshop on Foundations of Dependable and Secure Cyber-Physical Systems (2011)

On the Vulnerability of Hardware Hash Tables to Sophisticated Attacks

Udi Ben-Porat[1], Anat Bremler-Barr[2], Hanoch Levy[3], and Bernhard Plattner[1]

[1] Computer Engineering and Networks Laboratory, ETH Zurich, Switzerland
{ehudb,plattner}@tik.ee.ethz.ch
[2] Computer Science Dept., Interdisciplinary Center, Herzliya, Israel
bremler@idc.ac.il
[3] Computer Science Dept., Tel-Aviv University, Tel-Aviv, Israel
hanoch@cs.tau.ac.il

Abstract. Peacock and Cuckoo hashing schemes are currently the most studied hash implementations for hardware network systems (such as NIDS, Firewalls, etc.). In this work we evaluate their vulnerability to sophisticated complexity Denial of Service (DoS) attacks. We show that an attacker can use insertion of carefully selected keys to hit the Peacock and Cuckoo hashing schemes at their weakest points. For the Peacock Hashing, we show that after the attacker fills up only a fraction (typically $5\% - 10\%$) of the buckets, the table completely loses its ability to handle collisions, causing the discard rate (of new keys) to increase dramatically ($100 - 1,800$ times higher). For the Cuckoo Hashing, we show an attack that can impose on the system an excessive number of memory accesses and degrade its performance. We analyze the vulnerability of the system as a function of the critical parameters and provide simulations results as well.

1 Introduction

Modern high speed networks pose a challenge for routers, Firewalls, NIDS (Network Intrusion Detection System) or any other network devices that have to route, measure or monitor a network without slowing it down. Such network hardware elements are highly preferable targets for DDoS (Distributed Denial of Service) attacks since their failure can severely slow the network and, in the case of security systems, their failure can allow an attacker to conduct an attack on a critical system they are meant to protect. Equipped with knowledge about how the system works, an attacker can perform a low-bandwidth sophisticated DDoS attack, targeting weak points in the system, rather than just flooding it (which takes more efforts and can be detected and countered more easily).

For example, Crosby and Wallach [2] demonstrated attacks on Open Hash table implementations in the Squid web proxy and in the Bro intrusion detection system. They showed that an attacker can design an attack that achieves worst case complexity of $O(n)$ elementary operations per insert operation (instead of the average case complexity of $O(1)$), causing, for example, the Bro server to drop 71% of the traffic (without increasing the volume of the traffic).

R. Bestak et al. (Eds.): NETWORKING 2012, Part I, LNCS 7289, pp. 135–148, 2012.

In another example, Smith et al. [1] describe a low bandwidth sophisticated attack on a NIDS system, in which the attack disables the NIDS of the network by exploiting the behavior of the rule matching mechanism and sending packets which require very long inspection times.

Hash tables play an important role in the operations of the most important and time consuming tasks these systems have to perform. Using hashing techniques which allow constant operation complexity is therefore highly desirable. Multiple-choice Hash Tables (MHT), and in particular Peacock [3] and Cuckoo [4] Hashing, are easy to implement and are currently the most efficient and studied implementations for hardware network systems such as routers for IP lookup (for example [5], [6] and [7]), network monitoring and measurement (for example, [16]) and Network Intrusion Detection/Prevention Systems (NIDS/NIPS) [8]. For more information about hardware-tailored hash tables we recommend the recent survey by Kirsch et al. [9].

A Peacock hash table consists of a large main table (which typically holds 90% of the buckets) and a series of additional small sub-tables where collisions caused during insertions are resolved. Its structure is based on the observation that only a small fraction of the keys inserted into a hash table collide with existing keys (that is, hashed into an occupied bucket) and even a smaller fraction will collide again, etc. These backup tables are usually small enough for their summary (implemented by bloom filters) to be saved on fast on-chip memory which dramatically increases the overall operation performance in Peacock Hashing.

A Cuckoo Hashing is made of two (or more) sub-tables of the same size. Every key can be placed in one bucket (to which it hashes) in each sub-table. When a key k finds all its buckets occupied, one of the keys residing in those buckets is then moved to one of its alternate locations to free the bucket for k. Cuckoo Hashing, therefore, allows achieving a higher table utilization than that achieved by alternative MHT schemes that do not allow moves, while maintaining $O(1)$ amortized complexity of an Insert operation (although the complexity of a single Insertion is not bounded by a constant).

In this work we expose the weak points of the Peacock and Cuckoo Hashing and the system parameters that affect them. To evaluate the vulnerability of Peacock and Cuckoo we refer to [10] which observed that an attack on a hash table data structure can damage the performance of the system in two ways: 1. *In-attack damage* - Insertions of keys that require excessive number of memory accesses; 2. *Post-attack damage* - Insertion of keys that are placed in the table in a way that causes future insertions of keys to take excessive memory accesses and/or reduce their probability to find an empty bucket. Using this classification, we show that Peacock is resilient against in-attack damage and explain how such an attack can be countered easily. On the other hand we show that it is vulnerable to post-attack damage. We propose a sophisticated attack that can dramatically increase the discard probability of a newly inserted key after the attack has ended. We show that after the attacker inserts keys into the table, it brings the table into an irreversible state in which the discard probability for a newly inserted key can be 100 to 1, 800 times higher than the discard probability after the same amount of keys are inserted by regular users.

For Cuckoo hashing we explain why post-attack damage is irrelevant and analyze its in-attack vulnerability. We show that an attacker can slow the system by inserting keys that require 4 times more memory accesses than regular keys in a typical settings. We further analyze the vulnerability with respect to two key design parameters – the number of sub-tables and the number of moves allowed per insertion; we show that while the utilization in the table increases with either of these parameters, the vulnerability decreases with the former while increases with the latter. In addition to mathematical analysis, we also provide simulation results for a use case in which a system designer plans to design Cuckoo and Peacock hash tables which comply with the same requirements. In addition, we discuss the feasibility of the attack by evaluating the complexity of finding keys suitable for a sophisticated attack and show for both Peacock and Cuckoo Hashing that high number of sub-tables makes it harder for the attacker to find suitable keys.

The structure of the rest of the work is as follows: In Section 2 we explain the nature of sophisticated attacks against hash tables and the Vulnerability metric used in this work. Then, the main body of the work consists of sections 3 and 4 which are dedicated to the Peacock and Cuckoo Hashing, respectively. Both sections have a similar structure. Each is divided into five parts covering the following topics: 1. The hashing algorithm; 2. Attack strategy; 3. Feasibility of the attack; 4. Vulnerability analysis and simulation results and 5. The resilience to in-attack (Peacock) or post-attack (Cuckoo) damage. Finally, Section 5 concludes the key results. In addition, a glossary of the key notations used throughout the work can be found at the Appendix.

2   Sophisticated Attacks on Multiple Hash Tables

In multiple hash table schemes, such as Peacock and Cuckoo Hashing, every key can be placed only in a small fraction of the buckets. While it allows performing Search and Delete operation with no more than a predefined constant number of memory references, it also poses a challenge: As the load in the table grows, both the *insertion complexity* (measured by the number of probed buckets) and the *discard probability* - the probability for an inserted key to not find an available bucket to be stored in - increases. In order to keep these variables bounded by acceptable values, the utilization (maximal load) in the table is limited. Therefore, a simple flooding attack where the attacker simply inserts a large number of (random) keys cannot degrade the system performance beyond its acceptable limits. Note that a flooding attack can be handled by forwarding the keys to another table/device or by blocking the attacker since it can then be detected. Using knowledge about the table, an attacker can perform a *sophisticated* attack that degrades the system performance beyond its acceptable values (with which it was designed to comply) using just a small number of keys and hence avoid reaching the maximal load in the system.

The vulnerability metric we use in this work has been proposed in [10] and is defined as the maximal performance degradation (damage) that malicious users can inflict on the system using a specific amount of resources (budget) normalized by the performance degradation attributed to regular users using the same amount of resources. Formally, according to [10], the *effectiveness* of an attack is defined by

$$E_{st}(budget = K) = \frac{\Delta Perf(M_{st}, K)}{\Delta Perf(R, K)}, \tag{1}$$

where $\Delta Perf(M_{st}, K)$ and $\Delta Perf(R, K)$ are the performance degradations caused by inserting additional K keys to the table (in the context of hash tables) by malicious and regular users, respectively, where st is the attack strategy used by the attacker. Then, the Vulnerability V of a system is defined by the effectiveness of the strategy that causes the maximal damage:

$$V(budget = K) = max_{st}\{E_{st}(K)\}. \tag{2}$$

Therefore, when an attack strategy is not proved to be the optimal, its effectiveness is considered as a lower bound for the vulnerability of the system.

Note that in order to perform the sophisticated attacks analyzed in this work, the attacker is assumed to gain knowledge of the structure of the table (number of tables and their sizes, but not how many keys are already stored in the table and where). In addition, the attacker is assumed to be able to compute or guess the hash values of keys. This knowledge can be achieved by reverse engineering of similar products acquired by the attacker, various guessing methods or due to the use of open source algorithms. For more information see [10].

3   Peacock Hashing

3.1   Insertion Algorithm

In Peacock Hashing [3] the buckets are divided into d sub-tables $\{T_i\}_{i=1}^{d}$ and d corresponding hash functions $\{h_i\}_{i=1}^{d}$. The sizes of the sub-tables follow a decreasing geometric sequence $M_{i+1} = M_i/r$ where r is the proportion between the table sizes[1]. The first sub-table T_1 is called the *main table*, while the rest are called the *backup tables*. T_1 is the largest table and it is where the insertion algorithm first tries to store a key. Every backup table handles the collisions in the sub-table that precedes it. The insertion algorithm probes the sub-tables $\{T_i\}_{i=1}^{d}$ one after the other until finding a table where the bucket to which the key hashes is free. In the rare case where a key cannot find its place in any of the tables - it is dropped. In addition, a summary of the keys stored in every backup table is maintained (implemented by Bloom filters stored on the on-chip memory). It is used to avoid checking all the sub-tables when making sure an inserted key does not already exist in the table. Note that due to lack of space,

---

[1] In [3] the authors recommended to use $r = 10$.

in this work we follow the original Peacock and Cuckoo models and exclude the case in which a bucket can hold more than one key; this case is discussed in a technical report [11].

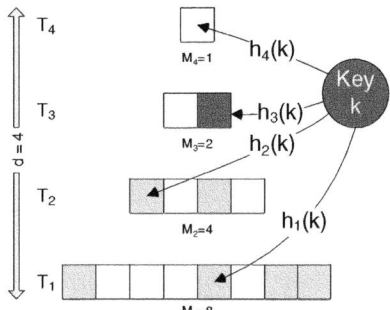

Fig. 1. An insertion of key k into a small Peacock hash table with $r = 2$ and $d = 4$. Gray buckets are occupied with existing keys and white buckets are free. The black bucket marks the bucket where k is finally placed.

After a long series of insertions and deletions (of keys) from the hash table, it becomes unbalanced. That is, the load in the smaller sub-tables is higher than in the bigger tables and this increases the discard probability of a new insertion [3]. This state can be prevented by *re-balancing* the table after a key is deleted as follows. After a key was deleted from bucket b in T_i, if there is a key k that is stored in $T_{j>i}$ such that $h_i(k) = b$, then k is moved back from T_j to the now-free bucket in T_i. The attack we propose and analyze below brings the hash table to an unbalanced state that cannot be solved by re-balancing (unlike a natural unbalanced state that occurs over time).

3.2   Post-Attack Damage: Attack on Peacock Hashing

As already explained, [3] showed that when the keys are concentrated in the backup tables (the table is unbalanced) the discard probability increases. The malicious user can artificially create an extreme case of this scenario by flooding the backup tables. A simple example of such an attack can be done by inserting K keys that all hash into the same bucket b at T_1. Every inserted key (except possibly for the first one) will collide with an existing key in b and then, according to the insertion algorithm, will be rehashed into a bucket in one of the backup tables. After the attack has ended the table is unbalanced, because the keys inserted by the malicious user are concentrated in the backup table while almost none are in the main table. This causes *Post-Attack* damage, measured by the increase in the discard probability.

Normally, Peacock hash table maintains a desired low discard probability by limiting the maximal load in the table. Nevertheless, an attacker using the above

sophisticated attack algorithm can cause the discard probability to exceed its
desired value by inserting only a small number of keys that do not cause the table
to reach its maximal load. Unlike a Peacock table which became unbalanced
naturally, a re-balancing routine cannot bring items back into the main table
after such an attack, because the bucket to which they all hash in T_1 can contain
only one of them but not all. So not only the table becomes unbalanced, there
is also no way to re-balance it until the keys are flushed out from the table or
the table is reconstructed with a new set of hash functions[2].

Denote the set of buckets to which a key is hashed in all sub-tables as the
pool of the key in the table. The insertion algorithm discards a new key only if
all the buckets in its pool are already occupied. Every key is hashed into one
bucket in each table, therefore, a bucket b in a table of size M_i, belongs to the
pools of $1/M_i$ of the keys in the key space. In simple words, a key placed in
a high table T_i "gets in the way" of more potential keys than if it was placed
in a bucket in a lower table T_j ($j < i$) (since $1/M_i > 1/M_j$). Therefore, the
basic idea behind the attack is to insert keys that will be placed in the upper
tables. Following is the formal description of the attack algorithm, generalizing
the simple attack described earlier. An attack in depth j is an attack to which
the attacker inserts keys that not only hash into the same bucket in T_1, but also
hash into the same buckets in $T_2, ..., T_j$. This will lead to the flooding of the
most upper tables $T_{j+1}, ..., T_d$. That is, the simple attack we described above is
an attack in depth 1. Due to lack of space, the complexity of finding the keys
is excluded from this work (and can be found in [11]). Note just that for large
tables (M) and for deeper attacks (larger j) - it is harder for the attacker to find
the keys for the attack.

3.3   Post-Attack Damage: Vulnerability of Peacock Hashing

We use the Vulnerability factor (described in Section 2) to measure the proportion between the increase in the discard probability caused by additional keys
inserted by an attacker and regular users. Let DP_I, DP_R and DP_A ('I' - Initial,
'R' - Regular, 'A' - Attacker) be the expected discard probabilities of newly
inserted keys at the following states: 1. DP_I - when the load in the table is α,
before any additional key is inserted; 2. DP_R - after K regular (random) additional keys were inserted; 3. DP_A - after K additional keys were inserted by a
sophisticated attacker. Note that due to their length, we exclude from this work
the proofs and the full discussion of the following claims and they can be found
in our technical report [11].

Lemma 1. *The drop probability after regular key insertion is approximated by*

$$DP_R(K) \sim (\alpha + R^{IN}/M)^d, \qquad (3)$$

where $R^{IN} = \sum_{s=1}^{K} p_s$, $p_1 = 1 - \alpha^d$ *and* $p_i = 1 - (\alpha + \sum_{s=1}^{i-1} p_s/M)^d$.

---

[2] Which if possible at all, will undoubtedly consume a huge amount of resources.

Lemma 2. *The drop probability after a sophisticated attack is approximated by*

$$DP_A(K) \sim [\prod_{l=1}^{j}(\alpha + (1 - \alpha)/M_l)](\alpha + A^{IN}/M')^{d'}, \tag{4}$$

where $A^{IN} = \sum_{s=1}^{K'} p'_s$, $K' = K - \lfloor(1 - \alpha)j\rfloor$, $M' = \sum_{i=j+1}^{d} M_i$, $d' = d - j$, $p'_1 = 1 - \alpha^{d'}$ *and* $p'_i = 1 - (\alpha + (\sum_{s=1}^{i-1} p'_s)/M')^{d'}$.

Theorem 1. *The vulnerability of the discard probability is given by*

$$V_{DP}(K) = \frac{DP_A(K) - DP_I}{DP_R(K) - DP_I}, \tag{5}$$

where $DP_I = \alpha^d$.

R^{IN} (Eq. 3) and A^{IN} (Eq. 4) can be roughly described as the number of keys that remain in table after the K keys were inserted by regular users and an attacker, respectively. Practically, the values of R^{IN} and A^{IN} are very close to each other and to K. The major factor that makes the value of $DP_A(K)$ (Eq. 4) significantly higher than $DP_R(K)$ (Eq. 3) is that R^{IN} is divided by M while A^{IN} is divided only by M'. This is because keys inserted by attacker are spread among the M' of the attacked backup tables while keys inserted by regular users are spread among all the M buckets in the table. Since $M_i = r^{d-i}$, $M' = \sum_{i=j+1}^{d} M_i$ is only a small fraction of $M = \sum_{i=1}^{d} M_i$ and this is the major factor behind the differences between $DP_A(K)$ and $DP_R(K)$ as seen in Figure 2(a).

Evaluation of the Vulnerability and the Analysis. To evaluate the vulnerability of the system as a function of the system parameters and to evaluate the quality of the approximations (lemmas 1 and 2) we next conduct a set of simulations. The simulations we conducted follow a scenario in which the system designer examines the option of using Peacock Hashing for hardware that can support approximately 10^5 buckets in the table. Building a table consisted of $d = 5$ sub-tables with sub-tables proportion of $r = 10$ results in a table of size $M = 11,111$ buckets. As mentioned before, in Peacock Hashing (as well in Cuckoo Hashing and other modern hashing schemes) the probability for a key to be dropped during an insertion increases with the load in the table. Therefore, the maximal load in the table is decided by the *Acceptable Loss Fraction*, that is, the maximal percentage of inserted keys that the system can afford to lose. It is important to note that *Discard Probability* (Figure 2(a)) that is used to measure the vulnerability and *Loss Fraction* that is used to set the maximal load are two different metrics. *Discard Probability* measures the probability to drop an inserted regular key *after* additional were inserted by malicious or regular users while *Loss Fraction* measures the percentage of the inserted (regular) keys that are dropped during an insertion and is used to set the maximal load of the table. In this example, we assume that the desired maximal *Loss Fraction* allowed is 1%. Our simulations showed that such a Peacock table with $d = 5$, $r = 10$ and $M = 11,111$ is suitable for the insertion of up to $0.3M = 3,333$ keys (the table utilization is 30%) before exceeding loss fraction of %1.

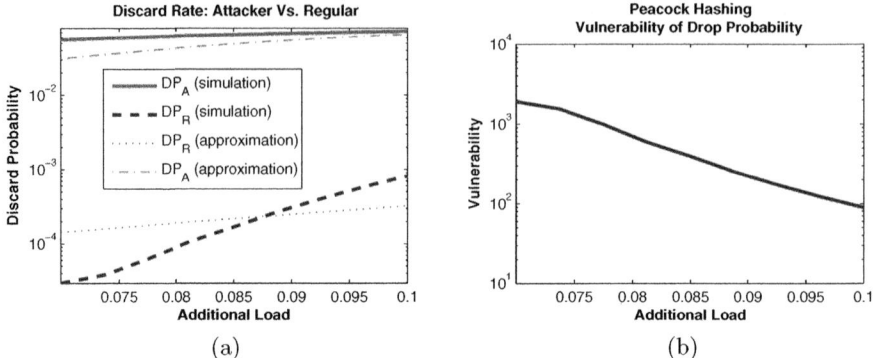

Fig. 2. Figure (a): The discard probability after insertions by regular (DP_R) and malicious (DP_A) users as a function of the load they add (K/M). Figure (b): The vulnerability simulation results (V) of a table with proportion $r = 10$ and $d = 4$ with an existing load of 10% as a function of the additional load.

In the simulations, the attack was conducted on a table with an existing load $\alpha = 0.1$ and the attack by the malicious user is in depth $j = 1$. Recall that an attack in depth $j = 1$ on target the upper 4 tables that hold together only $1,111/11,111 = 9.9\%$ of the buckets that some of which are already occupied prior the attack. Therefore, an attacker would insert no more than 10% of additional load. Therefore, the range of the x-axis in Figures 2(a) and (b) was chosen accordingly. Note that the additional 0.1 load together with the existing load $\alpha = 0.1$ does not exceed the maximal load in the table which is 0.3. We can see in Figure 2(a) that when the additional load is 7%, the discard probability remains very low (below 0.1%), while the sophisticated attack causes the discard probability to increase to values between 5.5% (when the additional load is 0.07) and 7.2% (when the additional load is 0.1), a discard probability that is achieved by regular insertions only with load which is 3.5 times larger. In addition we can see that, as expected, the approximation curves in Figure 2(a) reflect the behavior of DP_R and DP_A although do not imitate them precisely (see [11] for more details).

We can see a significant difference in the discard probability after a malicious attack and after regular insertions. This difference is expressed by the extremely high vulnerability values, depicted in Figure 2(b), where the results show that the discard probability after the attack has ended caused by the attacker is between 100 and 1,800 times larger than the discard probability after the insertion of the same load of keys by regular users. This drives the discard probability of the hash table far beyond the discard probability in which it is assumed to be operating. This result emphasizes the fundamental vulnerability of Hashing schemes, such as the Peacock Hashing, that dedicate specific range of buckets (the upper sub-tables) for collision resolution.

3.4 Resilience to In-Attack Damage (and Improving Performance Using Bitmaps)

As already mentioned in Section 1, we analyze an attack focused at creating *post-attack* damage. We avoided analyzing attacks aiming at *in-attack* damage by inserting keys that require excessive number of memory accesses during the attack, since we believe such attacks can be countered easily. Our suggestion is to keep a bitmap summary of the occupied buckets for every *backup* table. Since the total size of the backup tables is small (about 10% of M when $r = 10$), these bitmaps are compact enough to be stored in the fast on-chip memory. Hence, the complexity of accessing a key is negligible. Then, when handling an insertion of a new key which cannot find its place in the main table, its final bucket (in a backup table) can be found directly by checking its hash values against the fast bitmap summary. This way, no insertion has to probe more than two buckets (one in the main table, and one in the final destination). Note that it will not cause a mistaken insertion of a key that already exists, since (as mentioned earlier) in addition to the bitmap summary, there are also on-chip *key* summaries (commonly implemented in bloom filters [3]) which are used to make sure the key does not already exist in the sub-tables (before checking the bitmap summary to locate a free bucket for the key).

4 Cuckoo Hashing

4.1 Algorithm Description

According to the original definition by Pagh and Rodler [4] a Cuckoo hash table is made of two sub-tables, equal in size. Every key k hashes into one bucket in each of them using two hash functions $h_1(k)$, $h_2(k)$. If during an insertion, both of the possible buckets are occupied, the key is placed in one of them, causing the key placed in the occupied bucket to *move* to its alternative bucket in the same manner. Therefore, every insertion of a new key consists of a series of one or more *moves*. The complexity of a new insertion is measured by the number of moves it triggers. The expected complexity of an insertion is proved to be bounded by $O(1)$ as long the maximal capacity of the table is not reached. Figure 3 depicts an example of an insertion of k_2 into a Cuckoo hash table with 4 sub-tables. When k_2 is inserted, the buckets into which it hashes (dark left arrows) are all occupied. Then, the insertion algorithm chooses to eject k_1 (from T_4) and place k_2 in its place. k_1 is then relocated to an alternative free bucket to which it hashes in T_3. Note that as already explained in Section 3.1, we discuss the model in which a bucket can hold only one key.

For the original Cuckoo hash table consisting of two sub-tables, Pagh and Rodler [4] showed that the complexity of an insertion is $O(1 + 1/\epsilon)$ where $M = 2N(1 + \epsilon)$, M is the total number of buckets (in both tables) and N is the maximal number of keys the table is meant to hold. In [13] Fotakis, Pagh et al. generalized Cuckoo Hashing to *d-ary Cuckoo Hashing* where $d \geq 2$ sub-tables are used. They showed how N keys can be stored in $M = (1 - \epsilon)N$ buckets

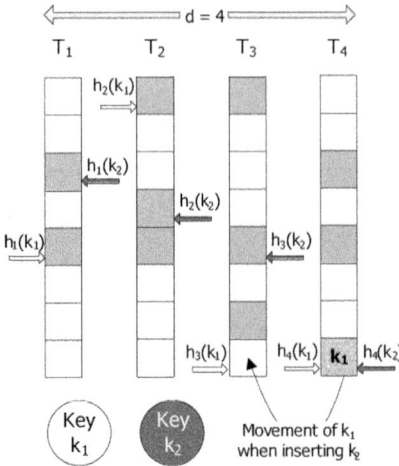

Fig. 3. A move of key k_1 during the insertion of k_2 into a Cuckoo hash table with 4 sub-tables. The white (right) and the dark (left) arrows mark the hash values of k_1 and k_2 in the different sub-tables, respectively.

for any constant $\epsilon > 0$. They showed that Search and Delete operations take[3] $O(ln(1/\epsilon))$ and proved the generalized Cuckoo Hashing has a constant amortized insertion time. Following [13], Frieze and Mitzenmacher [14] suggested a more efficient insertion method with a polylogarithmic upper bound. In later studies (such as [14], [13] and [15]) different insertion algorithms have been proposed. They mainly differ in the way they choose the key that will be relocated (in order to free a bucket for an inserted key that all his possible d locations are occupied). Note that the attack efficiency is independent of the way the key (to be moved) is chosen, hence we do not discuss here the variations of the insertion algorithm and for our work one can assume the moved key is chosen randomly. In addition, since the exact insertion complexity of the various insertion algorithms is not fully analyzed, we use the fact that it is proved to take an amortized $O(1)$ time when we approximate the vulnerability. In addition, we use simulations to give precise results for selected examples.

4.2 In-Attack Damage: Attack on Cuckoo Hashing

The basic idea behind the attack is to insert K keys that all hash into the same small set of buckets B, such that $K > |B|$. Except for the first $|B|$ keys, every attack key causes an insertion loop in which every move triggers another. Formally, the attacker inserts K keys such for every key k, $\{h_i(k)\}_{i=1}^{d} \subset B$ where B is a set of buckets such that $K > |B|$. Such attack creates insertion loops and cause the insertion algorithm to require excessive number of memory accesses. Note that the size of B can be is as low as d (when B contains exactly one bucket

---

[3] Their experiments showed that 4 probes suffice for $\epsilon \approx 0.03$.

from each sub-table) but a large bucket set B allows the attacker to find keys for the attack more easily. Note that a general algorithm that will detect every possible loop can be proved impractical, especially in hardware. Therefore, the most popular approach to address this issue is to limit the number of moves to a predefined fixed value W since their vast majority is very low[4].

In our technical report [11] we describe the algorithm used by the attacker to find the keys for the attack and its complexity is analyzed. Our main result on this subject is that it is harder to find Cuckoo attack keys than to find the Peacock attack keys[5]. However, on the other hand, in contrast to the Peacock attack, the attacker can use the same keys over and over again and sustain his attack until it is mitigated, if at all.

4.3   In-Attack Damage: Vulnerability of Cuckoo Hashing

Using the vulnerability measure described in Section 2, we now evaluate the system vulnerability due to new-key insertion complexity (measured by the number of moves it triggers). Let IC_R and IC_A ('R' - Regular, 'A' - Attacker) denote the overall number of operations performed during the insertion of K keys, by the attacker and by regular users, respectively. Note that regardless of the strategy of the attacker, the vulnerability cannot exceeds W ($V_{IC(K)} = \frac{IC_A(K)}{IC_R(K)} \leq W$) since trivially $IC_R(K) >= K$ and since the system enforces $IC_A(K) <= KW$.

Theorem 2. *The system vulnerability due to insertion complexity is*

$$V_{IC(K)} = \frac{IC_A(K)}{IC_R(K)} = |B|/K + (1 - |B|/K)\frac{W}{c}, \tag{6}$$

where $c \geq 1$ is the average time complexity of a regular key insertion and B is the group of buckets into which all the attack keys hash.

Proof. Since the insertion time of a random regular key has an amortized $O(1)$ complexity we can conclude that $IC_R(K) = Kc$ where $c \geq 1$ is a constant (very close to 1 in practice) denoting the insertion complexity of a random key. The first $|B|$ keys inserted by the attacker might not cause insertion loops since they all can possibly find an empty bucket. Therefore the complexity of their insertion is $|B|c$. Each of the remaining $K - |B|$ keys inserted by the attacker triggers an insertion loop that is terminated only after W moves. Then $IC_A(K) = |B|c + (K - |B|)W$. To conclude, the values of $IC_R(K)$ and $IC_A(K)$ lead to the result in Eq. 6. □

Evaluation of the Vulnerability and the Analysis. In the same way as we did for Peacock Hashing, we aim to construct (and then attack) hash table that can hold up to 3333 keys with an *Acceptable Loss Fraction* of 1%. As a

---

[4] W suppose to be set to $alog(n)$ where n is the number of keys the table can hold and a is appropriately chosen constant [14].

[5] Since keys have to hash to the same buckets in all sub-tables.

system designer would do, we used simulations to measure the loss fraction with different values of d, W and M and decided to use a system with $d = 3$, $W = 4$ and $M = 6000$. Note that higher values of d and W allow higher utilization of the table[6]. That is, smaller value of M is required in order to accommodate up to 3333 keys with the required loss fraction.

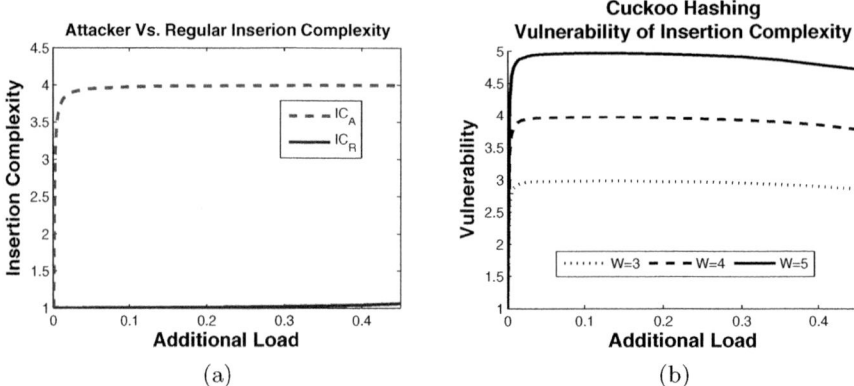

Fig. 4. Plot (a): The average insertion complexity of keys inserted by an attacker $IC_A(K)/K$ and a regular user $IC_R(K)/K$ in a system with $W = 4$. Plot (b): The vulnerability $(V = IC_A(K)/IC_R(K))$ as a function of the additional load (K/M). In both plots the existing load is $\alpha = 0.1$.

Figures 4(a) and 4(b) depict the simulation results on a table with (arbitrarily chosen) existing load of $\alpha = 0.1$. The x-axis in both figures corresponds to the additional load of keys that was added to the table. The x-axis ends at 0.45 since the maximal allowed load in the table is $3333/6000 = 0.55\%$ (and it already contains 0.1).

According Theorem 2 the vulnerability is mainly defined by W/c (since $|B|/K$ is expected to be very small, especially when $|B| = d$). Note that using $c = 1$ gives an upper bound on the estimated vulnerability which is then approximated by W. As we can see in Figure 4(a), the lower curve which depicts $IC_R(K)/K$ (the average complexity of regular insertion) - c is indeed very close to 1 in the simulation results. Therefore, as also shown in Figure 4(b), setting the moves limit W practically decides the vulnerability of the system. In addition, one can observe that, unlike Peacock Hashing, in Cuckoo Hashing the size of the table M has no role in the vulnerability (Theorem 2).

The following table summarize the results so far, including those from the attack feasibility analysis which is excluded from this paper and can be found in a technical report [11]. The arrows marks how different properties of a Cuckoo hash table are affected when increasing W, d and M. ⇑, ⇓ and ⇔ mark that a property is increased, decreased or do not chance (respectively).

---

[6] However, high values of d and W also increase the complexity of the different operations in the table.

	High W	High d	High M
Table Utilization	⇑	⇑	⇓
Vulnerability	⇑	⇓	⇔
Attack Feasibility	⇔	⇓	⇓

As explained above, from the efficiency point of view the system designer has to choose between higher complexity of the hash operations (high values of d and W) and lower utilization (large M). Our work, shows through the above analysis and simulation results the *security implications* of setting W and d. The results show that W is a key factor in the vulnerability of Cuckoo Hashing. Therefore, from the vulnerability point of view, it is preferable to increase the number of sub-tables d in order to keep the moves limit W as low as possible. Increasing d not only decreases the vulnerability but also decrease the feasibility by forcing the attacker to invest more effort in finding a suitable attack keys set (see [11]).

4.4   Resilience to Post-Attack Damage

There is no general way to cause post attack damage (In contrast to Peacock Hashing, see Section 3.2) since there is no specific layout of elements in the table that Cuckoo is vulnerable to more than other; This is true since it is impractical to assume the attacker knows where new keys will hashed to after the attack has ended.

5   Summary

In this work we exposed the weak points of the Peacock and Cuckoo Hashing. We showed that Peacock is resilient against in-attack damage. Nevertheless, we showed that it is highly vulnerable to an attack that aims at driving the system to high post-attack discard rates; such an attack can increase these rates by a factor of $100 - 1,800$ in comparison to normal behavior. For Cuckoo hashing we showed that an attacker can slow the system by inserting keys that require 4 times more memory accesses than regular keys in a typical settings. We also provided simulation results for a use case in which a system designer plans to design a Cuckoo and Peacock hash tables which comply with the same requirements.

References

1. Smith, R., Estan, C., Jha, S.: Backtracking Algorithmic Complexity Attacks Against a NIDS. In: Proceedings of ACSAC Annual Computer Security Applications Conference (2006)
2. Crosby, S., Wallach, D.: Denial of Service via Algorithmic Complexity Attacks. In: Proceedings of USENIX Security Symposium (2003)

3. Kumar, S., Turner, J., Crowley, P.: Peacock Hash: Fast and Updatable Hashing for High Performance Packet Processing Algorithms. In: Proceedings of IEEE INFOCOM (2008)
4. Pagh, R., Rodler, F.: Cuckoo Hashing. Journal of Algorithms (2001)
5. Mitzenmacher, M., Broder, A.: Using Multiple Hash Functions to Improve IP Lookups. In: Proceedings of IEEE INFOCOM (2000)
6. Song, H., Dharmapurikar, S., Turner, J., Lockwood, J.: Fast Hash Table Lookup Using Extended Bloom Filter: An Aid to Network Processing. In: Proceedings of ACM SIGCOMM (2005)
7. Waldvogel, M., Varghese, G., Turner, J., Plattner, B.: Scalable High Speed IP Routing Lookups. In: Proceedings of ACM SIGCOMM (1997)
8. Thinh, T., Kittitornkun, S.: Massively Parallel Cuckoo Pattern Matching Applied for NIDS/NIPS. In: Proceedings of IEEE DELTA (2010)
9. Kirsch, A., Mitzenmacher, M., Varghese, G.: Hash-Based Techniques for High-Speed Packet Processing. Algorithms for Next Generation Networks. Springer (2010)
10. Ben-Porat, U., Bremler-Barr, A., Levy, H.: Evaluating the Vulnerability of Network Mechanisms to Sophisticated DDoS Attacks. In: Proceedings of IEEE INFOCOM (2008)
11. Ben-Porat, U., Bremler-Barr, A., Levy, H., Plattner, B.: On the Vulnerability of Hardware Hash Tables to Sophisticated Attacks. Technical Report (2011), http://www.faculty.idc.ac.il/bremler/
12. Kirsch, A., Mitzenmacher, M., Wieder, U.: More Robust Hashing: Cuckoo Hashing with a Stash. In: Halperin, D., Mehlhorn, K. (eds.) ESA 2008. LNCS, vol. 5193, pp. 611–622. Springer, Heidelberg (2008)
13. Fotakis, D., Pagh, R., Sanders, P., Spirakis, P.: Space Efficient Hash Tables with Worst Case Constant Access Time. In: Alt, H., Habib, M. (eds.) STACS 2003. LNCS, vol. 2607, pp. 271–282. Springer, Heidelberg (2003)
14. Frieze, A., Melsted, P., Mitzenmacher, M.: An Analysis of Random-Walk Cuckoo Hashing. In: Dinur, I., Jansen, K., Naor, J., Rolim, J. (eds.) APPROX and RANDOM 2009. LNCS, vol. 5687, pp. 490–503. Springer, Heidelberg (2009)
15. Kirsch, A., Mitzenmacher, M.: The Power of One Move: Hashing Schemes for Hardware. IEEE/ACM Transactions on Networking 18(6), 1752–1765 (2010)
16. Estan, C., Keys, K., Moore, D., Varghese, G.: Building a Better NetFlow. In: Proceedings of ACM SIGCOMM (2004)

A   Appendix

Glossary of Notations	
T_i	Sub-table # i
d	Number of sub-tables
M_i	Number of buckets in T_i ($M_i = |T_i|$)
M	Total number of buckets ($M = \sum_{i=1}^{d} M_i$)
α	The existing load (keys/buckets) in the table (prior an attack)
K	Number of additional keys inserted by a malicious/regular user
r	$r = M_i/M_{i+1}$ (in Peacock Hashing)
j	Attack depth (in Peacock Hashing)
W	The maximal number of moves allowed (in Cuckoo Hashing)

Degree and Principal Eigenvectors in Complex Networks

Cong Li, Huijuan Wang, and Piet Van Mieghem

Faculty of Electrical Engineering, Mathematics and Computer Science,
Delft University of Technology, P.O. Box 5031, 2600 GA Delft, The Netherlands
{Cong.Li,H.Wang,P.F.A.VanMieghem}@tudelft.nl

Abstract. The largest eigenvalue λ_1 of the adjacency matrix powerfully characterizes dynamic processes on networks, such as virus spread and synchronization. The minimization of the spectral radius by removing a set of links (or nodes) has been shown to be an NP-complete problem. So far, the best heuristic strategy is to remove links/nodes based on the principal eigenvector corresponding to the largest eigenvalue λ_1. This motivates us to investigate properties of the principal eigenvector x_1 and its relation with the degree vector. (a) We illustrate and explain why the average $E[x_1]$ decreases with the linear degree correlation coefficient ρ_D in a network with a given degree vector; (b) The difference between the principal eigenvector and the scaled degree vector is proved to be the smallest, when $\lambda_1 = \frac{N_2}{N_1}$, where N_k is the total number walks in the network with k hops; (c) The correlation between the principal eigenvector and the degree vector decreases when the degree correlation ρ_D is decreased.

Keywords: networks, spectral radius, principal eigenvector, degree, assortativity.

1 Introduction

Dynamic phenomena occurring on networks are affected by the structure of networks, e.g., the absence of epidemic thresholds in large scale free networks [2][3][6], the effect of the degree correlations on the percolation of networks [8]. The largest eigenvalue $\lambda_1(A)$ of the adjacency matrix A, called the spectral radius of the graph, has been shown to play an important role in dynamic processes on graphs, such as SIS (susceptible-infected-susceptible) virus spread [12] and the Kuramoto type of synchronization process of coupled oscillators [11] on a given network topology. For instance, in a SIS spreading model, the epidemic threshold $\tau_c \simeq \frac{1}{\lambda_1(A)}$ separates two different phases of a dynamic process on a network: if the spreading rate τ is above the threshold, the infection spreads and becomes persistent in time; where $\tau < \tau_c$, the infection dies out exponentially fast [10][12]. In the past decade, researches have focused on how topological changes, such as link (or node) removal, may alter the spectral radius. Milanese *et al.* [7] studied the dynamical importance of the structural perturbation by

R. Bestak et al. (Eds.): NETWORKING 2012, Part I, LNCS 7289, pp. 149–160, 2012.
© IFIP International Federation for Information Processing 2012

removing one node or link. Van Mieghem *et al.* [15] have proved that to minimize the largest eigenvalue by removing a set of links or nodes is a NP-hard problem and have shown that the best strategy so far is based on the components of the principal eigenvector x_1, which underlines the importance of the principal eigenvalue in characterizing the influence of link/node removal on the spectral radius. Our main objective is to investigate the topological meaning of x_1, which has been rarely studied. Especially, we aim to understand the relation between x_1 and the degree vector/sequence[1] d, the computationally simplest and mostly studied property of a network.

The degree correlation, also called the assortativity ρ_D is computed as the linear correlation coefficient of the degree of nodes connected by a link. It describes the tendency of network nodes to connect preferentially to other nodes with either similar (when $\rho_D > 0$) or opposite (when $\rho_D < 0$) properties i.e. degree [9]. The assortativity was widely studied after it was realized that the degree distribution alone provides an insufficient characterization of complex networks. Networks with the same degree distribution may still differ significantly in various topological features. Degree-preserving rewiring [13] allows us to either increase or decrease the assortativity of a network without changing the degree of each node. The relation between the principal eigenvector and the degree vector is systematically investigated in networks with various degree distributions and degree correlations.

Section 2 illustrates the importance of the principal eigenvector in characterizing the influence of link/node removal on the spectral radius by two key theories developed in our early work and further simulations. Subsequently, we explore the properties of the principal eigenvector and the relation between the (normalized) degree vector and the principal eigenvector in networks with different degree correlation and with the degree distribution derived from the Erdös-Rényi random graphs[2] [4], the Bárabasi-Albert graphs[3] [1], and real-world networks (see Section 3). Our major contributions are: (a) the average of the components in the principal eigenvector $E[x_1]$ is shown and explained to decrease with the assortativity ρ_D; (b) the difference between the principal eigenvector and the degree vector is proved to be the smallest, when $\lambda_1 = \frac{N_2}{N_1}$, where N_k is the total number of walks with k hops in a network and (c) the correlation between principal eigenvector and the degree vector decreases as the assortativity ρ_D is decreased. These finds provide essential inspiration on when the degree vector well approximates the principal eigenvector. Finally, we illustrate the possibility to approximate the principal eigenvector based strategy to minimize the largest

---

[1] The degree vector/sequence is composed of the degree of each node, following the same ordering as the principal eigenvector.

[2] An Erdős-Rényi random graph can be generated from a set of N nodes by randomly assigning a link with probability p to each pair of nodes.

[3] A Bárabasi-Albert graph starts with m nodes. At every time step, we add a new node with m links that connect the new node to m different nodes already present in the graph. The probability that a new node will be connected to node i in step t is proportional to the degree $d_i(t)$ of that node. This is referred to as preferential attachment.

eigenvalue by removing links/nodes by its corresponding degree based strategy (see Section 4), which can be well explained by the findings in early sections.

2   The Decrease of the Spectral Radius

We consider a network as a graph $G = (\mathcal{N}, \mathcal{L})$, where $\mathcal{N}$ is the set of nodes and $\mathcal{L}$ is the set of links. The number of nodes is denoted by $N = |\mathcal{N}|$ and the number of links is represented by $L = |\mathcal{L}|$. The graph G can be represented by the $N \times N$ adjacency matrix A, consisting of elements a_{ij} that are either one or zero depending on whether there is a link between nodes i and j. The eigenvalues of the adjacency matrix are ordered as $\lambda_N \leq \lambda_{N-1} \leq \cdots \leq \lambda_1$, where λ_1 is the spectral radius and the corresponding eigenvector x_1 is called the principal eigenvector. Let $\mathcal{L}_m$ (or $\mathcal{N}_m$) denote the set of the m links (or nodes) that are removed from G, and $G_m(\mathcal{L}) = G \backslash \mathcal{L}_m$ (or $G_m(\mathcal{N}) = G \backslash \mathcal{N}_m$) is the resulting graph after the removal of m links (or nodes) from G. We denote the adjacency matrix of $G_m(\mathcal{L})$ (or $G_m(\mathcal{N})$) by $A_m(\mathcal{L})$ (or $A_m(\mathcal{N})$), which is still a symmetric matrix.

Theorem 1. *For any graph G and graph $G_m(\mathcal{L}) = G \backslash \mathcal{L}_m$, by removing m links from G, it holds that*

$$2 \sum_{l \in \mathcal{L}_m} (w_1)_{l^+} (w_1)_{l^-} \leq \lambda_1(A) - \lambda_1(A_m(\mathcal{L})) \leq 2 \sum_{l \in \mathcal{L}_m} (x_1)_{l^+} (x_1)_{l^-} \quad (1)$$

where x_1 and w_1 are the principal eigenvectors of A and $A_m(\mathcal{L})$ corresponding to the largest eigenvalues $\lambda_1(A)$ and $\lambda_1(A_m(\mathcal{L}))$, respectively, and where a link l joins the nodes l^+ and l^-.

Proof. [15]   □

The decrease of the largest eigenvalue $\lambda_1(A) - \lambda_1(A_m(\mathcal{L}))$ tends to be larger if the upper bound $2 \sum_{l \in \mathcal{L}_m} (x_1)_{l^+} (x_1)_{l^-}$ is larger. This motivates the principal eigenvector strategy to minimize the largest eigenvalue: removing the set of links that maximizes $2 \sum_{l \in \mathcal{L}_m} (x_1)_{l^+} (x_1)_{l^-}$. Moreover, when only one link is removed, removing the link with the maximum $(x_1)_{l^+} (x_1)_{l^-}$, maximizes not only the upper bound of (1), but likely the lower bound as well, since w_1 is close to x_1 in this case. This eigenvector strategy performs almost optimally in this situation.

Theorem 2. *For any graph G and graph $G_m(\mathcal{N}) = G \backslash \mathcal{N}_m$, by removing m nodes from G, it holds that*

$$0 \leq \lambda_1(A) - \lambda_1(A_m(\mathcal{N})) \leq 2 \sum_{n \in \mathcal{N}_m} (x_1)_n^2 \lambda_1(A) - \sum_{j \in \mathcal{N}_m} \sum_{i \in \mathcal{N}_m} a_{ij}(x_1)_i(x_1)_j \quad (2)$$

where x_1 is the principal eigenvectors of A corresponding to the largest eigenvalues $\lambda_1(A)$. In particular, if $m = 1$, then

$$0 \leq \lambda_1(A) - \lambda_1(A_1(\mathcal{N})) \leq 2(x_1)_n^2 \lambda_1(A) \quad (3)$$

where n is the node removed.

Proof. [5]   □

Theorem 2 implies that the decrease of spectral radius by removing a node or a set of nodes is strongly related to the principal eigenvector components corresponding to the removed nodes. Motivated by Theorem 2, the eigenvector based one node removal strategy to minimize the largest eigenvalue simply removes the node with the largest principal eigenvector component $(x_1)_n$.

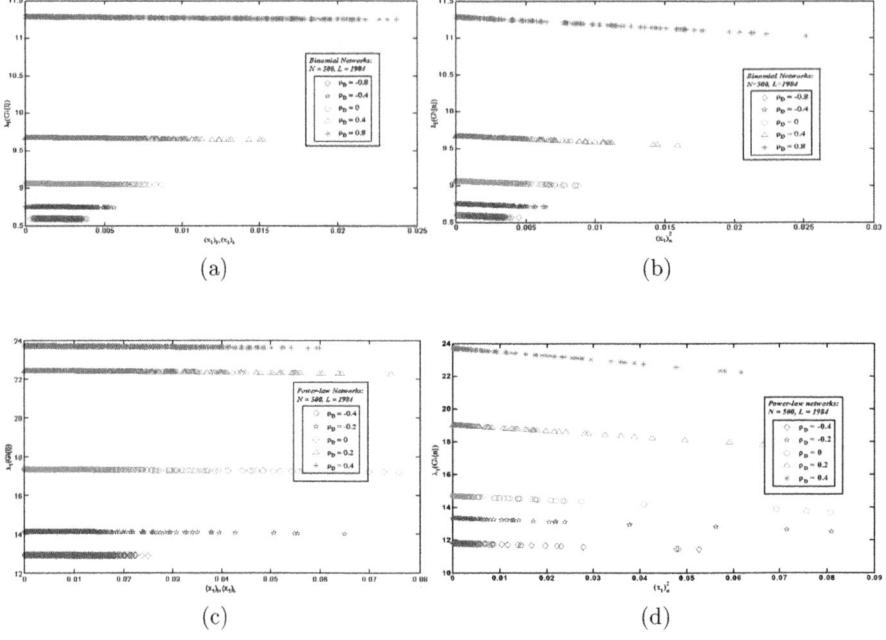

Fig. 1. The spectral radius of graphs by removing a link (or node) as a function of corresponding components in principal eigenvector (a), (b)in Binomial graphs,(c), (d) in power-law graphs

We perform further simulations to illustrate the importance of the principal eigenvector components in characterizing the influence of the link/node removal on λ_1. We deduce networks with different assortativities but with a given degree vector, which may follow a binomial or power-law degree distribution. Upon each network, we try all possible one link (or node) removal and examine the largest eigenvalue $\lambda_1(G\backslash(l))$ (or $\lambda_1(G\backslash(n))$) after removing one link (or node) as a function of $(x_1)_{l+}(x_1)_{l-}$ (or $(x_1)_n^2$) corresponding to the link (or node) removed. By the Perron-Frobenius theorem [14], all components of x_1 and w_1 are non-negative (positive if the corresponding graph is connected). Interestingly, $\lambda_1(G\backslash(l))$ (or $\lambda_1(G\backslash(n))$) decreases linearly as a function of increasing $(x_1)_{l+}(x_1)_{l-}$ (or $(x_1)_n^2$), as shown in Fig. 1. In other words, the spectral radius will be decreased more if the link (or node) removed has a larger $(x_1)_{l+}(x_1)_{l-}$ (or $(x_1)_n^2$).

3   Relation between the Principal Eigenvector and the Degree Vector

In view of the importance of the principal eigenvector in characterizing the influence of link/node on the spectral radius, in this section, we explore how the average $E[x_1]$ as well the variance of x_1 changes with the assortativity ρ_D when the degree vector, which may follow the degree distribution of network models or of real-world networks, remains the same. Moreover, we explore the difference and the linear correlation coefficient between the principal eigenvector and the degree vector, the simplest and mostly studies network metric, which as well provides important insights on under which condition the degree vector/sequence well approximates the principal eigenvector.

3.1   Properties of the Principal Eigenvector

Two types of degree distributions have been so far widely studied: the binomial and power-law degree distribution. The binomial degree distribution is a characteristic of an Erdős-Rényi random graph $G_p(N)$, which has N nodes and any two nodes are connected independently with a probability p. Such a random construction leads to a zero assortativity as proved in [13]. However, the class of graphs $G(N, p)$ with the same binomial degree distribution $\Pr[D_G = k] = \binom{N-1}{k}p^k(1-p)^{N-1-k}$ as Erdős-Rényi random graphs $G_p(N)$ and obtained, for instance, by degree-preserving rewiring feature an assortativity that may vary within a wide range. The power-law degree distribution $\Pr[D = k] = ck^{-\alpha}$, where $c = 1/\sum_{k=1}^{N-1}k^{-\alpha}$ has been widely observed in real-world networks. Similarly, graphs with a given power-law degree distribution, for example, generated by the Barabási-Albert power model [1] can be altered by the degree-preserving rewiring to obtain different assortativity.

We explore the principal eigenvector components (see Figure 2) as well as its average $E[x_1]$ (see Figure 3) in graphs with the same degree distribution (i.e. binomial or power-law) but with different assortativities ρ_D obtained by degree-preserving rewiring. Figure 2 shows that the variance of the principal eigenvector increases with assortativity ρ_D. Furthermore, as shown in Figure 3, $E[x_1]$ decreases with the increase of assortativity ρ_D. Similarly, we consider a set of 11 real-world networks. We apply degree-preserving rewiring to each real-world network to derive network instances with different assortativity. In other words, we derive a class of networks that possess the same degree distribution as a real-world network but different assortativities. Interestingly, we observe the same, $E[x_1]$ decreases with increasing assortativity (see Figure 3(b)).

The decrease of $E[x_1]$ and the increase of the variance of the principal eigenvector components with increasing assortativity can be qualitatively explained as follows. As defined, the principal eigenvector x_1 corresponds to the largest eigenvalue λ_1 follows

$$\lambda_1(x_1)_j = \sum_{q=1}^{N} a_{jq}(x_1)_q, \qquad (4)$$

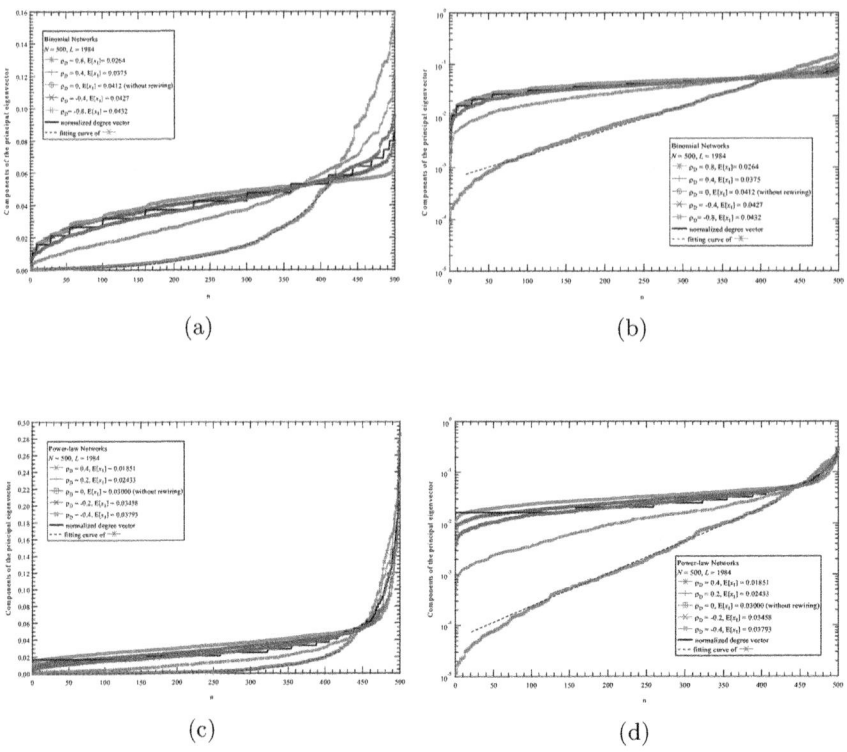

Fig. 2. The components of the principal eigenvector in increasing order. Images (a) (linear) (b) (semilogarithmic) are binomial graphs with different assortativity. Images (c) (linear) (d) (semilogarithmic) are power-law graphs with different assortativity.

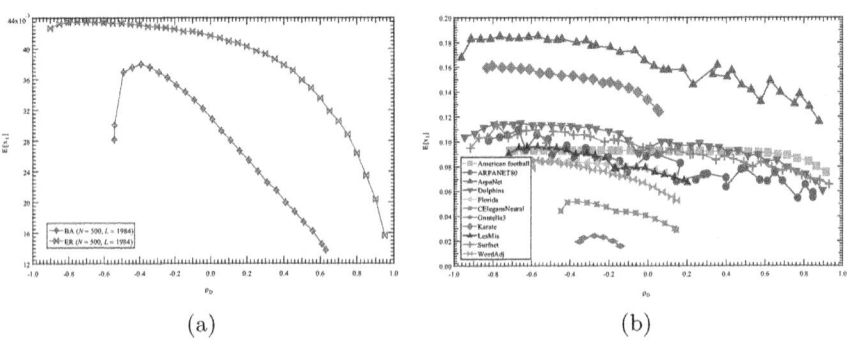

Fig. 3. The average of the components of the principal eigenvector versus the assortativity. (a) in binomial and power-law graphs (b) in network instances derived from real-world networks via degree-preserving rewirings.

where $a_{jq} = 1$ if q is a neighbor of node j, or else $a_{jq} = 0$. The j-th component of the principal eigenvector $(x_1)_j$ tends to be large if node j has a large degree (number of neighbors) or if the components corresponding to its neighbors are large. When ρ_D is large, high degree nodes prefer to link with other high degree nodes. In this case, a high degree node possesses a large number of neighbors, whose corresponding eigenvector components are again likely to be large, whereas a low degree node connects to a small number of neighbors, whose corresponding components tend to be small. Both a large variance in degree and a large assortativity ρ_D contribute to a large variance $Var[x_1]$ of the principal eigenvector x_1. This explains why variance $Var[x_1]$ of x_1 increases with ρ_D and with a given assortativity, the power-law graphs have a larger $Var[x_1]$ than the binomial graphs (see Figure 2). Furthermore, since $Var[X] = E[X^2] - (E[X])^2$ and $x_1^T x_1 = 1$,

$$E[x_1] = \sqrt{\frac{1}{N} - Var[x_1]}, \tag{5}$$

Correspondingly, both a large variance in degree and a large assortativity ρ_D contribute to a small $E[x_1]$ of the principal eigenvector x_1. Hence, $E[x_1]$ decreases with increasing ρ_D and tends to be smaller when the degree variance is larger. Moreover, considering Eq. (5), we can deduce the upper bound $E[x_1] \leq \frac{1}{\sqrt{N}}$.

Figure 2 compares as well the principal eigenvector x_1 with the normalized degree vector $\bar{d} = \frac{d}{\sqrt{d^T d}}$ in binomial graphs and power-law graphs ($N = 500$, $L = 1984$) with different assortativities. The components of x_1 and $\bar{d}$ are plotted in the order of increasing magnitude. The difference between x_1 and $\bar{d}$ is affected by ρ_D, which will be further explored in the following part.

3.2   Relation between Degree Vector and Principal Eigenvector

In this section, we investigate the relation between the principal eigenvector and the degree vector by their difference and linear correlation coefficient. The degree vector has to be first normalized to quantify its difference with the principal eigenvector. We propose two scalings of the degree vector $\bar{d} = \frac{d}{\sqrt{d^T d}}$ and $\tilde{d} = \frac{\alpha}{\lambda_1} d$, where α is a constant. The corresponding *difference vector* between x_1 and the scaled degree vector is $w = x_1 - \frac{d}{\sqrt{d^T d}}$ and $y = x_1 - \frac{\alpha}{\lambda_1} d$, respectively. The overall difference can be quantified by either the relative difference $u^T w$ (or $u^T y$) or the absolute difference $w^T w$ (or $y^T y$), actually, the square sum or the sum of the components in the difference vector respectively. The first scaling of the degree vector $\bar{d} = \frac{d}{\sqrt{d^T d}}$ aims to obtain the same norm for the the degree vector and the principal eigenvector: $\sqrt{d^T d} = \sqrt{x_1^T x_1} = 1$. The other $\tilde{d} = \frac{\alpha}{\lambda_1} d$ is motivated by $(x_1)_j = \frac{1}{\lambda_1} \sum_{r=1}^{N} a_{jr} (x_1)_r \leq \frac{d_i}{\lambda_1}$ and the constant α is determined (see Theorem 3) as the one minimize the absolute difference $y^T y$. Note that both linear scalings of the degree vector will not change the linear correlation coefficient between the principal eigenvector and the degree vector.

Theorem 3. *The absolute difference $w^T w$ (or $y^T y$) between the principal eigen-vector and the degree vector is the smallest ($w^T w = 0$ or $y^T y = 0$) when the spectral radius follows $\lambda_1 = \frac{N_2}{N_1}$, where N_k is the total number of k hop walks between any two nodes which can be the same.*

Proof. The absolute difference

$$w^T w = (x_1 - \frac{d}{\sqrt{d^T d}})^T (x_1 - \frac{d}{\sqrt{d^T d}}) = x_1^T x_1 - 2 \frac{d^T x_1}{\sqrt{d^T d}} + \frac{d^T d}{(\sqrt{d^T d})^2} = 2 - 2 \frac{d^T x_1}{\sqrt{d^T d}}.$$

(6)

Moreover, the generalized form of (4) for the k-th largest eigenvalue λ_k and the corresponding eigenvector x_k follow $(x_k)_j = \frac{1}{\lambda_k} \sum_{r=1}^{N} a_{jr} (x_k)_r = \alpha \frac{d_j}{\lambda_k} - \frac{1}{\lambda_k} \sum_{r=1}^{N} a_{jr} (\alpha - (x_k)_r)$, we will determine α so that $y_k = x_k - \frac{\alpha}{\lambda_k} d$ has minimum norm. Hence,

$$y_k^T y_k = \left(x_k - \frac{\alpha}{\lambda_k} d \right)^T \left(x_k - \frac{\alpha}{\lambda_k} d \right) = 1 - 2 \frac{\alpha}{\lambda_k} d^T x_k + \frac{\alpha^2}{\lambda_k^2} d^T d,$$

(7)

is minimized with respect to α if $-\frac{2}{\lambda_k} d^T x_k + 2 \frac{\alpha}{\lambda_k^2} d^T d = 0$ or $\frac{\alpha}{\lambda_k} = \frac{d^T x_k}{d^T d}$. Let $y = y_1$, we obtain

$$y^T y = 1 - \frac{(d^T x_1)^2}{d^T d},$$

(8)

using the α derived in the last step. In both Eq. (6) and Eq. (8), $w^T w = 0$ and $y^T y = 0$ if $d^T x_1 = \sqrt{d^T d}$. In other words, when the principal eigenvector is proportion to degree vector, $w = \mathbf{0}$ (or $y = \mathbf{0}$). Since $Ax_1 = \lambda_1 x_1$, $d^T x_k = \lambda_1 u^T x_1$. The condition $d^T x_1 = \sqrt{d^T d}$ implies

$$\lambda_1 u^T x_1 = \sqrt{d^T d} = \sqrt{N_2}$$

where $N_2 = d^T d$. Since $x_1 = \frac{d}{\sqrt{d^T d}}$, and $u^T d = N_1$, Lemma 3 follows.   □

Notice that in some approximate mean-field models for virus spreading [10], $\tau_c \sim \frac{N_1}{N_2} = \frac{1}{\lambda_1}$. Furthermore, $w^T w = 0$ (or $y^T y = 0$) is a special case of $u^T y_k = 0$, when $\lambda_1 = \frac{N_2}{N_1}$.

The relative difference $w^T u = u^T x_1 - \frac{d^T u}{\sqrt{d^T d}}$ ($y^T u$) is zero when the absolute difference is zero. We explore the relative difference in general cases by considering the binomial graphs as an example. The sum of the principal eigenvector $u^T x_1$ and the relative difference $w^T u$ as a function of the assortativity are shown in Figure 4 to follow exactly the same trend, since the degree of each node, thus, $\frac{d^T u}{\sqrt{d^T d}}$ remains the same when we change the assortativity by degree-preserving rewiring. When the assortativity $\rho_D = 0$, the binomial graphs are actually Erdős-Rényi random graphs, for which $\lambda_1 \simeq \frac{N_2}{N_1}$ when the network size is large [14]. Hence, both the absolute and relative difference are zero when the assortativity is around zero. The sum of the principal eigenvector $u^T x_1$ decreases with the assortativity ρ_D, as explained in Subsection 3.1.

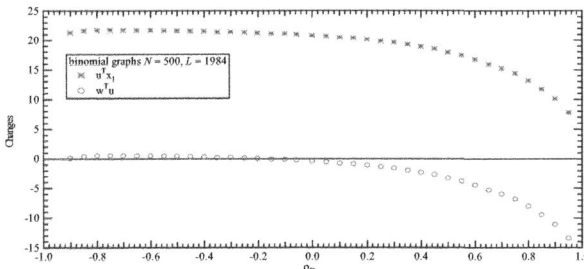

Fig. 4. The difference between the principal eigenvector and degree vector as a function of the assortativity

3.3 Correlation between the Principal Eigenvector and the Degree Vector

Recall that so far the best strategy to minimize the spectral radius by links/nodes removal is based on the principal eigenvector. When the correlation $\rho(x_1, d)$ between the principal eigenvector and the degree vector is positively strong, we may use the degree vector instead of the principal eigenvector to determine which links/nodes to remove, which will be further illustrated in Section 4. Here, we investigate the linear correlation coefficient $\rho(x_1, d)$ between the principal eigenvector and the degree vector as a function of ρ_D. Linear scaling of the degree vector will not change the linear correlation coefficient. Hence, we consider the original degree vector. When the absolute difference between the principal eigenvector and the scaled degree vector is zero, the principal eigenvector is proportion to degree vector. In this case, $\rho(x_1, d) = 1$, which seldom occurs in real-world networks. A strong positive correlation, not necessarily to be one, is already interesting with respect to approximate the eigenvector strategy by the corresponding degree vector strategy in minimizing the spectral radius.

Figure 5(a) depicts that $\rho(x_1, d)$ is mostly positively strong in the Erdös-Rényi random graphs and Bárabasi-Albert graphs. However, $\rho(x_1, d)$ decreases dramatically when the assortativity is decreased, actually around the minimal assortativity. Similarly, we derive networks with different assortativities by applying degree preserving rewiring to each of the 11 real-world networks. As in Figure 5(b), We are interested in how $\rho(x_1, d)$ changes with the assortativity ρ_D in real-world networks. Figure 5(b) illustrates that, the correlation $\rho(x_1, d)$ creases as the assortativity is decreased, especially around the minimal assortativity, which is the same as observed in network models. In the simulations of both network models and real-world networks, the most evident decrease is observed in networks with a power-law degree distribution such as the C. elegans neural network, the Gnutella 3 network and the WordAdj network.

These observations can be explained similarly as we explain the average/variance of the principal eigenvector versus assortativity in Section 3.1. In general, if a node has a large degree, its corresponding principal eigenvector component tends to be large even when the assortativity is zero, due to (4). A large positive assortativity implying large (or small) degree nodes tend to connect to other large

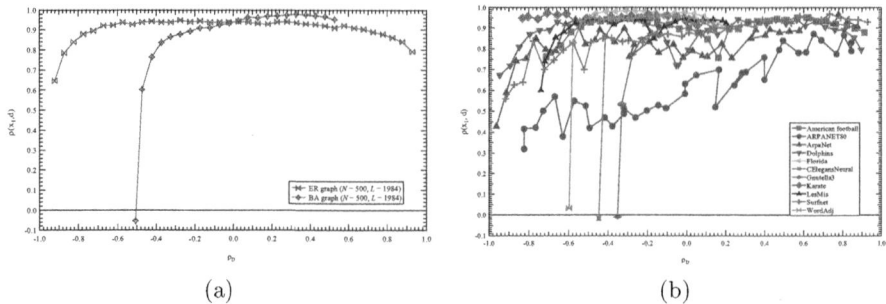

(a)   (b)

Fig. 5. The linear correlation coefficient between the degree vector and the principal eigenvector as a function of the assortativity (a) in both binomial graphs (red marks and line) and power-law graphs(blue marks and line); (b) in network instances derived from real-world networks via degree-preserving rewirings.

(or small) degree nodes, further enforces a large degree node to have more likely a even larger principal eigenvector component compared to a small assortativity. Hence, a negative assortativity will weaken the correlation $\rho(x_1, d)$. Note that the correlation coefficient is not necessarily the maximum at the maximal assortativity as shown in Figure 5, because here we examine the linear correlation coefficient but not the rank correlation.

4   Application: Degree *vs.* Principal Eigenvector Strategy in Minimizing the Spectral Radius

In this section, we illustrate the possibility to replace the principal eigenvector strategy by the degree vector in minimizing the spectral radius λ_1 via an example of node removal in power-law networks with different assortativities. As mentioned in Section 2, so far the best node removal strategy removes the node with the largest principal eigenvector component $(x_1)_j$. A widely applied strategy to minimize λ_1 by removing m nodes (a) removes the set of m nodes with the highest component in the principal eigenvector of the original graph. The corresponding degree vector strategy (b) removes the set of m nodes with the highest degree in the original graph. We compare these two strategies in removing $m \in [1, 200]$ nodes in graphs with positive, zero and negative assortativity (see Fig. 6) but with the same power-law degree distribution as in Fig. 5(a).

Figure 6 shows that the decreases of λ_1 by removing nodes with strategy (a) and (b) are almost same when ρ_D is large. The eigenvector strategy (a) decreases the spectral radius more thus performs better than the degree vector strategy (b) when the assortativity is small. When the assortativity is large, the degree vector is positively and strongly correlated with the principal eigenvector. In such a case, the degree vector strategy, the simplest to compute, well approximates the principal eigenvector strategy in minimizing the spectral radius.

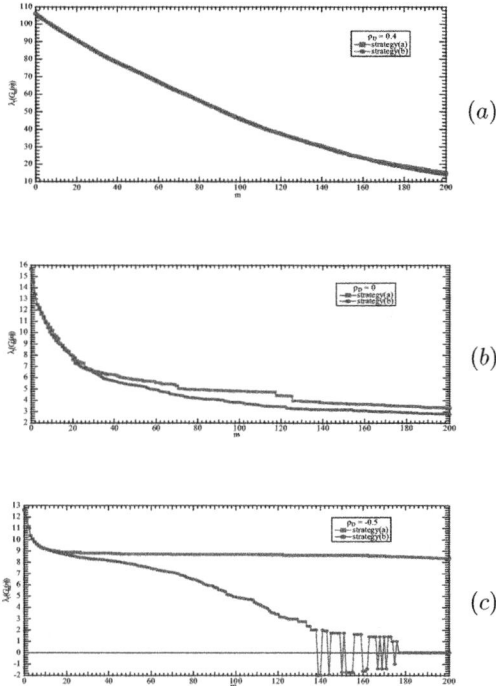

Fig. 6. The decrease of the spectral radius by successively removing m nodes in power-law networks. The square and circle dot dash lines show the decrease of the spectral radius by strategies (a) and (b) separately.

5   Conclusions

The principal eigenvector is essential in characterizing the influence of link/node on the spectral radius, whereas its topological meaning is far from well understood. This work, via both theoretical analysis and systematic simulations, contributes to the following aspects: (a) the average $E[x_1]$ (or variance) of the principal eigenvector is shown and explained to decrease (or increase) with the assortativity ρ_D; (b) the difference between the principal eigenvector and the degree vector is proved to be the smallest, when $\lambda_1 = \frac{N_2}{N_1}$ and (c) we illustrate and explain why the correlation between principal eigenvector and the degree vector decreases as ρ_D is decreased. In general, both a large variance (heterogeneity) in nodal degree and a large degree correlation (homogeneity in connection) contribute to a large average and a small variance of the principal eigenvector and a strong correlation between the degree and the principal eigenvector. As a straightforward application of these finds, we illustrate that when the assortativity is large, we could approximate the well performance principal eigenvector based strategy (to minimize λ_1 by removing links/nodes) by the corresponding degree vector, which is the simplest network property to compute.

References

1. Barabasi, A.-L., Albert, R.: Emergency of Scaling in Random Networks. Science 286, 509–512 (1999)
2. Barthélemy, M., Barrat, A., Pastor-Satorras, R., Vespignani, A.: Velocity and Hierarchical Spread of Epidemic Outbreaks in Scale-Free Networks. Phys. Rev. Lett. 92, 178701 (2004)
3. Boguñá, M., Pastor-Satorras, R., Vespignani, A.: Absence of Epidemic Threshold in Scale-Free Networks with Degree Correlations. Phys. Rev. Lett. 90, 028701 (2003)
4. Erdös, P., Rényi, A.: On Random Graphs. I. Publicationes Mathematicae 6, 290–297 (1959)
5. Li, C., Wang, H., Van Mieghem, P.: Bounds for the spectral radius of a graph when nodes are removed, accepted, Linear Algebra and its Applications
6. May, R.M., Lloyd, A.L.: Infection dynamics on scale-free networks. Phys. Rev. E 64, 066112 (2001)
7. Milanese, A., Sun, J., Nishikawa, T.: Approximating spectral impact of structural perturbations in large networks. Physical Review E 81, 046112 (2010)
8. Newman, M.E.J.: Assortative Mixing in Networks. Physic Rev. Lett. 89, 208701 (2002)
9. Newman, M.E.J.: Mixing patterns in networks. Phys. Rev. E 67, 026126 (2003)
10. Pastor-Satorras, R., Vespignani, A.: Epidemic dynamics and endemic states in complex networks. Phys. Rev. E 63, 066117 (2001)
11. Restrepo, J.G., Ott, E., Hunt, B.R.: Onset of synchronization in large networks of coupled oscillators. Physical Review E 71, 036151, 1–12 (2005)
12. Van Mieghem, P., Omic, J., Kooij, R.E.: Virus spread in networks. IEEE/ACM Transactions on Networking 17(1), 1–14 (2009)
13. Van Mieghem, P., Wang, H., Ge, X., Tang, S., Kuipers, F.A.: Influence of assortativity and degree-presreving rewiring on the spectra of networks. The European Physical Journal B, 643–652 (2010)
14. Van Mieghem, P.: Graph Spectra for Complex Networks. Cambridge University Press, Cambridge (2011)
15. Van Mieghem, P., Stevanović, D., Kuipers, F., Li, C., van de Bovenkamp, R., Liu, D., Wang, H.: Optimally decreasing the spectral radius of a graph by link removals. Physical Review E 84, 016101 (2011)

Resilient Virtual Network Design
for End-to-End Cloud Services

Isil Burcu Barla[1,2], Dominic A. Schupke[1], and Georg Carle[2]

[1] Nokia Siemens Networks, Munich, Germany
{isil.barla.ext,dominic.schupke}@nsn.com
[2] University of Munich, Munich, Germany
carle@in.tum.de

Abstract. Network virtualization with combined control of network and IT resources enables network designs for end-to-end cloud services with latency and availability guarantees. Even though providing such QoE guarantees is of high importance for cloud services, it is mostly not possible today if the services traverse different domains. To addresses this problem, firstly, we introduce novel resilient design methods for virtual networks minimizing the cost or the latency of the virtual network. We realize the routing of the services and the mapping of the virtual network simultaneously. Secondly, we provide two fundamental cloud connection architectures, which provide end-to-end resilience for cloud services in the presence of both network and datacenter failures. Using extensive simulations, we evaluate the performance of the proposed architectures in terms of cost of the virtual networks and maximum end-to-end delay that they can guarantee for cloud services.

Keywords: network virtualization, resilience, cloud services, latency.

1 Introduction

Cloud services are more and more utilized by businesses and for private applications, where latency and availability guarantees are of high importance especially for business critical applications. The performance of the cloud services is the key metric to measure the acceptability of that service by the end-users and hence it directly impacts the revenue for service providers [1]. Moreover, resilience of the cloud services in the presence of both *DataCenter* (DC) and network failures is as well a key point especially for the "cloudified" businesses, where a DC failure might cause a service outage in the range of days.

Cloud providers are aware of these problems and try to address them by offering service level agreements to their customers. However, these agreements only cover the performance and connectivity inside the cloud and exclude the telecommunication networks, which might actually cause excessive latencies and even service outages. Thus, today it is essentially impossible to provide quality of experience guarantees in an end-to-end fashion for cloud services. One solution for this problem is using the concept of *Network Virtualization* with combined control for network and IT resources.

R. Bestak et al. (Eds.): NETWORKING 2012, Part I, LNCS 7289, pp. 161–174, 2012.

Network virtualization is proposed as a key enabler for the next generation networks and the future Internet [3,4]. Unlike current virtualization techniques like Virtual Private Networks (VPNs) and overlay networks, network virtualization enables operation of isolated network slices. A slice consists of isolated computational resources inside network and DC nodes, as well isolated network resources between them. In a virtual network environment new business models realizing different tasks are expected to be established and as a result, as well trading of virtual resources between them [5]. Note that these resources can be network resources and/or IT resources. In such an environment new control mechanisms and interfaces are necessary to realize the setup and operation of the *Virtual Networks* (VNets). For possible realizations of combined control of IT and network resources using virtualization, there are already several suggestions in the literature [17]. There are as well some commercial offers like from Amazon [6], where a virtual network is offered together with the cloud services. Still, such solutions currently lack redundancy and QoE guarantees, which are the main reasons for hesitation of businesses to adopt cloud solutions according to a survey done in 2011 with 3700 enterprises worldwide [7]. Therefore, in this paper we propose novel architectures to enable provisioning of cloud services with end-to-end availability and latency guarantees. In our network virtualization model, we define two business roles, namely the *Virtual Network Operator* (VNO), operating a VNet on the physical substrate, and the *Physical Infrastructure Provider* (PIP) owning the physical substrate.

A PIP is the owner of the physical infrastructure, which can be e.g. a transport network, wireless access network, IT resources like processing or storage units or any combination of them. The PIP is in the position to monitor all of its physical and virtual resources and has the knowledge of the usage and physical location of its virtual resources. Given a PIP with existing resources, its incentive would be optimizing the utilization of its resources to maximize its revenue according to a chosen strategy, e.g. minimizing the used capacity or load balancing, by allocating the virtual resources accordingly. In this paper we focus on the PIPs owning transport networks and IT resources. In the remainder of the paper, if there is a need for distinguishing, the PIPs owning only network resources will be called nPIPs and the ones having only DC or a combination of both will be called dcPIPs. A VNO can operate one or several VNets, which are mapped onto the physical infrastructure of possibly one or more PIPs. A VNet can consist of both virtual network and IT resources. The interfaces and information sharing between the VNO and the PIP would depend on their internal business models and the contract between them [2]. Without loss of generality, we assume that for the VNet setup the available virtual resources of the PIPs are advertised to the VNO. The VNO can negotiate with various PIPs and compute an optimal VNet according to its specific needs. These can be e.g. either minimizing the cost of the VNet or optimizing the VNet design to provide minimum latency for the running services.

In such a virtual network environment there are two fundamental resilience architecture options, namely resilience can be provided solely by the PIP, *PIP-Resilience*, or only by the VNO, *VNO-Resilience*. Applying resilience at different layers has several advantages and drawbacks. In [8], we address this question and compare the PIP and VNO-Resilience cases in a qualitative manner in terms of network resource usage, service level resilience adaptability and network setup and operation complexity. On the one hand, a VNO can utilize the available resources of different PIPs to reach an overall optimal design regarding both resilience and performance considerations. On the other hand, PIP-Resilience can offer a simpler signaling interface between the VNO and the PIP. It also provides lower system complexity in terms of the concurrent actions taken in case of a failure. Furthermore, in PIP-Resilience, the recovery action is transparent to the VNet and hence the virtual topology remains unchanged. Finally, the VNO-Resilience can offer VNet setup at service level granularity. Note that hybrid mechanisms are out of the scope of this paper due to our aim of investigating the resilience design choices affecting the delay performance and cost of the VNets.

In this paper, we propose novel solutions for end-to-end cloud services with availability and latency guarantees under both network and DC failures. We provide end-to-end solutions both for VNO and PIP-Resilience cases. Firstly we introduce novel resilient VNet design methods, which route the requested services and map the VNets onto the physical substrate simultaneously. We model our resilient VNet designs as two sets of *Mixed Integer Linear Problems* (MILPs) for VNO and PIP-Resilience cases by minimizing the delay of all the possible services in the VNet and the cost of the VNet, while providing network resilience, respectively. Afterwards, we combine these VNets with resilient cloud connection models for VNO and PIP-Resilience, which provide resilience in the presence of both physical network and DC failures for end-to-end cloud services. We evaluate the performance of our VNet designs in terms of cost and delay and we show their efficiency and applicability compared to traditional shortest paths mapping approaches. Finally, using extensive simulations we compare the two end-to-end solutions in terms of their delay performances. The remainder of the paper is organized as follows: Section 2 gives a short summary of the related work, in Section 3 the resilient VNet designs and in Section 4 the cloud connection models are introduced and their performances are evaluated. Finally, Section 5 concludes the paper with a discussion of the results and an outlook.

2   Related Work

Regarding the VNet design there are mainly two types of works in the literature. The first one is on routing the services in a VNet according to quality of service or availability requirements [9,10]. It is assumed that the VNet is already existing and mapped onto the physical substrate. However, in this paper we deal with the problem of designing a new VNet for a VNO according to the given requirements. Unlike the overlay or VPN services where the customer needs to pay only per usage, a VNO would need to pay for the setup and maintenance of its VNet.

Therefore, it is very important from the beginning to design a cost-efficient VNet, which can offer the required service quality.

The second type of work in the literature offer solutions for the mapping of a VNet onto the physical substrate [11,12]. Moreover, there are as well proposed algorithms for mapping survivable VNets [13,14]. However, all of these works assume that the virtual topology is already given and in case of survivable mapping, the VNet has to be even bi-connected so that the mapping can be realized at all. Our work does not have such a limitation and designs a VNet with the given requirements. In [16], the authors deal with the VNet design problem for the case of the overlay networks by minimizing the cost of the overlay network. However, they do not consider resilience and use direct shortest path mappings. In [15], resilient VPN designs are realized but again assuming direct mapping of the virtual links on shortest physical paths. We extend this approach by allowing several mapping choices for a virtual link and show that our approach outperforms the shortest path mapping model in terms of feasibility and delay performance. Thus, there is extensive work available for mapping a given VNet on the physical substrate and routing a set of services in a VNet, which is already mapped onto the physical substrate. However, to the best of our knowledge this is the first paper, which considers both mappings simultaneously to realize cost-efficient and latency-optimized resilient VNet designs.

Finally, we provided a qualitative comparison of PIP and VNO-Resilience cases in a virtual network environment in [8]. However, to the best of our knowledge, this is the first paper proposing resilient VNet designs and cloud connections for these two fundamental cases and conducting a quantitative study for the cost and latency evaluation of these models.

3   Resilient Virtual Network Design

In this section, the resilient VNet design models are introduced. It is assumed that a set of requested services and the physical network are provided. The models are given in the form of MILPs and the optimization is performed for minimizing the maximum latency or the cost of the VNet. The first subsection introduces the simple model without any resilience considerations. In the following subsections VNO-Resilience and PIP-Resilience models are introduced. Finally, in Subsection 3.4 we evaluate the performance of the different models using different parameter settings and comparing to *Shortest Path Mapping* (SPM) and *SPM with Additional Nodes* (SPMwAN) models, where each virtual link is directly mapped onto the physical shortest path between its end-nodes.

3.1   Simple Model without Resilience

In the *Simple Model* (SM) the virtual links are mapped onto single paths in the physical network and the services are routed in the VNet on $i \in \{1, .., r\}$ routes. In SM, the *service nodes*, i.e. the end-nodes of the given services, are directly used as the virtual nodes of the resulting VNet. The virtual links have k-shortest paths[1] mapping possibilities. However, to maintain linearity instead of using one virtual link with several possible mappings, we generate a new virtual link for each mapping and add it to the list of all the possible virtual links. The result of the optimization problem is a VNet, which consists of only the links and nodes that are used to route any of the given services. In the following, the sets, parameters and variables used in the MILPs are briefly introduced.

- *Sets:*
    - V: Set of the all virtual node candidates
    - L: Set of the all virtual link candidates
    - D: Set of the requested services
    - E_l: Set of the endpoints of link $l \in L$
    - N: Set of virtual links $(j, k) \in L^2$, which share at least one physical edge

- *Parameters:*
    - b_d: Requested bandwidth for the service $d \in D$
    - c_d: Requested node resources for the service $d \in D$
    - s_l: Physical length of link $l \in L$
    - λ_l: Fixed setup cost for having a new link $l \in L$
    - θ_l: Setup cost per unit capacity for link $l \in L$
    - μ_v: Fixed setup cost for having a new node $v \in V$
    - η_v: Setup cost per unit capacity for node $v \in V$

- *Variables*
    - $\beta_{i,d,l}$: Binary variable taking the value of 1 if the link $l \in L$ is used for the ith route of the demand $d \in D$, 0 otherwise
    - $\delta_{i,d,v}$: Binary variable taking the value of 1 if the node $v \in V$ is used for the ith route of the demand $d \in D$, 0 otherwise
    - γ_l: Binary variable taking the value of 1 if the link $l \in L$ is in the resulting VNet, 0 otherwise
    - α_v: Binary variable taking the value of 1 if the node $v \in V$ is in the resulting VNet, 0 otherwise
    - $u_l \in [0, \infty]$: Used capacity on link $l \in L$
    - $\omega_v \in [0, \infty]$: Used capacity on node $v \in V$

The constraints for SM are given in the following. Eq. (1) is the link-flow constraint. Eq. (2) makes sure that a node is flagged as "used" for a service if it is the source or the target of that service. Eq. (3) and (4) state that a virtual link or node is part of the resulting VNet if it carries the traffic of any service, respectively. Finally, Eq. (5) and (6) are the constraints for link and node capacity, respectively.

$$\sum_{l:v \in E_l} \beta_{i,d,l} = \begin{cases} 1 & \text{if } v = s \text{ or } v = t \\ 2\delta_{i,d,v} & \text{otherwise} \end{cases} \quad \forall d = (s,t) \in D, \ v \in V, \ i \in \{1,..,r\}$$

$$\tag{1}$$

---

[1] When all simple paths between two nodes are listed in ascending order according to their lengths, k-shortest paths between these two nodes are the first k paths in the list.

$$\delta_{i,d,v} = 1 \quad \forall d = (s,t) \in D, \ \forall v \in (s,t), \ i \in \{1,..,r\} \tag{2}$$

$$\gamma_l \geq \beta_{i,d,l} \quad \forall l \in L, \ \forall d \in D, \ \forall i \in \{1,..,r\} \tag{3}$$

$$\alpha_v \geq \delta_{i,d,v} \quad \forall v \in V, \ \forall d \in D, \ \forall i \in \{1,..,r\} \tag{4}$$

$$u_l \geq \sum_{i \in \{1,..,r\}} \sum_{d \in D} \beta_{i,d,l} \, b_d \quad \forall l \in L \tag{5}$$

$$\omega_v \geq \sum_{i \in \{1,..,r\}} \sum_{d \in D} \delta_{i,d,v} \, c_d \quad \forall v \in V \tag{6}$$

There are two objective functions defined for different optimization objectives, namely VNet cost minimization and delay minimization. Note that the cost of the VNet constitutes of the link cost and the node cost, where each of them has again two parts, namely the fixed setup cost for having a new link or node in the VNet and the capacity dependent cost depending on the requested capacity on that link or node. To achieve simplicity in the PIP-VNO business relationships, a linear cost model is assumed. In cost minimization the total cost of the VNet and in propagation delay minimization the total length of the routes for each service are minimized. We only consider the propagation delay in the physical path as the latency metric for a service since the network is designed for normal load conditions. Thus, the queueing delay is negligible and the main latency is caused by the propagation of the signal over physical distances. Expressions (7) and (8) show the objective functions for VNet cost minimization and delay minimization of the services, respectively.

$$\min \ \left(\sum_{l \in L} (\lambda_l \gamma_l + \theta_l u_l) + \sum_{v \in V} (\mu_v \alpha_v + \eta_v \omega_v) \right) \tag{7}$$

$$\min \ \sum_{d \in D} \sum_{i \in \{1,..,r\}} \sum_{l \in L} \beta_{i,d,l} \, s_l \tag{8}$$

3.2   VNO-Resilience

For VNO-Resilience, 1:1 protection routing is used in the virtual layer, where the working and protection paths of a service have to be physically disjoint. Hence, the number of the routes r is 2. To provide resilience additional diversity constraints are introduced to the model. The constraint given in (9) ensures that the virtual working and protection paths of a service do not contain any two virtual links, which share common edges in the physical layer. Equation (10) provides node-diversity, where the working and protection paths are not allowed to share any nodes other than the end-nodes. In case of a physical link or node failure, the affected services are re-routed by the VNO on their pre-calculated protection paths.

$$\beta_{1,d,j} + \beta_{2,d,k} \leq 1 \quad \forall d \in D, \ (j,k) \in N \tag{9}$$

$$\delta_{1,d,j} + \delta_{2,d,k} \leq 1 \quad \forall d = (s,t) \in D, \ (j,k) \in V \setminus \{s,t\} \tag{10}$$

3.3   PIP-Resilience

In the case of PIP-Resilience, providing resilience is the responsibility of the PIP(s). The services are routed on single paths in the VNet layer, where each virtual link is mapped on two disjoint physical paths in the physical layer. The disjointness criteria can be defined as link-disjoint or node-disjoint. For PIP-Resilience, SM is directly applied where the number of virtual routes is set to 1. However, instead of k-shortest physical path mapping for the virtual links, k-shortest disjoint path pairs[2] mapping is used. Therefore, the VNO sees only a simple network, which is protected in the physical layer. The re-routing in case of a failure is realized in the physical layer by the corresponding PIP, i.e. the virtual topology remains unchanged and ideally the services are not disrupted.

3.4   Performance Evaluation of the Resilient Virtual Network Design Options

In this section the two proposed models are compared in terms of the VNet cost and maximum service delay they provide for different cost parameters and optimization functions. Moreover, the performance of the models is evaluated against the SPM and SPMwAN models, where both of them use direct shortest path mapping. As the resilience strategy they both utilize the VNO-Resilience, i.e. the services are routed on two virtual paths, which are physically disjoint. In SPM, like in the VNO and PIP-Resilience cases, the virtual node set consists of the service nodes. In SPMwAN, however, the virtual node set is extended, where the VNO can use as well additional virtual nodes for routing purposes, which are not the source or target of any of the services. Similarly, the virtual link set is as well extended to cover the possible links between all the node pairs in the new node set.

For the performance evaluation, we used two test networks, namely the NobelUS and NobelEU [18] networks. For VNet generation, first, we select a certain number of service nodes randomly from the physical network, where there is uniform demand between all of them. Then we solve the optimization problem for different resilience models, where each of them result in a different VNet. They are then compared regarding their delay/cost performances until a confidence level of 95% and ±5% confidence interval is reached. Link diversity option is used for the simulations. However, our results show that node diversity option results in comparable delay and cost values as link-diversity. To evaluate the effect of different cost factors and optimization functions on the resulting cost and delay, we distinguish between seven cases as shown in Fig.1a. For this analysis the NobelUS network is used. Results for 3-node VNets are shown due to their significance and applicability in real life scenarios. The cost and delay differences shown in the figure are the relative differences of the two models, which are calculated by taking the difference of PIP and VNO-Resilience value and

---

[2] K-shortest disjoint path pairs are the first k disjoint path pairs when all the disjoint path pairs are listed in ascending order according to their total length.

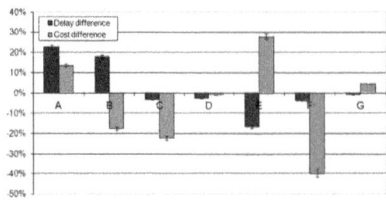

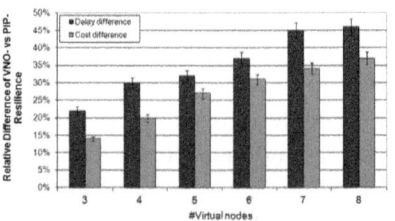

(a) Effect of different cost and optimiza-   (b) VNet Design for VNO-Resilience vs
tion function settings denoted by A-G   PIP-Resilience (cost optimization)

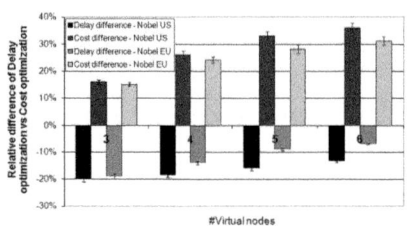

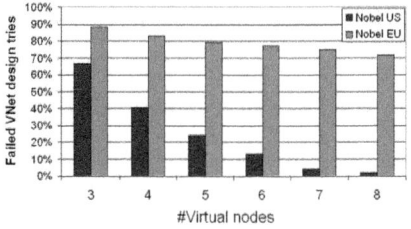

(c) Delay Minimization vs. Cost Mini-   (d) % of failed VNet designs with SPM
mization

Fig. 1. VNet design performance comparisons for VNO-Resilience, PIP-Resilience and
Shortest Path Mapping (SPM)

dividing it to the VNO-Resilience value. In cases A-F, cost optimization is used
and the cost factors for link and node costs are varied. The link cost can be
defined as a fixed value or might depend on the length of the physical path it
is mapped on. In the latter case, we use the flag "length." The cases are defined as a quadruples; {the fixed link setup cost, the capacity dependent link
setup cost, the fixed node setup cost, the capacity dependent node setup cost}.
The cases are defined as A={length,length,1,1}, B={length,length,2000,2000},
C={1,1,1,1}, D={1,100,1,1}, E={100,1,1,1} and F={1,1,100,100}. The cases are
chosen to investigate the effect of each individual cost component in case of fixed
and length-dependent cost factors. In case B, the node cost factor is taken as
2000, which is a value in the range of average virtual length link for the used
test network. Note that the length corresponds to the total physical length of
the virtual link, i.e. in VNO-Resilience it is the length of the single physical path
and in PIP-Resilience it is the sum of the lengths of the two disjoint physical
paths for each virtual link. Hence, the protected virtual links are in general more
expensive than the unprotected ones. Similarly, for fixed link cost values we introduce a resilience cost factor for PIP-Resilience. Its value is taken as 2 for the
simulations. Finally, in cases C-F we investigate the effect of the cost component
with the weight 100, where the rest is kept minimum.

Cases B,C and E show that when the node cost is in the range of the link
cost or higher, VNO-Resilience results in higher VNet cost compared to PIP-
Resilience. This effect is caused by the higher virtual node capacity usage in

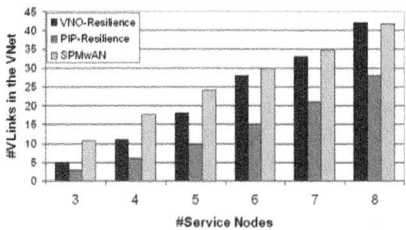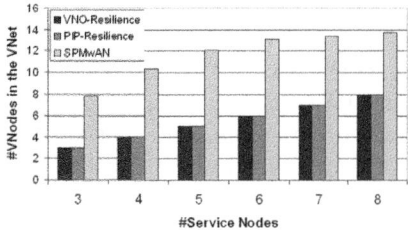

(a) Comparison of the number of virtual nodes

(b) Comparison of the number of virtual links

Fig. 2. VNet design performance comparisons for VNO-Resilience, PIP-Resilience and Shortest Path Mapping with Additional Nodes

VNO-Resilience due to the two-paths routing in the VNet. In cases A and B, the cost of the link depends on the physical length of the link and hence cost optimization is aligned with delay optimization. In these cases, VNO-Resilience results in 20% lower delay than PIP-Resilience. Wherever the delay optimization function is used, the VNO and PIP-Resilience result in comparable delay and cost values. However, if we compare the results of delay optimization and cost optimization for VNO-Resilience with the same cost factors as in case A, it is observed that the delay minimization option results always in lower delay but higher VNet cost. Increasing the number of the service nodes, decreases the delay difference of the two optimization functions but increases the cost difference as shown in Fig.1c. Hence, the appropriate optimization function should be chosen according to both the number of service nodes and the cost factors.

In the remainder of the results the cost factors are always taken as in case A. For cost optimization the delay and cost differences of PIP and VNO-Resilience increases with increasing VNet size as shown in Fig.1b. As can be seen for larger VNets, a VNet with PIP-Resilience costs on average 35% more than a VNet with VNO-Resilience. Moreover, PIP-Resilience results in 45% higher VNet-latency than VNO-Resilience for these settings. These results are obtained for the NobelUS network.

In SPM direct shortest path mapping is used and hence it is not always possible to find disjoint paths to route the services and the design cannot be performed. Fig.1d shows the ratio of the VNet design tries, which failed to find a solution during the simulations. Note that with increasing VNet size, the probability to find a solution for SPM is increasing. However, for NobelEU network, even for 8-nodes VNets in over 70% of the tries, no solution could be found. Moreover, even if a solution is found for SPM, it always results in higher maximum delay compared to VNO-Resilience. This difference decreases with increasing VNet size but is still over 20% for 8-nodes VNet on the test network NobelUS.

SPMwAN results in comparable latency as the VNO-Resilience and solves the problem faced by SPM. However, firstly it less scalable for larger physical networks. For our test networks, VNO and PIP-Resilience simulations find a solution in a time interval of seconds but for SPMwAN, the simulation lasts for several minutes or even hours. Second, the resulting VNet has more virtual links and nodes compared to the PIP and VNO-Resilience cases as shown in Fig.2a and 2b. The virtual link numbers of SPMwAN and VNO-Resilience come closer for higher service node numbers. However, SPMwAN always has a higher number of nodes independent of the VNet size. Hence, especially for a high node cost factor, the network cost is drastically higher for SPMwAN.

4   Enhanced Datacenter Connection Models

In this section, the DC connection models for VNO and PIP-Resilience cases are introduced. In both cases, the VNet is connected to one primary and one backup DC to serve all the cloud services within the VNet. The design aim of both models is providing resilience in presence of both network and DC failures.

4.1   VNO-Resilience

In VNO-Resilience, all elements of the DC connection model, namely the DCs and the links and nodes connecting them to the VNet, are chosen by the VNO. To provide resilience in the presence of DC failures, the two DCs should be located in geographically disjoint locations. In our model, we divide the physical network into *availability regions*, where a failure in one region does not affect any other region. While choosing the DCs, the region information of the DCs should be provided to the VNO to guarantee disjointness. Moreover, to provide network resilience in case of single link failures, we choose the three virtual links, namely the two connecting the DCs with the VNet and the one connecting the two DCs, all mutually disjoint. Hence, the physical disjointness information of the available virtual links should be also provided to the VNO by the PIPs.

Fig.3a shows the DC connection model for VNO-Resilience. In normal operation, the services are routed via the link l_p to the *Virtual Machine* (VM) located in the primary DC. In case of a failure in the primary DC, the services will be rerouted to the backup DC using the second connection node and the backup link l_b. Similarly, in case of a failure on l_p, the traffic is rerouted on l_b to the backup DC. The link between the two DCs, l_c, is established for synchronization, data migration and failure routing purposes. In the special case, where both l_p and the backup DC fail simultaneously, the primary DC can be still reached using the path l_b and l_c. This case will be referred to as the *worst-case scenario*. Note that, the VNO can choose the two DCs from the same or different dcPIP(s) to optimize the performance of the cloud services in terms of latency.

The VNO-Resilience model becomes non-scalable with increasing number of DCs due to the large number of possible DC-connection node combinations. Therefore, we introduce a heuristic, where the primary DC and its connection node, node 1, are chosen first according to the maximum end-to-end delay it provides. However, the path l_p is not fixed but rather a candidate path list is created holding the k-shortest paths between the primary DC and node 1. Afterwards, the backup DC and its connection node, node 2, are chosen to

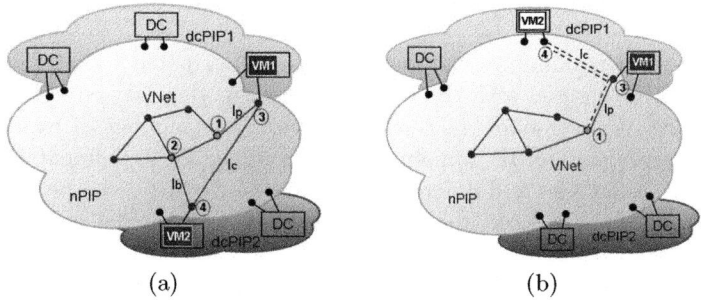

(a)   (b)

Fig. 3. Datacenter Connection Models: (a) VNO-Resilience, (b) PIP-Resilience

minimize the end-to-end delay considering both the VNet delay and the routing on l_p, l_b and l_c, where all of these links are mutually physically disjoint. The end-to-end delay performance difference of the optimal case and the heuristic remain in $\pm 5\%$ interval for the NobelUS network with different DC and dcPIP settings and it is hence negligible.

4.2   PIP-Resilience

In the PIP-Resilience case, as shown in Fig.3b, the connection to the VNet is established over one virtual link. The primary DC with the connection link, l_p, is chosen by the VNO. In this model, providing resilience is the responsibility of the PIPs. Therefore, l_p is mapped in the physical network on two disjoint paths. Moreover, the dcPIP, owning the primary DC is responsible to provide resilience against DC failures and, thus, provides the connection to a DC it chooses from its own domain, which serves as the backup DC. Note that this connection link, l_c, has to be as well mapped onto two disjoint physical paths. This link might be owned directly by the dcPIP if it has relevant resources or has to be leased from an nPIP to connect the two DCs. In PIP-Resilience model, l_c and the backup DC are not visible to the VNO and all the recovery actions taken in case of any network or DC failure are transparent to the VNO. In case of a DC failure, the services are redirected to the backup DC and in case of a network failure, the protection paths are used to route the services in the physical layer.

For PIP-Resilience model to be realizable, each dcPIP has to own at least two DCs, which are located in geographically disjoint locations. For a dcPIP having a single DC, providing resilience is impossible. Moreover, the choice of the backup DC would depend on internal strategy of the dcPIP and on the contract with the VNO. The internal strategy of a dcPIP might be e.g. load balancing among its DCs or providing the highest performance for the services. Finally, on the one hand, for VNO-Resilience, the number of the virtual links and nodes, which have to be established and maintained is higher than the PIP-Resilience as shown in Fig.3a and 3b. On the other hand, for PIP-Resilience the virtual network links have to be mapped on two physically disjoint paths, where for VNO-Resilience single path mapping is sufficient.

4.3   Delay Performance Evaluation of the End-to-End System

The end-to-end maximum delay performance is evaluated by combining corresponding VNet designs with the DC connection models. We compare the delay performance of the models in terms of VNet size, number of available DCs, number of different dcPIP domains, location of the DCs, different DC connection model preferences and different failure cases. The DCs can be placed either randomly, with the option "random", or to obtain maximum distance between them, namely with the option "away." Note that for both cases, the DCs of a dcPIP are located in different availability regions. Finally, it is assumed that in PIP-Resilience, the dcPIP chooses the backup DC randomly from its domain. In the simulations random VNets are generated for the test networks with DCs located randomly on them. Note that the chosen test networks are realistic topologies covering large physical areas. This enables end-to-end resilience design even in case of disasters and makes the problem more interesting by possibly enabling having multiple PIPs. For each VNet and DC set, the cloud connections are designed using the two models and the maximum end-to-end latency observed in both cases is compared until the confidence level of 95% with ±5% confidence interval is reached for the result.

Fig.4a shows the effect of the number of the different dcPIP domains and number of DCs each domain possesses on the end-to-end maximum delay difference of PIP and VNO-Resilience. The results are obtained using 3-nodes VNets mapped on the NobelEU network with random DCs. For this simulation the DC location option is "away" and the same primary DC is used for PIP and VNO-Resilience. It is observed that the PIP-Resilience results always in higher end-to-end delay compared to VNO-Resilience and this difference increases with increasing number of dcPIPs. However, for a certain number of dcPIPs, increasing the number of DCs per dcPIP decreases the relative delay difference, since the dcPIPs' DC selection options increase as well.

The simulations performed with the NobelUS network show that if in PIP-Resilience the primary DC is selected freely to minimize the latency, the relative delay difference is decreased by 10% compared with the same primary DC selection scenario as shown in Fig.4b. Moreover, comparing Fig.4a and 4b, it is seen that a larger physical network results in higher relative delay difference. For NobelUS network, with 1 dcPIP and 2 DCs, the absolute maximum end-to-end round-trip delay of the PIP-Resilience is around 112 ms for random 3-nodes VNets. For NobelEU network, the maximum round-trip delay of 3-nodes VNets is 107 ms and of 5-nodes VNets 117 ms. However, the relative delay difference of the PIP and VNO-Resilience remains almost constant for different VNet sizes.

Finally, different DC location and protection options are compared using the NobelEU network and 3-nodes VNets as shown in Fig.4c. In all cases PIP-Resilience results in higher maximum delay compared to VNO-Resilience. This difference goes beyond 120% if more than 5 dcPIPs are available for the "away" DC location option. When the DCs are placed randomly, the relative delay difference is reduced by around 20%. Finally, in worst-case scenario, the relative delay difference is drastically decreased and reaches 40% for 10 dcPIPs.

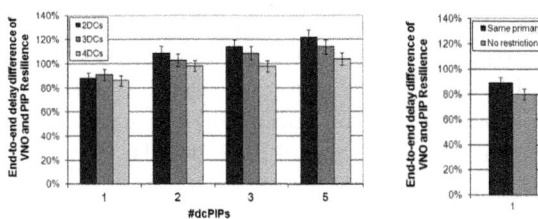 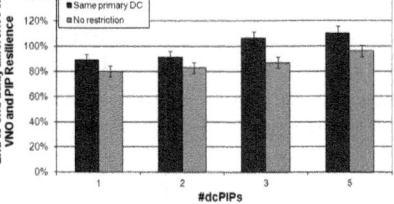

(a) Delay performance comparison (b) Effect of using the same primary DC
for different # DCs and dcPIPs

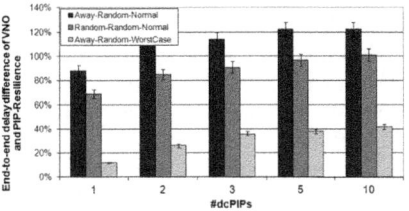

(c) Effect of random and worst-case
options

Fig. 4. Performance comparisons of DC connection models

5   Conclusion

In this paper, we propose novel solutions for enabling the provisioning of cloud
services with end-to-end availability and latency guarantees. The problem is separated into resilient Virtual Network (VNet) design and cloud connection design
parts, where each are of different nature but enable together end-to-end resilient
cloud services. First, we introduce two fundamental resilient VNet designs, one
at the Virtual Network Operator (VNO) layer, namely VNO-Resilience, and
the other one at the Physical Infrastructure Provider (PIP) layer, namely PIP-
Resilience, in form of mixed-integer linear problems. We show that the proposed
models outperform the models, where the mapping is done using shortest paths,
in terms of efficiency and applicability. With direct shortest path mapping, in
more than 80% of the cases no solution can be found for small VNets. Allowing
additional virtual nodes solves this problem but results in relatively higher cost
compared to the proposed models. Different cost factor values and optimization
functions are discussed and it is shown how the model decision should be made
according to the actual cost factor values and the cost vs. delay requirements.

In the second half of the paper, we introduce two DataCenter (DC) connection
models for the designed VNets, which allow end-to-end reliability in presence of
DC and network failures. We perform an end-to-end maximum delay performance analysis and our simulation results show that the relative delay difference
of the two models can reach 120% for the test networks and PIP-Resilience
results always in higher end-to-end delay compared to VNO-Resilience.

In this paper, we obtain the end-to-end system by designing the VNets and cloud connections sequentially. As future work, the designs can be optimized together for both parts to achieve maximum efficiency and performance. Moreover, in the delay-optimization MILP, we minimize the total delay for all the services within the network. Another option would be minimizing the maximum delay. Finally, capacity constraints can be added to the MILP models.

References

1. Greenberg, A., et al.: The Cost of a Cloud: Research Problems in Data Center Networks. In: ACM SIGCOMM CCR (January 2009)
2. Meier, S., et al.: Provisioning and Operation of Virtual Networks. In: Electronic Communications of the EASST, vol. 37 (March 2011)
3. Chowdhury, M.K., Boutaba, R.: A survey of network virtualization. Elsevier Computer Networks 54(5) (2010)
4. Tutschku, K., et al.: Network virtualization: Implementation steps towards the future internet. In: KiVS, Kassel, Germany (March 2009)
5. Papadimitriou, P., et al.: Implementing network virtualization for a future internet. In: 20th ITC Specialist Seminar, Hoi An, Vietnam (May 2009)
6. Amazon AWS, http://aws.amazon.com/directconnect/
7. Symantec, Virtualization and Evolution to the Cloud Survey (2011)
8. Barla, I.B., Schupke, D.A., Carle, G.: Analysis of Resilience in Virtual Networks. In: 11th Würzburg Workshop on IP: Joint ITG and Euro-NF Workshop Visions of Future Generation Networks (August 2011)
9. Anderson, D., et al.: Resilient overlay networks. In: Proceedings of 18th ACM Symposium on Operating Systems Principles (2001)
10. Li, Z., Mohapatra, P.: QRON: QoS-aware routing in overlay networks. IEEE Journal on Selected Areas in Communications 22(1), 29–40 (2004)
11. Zhu, Y., Ammar, M.: Algorithms for Assigning Substrate Network Resources to Virtual Network Components. In: Proc. IEEE INFOCOM (March 2006)
12. Chowdhury, N.M.M.K., Rahman, M.R., Boutaba, R.: Virtual Network Embedding with Coordinated Node and Link Mapping. In: IEEE INFOCOM (2009)
13. Modiano, E., Narula-Tam, A.: Survivable lightpath routing: A new approach to the design of WDM-based networks. IEEE J. Selected Areas in Communications 20, 800–809 (2002)
14. Rahman, M.R., Aib, I., Boutaba, R.: Survivable Virtual Network Embedding. In: Crovella, M., Feeney, L.M., Rubenstein, D., Raghavan, S.V. (eds.) NETWORKING 2010. LNCS, vol. 6091, pp. 40–52. Springer, Heidelberg (2010)
15. Maliosz, M., Cinkler, T.: Configuration of Protected Virtual Private Networks. In: DRCN 2001, Budapest (2001)
16. Kamel, M., Scoglio, C., Easton, T.: Optimal Topology Design for Overlay Networks. In: Akyildiz, I.F., Sivakumar, R., Ekici, E., Oliveira, J.C.d., McNair, J. (eds.) NETWORKING 2007. LNCS, vol. 4479, pp. 714–725. Springer, Heidelberg (2007)
17. Koponen, T., et al.: Onix: A Distributed Control Platform for Large-scale Production Networks. In: Proc. OSDI (October 2010)
18. Nobel US and Nobel EU Topologies, http://sndlib.zib.de

Dynamic Scaling of Call-Stateful SIP Services in the Cloud

Nico Janssens, Xueli An, Koen Daenen, and Claudio Forlivesi

Service Infrastructure Research Dept.
Alcatel-Lucent Bell Labs, Antwerp, Belgium
{nico.nj.janssens,xueli.an,koen.daenen,
claudio.forlivesi}@alcatel-lucent.com

Abstract. Many cloud technologies available today support dynamically scaling out and back computing services. The predominantly session-oriented nature and the carrier-grade requirements of telco services (such as SIP services) complicate the successful adoption of dynamic scaling in a telco cloud. This paper investigates how to enable dynamic scaling of these telco services in an effective manner, focusing in particular on call-stateful SIP services. First, we present and evaluate two protocols to transparently migrate ongoing sessions between call-stateful SIP servers. These allow to quickly shutdown call-stateful SIP servers in response to a scale back request, removing the need to wait until their ongoing calls have finished. Second, instead of responding to load changes in a reactive manner, this paper explores the potential value of pro-active resource provisioning based on call load forecasting. We propose a self-adaptive Kalman filter to implement short-term call load predictions and combine this with history-based predictions to anticipate future call load changes. We believe that session migration and call load forecasting are two important elements to safely reduce the operational expenditure (OpEx) of a cloudified SIP service.

Keywords: cloud, telecommunication, elasticity, dynamic scaling, session migration, load prediction, SIP.

1 Introduction

Cloud computing has gained substantial momentum over the past few years, fueling technological innovation and creating considerable business impact. Public, private or hybrid cloud infrastructure shortens customers' time to market (new hosting infrastructure is only a few mouse-clicks away), and promises to reduce their total cost of ownership by shifting the cost structure from higher capital expenditure to lower operating expenditure. One of the fundamental features of cloud computing is the ability to build dynamically scaling systems. Virtualization technologies (including XEN, KVM, VMware, Solaris and Linux Containers) facilitate computing services to automatically acquire and release resources. This enables to dynamically right-size the amount of allocated resources, instead of statically over-dimensioning the capacity of such services. Dynamic scaling thus enables to reduce operational costs and to gracefully handle unanticipated load surges, all without compromising the performance and correct functioning of the affected services.

R. Bestak et al. (Eds.): NETWORKING 2012, Part I, LNCS 7289, pp. 175–189, 2012.

Although the majority of existing (cloud) scaling solutions have been targeted at web and enterprise applications [10, 14, 18, 20], telco services can also benefit significantly from dynamic scaling. To guarantee carrier-grade service execution, telco operators typically over-provision the employed resources to handle sporadic unanticipated load surges (e.g. caused by events with a significant social impact) or anticipated load spikes (e.g. caused by New Year wishes). This reduces their resource utilization ratio and raises their operational cost.

This paper investigates the application of dynamic scaling in telco services, focusing in particular on call-stateful SIP servers. While stateless web applications or RESTful [4] web services can scale back immediately without breaking ongoing interactions, this is not the case for call-stateful SIP servers. Before removing a call-stateful SIP server from an elastic SIP cluster, one needs to ensure that all ongoing sessions processed by that server have ended[1]. To fully exploit the potential of dynamic scaling for call-stateful SIP services, this paper explains how to transparently migrate the processing of ongoing sessions to peer servers. Hence a call-stateful SIP server can be released quickly in response to a scaling back event.

Our second contribution builds upon the observation that successful adoption of dynamic scaling support for telco services highly depends on its ability to preserve the services' stringent availability requirements. Instead of responding to load changes in a reactive manner, this paper explores the value of pro-active resource provisioning based on call load forecasting. We observed that the daily call variations of a local trunk group adheres to recurring patterns. This allows to formulate load predictions (and the consequent decisions to increase or decrease the amount of virtual resources) from a history of load observations. To handle also sporadic unanticipated load surges that significantly diverge from these recurring patterns, we combine history-based forecasting (based on time series spanning multiple days) with limited look-ahead predictions (taking into account only on a few prior observations).

The remainder of this paper is structured as follows. Section 2 provides the required background information on SIP services and discusses related work. Section 3 elaborates on how to transparently migrate sessions between elastic SIP servers. Next, Section 4 presents our algorithms to predict call load variations and simulates a dynamically scaling communication service to evaluate these prediction algorithms. Finally, conclusions and future work are presented in Section 5.

2   Background and Related Work

SIP (Session Initiation Protocol) is an IETF-defined signaling protocol for creating, modifying and terminating sessions (including Internet telephone calls, multimedia distribution and multimedia conferences) between two or more remote participants over Internet Protocol (IP) networks. Although SIP is essentially a peer-to-peer protocol (more details on the protocol can be found in [6, 17]), a SIP telco service includes servers to help routing requests to a user's current location, to authenticate and authorize users

---

[1] An experimental analysis of Skype® usage [5] indicated that the average length of a Skype® call was 12m 53s, while the longest call lasted for 3h 26m. Although Skype® is not a SIP service, both technologies offer similar Voice over IP (VoIP) services.

for services, to implement provider call-routing policies and to provide extra features to users [17]. Such SIP servers can operate either in a stateless or stateful mode. If *stateless*, a SIP server processes each message as unrelated to any previous messages – hence simply forwarding SIP requests and responses straightaway. Because of their very nature, such stateless servers (like RFC3261 proxies) can be safely added to or removed from a server farm without compromising ongoing calls. A *stateful* SIP server, in contrast, remembers information about each incoming request and any request it sends as a result of processing incoming requests, and uses this information to affect the processing of future messages associated with that request [17]. In the remainder of this section we further clarify the difference between *transaction-stateful* SIP servers, retaining only transaction state, and *call-stateful* SIP servers, retaining both transaction and session state.

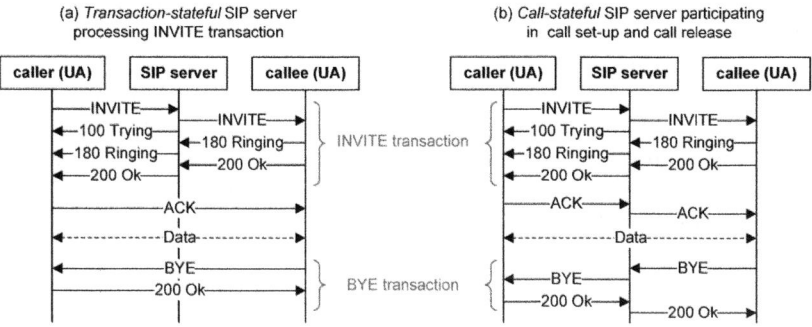

Fig. 1. Transaction-stateful vs. call-stateful SIP servers

Waiting until a stateful SIP server holds no more execution state (and therefore can be removed safely) is only suitable if this condition can be met in bounded time. This is the case, for instance, when stateful SIP servers retain *only transaction state* and *no session state*. As an example, Figure 1(a) illustrates a transaction-stateful SIP proxy participating solely in INVITE transactions. RFC 3261 [17] specifies that the default timeout window of an INVITE and a non-INVITE transaction equals 32 seconds and 4 seconds, respectively. Since ongoing transactions will be canceled if they did not complete after this timeout, a transaction-stateful SIP server reaches a safe removal state in bounded time after preventing the initiation of new transactions.

This is not the case for *call-stateful* SIP servers, such as back-to-back user agents (B2BUAs) or SIP proxies controlling middle boxes that implement firewall and NAT functions. As illustrated in Figure 1(b), call-stateful SIP servers retain session state (like dialogs) during the entire call – that is, from the initiating INVITE to the terminating BYE transaction. Since sessions (in contrast to transactions) typically do not complete in bounded time, waiting until all ongoing sessions have terminated before removing a call-stateful SIP server can significantly delay scaling back operations.

Related work has focused on various aspects of dynamic scaling. The work presented in [8, 11, 12] defines feedback loops to dynamically right-size the amount of provisioned resources. In [18], Seung et al. discuss how to scale enterprise applications over a hybrid cloud (including both private and public resources). The authors of [20], in turn, propose a dynamic provisioning technique for scaling multi-tier Internet applications. Furthermore, today's commercial public clouds (including Amazon Web Service®, Google App Engine® and Heroku®) offer support to automatically scale out and back cloud applications. To the best of our knowledge, however, none of these general purpose solutions take into account the session-oriented nature and the stringent availability requirements of telco applications.

More closely related to our work, [3] explains how IMS functionality can be merged and split among different nodes without disrupting the ongoing sessions or calls. An important difference with our work is that the presented IMS scaling solution is *not* transparent to the SIP UAs. To be precise, SIP UAs need to re-REGISTER and re-INVITE when the IMS functionality has been merged or split among different nodes. Since changing the behavior of SIP UAs complicates the successful adoption of elastic SIP services, our solution seeks to be transparent to both SIP UAs and (non-elastic) SIP servers that belong to different domains.

Finally, we briefly compare our work both with SIP session mobility and SIP handover (as discussed for instance in [2]). SIP session mobility targets the transfer of an ongoing session from one device to another, while SIP handover aims to preserve ongoing SIP sessions when roaming between different networks. In both cases, the UAs initiate and participate in the migration process. This paper, in contrast, presents a solution to migrate the processing of ongoing sessions between SIP servers transparently to the UAs.

3   SIP Session Migration

Safe and transparent migration of SIP sessions between servers can be decomposed into two sub-problems. First, *referential integrity* must be preserved while and after executing a session migration. Since call-stateful SIP servers by nature share their state with their clients (which can be phones as well as other SIP services), these clients are tightly coupled to a specific server during the entire call. Migrating the processing of that particular session to another server, therefore, requires preserving at all times the client's reference to the server that is actually processing its session.

Second, safe and transparent migration of session state from SIP SERVER A to B must be coordinated properly. First, SERVER A must be put into a quiescent execution state. Goudarzi and Kramer describe in [13] that such a quiescent execution state is reached when a service (1) is currently not involved in *ongoing* transactions, and (2) will not participate in any *new* transaction. When applying these prerequisites to call-stateful SIP services, a SIP server reaches a quiescent execution state once (1) all ongoing transactions that belong to the affected SIP dialogs have been completed or terminated, and (2) no new transactions will be started on that server unless to complete other ongoing transactions. The latter occurs, for instance, when a PRACK transaction (to exchange a provisional ACK request [16]) is executed as part of an ongoing INVITE transaction.

To put SERVER A into a quiescent execution state, all SIP requests creating new dialogs (such as INVITE and SUBSCRIBE requests) must be redirected to other SIP servers. Additionally, all requests starting new transactions on confirmed dialogs [17] processed by SERVER A (such as re-INVITE and BYE requests) must be buffered. All other messages, including requests that are sent within an early dialog [17] such as CANCEL and PRACK requests, need to be delivered to SERVER A in order to complete ongoing transactions. Once a quiescent execution state is reached, all remaining session state can safely be captured from SERVER A to be reinstated in SERVER B. Finally, intercepted messages must be released again, but should be redirected to SERVER B.

The remainder of this section presents two stateless "elasticity gateways" to preserve referential integrity while and after migrating SIP sessions. Additionally, we describe and evaluate two protocols coordinating the migration scenario presented above.

3.1 Architectural Overview

We developed a stateless SIP *Client Elasticity Gateway* (CEG) to decouple SIP User Agents (UAs) from the call-stateful SIP servers that process their calls (as illustrated in Figure 2). Configured as the UA's outbound proxy, a CEG conceals the elastic SIP servers from a UA by acting as a single SIP server. It includes load balancing support based on the weight and priority tags of DNS Service (SRV) records and can also be equipped with fail-over support to cope with SIP server crashes. Additionally, the SIP CEG terminates elasticity control messages originating from the elastic SIP cluster, hence concealing the dynamics of the elastic SIP cluster from the UA. We note that traditional load balancing support processes only incoming messages. Since the SIP CEG seeks to control all communication between the UAs and the elastic SIP cluster, it also forwards outgoing messages to the UA. Hence, the CEG can enforce the UA to send back responses to itself instead of to the actual SIP server that previously processed this message. This enables the CEG to transparently redirect requests and responses when the associated dialog has been migrated.

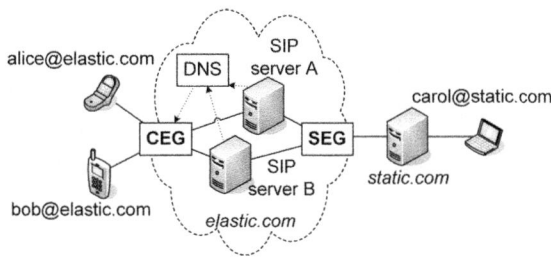

Fig. 2. Architecture of a dynamically scaling (call-stateful) SIP cluster

To achieve this behavior, the CEG combines stateless proxy functionality with registrar support. To enforce that all communication directed to a SIP UA passes the UA's CEG first, the CEG updates all REGISTER requests to replace the UA's contact address with its own - and thus publishes itself as a contact on the client's behalf. This way the CEG will intercept all SIP messages directed to the UA's contact address.

To dispatch these incoming messages to the target UA, the CEG can store the UA's original contact address locally, or it can encode this address in the updated contact header. Additionally, when processing non-REGISTER requests, the CEG adds a VIA header and a RECORD-ROUTE header [17] to the request to make sure responses and subsequent requests pass the CEG as well.

As a potential drawback of this architecture, one could argue that these CEGs may become the system's choke point – thus shifting the scaling problem of stateful SIP servers to another SIP component. We want to stress that the functionality of the CEG is stateless and very lightweight – in our current implementation a single CEG consumes approximately 10% of the CPU load a stateful SIP server consumes when processing the same amount of Calls Per Second (CPS). Furthermore, by deploying (multiple) CEGs close to the client[2] instead of only a few CEGs close to the elastic SIP servers, the CEGs have to meet less strict scalability and high-availability requirements. Since only a few UAs depend on their functionality, the impact of a failure is limited, as is the probability of the CEG to become a choke point.

In addition to the SIP CEG, a stateless SIP *Server Elasticity Gateway* (SEG) has been developed to decouple elastic SIP servers from peers that belong to different (non-elastic) domains (see Figure 2). The role of the SEG is similar to the CEG; it redirects incoming messages to the appropriate server when the associated dialog has been migrated, it terminates elasticity control messages originating from the elastic SIP servers, and it forwards outgoing messages to the target domain (hence concealing the elastic SIP server that actually processed this message and ensuring responses are sent back to the SEG). Although the objectives of the CEG and the SEG are similar, their implementation is slightly different. SEGs do not receive REGISTER requests, for instance, and must be registered with DNS to intercept requests sent to the domain. These and other implementation differences have been the main reason to distinguish between the CEG (which decouples elastic SIP servers from SIP UAs) and the SEG (which decouples elastic SIP servers from peers that belong to different domains).

Finally, we note that in contrast to this application layer solution, network technologies such as MIP [15] or dynamic NAT could be considered as well to transparently redirect messages between SIP servers. By holding together all SIP session migration functionalities at the application layer, however, we seek to limit dependencies to technologies that are not omnipresent – which complicates large scale deployment. Besides, we note that in addition to redirecting messages, these elasticity gateways also participate in the actual session migration process, as we further explain in the next section.

3.2   Migration Protocols

This section introduces two protocols to safely coordinate a session migration between two SIP servers. The first protocol (further referred to as the Gateway Intercept Protocol – GIP) imposes a quiescent execution state by intercepting messages on the elasticity gateways. Figure 3(a) illustrates the various steps of this protocol, exemplified with the migration of an ongoing session from SERVER A towards SERVER B. This migration starts by acquiring the dialog specifications of the affected session (see step 1 in Figure 3(a)), including the addresses of the CEGs and SEGs participating in this session.

---

[2] CEGs can be deployed on home gateways, femto-cells or as a separate service on the UAs.

Next, the Scaling Logic (coordinating the execution of the migration) instructs these CEGs and SEGs to intercept requests starting new transactions on a confirmed dialog (as explained in the beginning of Section 3) and to temporarily buffer these messages in a waiting queue (step 2). The elasticity gateways keep on forwarding all other messages to SERVER A such that any ongoing transaction can complete. Next, SERVER A must be monitored until all ongoing transactions have been completed or terminated (step 3). At this point, a quiescent execution state is reached and the remaining dialog state (as well as all other session state) can safely be transferred from SERVER A towards SERVER B[3] (steps 4 and 5). After the dialog state has been transferred, the CEGs and SEGs are instructed to release all intercepted requests and to redirect them to SERVER B (step 6). If this dialog migration is preceding a shutdown of SERVER A, the latter should also be prevented from receiving dialog-creating requests. This can be accomplished by deregistering SERVER A before executing the dialog migration (step 0). When using DNS to implement load balancing, for instance, removing the records associated with SERVER A prevents the arrival of new dialog-creating requests on that server.

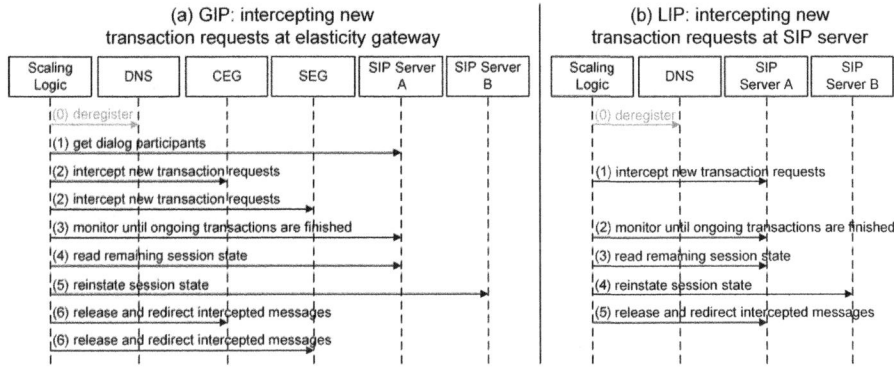

Fig. 3. SIP Session Migration Protocols

The second protocol (further referred to as the Local Intercept Protocol – LIP) builds upon the same principles of the previous one, but intercepts and buffers requests at the SIP server instead of the elasticity gateways. As illustrated in Figure 3(b), the protocol starts by instructing SERVER A to intercept requests starting new transactions on a confirmed dialog (see step 1 in Figure 3(b)). Next, SERVER A must be monitored until all ongoing transactions are completed or terminated (see step 2), which indicates that a quiescent execution state is reached and the remaining dialog state (as well as all other session state) can safely be transferred from SERVER A towards SERVER B (steps 3 and 4). After the dialog state has been migrated, SERVER A acts as stateless proxy, forwarding intercepted messages as well as messages that were still in transit during

---

[3] State migration to transfer dialogs can be implemented by extending the APIs of the affected servers to capture and reinstate state data, or by exploiting a service's high availability support that periodically stores state data to recover from failures (if present).

the migration towards SERVER B. We note that although the elasticity gateways are not involved in this session migration process, they are updated indirectly after the Scaling Logic added or removed server instances to/from DNS (step 0).

3.3 Experiments and Evaluation

In this section we compare the protocols presented above, focusing in particular on their *transaction interruption window*. This interval quantifies the maximum amount of time that requests starting new transactions may be buffered in the course of a migration. Delaying these messages for too long may cause redundant retransmissions or even cancel the affected transaction. To compare the potential transaction interruption window of both protocols, we benchmarked the migration of a single session between two elastic SIP servers in three different settings. A first benchmark was executed while our Scaling Logic (coordinating the execution of both protocols), a single CEG prototype and both elastic SIP servers[4] (SERVERS A and B) were deployed on a local private cloud platform. This scenario, depicted in Figure 4(a), is further referred to as LOCAL. To measure the impact of deploying a CEG outside this private cloud (which impacts the communication latency between the Scaling Logic and the CEG), we performed a second benchmark while the CEG was running on Amazon's EC2 cloud computing platform in Dublin, Ireland[5]. This scenario, illustrated in Figure 4(b), is further referred to as REMOTE CEG. Finally, to measure the impact of a session migration when the affected SIP servers are deployed on a hybrid cloud, we performed an additional benchmark with the CEG, Scaling Logic and one elastic SIP server (SERVER A) running on Amazon's EC2 data center in Dublin, while SERVER B was deployed on the private cloud platform in Antwerp. This scenario, depicted in Figure 4(c), is further referred to as HYBRID.

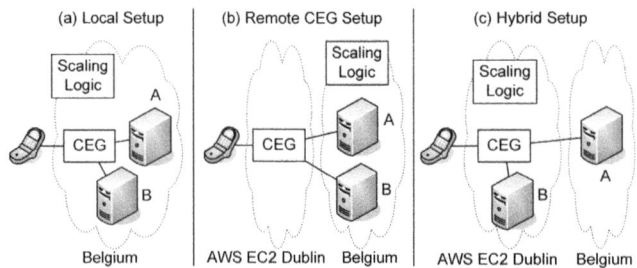

Fig. 4. SIP Session Benchmark Scenarios

For each scenario, we benchmarked 200 session migrations using a prototype implementation of both protocols presented above. The results of these benchmarks are depicted in Figure 5. We can deduce from Figure 5 that LIP has a smaller transaction interruption window than GIP. This can easily be explained by the fact that LIP

---

[4] The employed prototypes of the elasticity gateways and the elastic SIP servers are developed in Java, using the JAIN-SIP stack version 1.2.

[5] The measured average round-trip time between the private cloud platform located in Antwerp and the employed VMs from Amazon's EC2 located in Dublin was around 32 ms.

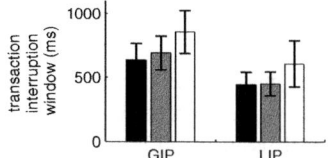

	GIP	LIP
Local Setup (black bars)	637 ms	446 ms
Remote CEG Setup (gray bars)	691 ms	452 ms
Hybrid Setup (white bars)	855 ms	607 ms

Fig. 5. Result from benchmarking the transaction interruption window of GIP and LIP

involves only the SIP servers participating in the migration, while GIP includes the elasticity gateways in the migration process as well (see step 2 and 6 in Figure 3(a)). A different deployment of the CEG only impacts the transaction interruption window when using GIP, for the same reason. Finally, to understand the impact of the measured transaction interruption windows one must take into account that a (transactional) SIP entity starts a retransmission timer with a default value of 500 ms when transmitting a message over an unreliable transport protocol. The results shown in Figure 5 indicate that the measured transaction interruption window of GIP may potentially cause a client's retransmission timer to expire once, resulting in a redundant retransmission of the buffered message. When benchmarking LIP, however, the measured transaction interruption window for LOCAL and REMOTE CEG turns out to be smaller than the default retransmission timeout. Hence, in the absence of significant communication delays between UA and CEG, LIP has the lowest probability to cause redundant message retransmissions. We also note that the actual transaction interruption window can be further reduced by optimizing our prototype implementation.

We conclude this section with some final remarks. First, the benchmark results discussed above may create the perception that LIP is in general a better solution than GIP to implement SIP session migration. This is indeed the case if we focus exclusively on the transaction interruption window of both protocols. One of the main benefits of GIP over LIP, however, is that it can be integrated more easily into existing SIP infrastructure. To apply GIP, existing SIP servers must (1) provide access to the state and specification of the servers' ongoing transactions and dialogs, and (2) enable reinstating the state of migrated dialogs. All remaining support to buffer and redirect messages is handled by separate elasticity gateways. When integrating LIP, in contrast, the affected SIP servers must accommodate this functionality as well.

Finally, instead of intercepting and buffering messages to safely migrate sessions, one can also exploit SIP message retransmissions when using an unreliable transport protocol. In this case, messages are not buffered but become discarded instead. This could be particularly useful when implementing GIP, as the benchmarks indicate that the execution of this protocol may potentially cause a retransmission of these buffered messages anyway.

4   Call Load Forecasting

Migration protocols enable to quickly shutdown call-stateful SIP servers in response to scale down requests, excluding the need to wait until these servers' ongoing calls have finished. In this section, we explore the potential value of pro-active resource provisioning to support dynamic scaling of telco services without compromising their stringent availability requirements.

The call capacity of conventional telephony systems is typically designed to meet the expected Busy Hour Call Attempts (BHCA). Figure 6 depicts the average amount of call attempts per 15 minutes, collected from a trunk group in Brussels from May 2011 until October 2011. Based on these data, we deduce that static peak load dimensioning in this case results into an average call capacity usage of only 50% when averaged over a day. This resource utilization ratio is even lower if the system needs to be dimensioned to handle sporadic unanticipated load surges, such as in case of natural disasters, or anticipated load spikes caused by events with a significant social impact.

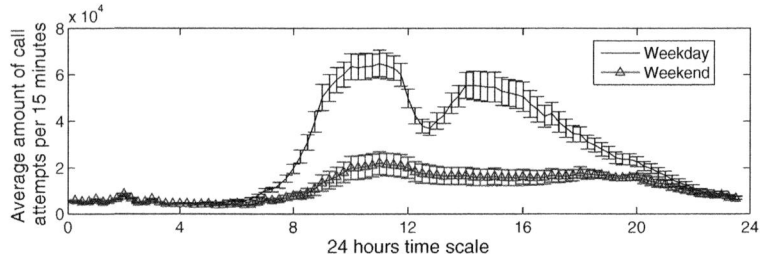

Fig. 6. Average number of call attempts/15 minutes, collected from a local trunk group

Although dynamic scaling enables telco services to optimize their resource utilization ratio, it also increases the risk to compromise their availability requirements. Insufficient resource provisioning to handle load raises, for instance, may cause SLA violations and increases the risk of losing customers. This section explores the potential value of *pro-active* scaling based on call load forecasting to preserve availability requirements while dynamically scaling a telco service.

4.1 Forecasting Algorithms

We present a lightweight limited look-ahead prediction algorithm to forecast short-term call load variations. Furthermore, we combine the short-term forecasting mechanism with history-based forecasting (based on measurements spanning multiple months) to improve the accuracy of call load forecasting by exploiting recurring load variation patterns.

History Based Predictions. Usage variations of communication systems typically represent iterative patterns, resulting from the end users' daily activity routines. The weekday call pattern shown in Figure 6, for instance, includes a peak in the morning around 10 am (when a business day has started for most employees) followed by another peak around 2 pm (after lunch time). Brown's statistical analysis of a telephone call center shows similar call patterns [1]. Additional to weekdays, recurring call patterns can also be observed on weekend days (although the amount of call attempts typically is much lower than on weekdays). These recurring patterns enable to predict call load expectations for each type of day based on a history of measurements. One possible technique to accomplish this involves the use of a Kalman filter, which is an established technique in control systems for noise filtering and state prediction [7]. For every time k, a Kalman filter is trained using a history of measurements collected on previous days at the same time. Based on today's measurements at time k, this Kalman filter can be used to estimate tomorrow's expected call load at the same time [9].

Limited Look-Ahead Predictions. An important limitation of history based predictions is the inability to handle irregular or unexpected events (such as natural disasters or popular sports events) triggering significant load surges. Short-term forecasting (also referred to as limited lookahead control) aims to cope with such unexpected surges by making predictions based solely on a set of recent measurements. Short-term forecasting is a well-studied subject [8,19]. In this paper, we propose a lightweight *Self-adaptive Kalman Filter* (SKF) to anticipate call surges without the knowledge of history data.

A Kalman filter is a recursive estimator. It estimates a new state $x(k+1)$ based on both the current measurement $z(k)$ and the estimation of the previous state $x(k)$. The call load $x(k+1)$ at time $k+1$ can be described as a linear equation $x(k+1) = Ax(k) + w(k)$ with a measurement $z(k) = Hx(k) + v(k)$, in which A represents the relation between the call load from the previous and the current time. $z(k)$ is the measured value at time k, which is related to the load state $x(k)$ multiplied with a factor H. The normally distributed variables w and v represent the process and measurement noise, respectively. Furthermore, we assume that w and v have zero mean and have variance Q and R, respectively. $\widehat{x}$ and $\widetilde{x}$ are defined as the a priori and a posteriori estimations, where $\widetilde{x}(k) = \widehat{x}(k) + K(k)(z(k) - H\widehat{x}(k))$, $\widehat{x}(k+1) = A\widetilde{x}(k)$ and $K(k)$ is the Kalman gain. $\widehat{P}$ and $\widetilde{P}$, in turn, are the a priori and posteriori estimation error variance, where $\widehat{P}(k+1) = A\widetilde{P}(k)A^T + Q$ and $\widetilde{P}(k) = (I - K(k)H)\widehat{P}(k)$. Taking all this into account, the Kalman gain is obtained as $K(k) = H\widehat{P}(k)(H^2\widehat{P}(k) + R)^{-1}$. Assuming that w and v have negligible influence on the system, the Kalman gain is dominated by H. We define that H is within the range $(0, 1]$. When H approaches 1, the system trusts the measurement more. When the system under-predicts the load, H should be decreased to proportionally increase the estimation for the next time. Hence, to enhance the accuracy of our prediction system, we propose to let H self-adapt according to the estimation error as

$$H(k+1) = \begin{cases} I_{e_k<0}\,(H(k) - \tau) + I_{e_k>e_{th}}\,(H(k) + \tau)\,, 0 \le H(k+1) \le 1 \\ 0 \qquad\qquad\qquad\qquad\qquad\qquad\qquad\qquad ,\, H(k+1) < 0 \\ 1 \qquad\qquad\qquad\qquad\qquad\qquad\qquad\qquad ,\, H(k+1) > 1 \end{cases} \qquad (1)$$

where $e_k = \widetilde{x}(k) - z(k)$ is the estimation error at time k. I_A is an indicator function returning value 1 if condition A is true, while otherwise value 0 is returned. As expressed in equation (1), when at time k the call load is under-predicted we increase the value of H at time $k+1$ with a small pre-defined value τ. Otherwise, if the call load is over-predicted, we decrease the value of H by τ. Reducing H will be more harmful than increasing H since under-provisioning resources might violate the service availability. Hence we decrease H only if the error is higher than a threshold e_{th}.

Hybrid Call Load Forecasting. By using the history call load measurements, we can calculate the mean ν and standard deviation σ for a certain time. In this section, we propose two algorithms that use this information to limit abnormal predictions resulting from limited look-ahead forecasting. Hence, we aim to further improve the prediction accuracy.

Hybrid Algorithm 1. At time $k - 1$, we perform both a limited look-ahead prediction $\hat{x}(k)$ and a history based forecasting $\overline{x}(k)$. At time k we calculate $e_s = \hat{x}(k) - z(k)$ and $e_l = \overline{x}(k)(1 + \sigma) - z(k)$. If $e_s < e_l$, the resulting prediction for time $k + 1$ relies exclusively on the limited look-ahead prediction, while otherwise the history based forecasting is used. The motivation of this algorithm is to give more credibility to the algorithm that has the lowest prediction error at the current time.

Hybrid Algorithm 2. If the previous measurement $z(k - 1)$ is within the range $(\overline{x}(k - 1)(1 - \sigma), \overline{x}(k - 1)(1 + \sigma))$, the predicted load for time k is set to $\overline{x}(k)(1 + \sigma)$. Otherwise, we fall back to a limited look-ahead prediction. This algorithm gives more credibility to the history data. The limited look-ahead prediction is adopted only if the measurement does not fall in the history range.

4.2 Safety Margin

Due to the intrinsic cost and risk of under-provisioning telco services, we apply a safety margin δ to the predicted call load x when calculating the amount of required resources. By provisioning enough resources to handle x times $(1 + \delta)$ call attempts, we seek to reduce the possibility of under-provisioning.

4.3 Evaluation

We simulated the behavior of a dynamically scaling communication service to evaluate the prediction algorithms presented above. This simulation uses real-life call attempt measurements, collected from a local trunk group during 66 weekdays (similar to the data depicted in Figure 6). The simulation implements a control function that periodically updates the number of server instances based on the call load prediction. The simulation assumes these instances can be removed quickly, for instance by using the session migration techniques presented in Section 3. To evaluate the forecasting algorithms, the simulation calculates the amount of server instances that are over-provisioned as well as the amount of missing instances to handle the current load (under-provisioning). We define over-provisioning as the ratio between (1) the amount of over-provisioned instances during a single day using call load forecasting and (2) the amount of over-provisioned instances to handle the BHCA of the same day. If the incoming call load is higher than the overall capacity of all provisioned instances, in contrast, a number of requests will be dropped (under-provisioning). We define the Successful Call Processing Rate (SCPR) as the ratio between (1) the total amount of processed call attempts and (2) the total amount of offered call attempts. Both parameters help to understand and evaluate the effectiveness of our forecasting algorithms.

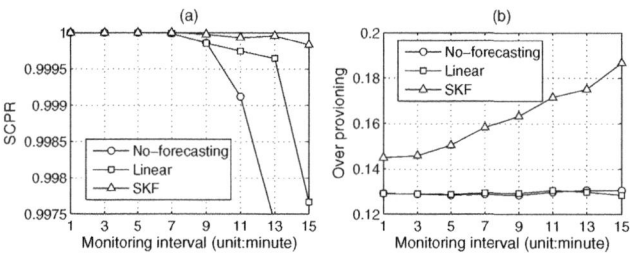

Fig. 7. Limited look-ahead predictions with varying monitoring interval. SKF has initial values $H(1) = 1, \tau = 0.1, \delta = 0.15, e_{th} = 100$.

First, we analyze the correlation between the monitoring interval and the SCPR when using limited look-ahead predictions. We compare our SKF algorithm with limited look-ahead predictions based on linear extrapolation and with a scenario that does no forecasting at all. As illustrated in Figure 7(a), using SKF results into the highest SCPR. Furthermore, for all tested monitoring intervals SKF achieves a SCPR > 99.95%. When the monitoring interval is smaller than 13 minutes, SKF can even achieve a SCPR > 99.99%. These experiments also indicate that SCPR > 99.99% could be achieved without call load forecasting if the monitoring interval is smaller than 8 minutes. Although using a small monitoring interval indeed enables the system to quickly respond to under or over-provisioning, frequently scaling may compromise the stability of the system as well as the overall OpEx reduction depending on the cost associated with every scaling action [11]. Additional to the SCPR, Figure 7(b) depicts the over-provisioning ratio of the tested limited look ahead prediction algorithms. Although SKF generates the highest over-provisioning, we can observe that when the monitoring interval is 15 minutes SKF safely reduces the provisioned call capacity to 18.66% of the capacity needed without dynamic scaling.

We also compare linear and SKF predictions with history-based Kalman predictions and width both hybrid call load forecasting algorithms. During these simulations the monitoring interval was set to 15 minutes. The measured SCPR and over-provisioning rate are depicted in Figures 8(a) and 8(b), respectively. From these results we can deduce that only SKF and Hybrid 1 can realize a SCPR above 99.9%, while Hybrid 1 generates less over-provisioning than SKF. To further compare these two algorithms, we depict their cumulative distribution function (cdf) in Figure 9 by using a different buffering ratio δ. It is easy to understand that increasing δ results in a higher SCPR. Based on this experiment we can also observe that the performance of SKF and Hybrid 1 is very similar for all tested δ values.

In the previous simulations we restricted the capacity of a single instance to 100 calls per minute due to the low BHCA of the employed trunk group measurements (around 5k calls per minute). When increasing this capacity to 500 calls per minute, we observe similar results in terms of SCPR and over-provisioning rate. The only difference worth mentioning relates to the safety margin δ. Since increasing the capacity of a single instance reduces the probability to cause resource under-provisioning, it also decreases

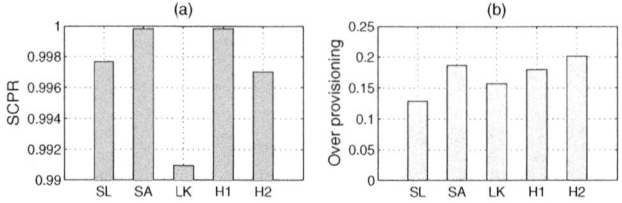

Fig. 8. Call load forecasting comparison, $\delta = 0.15$, SL: linear extrapolation, SA: SKF predictions with $e_{th} = 100$, LK: history-based predictions with Kalman filter, H1: Hybrid 1 with $\tau = 0.1$, H2: Hybrid 2 with $\tau = 0.1$

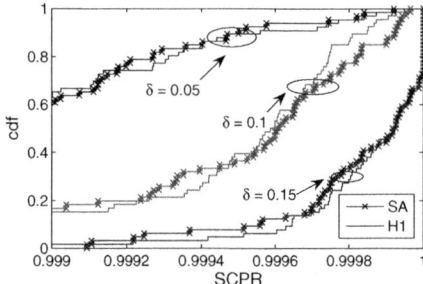

Fig. 9. CDF for SCPR comparison between SA and H1 by using different δ, $e_{th} = 100$

the impact of safety margin δ. Our simulations indicate that when the instance capacity is increased to 500 calls per minute, δ can be set to 0 to achieve similar results as shown in Figure 8 (which uses $\delta = 0.15$).

5   Conclusion and Future Work

In this paper, we investigate the feasibility of applying dynamic scaling to cloudified telco services. We present and evaluate two protocols for transparently migrating ongoing sessions between call-stateful SIP servers. This enables to quickly release a server in response to a scale down request, instead of unnecessarily wasting resources by waiting until all ongoing sessions on that server have ended. Additionally, we propose a self-adaptive Kalman filter to implement limited look-ahead call load predictions and combine this with history-based Kalman predictions to reduce the amount of resource over-provisioning. We believe that both techniques enable to reduce the OpEx of a cloudified SIP service and to increase the resource utilization ratio of a telco cloud provider without compromising service availability.

Future work focuses on how to protect a dynamically scaling SIP service against malicious load surges. Additionally, we are studying the influence of server capacity variations caused by the underlying virtualization technology on the employed SIP scaling feedback system. By combining these results with the findings presented in this paper, we seek for dedicated SIP scaling solution to optimize the amount of employed cloud resources in a safe manner.

References

1. Brown, L., Gans, N., Mandelbaum, A., Sakov, A.: Statistical analysis of a telephone call center: a queueing science perspective. Journal of the American Statistical Association (2005)
2. Chen, M.X., Wang, F.J.: Session mobility of sip over multiple devices. In: Proceedings of the 4th International Conference on Testbeds and Research Infrastructures for the Development of Networks & Communities, TridentCom 2008, pp. 23:1–23:9 (2008)
3. Dutta, A., Makaya, C., Das, S., Chee, D., Lin, J., Komorita, S., Chiba, T., Yokot, H., Schulzrinne, H.: Self organizing IP multimedia subsystem. In: Proc. of the 3rd IEEE Int. Conf. on Internet Multimedia Services Architecture and Applications, IMSAA 2009, pp. 118–123 (2009)
4. Fielding, R.T.: Architectural Styles and the Design of Network-based Software Architectures. Ph.D. thesis, University of California, Irvine (2000)
5. Guha, S., Daswani, N., Jain, R.: An experimental study of the skype Peer-to-Peer VoIP system. In: 5th International Workshop on Peer-to-Peer Systems. Microsoft Research (2006)
6. Hilt, V., Widjaja, I.: Controlling overload in networks of SIP servers. In: IEEE International Conference on Network Protocols, ICNP 2008, pp. 83–93 (October 2008)
7. Kalman, R.E.: A new approach to linear filtering and prediction problems. Transactions of the ASME, Journal of Basic Engineering, 35–45 (1960)
8. Kusic, D., Kephart, J.O., Hanson, J.E., Kandasamy, N., Jiang, G.: Power and performance management of virtualized computing environments via lookahead control. In: Proceedings of the 2008 International Conference on Autonomic Computing, ICAC 2008, pp. 3–12 (2008)
9. Li, J., Moore, A.: Forecasting web page views: methods and observations. Journal of Machine Learning Research 9, 2217–2250 (2008)
10. Lim, H.C., Babu, S., Chase, J.S., Parekh, S.S.: Automated control in cloud computing: challenges and opportunities. In: Proceedings of the 1st Workshop on Automated Control for Datacenters and Clouds, ACDC 2009, pp. 13–18 (2009)
11. Lin, M., Wierman, A., Andrew, L.L.H., Thereska, E.: Dynamic right-sizing for power-proportional data centers. In: Proceedings IEEE INFOCOM 2011, pp. 1098–1106 (2011)
12. Mao, M., Li, J., Humphrey, M.: Cloud auto-scaling with deadline and budget constraints. In: 11th IEEE/ACM International Conference on Grid Computing, Grid 2010 (2010)
13. Moazami-Goudarzi, K., Kramer, J.: Maintaining node consistency in the face of dynamic change. In: Proceedings of ICCDS 1996, pp. 62–69 (1996)
14. Padala, P., Shin, K.G., Zhu, X., Uysal, M., Wang, Z., Singhal, S., Merchant, A., Salem, K.: Adaptive control of virtualized resources in utility computing environments. SIGOPS Oper. Syst. Rev. 41, 289–302 (2007)
15. Perkins, C.: RFC 5944:IP Mobility Support for IPv4, Revised (2010)
16. Rosenberg, J., Schulzrinne, H.: RFC 3262:Reliability of Provisional Responses in Session Initiation Protocol, SIP (2002)
17. Rosenberg, J., Schulzrinne, H., Camarillo, G., Johnston, A., Peterson, J., Sparks, R., Handley, M., Schooler, E.: RFC 3261: SIP: Session Initiation Protocol (2002)
18. Seung, Y., Lam, T., Li, L.E., Woo, T.: Cloudflex: Seamless scaling of enterprise applications into the cloud. In: Proceedings IEEE INFOCOM 2011, pp. 211–215 (April 2011)
19. Trudnowski, D.J., McReynolds, W.L., Johnson, J.M.: Real-time very short-term load prediction for power-system automatic generation control. IEEE Tran. Control Systems Technology 9(2), 254–260 (2001)
20. Urgaonkar, B., Shenoy, P., Chandra, A., Goyal, P.: Dynamic provisioning of multi-tier internet applications. In: Proceedings of ICAC 2005, pp. 217–228 (June 2005)

Remedy: Network-Aware Steady State VM Management for Data Centers

Vijay Mann[1,*], Akanksha Gupta[1], Partha Dutta[1], Anilkumar Vishnoi[1],
Parantapa Bhattacharya[2], Rishabh Poddar[2], and Aakash Iyer[1]

[1] IBM Research, India
{vijamann,avishnoi,akguptal,parthdut,aakiyer1}@in.ibm.com
[2] Indian Institute of Technology, Kharagpur, India
parantapa@cse.iitkgp.ernet.in, rishabh.pdr@gmail.com

Abstract. Steady state VM management in data centers should be network-aware so that VM migrations do not degrade network performance of other flows in the network, and if required, a VM migration can be intelligently orchestrated to decongest a network hotspot. Recent research in network-aware management of VMs has focused mainly on an optimal network-aware initial placement of VMs and has largely ignored steady state management. In this context, we present the design and implementation of Remedy. Remedy ranks target hosts for a VM migration based on the associated cost of migration, available bandwidth for migration and the network bandwidth balance achieved by a migration. It models the cost of migration in terms of additional network traffic generated during migration.

We have implemented Remedy as an OpenFlow controller application that detects the most congested links in the network and migrates a set of VMs in a network-aware manner to decongest these links. Our choice of target hosts ensures that neither the migration traffic nor the flows that get rerouted as a result of migration cause congestion in any part of the network. We validate our cost of migration model on a virtual software testbed using real VM migrations. Our simulation results using real data center traffic data demonstrate that selective network aware VM migrations can help reduce unsatisfied bandwidth by up to 80-100%.

Keywords: Network aware VM migration, VM migration traffic modeling.

1 Introduction

As more and more data centers embrace end host virtualization and virtual machine (VM) mobility becomes commonplace, it is important to explore its implications on data center networks. VM migration is known to be an expensive operation because of the additional network traffic generated during a migration, which can impact the network performance of other applications in the network, and because of the downtime that applications running on a migrating VM may experience. Most of the existing work

---

* Corresponding author.

R. Bestak et al. (Eds.): NETWORKING 2012, Part I, LNCS 7289, pp. 190–204, 2012.

on virtual machine management has focused on CPU and memory as the primary resource bottlenecks while optimizing VM placement for server consolidation, dynamic
workload balance, or disaster recovery [2, 10, 16]. They do not take into account network topology and current network traffic. However, steady state management of VMs
should be network-aware so that migrations do not degrade network performance of
other flows in the network, and if required, a VM migration can be intelligently orchestrated to decongest a network hotspot. If the trigger to migrate one or more VMs comes
from the network itself, it should ensure that the number of VM migrations are kept
low. Recent research [11] [7] [14] [13] in network-aware VM management has focused
mainly on an optimal network-aware initial placement of VMs and has largely ignored
steady state management.

In this context, we present the design and implementation of Remedy: a system for
network-aware steady state management of VMs. To the best of our knowledge, Remedy is the first system to perform network-aware management of VMs so as to minimize
the cost of VM migration over a long-term. We make the following contributions in this
paper:

- We model the cost of VM migration in terms of total traffic generated due to migration.
- We present a target selection heuristic (TSH) that ranks target hosts for a VM migration based on the associated cost of migration, available bandwidth for migration
  and the network bandwidth balance achieved by a migration.
- We implement Remedy as an OpenFlow [12] controller application that can detect the most congested links in a data center network, identify the flows that pass
  through those links, and select the best flows to be rerouted that can resolve network congestion.[1] It then uses an intelligent VM selector heuristic (VSH) to select VMs that are part of these flows based on their relative cost of migration and
  then migrates these VMs in a network-aware manner to decongest these links. Our
  target selection heuristic (TSH) ensures that neither the migration traffic nor the
  flows that get rerouted as a result of migration cause congestion in any part of the
  network.
- We also implement our prototype in the VMFlow simulation framework [11] from
  our earlier work. Our simulation results using real traffic data from a data center
  demonstrate that selective network aware VM migrations can help reduce unsatisfied bandwidth by up to 80-100%.
- We evaluate the Remedy OpenFlow prototype on a virtual software testbed. Our
  results validate our cost of migration model.

The rest of this paper is organized as follows. Section 2 presents a description of the
problem and describes our cost of migration model. Section 3 describes the design and
architecture of Remedy. We provide an experimental evaluation of various aspects of
Remedy in Section 4. An overview of related research is presented in Section 5. Finally,
we conclude the paper in Section 6.

---

[1] Remedy can possibly be implemented over other network monitoring and control frameworks.
We chose OpenFlow for ease of implementation.

2   Online Minimum Cost Network Utilization Using VM Migration

In this section we formulate the main optimization problem that need to be solved for network-aware steady state management of VMs. Since the arrival, departure and change of network demands is not known in advance, the problem is an online one.

Context: The data center network is modeled as a directed graph $G(V, E)$, where each link e in E has a capacity $C(e)$. There are two types of vertices in V: the hosts and the network switches. The edges represent the communication links. Each host has a processing capacity.[2] A network demand is defined by the three parameters, its source, destination and its rate.

Configuration: At any given point of time, the configuration of the system consists of the set of VMs (with their current processing requirement), the set of existing demands (with their current rate), the *placement* Π that maps VM to hosts, and the routing RT that maps demands to paths (in G). (A flow consists of a demand along with its routing path.) The routing is unsplittable, i.e., all the traffic for a demand is sent using a single flow over a single path. Note that multiple VMs can be mapped to a single host.

The load of a link is the sum of the rates of all flows that are routed over that link. The utilization (also called congestion) of a link is defined as the ratio of the load on the link to its capacity. We say that network has α-*utilization* in a configuration, if the utilization of every link is at most α.

Constraints: A configuration should satisfy the capacities of all the hosts and the links. In other words, the sum of the processing requirements of all VMs that are placed on a host should be less than or equal to the capacity of the host, and the sum of the rates of all flows routed over a link should be less than or equal to the capacity the link.

Migration Cost: Migrating a VM generates network traffic between the source and the destination hosts of the migration, where the amount of traffic depends on the VMs image size, its page dirty rate, the migration completion deadline and the available bandwidth. We model the cost of a migration as the total traffic generated due to the migration. In this paper, we model the migration cost of VMware vMotion [3]. Note that, even though the model presented in this paper is for VMware vMotion, our techniques are generic enough to be applicable for most live migration systems that are based on iterative pre-copy technique. In particular, the model presented in this work can be extended by making minor modifications to the stop conditions that vary across different hypervisors.

VMware vMotion[3] is a pre-copy live migration technique. The conditions for stopping the pre-copy cycle are as follows:

1. The current dirty memory can be transmitted in T milliseconds (i.e., dirty memory is small enough to start a stop-copy of at most T millisecond). Here, T is a user setting, and it is called the switchover goal time.

---

[2] Although, storage and memory requirements/capacities of the VMs and the hosts are not considered in this paper, our solution can be extended for these requirements.

[3] For ease of presentation, we use the terms vMotion and VM migration interchangeably.

2. vMotion did not make *enough progress* in the previous pre-copy cycle. vMotion measures the progress of a migration by the difference in dirty memory before and after a pre-copy cycle. This difference should at least be X MBs, where X is a user setting. We call X the minimum required progress amount.

In case vMotion did not make enough progress, and the current dirty memory cannot be transmitted in 100 seconds, the migration fails. In the following Theorem, we derive the equations for the number of pre-copy cycles, and for the total traffic generated during a migration, under the assumption that the following parameters are given constants: (1) the memory size M of a VM, (2) the page dirty rate R of a VM and (3) the bandwidth of the link used for migration L.

Theorem 1. *The number of pre-copy cycles* $n = \min\left(\left\lceil \log_{R/L} \frac{T \cdot L}{M} \right\rceil,\right.$ $\left.\left\lceil \log_{R/L} \frac{X \cdot R}{M \cdot (L-R)} \right\rceil\right)$, *and the total traffic generated by the migration* $N = M \cdot \frac{1-(R/L)^{n+1}}{1-(R/L)}$.

Proof. Let N_i denote the traffic on the ith pre-copy cycle. Taking into account the page dirty rate R and the bandwidth L used for migration, we see that the first iteration will result in migration traffic equal to the entire memory size $N_0 = M$, and time $\frac{M}{L}$. During that time, $M\frac{R}{L}$ amount of memory becomes dirty, and hence, the second iteration results in $N_2 = M\frac{R}{L}$ amount of traffic. Extending this argument it is easy to see that $N_i = M\left(\frac{R}{L}\right)^{i-1}$ for $i \geq 1$. Thus if there are n pre-copy cycles and one final stop-copy cycle, the total traffic generated by the migration is $N = \sum_{i=1}^{n+1} N_i = M \cdot \frac{1-(R/L)^{n+1}}{1-(R/L)}$. We now derive an equation for n.

Suppose that the pre-copy stops after n cycles. Then one of the following two conditions must hold true, at the end of cycle n. (1) From the first pre-copy stopping condition we have, $M\left(\frac{R}{L}\right)^n < T \cdot L$, which in turn implies $n < \log_{R/L} \frac{T \cdot L}{M}$. (2) From the second pre-copy stopping condition we have, $M\left(\frac{R}{L}\right)^{n-1} - M\left(\frac{R}{L}\right)^n < X$, which in turn implies $n < \log_{R/L} \frac{X \cdot R}{M \cdot (L-R)}$.

Thus, from the above two conditions, the number of pre-copy cycles $n = \min\left(\left\lceil \log_{R/L} \frac{T \cdot L}{M} \right\rceil, \left\lceil \log_{R/L} \frac{X \cdot R}{M \cdot (L-R)} \right\rceil\right)$.

We also note that the time $W(L)$ spent on the stop-copy transfer for a given migration bandwidth L, is given by $W(L) = \frac{M}{L} \cdot \left(\frac{R}{L}\right)^n$. ∎

We now describe the online problem that we address to perform network-aware steady state VM management.

Problem: Due to possibly under-provisioned links and time-varying demands, the network utilization may exceed a desired threshold over time. In this problem, we aim to reduce the network utilization by migrating VMs while incurring minimum migration cost. More specifically, we define the online α-Minimum Migration Cost Network Utilization (α-MCNU) problem as follows.

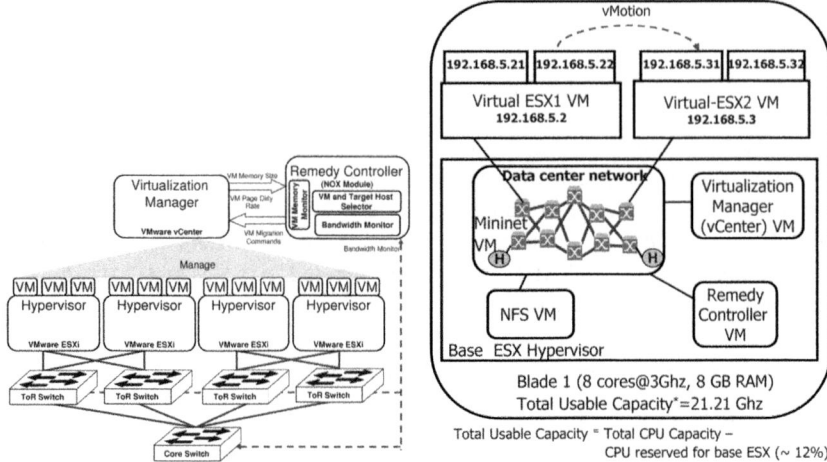

(a) Remedy Architecture - current imple- (b) Remedy virtual emulation environment:
mentation with VMware and NOX Open- emulates 2 physical hosts with 4 mobile
Flow controller   VMs using nested virtualization

Fig. 1. Remedy Architecture and Virtual Emulation Environment

During the data center operation, if there is a change in the demands (in particular, if some demand is created, terminated, or its rate is changed significantly), then the system may introduce one or more VM migrations. Each migration has an upper bound on its completion time. In the α-MCNU problem, given an initial configuration (placement and routing), and the arrival, departure and change of demands over time (which is not known in advance), we need to select a sequence of VM migrations such that the network utilization remains at most α, except during the period of a migration, while ensuring that the cost of migration is minimized. Even the offline version of MCNU problem, where all demand changes are known in advance, is NP-Hard.[4]

Theorem 2. *The offline version of MCNU problem is NP-Hard.*

We present our heuristics for the MCNU problem in section 3.2.

3   Architecture and Design

In this section we present the architecture and design of Remedy, a system for network-aware steady state VM management (refer Figure 1(a)). Remedy has been implemented in Python as a NOX [9] module, which is one of the popular OpenFlow controllers currently available. An OpenFlow controller is connected to all OpenFlow enabled switches and routers and communicates with these switches through a control channel using the OpenFlow control protocol. Remedy has three main components: (1) Bandwidth Monitor, (2) VM and Target Selector and (3) VM Memory Monitor. We now describe each of the above components in detail in the following subsections.

---

[4] We omit the proof due to lack of space.

3.1 Bandwidth Monitor

OpenFlow provides flow monitoring capabilities in the form of per flow, per port and per queue statistics (using OFPST_FLOW, OFPST_PORT and OFPST_QUEUE request types, respectively as per OpenFlow 1.1 specification). Remedy Bandwidth Monitor first detects all links by using the "discovery" module in NOX. Each link is represented as a 4 tuple: `<Start SwitchID, Start PortNumber, End SwitchID, End PortNumber>`.

Remedy controller polls for flows statistics and port statistics by periodically sending flowStats (OFPST_FLOW) and portStats (OFPST_PORT) requests to all the switches in the network. It uses the port statistics to find out the transmitted bytes from each port on each switch in the current polling interval and sorts this list in decreasing order to find the most busy ports in the network. For each port in the busy ports list, it then uses the flow statistics to find out the number of bytes recorded for each flow on that port in the current monitoring interval. Note that a flow represents a network demand and is uniquely identified through source and destination IP addresses (we ignore port numbers since our eventual goal is to find VMs that contribute the most to a link's utilization). This list of flows on each port is also sorted in decreasing order to find the heaviest flows on each port in the network. Since a port may appear on either end of a link, the list of busy ports is used to calculate the traffic on each link and a list of busy links is created. Similarly, the list of heaviest flows on a port is used to calculate the heaviest flows on each link, and their contribution to the overall link utilization is calculated.

Remedy uses an input headroom value to determine a link utilization value beyond which a link is considered congested. For example, if the input headroom value is 20, then a link is considered congested when its utilization crosses 80%. Since a VM may be part of multiple flows either as a source or a destination, the sorted list of heaviest flows on each congested link is used to determine the VMs which are top contributors to a link's utilization. This list of VMs is passed on to the VM and Target Host Selector.

3.2 VM and Target Host Selector

The VM and Target Host Selector uses two methods, the VM selector heuristic (VSH) and the target selector heuristic (TSH), to select a VM and the target host for migrating the selected VM, in order to control network utilization. These two heuristics together addresses the MCNU problem with $\alpha = 1$ (as defined in Section 2). In the case where the VM to be migrated is already provided as input (either by a user or by an existing VM placement engines which consider CPU and memory statistics to migrate a VM for server consolidation, workload balance or fail-over), Remedy uses only the TSH. We now describe these two heuristics in detail. (See Figure 2 for pseudocode.)

VM Selector Heuristic (VSH): VM selector heuristic uses the list of top contributor VMs for each congested edge and selects the VMs to be migrated based on number of communicating neighbors of a VM, total input/output traffic of a VM and an approximate cost of migrating a VM. As discussed in section 2, the cost of migrating a VM (in terms of additional network traffic generated) depends on the memory size of

```
1:  function VSH
2:      rank VMs in increasing order of their approx migration cost = 120% of VM size
3:      rank VMs in the increasing order of their number of communicating neighbor
4:      rank VMs in the increasing order of their total input/output traffic
5:      return VM with lowest sum of the three ranks

6:  function TSH (VM v)
7:      H ← set of all hosts with enough spare CPU capacity to place v; H_f ← ∅
8:      calculate available b/w from current host of v to each host in H
9:      for each host h in H do
10:         calculate the migration traffic and migration feasibility of VM v to host h following Sec. 2
11:         if migrating v to h is feasible then
12:             add h to H_f
13:             calculate available b/w for each flow of v when re-route to host h
14:             calculate normalized weighted sum of available b/ws calculated in previous step
15:     return a ranked list of hosts in H_f in the decreasing order of their normalized weighted sum
16:
```

Fig. 2. Pseudocode for VSH and TSH

the VM, its page dirty rate, the available bandwidth for migration, and some other hypervisor specific constants. Since, the available bandwidth for migration can be known only when the target host for a migration has already been determined, we use 120% of memory size of a VM as a rough indicator of the additional network traffic that would be generated on migrating a particular VM. VMs are first sorted in increasing order by this approximate cost of migration and then by their number of communicating neighbors (in increasing order) and finally by their total input/output traffic. The rank of each VM in these three sorted lists is noted and added together. The VM which has the lowest sum of ranks is selected as the VM to be migrated.

Target Selector Heuristic (TSH): Given a VM to migrate, target selector heuristic ranks target hosts for that VM based on the associated cost of migration, available bandwidth for migration and the network bandwidth balance achieved by a migration. In a typical data center, a physical host has many VMs running on it, which are connected to physical network interfaces through a virtual switch (or vSwitch) inside the hypervisor. Physical hosts are directly connected to a ToR switch, which in turn is connected to one or more aggregate switches. The aggregate switches then connect to a layer of core switches. In some data centers, the aggregate and core switches may form a single layer. TSH evaluates all end hosts that have free CPU capacity. For all such hosts, it finds the path from the current host and evaluates the available bandwidth. Note that, Remedy directly calculates the available bandwidth on a path using the flow statistics collected by Bandwidth Monitor from all switches. Thus, Remedy does not need to use any available bandwidth estimation tools. Based on the observed bandwidth, Remedy calculates the cost of migration as per the cost of migration model described in Section 2 (which is in terms of additional network traffic that will be generated during migration). This additional network traffic is added to existing traffic on the chosen path and the resultant available bandwidth is calculated to ensure that migration is feasible. If migration is feasible, TSH then calculates the available bandwidth for all network demands or flows that this VM is a part of and which will get rerouted as a result of this

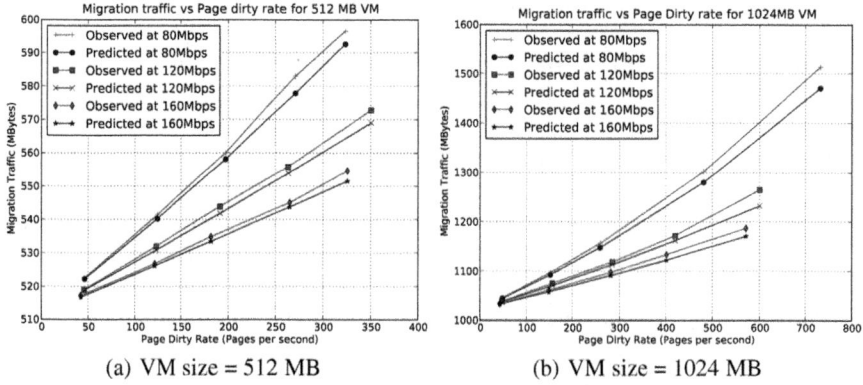

(a) VM size = 512 MB   (b) VM size = 1024 MB

Fig. 3. Cost of migration model predicts migration traffic to within 97% accuracy

migration. TSH calculates a normalized weighted sum of these available bandwidths (each demand from a VM is first normalized with respect to the largest demand for that VM, and then multiplied by the available bandwidth on its path - this step is repeated for all demands for a given VM, and these are added up across all network demands of that VM). This normalized weighted sum of available bandwidths is used to create a ranked list of all destination hosts to which a migration is feasible. Intuitively, this normalized weighted sum of available bandwidths favors large flows over small flows and the heuristic attempts to create the best possible available bandwidth balance after migration. The Remedy System tries to migrate the selected VM to the destination host one by one, according to the above ranked list, until a migration is successful.

3.3   VM Memory Monitor

VM Memory Monitor fetches the memory size and the current average page dirty rate for each VM. The settings for memory size are usually exposed by the virtualization manager. We use VMware ESXi hypervisor as our virtualization platform that support live VM migration in the form of VMware "vMotion". VMware ESXi hypervisor has a remote command line interface (RCLI) [4], that has commands to fetch memory size and also to initiate a migration. However, it does not expose a direct metric to measure page dirty rate. We currently rely on VMware ESXi logs to find out page dirty rates.

4   Experimental Evaluation

We first validated the correctness of our system on an innovative virtual software environment that leverages nested system virtualization and software switches. In order to analyze the effects of VM selector and target selector heuristics at scale, we implemented these heuristics in the VMFlow simulation framework from our earlier work [11] and conducted simulations using traffic data from a real data center. We now describe these experiments in detail.

4.1 Validating Remedy in an OpenFlow Network

Virtual Software Testbed: We created an innovative virtual emulation environment that gave us full control over the network as well as VM mobility across hosts. We leveraged nested server virtualization in VMware ESXi server [5] to create VMs that run hypervisor software within a single physical machine. Our virtual testbed that emulates two hosts (each with two VMs) inside a single physical blade server is shown in Figure 1(b). The various VMs that run the hypervisor software (we refer to them as virtual hypervisors) represent multiple virtual hosts. Virtual hypervisor or virtual hosts are connected to each other through an OpenFlow software switch (version 1.1) or a Mininet [1] data center network with multiple OpenFlow software switches running inside a VM. Mininet uses process-based virtualization and network namespaces to rapidly prototype OpenFlow networks. Simulated hosts (as well as OpenFlow switches) are created as processes in separate network namespaces. Connectivity between hosts and switches, and between switches is obtained using virtual interfaces and each host, switch and the controller has a virtual interface (half of a "veth" pair). We connect two real hosts (in this case, the virtual ESXi hosts) on two interfaces of the Mininet VM such that they are connected to the data center network created inside the Mininet VM. We run the Remedy controller on top of NOX as a remote controller inside another VM and this is also connected to the Mininet data center network. Shared storage for these virtual hosts is provided by shared NFS service also running inside a VM and connected to the Mininet data center network. Virtualization manager (VMware vCenter) runs inside another VM. Direct wire connectivity between these various VMs (OpenFlow or Mininet VM, Controller VM, NFS VM, vCenter VM and Virtual Hypervisor VMs) is achieved through various vSwitches on the base hypervisor. We also created VMs on top of the virtual hypervisors (or virtual hosts) that can be migrated across various hosts. While, the virtual testbed provides only an emulation environment and can not be used for any scalability or absolute performance comparison, it still proves to be a valuable tool to verify the correctness of our system using real VM migrations on VMware ESXi hypervisor, and OpenFlow capable software switches.

Results: In order to validate our cost of migration model described in section 2, we developed a simple page modification microbenchmark. This microbenchmark, first allocates large chunks of page aligned memory. It then randomly writes to different pages in a loop and executes busy loop for a predetermined count based on the specified page dirty rate. The page modification benchmark is executed with different page dirty rates on one of the VMs hosted on the virtual ESXi server. The two virtual ESXi servers were connected via an OpenFlow switch VM for this experiment. This VM was then migrated to another virtual ESXi at different link bandwidths. To create different link bandwidths, a queue was created on the output port of the OpenFlow switch and the minimum rate on that queue was set to the desired value. Remedy controller installed a rule to enqueue all migration traffic (identified through TCP port numbers used by "vMotion") to this queue and all other traffic was sent through the default queue. The total migration traffic was monitored inside the Remedy Controller, as well as verified through VMware ESXi logs. The page dirty rates were also verified through VMware ESXi logs. The results are shown in Figure 3 for two VM sizes - 512 MB and 1024 MB.

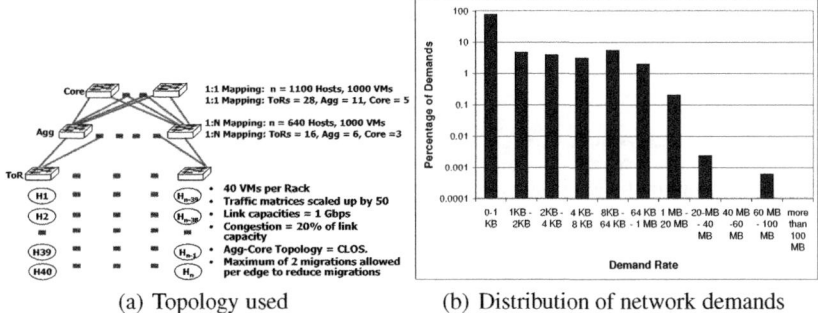

(a) Topology used   (b) Distribution of network demands

Fig. 4. VMFlow simulations to evaluate the VM selector and Target selector heuristics

In both the cases, our model predicts the total migration traffic to within 97% accuracy for all page dirty rates and link bandwidths.

4.2 Validating the Effectiveness of VM and Target Selector Heuristics at Scale through Simulations

In this section we evaluate the VM selector and target selector heuristics at scale through simulations in the VMFlow simulation framework [11]. Note that in all these simulations, the cost of migration was assumed to be 1.2 times the memory size of a VM as mentioned in section 3. Deadline for migration completion time was assumed to be equal to the monitoring interval for our input traffic matrices (15 minutes). The memory sizes for all VMs were assumed to be normally distributed with a mean of 1 GB and standard deviation of 512 MB. With these settings, the goal of the simulation experiments is to evaluate the effectiveness of VM and target selector heuristics at scale in achieving network balance and their ability to satisfy bandwidth demand.

VMFlow Simulator: VMFlow simulator simulates a network topology with VMs, switches and links as a discrete event simulation. Each host is assumed to contain one or more VMs. VMs are either randomly mapped to all hosts (in case of 1:1 host to VM mapping) or they are mapped based on their CPU requirements (in case 1:N host to VM mapping). VMs run applications that generate network traffic to other VMs, and VMs can migrate from one node to the other. At each time step, network traffic generated by VMs (denoted by an entry in the input traffic matrix) is read and the simulator attempts to satisfy these demands in decreasing order (largest demands are satisfied first). For a given demand, it calculates the corresponding route between the hosts based on an already used path or the shortest path in the network topology, while ensuring that available link capacities can satisfy the given network demand. In order to simulate congestion, we modified the simulator so that it can take more demands than available link capacities. For the results in this paper, we simulated 20% congestion where each link can take 20% more traffic than its rated capacity. The simulator then executes an ongoing optimization step in which it detects congestion, finds the set of VMs to migrate to decongest the network using VM selector heuristic (VSH) and selects the appropriate

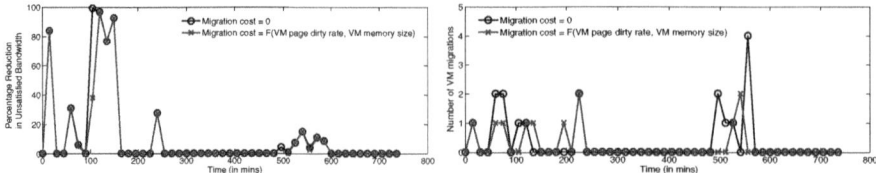

(a) Reduction in unsatisfied bandwidth due to (b) Improvement in number of migrations due
network-aware steady state VM management   to migration cost-awareness

Fig. 5. Results from VMFlow simulator with 1000 VMs and 1:1 host to VM mapping using data
from a real data center over a 12 hour period: Results illustrate benefits of using network-aware
steady state VM management - unsatisfied bandwidth reduces up to 100% at some time-steps
in the simulation, adding migration cost-awareness to the heuristics reduce the total number of
migrations required to achieve this benefit from 17 to 11

set of destinations for each VM using the target selector heuristic (TSH). At each time
step of the simulation, we compute the total bandwidth demand that can not be satisfied, the number of migrations required and link utilizations. A new input traffic matrix
that represents the network traffic at that time instance, is used at each time step of the
simulation.

Network Topology and Input Traffic Matrices: The simulator creates a network
based on the given topology. Our network topology consisted of 3 layers of switches: the
top-of-the-rack (ToR) switches which connect to a layer of aggregate switches which in
turn connect to the core switches. Each ToR has 40 hosts connected to it. We assume a
total of 1000 VMs and 1100 hosts for the 1:1 VM to host mapping scenario, and 640
hosts for 1:N host to VM mapping scenario. The topology used for these simulations is
shown in Figure 4(a).

To drive the simulator, we needed a traffic matrix (i.e. which VM sends how much
data to which VM). We obtained data from a real data center with 17,000 VMs with
aggregate incoming and outgoing network traffic from each VM. The data had snapshots
of aggregate network traffic over 15 minute intervals spread over a duration of 5 days.
Our first task was to calculate a traffic matrix out of this aggregate per VM data. Given
the huge data size, we chose data for a single day. We used the simple gravity model to
create a traffic matrix out of this aggregate per VM data (as described in [11]). As noted
in [11], these traffic matrices did not show much variation over time. Note that traffic
matrices generated by simple gravity model tend to be dense and traffic is distributed
over all the VMs in some proportion. This results in a large number of very low network
demands. In order to make these demands significant so that they can cause congestion in
the network, we used a scale-up factor of 50 (i.e. each entry in the input traffic matrix was
multiplied by 50) for the traffic matrices used in 1:1 host to VM mapping experiment and
scale up factors of 75 and 125 for the traffic matrices used in 1:N host to VM mapping
experiments. A distribution of network demands for one of the traffic matrices (used
for one time step in simulation) after applying a scale factor of 50 is shown in Figure
4(b). This matrix had 1 million total demands out of which only 20% are zero demands.
Maximum demand size was 65.91 MB/s and the mean demand size was 13 KB/s.

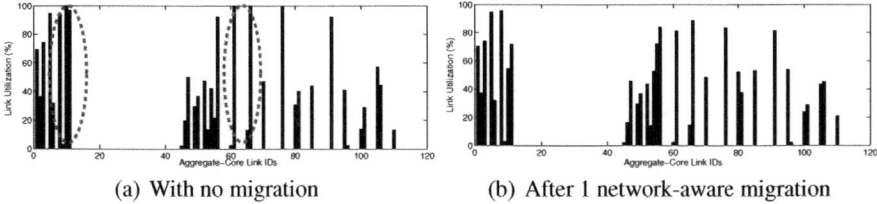

(a) With no migration   (b) After 1 network-aware migration

Fig. 6. Agg-Core layer link utilizations from VMFlow simulator with 1000 VMs and 1:1 host to VM mapping using data from a real data center at time step 2 (point 2 in Figure 5(a)). Results illustrate benefits of a single network-aware VM migration - links that were congested and hence could not take on additional bandwidth demands (in the no migration case) get decongested through 1 network-aware VM migration.

1:1 Host to VM Mapping Simulation Results

We first ran the simulation for 1:1 host to VM mapping. Since, the input traffic matrices are already dense (much more than a real data center traffic matrix), 1:1 host to VM mapping is useful to show the benefits of our heuristics. We conducted two sets of simulations:

- The first set of simulations, assumed that VM migrations were of 0 cost and the only objective was to decongest a given edge (with the constraint that only two migrations were allowed per congested edge).
- During our second set of simulations, we assigned a cost of VM migration based on the cost of migration model given in section 2 that uses memory size of a VM, its page dirty rate, and link bandwidth to predict the additional traffic generated on the network.

Simulation results for 1:1 host to VM mapping are shown in Figure 5(a) and in Figure 5(b). Results illustrate benefits of using network-aware steady state VM management. Unsatisfied bandwidth reduces up to 100% at some time-steps in the simulation (refer Figure 5(a)) while adding migration cost-awareness to the heuristics reduces the total number of migrations required to achieve this benefit from 17 to 11 (refer Figure 5(b). Link utilizations for the aggregate-core layer for one of the time steps (time step 2 in Figure 5) are given in Figure 6(a) for the no migration base case, and in Figure 6(b) for the 0 cost network aware migration case. Note that the links that were congested (marked by circles) and hence could not take on additional bandwidth demands (in the no migration case) get decongested through a single network-aware VM migration.

1:N Host to VM Mapping Simulation Results

In this section, we present the simulation results for 1:N host to VM mapping scenario. We assumed a fixed host CPU capacity of 4 GHz and generated normally distributed random CPU demands for each VM with a mean demand of 1.6 GHz and standard deviation of 0.6 GHz.

VMs are mapped based on their CPU requirements. 1000 VMs were mapped onto 574 hosts and the remaining hosts remained idle. While, choosing the best host for a VM under a given ToR switch, the host with the maximum available link bandwidth to

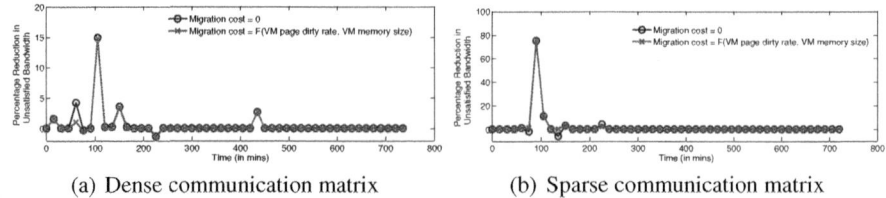

(a) Dense communication matrix   (b) Sparse communication matrix

Fig. 7. Results from VMFlow simulator with 1000 VMs and 640 hosts (1:N host to VM mapping, 1000 VMs were mapped to 574 hosts using CPU data) using data from a real data center over a 12 hour period: Reduction in unsatisfied bandwidth is lower (up to 15%) for dense communication matrix generated by simple gravity model and goes up to 75% for a sparse matrix

the ToR switch is chosen, provided that the host has enough spare CPU capacity to host the VM. Traffic matrices were scaled up by a factor of 75.

In this case again we conducted two sets of simulations - one which assumed a 0 migration cost, and the other that calculated a cost of migration in terms of additional network traffic generated as per our cost of migration model. Simulation results for 1:N host to VM mapping are shown in Figure 7(a). Unsatisfied bandwidth reduces by up to 15% at some time steps in the simulation. However, the gains here are modest, as compared to 1:1 mapping scenario, since the traffic matrix is already dense and as more VMs are packed into a single host, the network traffic matrix becomes denser, reducing the ability to move VMs freely. However, real data center traffic matrices are much less dense (as compared to the traffic matrices generated by simple gravity model that we used), and hence we expect the gains to be much higher in real data centers. To verify this, we generated sparse matrices by randomly making half of the demands zero, while increasing the scale factor for traffic matrices from 75 to 125. Simulation results for the sparse matrices are given in Figure 7(b). Unsatisfied bandwidth now reduces by up to 75%. Adding migration cost-awareness to the heuristics reduces the total number of migrations required in both the scenarios (from 11 to 8 in the dense communication matrix, and from 15 to 13 in the sparse communication matrix scenario). Note that at one time step in dense communication matrices, the unsatisfied bandwidth increases resulting in negative reduction value. We believe this also happened primarily due to a dense traffic matrix, where the added congestion in the network resulted in the simulator trying to satisfy larger demands and failing, when it could have satisfied smaller demands.

5   Background and Related Work

Recently there has been significant amount of work on VM placement and migration. The related research work can be broadly classified as follows.

5.1   Network-Aware Placement of VMs

Our prior work on network-aware initial placement of VMs - VMFlow [11], proposed a greedy heuristic for initial placement of VMs to save network power while satisfying more network demands. As one of the contributions of this paper, VMFlow simulator has been enhanced with the VM selector heuristic (VSH) and TOR selector heuristic (TSH) to incorporate network-aware steady state management.

The online minimum congestion mapping problem studied in [7] by Bansal et al. is close to our work: it considers placing a workload consisting of network traffic demands and processing demands on an underlying network graph (substrate) while trying to minimize the utilization over all links and nodes. The paper [7], however, differs from our work in two crucial aspects. First, in [7], once a workload is placed (i.e., its node and link assignment are finalized by the algorithm), this assignment cannot be changed during placement of subsequent workloads. On the other hand, migrating the source and destination of a network demand (which corresponds to VMs in our setting) is the primary technique used in this paper. Second, we consider the cost of VM migration, and ensure that the migration is completed in a timely manner, which are not considered in [7]. Sonnek et al. [14] propose a technique that monitors network affinity between pairs of VMs and uses a distributed bartering algorithm coupled with migration to dynamically adjust VM placement such that communication overhead is minimized. Meng et al. [13] consider heuristics for VM placement in a data center to optimize network traffic while minimizing the cost of VM migration. Although their primary algorithm is presented in an offline setting, it can be executed periodically to modify the VM placements according to the time-varying network demands. However, unlike our paper, their placement algorithm does not consider minimizing the cost over a sequence of network demand changes, but only tries to minimize the cost of each individual step. Moreover, the solution presented by [13] does not specify the routing path for the traffic demands, which is handled by our work. Finally, compared to [13], our solution uses a more detailed model for the cost and time required for the VM migrations.

5.2   Cost-Aware Migration of VMs and Prediction of Migration Performance

Most of the existing research [10] [8] [6] on cost-aware migration of VM analyzes the impact of a live VM migration on the workload that is executing on the migrating VM itself. Breitgand et al. [8] model the total cost of live in-band migration, and present an optimal bandwidth allocation algorithm for in-band live migration such that the cost of migrating a VM, in terms of the down time experienced by the migrating VM, is minimum. However, in a shared data center network, a VM migration also affects the performance of other hosts in the network, which is the focus of our work. Jung et al. [10] present a holistic optimization system that considers various costs of migration, such as, application performance degradation of the migrating VM and power trade offs. Sherif et al. [6] present two simulation models that are able to predict migration time to within 90% accuracy for both synthetic and real-world benchmarks running on Xen hypervisor. Stage et al. [15] propose a complementary scheme in which network topology aware scheduling models are used for scheduling live VM migrations such that effect of VM migrations on other nodes in the network is reduced.

6   Conclusion

In this paper, we presented the design and implementation of Remedy - a system for network aware steady state VM management that ranks target hosts for a VM migration based on the associated cost of migration, available bandwidth for migration and the network bandwidth balance achieved by a migration. Our simulation results using real data center traffic data demonstrated that selective network aware VM migrations, as proposed in Remedy, can help reduce unsatisfied bandwidth by up to 80-100%.

As part of ongoing and future work, we are currently working on implementing Remedy in a real data center with state-of-the-art network equipment. Future extensions to Remedy include a decentralized design for monitoring and an analysis engine to analyze network demands and link utilizations using auto regressive moving average model to verify whether a given demand or link utilization is stable or a transient spike. This will help us make more informed decisions to tackle long term congestion. Finally, we also plan to work on extending the cost of migration model to other hypervisors such as Xen and KVM.

References

1. Mininet: Rapid prototyping for software defined networks,
   http://yuba.stanford.edu/foswiki/bin/view/OpenFlow/Mininet
2. VMware Server Consolidation,
   http://www.vmware.com/solutions/consolidation/
3. VMware vMotion: Live migration of virtual machines,
   http://www.vmware.com/products/vmotion/overview.html
4. VMWare vSphere command line interface,
   http://www.vmware.com/support/developer/vcli/
5. VMware vSphere Hypervisor, http://www.vmware.com/products/vsphere-hypervisor/overview.html
6. Akoush, S., Sohan, R., Rice, A., Moore, A., Hopper, A.: Predicting the performance of virtual machine migration. In: IEEE MASCOTS (2010)
7. Bansal, N., Lee, K., Nagarajan, V., Zafer, M.: Minimum congestion mapping in a cloud. In: ACM PODC (2011)
8. Breitgand, D., Kutiel, G., Raz, D.: Cost-aware live migration of services in the cloud. In: SYSTOR 2010 (2010)
9. Gude, N., Koponen, T., Pettit, J., Pfaff, B., Casado, M., McKeown, N., Shenker, S.: NOX: Towards an Operating System for Networks. In: ACM SIGCOMM CCR (July 2008)
10. Jung, G., Hiltunen, M.A., Joshi, K.R., Schlichting, R.D., Pu, C.: Mistral: Dynamically managing power, performance, and adaptation cost in cloud infrastructures. In: IEEE ICDCS (2010)
11. Mann, V., Kumar, A., Dutta, P., Kalyanaraman, S.: VMFlow: Leveraging VM Mobility to Reduce Network Power Costs in Data Centers. In: Domingo-Pascual, J., Manzoni, P., Palazzo, S., Pont, A., Scoglio, C. (eds.) NETWORKING 2011, Part I. LNCS, vol. 6640, pp. 198–211. Springer, Heidelberg (2011)
12. McKeon, N., et al.: OpenFlow: Enabling Innovation in Campus Networks. In: ACM SIGCOMM CCR (April 2008)
13. Meng, X., Pappas, V., Zhang, L.: Improving the Scalability of Data Center Networks with Traffic-aware Virtual Machine Placement. In: IEEE INFOCOM (2010)
14. Sonnek, J., Greensky, J., Reutiman, R., Chandra, A.: Starling: Minimizing communication overhead in virtualized computing platforms using decentralized affinity-aware migration. In: IEEE ICPP (2010)
15. Stage, A., Setzer, T.: Network-aware migration control and scheduling of differentiated virtual machine workloads. In: ICSE CLOUD Workshop (2009)
16. Verma, A., Ahuja, P., Neogi, A.: pMapper: Power and Migration Cost Aware Application Placement in Virtualized Systems. In: Issarny, V., Schantz, R. (eds.) Middleware 2008. LNCS, vol. 5346, pp. 243–264. Springer, Heidelberg (2008)

Building a Flexible and Scalable Virtual Hardware Data Plane

Junjie Liu[1,2], Yingke Xie[1], Gaogang Xie[1], Layong Luo[1,2], Fuxing Zhang[1,2], Xiaolong Wu[1,2], Qingsong Ning[1,2], and Hongtao Guan[1]

[1] Institute of Computing Technology, Chinese Academy of Sciences, Beijing, China
{liujunjie,ykxie,xie,luolayong,zhangfuxing,wuxiaolong,
ningqingsong,guanhongtao}@ict.ac.cn
[2] Graduate University of Chinese Academy of Sciences, Beijing, China

Abstract. Network virtualization which enables the coexistence of multiple networks in shared infrastructure adds extra requirements on data plane of router. Software based virtual data plane is inferior in performance, whereas, hardware based virtual data plane is hard to achieve flexibility and scalability. In this paper, using FPGA (Field Program Gate Array) and TCAM (Ternary Content Addressable Memory), we design and implement a virtual hardware data plane achieving high performance, flexibility and scalability simultaneously. The data plane uses a 5-stage pipeline design. The procedure of packet processing is unified with TCAM based rule matching and action based packet processing. The hardware data plane can be easily configured to support multiple VDP (Virtual Data Plane) instances. And in each VDP instance, the pattern of packet processing can be flexibly configured. Also, it can achieve seamless migration of VDP instance between software and hardware. The hardware data plane also provides a 4-channel high-performance DMA engine which largely reduces packet acquisition overhead on software. So that software can be more involved in customized packet processing.

Keywords: Network Virtualization, Virtual Data Plane, FPGA.

1 Introduction

Network virtualization has been thought as a method to experiment with innovative protocols and a feasible way to immigrate to future network [1–3]. Network virtualization virtualizes network resources, such as node, link, and topology. It enables heterogenous networks to run parallel and guarantees the isolation of each subnetwork. A router which supports network virtualization is required to fulfill various goals - scalability, flexibility, programmability, manageability, isolation and so on [5]. The router should virtualize its resources and enable multiple virtual router instances run on it with fair isolation. Each virtual router instance should be configured flexibly to process packets in their patterns. And the number of virtual router instances shouldn't be constant.

Data plane is a critical part in router architecture. It is responsible for forwarding decision, packet processing and output link scheduling [6]. It has to process

R. Bestak et al. (Eds.): NETWORKING 2012, Part I, LNCS 7289, pp. 205–216, 2012.

millions of packets per second. For data plane in virtual router, it has to achieve virtualization, scalability, flexibility, and isolation. Those requirements can be easily achieved using sophisticated host virtualization technology [4, 10, 11, 16] (e.g. Xen). However, the performance is far from enough to satisfy practical scenario. Even after high performance I/O virtualization [21], virtual machine supervisor modification and kernel-level optimization, the performance is still inferior to raw linux performance [11].

For better performance, the data plane has been taken out from general purpose processor to platforms like NP (Network Processor), FPGA (Field Program Gate Array) or even GPU.

NP is specifically designed for network applications with pipeline of processors and parallel processing features. Nevertheless, the hardship of NP programming leads to the decline of NP usage in academic community. On the other hand, due to the popularization of NetFPGA [13], a FPGA based platform designed for network processing, FPGA is gaining popularity. Its high performance feature and the easiness of programming using HDL(Hardware Description Language) are favored by many researchers. Besides FPGA, there is a rising interest on using GPU for network processing. GPU has numerous processers and high I/O bandwidth which makes it a perfect platform for FIB (Forwarding Information Base) lookup [22]. However, the lack of platform specifically designed for network processing limits the use of GPU.

In this paper, we use FPGA to implement a virtual hardware data plane. To achieve high performance classification, we use TCAM (Ternary Content Addressable Memory) in our platform. TCAM is a special memory addressed by content. It is widely used in high-performance routers and other network devices. It stores 0, 1, or x (don't care). Each lookup of TCAM returns the address of the first entry matching the content in determined time.

The rest of the paper is organized as follows. In Section 2 we discuss some related work. By analyzing their pros and cons, we conclude our philosophies of building our data plane. In Section 3, we brief on the architecture of the hardware platform we used. And in Section 4 we show how our hardware data plane is designed and implemented. We show the overview of the design in Section 4.1, and elaborate each processing stage separately in Section 4.2. We discuss the isolation problem in Section 4.3. And in Section 4.4, we show the reference design of software for our virtual hardware data plane. In Section 5, we evaluate our hardware data plane in throughput, latency and scalability. Finally, we conclude our work in Section 6.

2   Related Work

There are already several works trying to build virtual data plane on FPGA. Anwer M. B. and Feamster [14] build eight identical virtual router instances in NetFPGA. Each instance possesses fixed forwarding table and forwarding logic. The design is preeminent in performance and isolation. But it only supports constant virtual router instances with fixed hardware resources. There is no flexibility or scalability. While their further work SwitchBlade [7] is designed in a completely different way. It pipelines and modularizes the packet processes which can be easily deployed customized protocol on programmable hardware like FPGA. The platform enables customized process to be implemented in hardware. However, the design fails to consider scalability issue.

To explore scalability, D. Unnikrishnan et al. [8] propose a scalable virtual data plane. It implements virtual router instances both in FPGA and in software which achieves better scalability. They also propose a method of dynamic FPGA logic reconfiguration. So that virtual router instance can be migrated from software to hardware and vice versa by reconfiguring FPGA. It resolves the problem of scalability in a degree, and the performance of each virtual data plane can be changed. Nevertheless, the reconfiguration of the hardware which disables the whole hardware data plane compromises isolation. In their later work [17], the reconfiguration is done using partial reconfiguration feature of FPGA. Based on this method, the reconfiguration can be done in less time without interfere other VDP (Virtual Data Plane) instances in hardware. However the number of VDP instances implemented on hardware is still constrained.

Another work need to be mentioned is OpenFlow [18]. Although OpenFlow itself does not concerns about network virtualization in its data plane, it has a relative simple yet flexible data plane. The processing of packet can be easily defined and configured using action based processing. FlowVisor [20] tries to implement virtualization in OpenFlow with a middle layer virtualizing the control plane. However, it only virtualizes the control plane, leaving the data plane unchanged.

Based on the analysis above, we have developed our philosophies on the construction of virtual hardware data plane. Firstly, the data plane should be pipelined and modularized. The pipeline will increase the performance of the hardware data plane. While modularity will facilitate the implementation of new features into the data plane. Secondly, hardware and software cooperation should be intensified. The resource in FPGA is limited comparing to increasingly powerful multicore processor in the host. The cooperation with software will boost scalability as well as programmability of the data plane. Thirdly, a unified action-based processing should be adopted to increase flexibility and utility of resources. The isolation of FPGA resources in virtual data plane not only constrains scalability but also deteriorates flexibility. Meanwhile, action-based processing saves hardware resources and increases flexibility.

3   Hardware Design

Our hardware data plane is built based on our own NAC (Network Acquisition Card) hardware platform. Figure 1 shows the architecture of NAC hardware platform. Unlike NetFPGA platform [13] which has quite scarce resources and is designed for prototype implementation and algorithm verification, NAC platform aims for practical deployment. The platform is centered with a Xilinx Virtex-5 LX155T FPGA with sufficient logic resource. The FPGA chip integrates a 8-lane endpoint block for PCI Express 1.0a providing an unidirectional data rate up to 20Gbps. It has 4 Gigabit Ethernet SFP interfaces, so that different SFP modules can be plugged in for different types of service access. Another important component is the on-board IDT75P42100 TCAM chip with a capability of 2.304Mbit. For large-volume memory storage, a 9MB QDRII SRAM chip is added to the hardware platform. Currently, this platform has been used in enterprise network as hardware accelerator in many practical systems such as intrusion detection system.

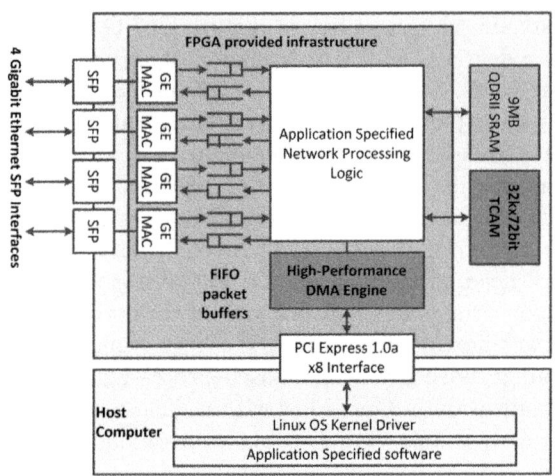

Fig. 1. Architecture of NAC hardware platform

4   Data Plane Design and Implementation

4.1   Design Overview

The data plane we designed is based on the philosophies we discussed in Section 2. The design overview of the data plane is showed in Figure 2. The data plane pipeline is made up of 5 stages - VDP mapping, action matching, action processing, software processing, and forwarding.

Packets are processed as follows. Firstly, packets are gathered from the ethernet interfaces and mapped to their VIDs (Virtual data plane ID) in VDP mapping stage. Then, packets along with their VIDs are used to obtain the action result in action matching stage. The action result contains the information how the packet will be processed. Afterward, packets are processed by function modules in action processing stage according to their action results. Packets are either forwarded to the software processing stage for further process or sent to the forwarding stage for output scheduling. Software processing stage acts as an extension of hardware data plane which implements slow data plane. As the software is highly programmable, the programmability of our data plane expanded.

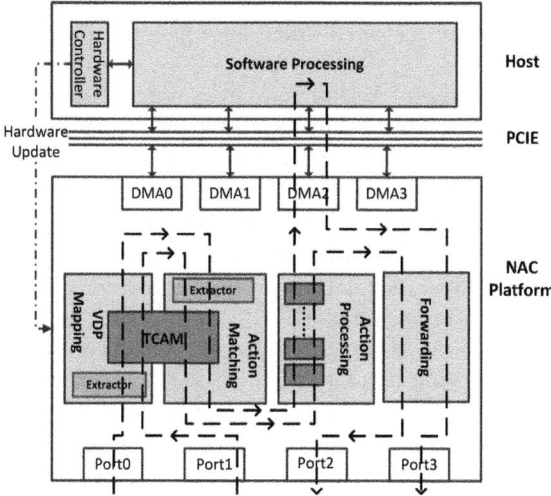

Fig. 2. Design overview of hardware data plane

In the implementation, we extracts 10-tuple packet information for rule matching in VDP mapping and action matching. The 10-tuple packet descriptor is similar to the 10-tuple flow header used in OpenFlow [18]. Hence, some innovative protocols can not be directly processed by the hardware data plane. However, we believe that in the near feature, TCP/IP stack based traditional packets will still dominate the traffic. And the minority of the traffic can be processed by software in our data plane. As the data plane is well integrated with software processing, innovation can be supported by software processing stage. What's more, the modularity of the hardware data plane enables the data plane to be reprogrammed to support new features.

4.2 Process Stages

VDP Mapping Stage. The VDP mapping stage contains two modules - information extracting and VDP mapping. Packet information is extracted from each packet for VDP mapping and used to match predefined VDP mapping rules in TCAM which returns the VID of the packet. Every VDP instance can install several mapping rules when it is created. These mapping rules are used to describe packets that belongs to the VDP instance.

The 10-tuple packet information is extracted in the implementation. It covers packet information from link layer to transport layer. Hence, VDP instance can be flexibly configured to implement different layers of virtualization [19]. For example, a layer 2 virtualized virtual data plane can be deployed using VDP mapping rules that only concerns source MAC address field, destination MAC address field and VLAN ID field. While for network layer virtualization, mapping rules should not care about fields other than source and destination IP addresses.

What's more, a default VDP mapping rule is used to match all the unmatched packets to a default VDP instance which is implemented in software.

Action Matching Stage. Action matching stage also includes information extracting and action matching. The extracted fields can be different from that in the VDP mapping stage. The extracted packet information along with the VID is used to match action matching rules in TCAM. Action matching rules are made up of extracted packet fields and VID field and installed by each VDP instance. So that packets in one VDP instance wouldn't match rules in another. And within each VDP instance, rules can be flexibly configured.

In the implementation we reuse the 10-tuple packet information for action matching. Every match of action matching rule returns an index of action result using which an action result can be obtained. Action result contains information of how the packet will be processed in action processing stage. There is a default rule for each VDP instance that match all unmatched packets. Packet that fails to match explicit rules will hit the default rule and be processed according to the default action result. Usually, the default action result can be set as sending packets to the software processing stage. With more explicit rules in hardware, the VDP instance will have better performance. So the performance of VDP instance can be configured by changing the number of explicit rules in TCAM. What's more, by replacing first rule with the default rule, a VDP instance will seamless migrate from hardware to software.

Action Processing Stage and Forwarding Stage. Packets along with their action results are sent to action processing stage. In this stage, packets are processed by a pipeline of function modules according to their action results. Each function module is responsible for a specific processing function. As this stage is modularized, more modules as well as corresponding fields in action result can be added to support more patterns of packet processing in hardware.

In the implementation, an action result contains a bitmap field and various value fields. Bitmap field has several bits: port-valid bit, change-MAC bit, TTL bit, forward bit, and drop bit. While the value fields are port ID field, source and destination MAC addresses field, and DMA ID field. Port-valid bit implies the validity of port ID field through which packets will be sent. Change-MAC bit is asserted when packets should change their MAC addresses with the ones in source and destination MAC addresses field. The TTL field indicates the decrease of the TTL field in packets and recalculation of the checksum in IP header. If forward field is asserted, packet should be directly forwarded to the given output port, otherwise they will be sent to the host through the DMA channel in DMA ID field. And drop bit denotes whether those packets should be dropped or not.

The combination of those fields results in various patterns of actions. For example, host forwarding action, with the forward bit deasserted and DMA ID given, can be used as the action result of default matching rule for each VDP instance. Packets that obtain host forwarding action are sent to host through the given DMA channel.

Forwarding stage is responsible for packet output scheduling. Packets and their output port IDs from either action processing stage or software processing stage are sent to this stage. Then they are scheduled and forwarded to the their coordinating output ports.

DMA Engine and Software Processing Stage. In our work, a high-performance DMA engine is adopted in the hardware data plane. The DMA engine directly sends packet data to the reception ring buffer in host memory without interrupting CPU. While the software polls data from the ring buffer all the time and assembles the packets. One a reverse way, the software writes packet data to the transmission ring buffer in host memory and changes the buffer pointers in the hardware for notification. Then the hardware automatically requests those data from host memory and assembles packets in the hardware. In this way, the interrupt rate of system dramatic drops eliminating frequent context switching. What's more, the packets can be processed in a batched fashion further improving the performance for software processing.

Additionally, the DMA engine implements 4 independent channels for packet reception and transmission. That means there are 4 independent reception and transmission ring buffers respectively in host memory. This provides an alternative way of I/O virtualization. One of the channels can be used by supervisor for distributing packets to correspondent virtual router instances. Other three channels can be dedicated to virtual router instances which requires high-performance I/O throughput.

Packets that can not be processed in hardware, due to the limitation of TCAM entries or the absence of customized processing modules, will be sent through the DMA channel to software. The software processing stage acts as an extension of hardware data plane which adding programmability to the data plane. The high-performance DMA engine we used in this design also boosts the performance of software processing.

4.3 Resource Isolation

Virtualization requires the share of physical resources, while isolation demands the elimination of interaction. For software, the resources needed to be isolated are CPU time, allocation and access of memory, hard drive space, and so on. The interrupt due to packet reception, i.e. I/O fairness, also should be taken into consideration [12]. For work in [14], the isolation is accomplished by separation of logic resources. However, in our hardware data plane, it no longer applies.

As described above, all the VDP instances in hardware data plane share the pipeline. We think that the unified pipeline is fast enough to handle all packets. With data width of 128 bit and a synchronous clock of 200MHz, the hardware is capable of handling packets in 25.6Gbps which is far more than maximal packet rate in this platform. So packets can peacefully time-share the processing modules. The major problem is the share of TCAM. TCAM is used to store mapping rules and action rules for all VDP instances. Every instance may need several mapping rules and action rules. The management of TCAM should be easy to allocate and deallocate TCAM entries to VDP instances, and provides fair access to TCAM.

We use a slice-based TCAM management in our work. The TCAM is divided into two parts - VDP mapping rules and action matching rules. TCAM entries are partitioned into slices. The allocation and deallocation of TCAM entries is based on slices. The supervisor maintains the usage information of all slices - whether a slice is used or not and which VDP instance does a slice belong to. The supervisor also acts as an agent translating TCAM operations of virtual router instances to the physical TCAM operations. Another benefit is the potential for TCAM power consumption reduction which is critical in every scenario using TCAM. Before we send the request for a TCAM lookup, we already know which slices will be matched, either VDP mapping slices or action matching slices for a certain VDP instance. So we can only select blocks which contains those slices to perform the match instead of using the whole TCAM.

4.4   Reference Software Architecture

To facilitate the configuration of the virtual data plane, we provide two sets of API. A higher level of API can be used by control plane of virtual router instance to manage its VDP mapping rules and action matching rules logically. The lower level of API is used by the supervisor to manage TCAM allocation and perform physical TCAM entries modification.

The Pearl Project [9], aiming to build a programmable virtual router plat-form, uses the virtual hardware data plane described in this paper. The software architecture of the Pearl Project is showed in Figure 2. The substrate layer deals with distribution and isolation of packets as well as hardware management. LXC, a light-weight OS-level virtualization technique, is used to contain control plane and slow data plane of a virtual router instance. Three of the four DMA chan-nels are dedicated to three virtual machines respectively, and the last one is shared by the rest virtual machines. And a VM (Virtual Machine) manager is used to manage both software virtual machines in software and VDP instances in hardware.

5   Evaluation

We evaluate our work from three perspectives - (1) throughput, (2) latency, and (3) scalability. We try to demonstrate that the hardware data plane is scalable and high-performance. We use a server to host our hardware platform. It has a 3.5GHz 64-bit 8-core Intel Xeon processor, 8 GB DDR2 RAM. The OS system is a 2.6.35.9 kernel version Ubuntu. We use Spirent TestCenter [23], a powerful Ethernet testing device, to generate packets for the test.

There are two processing paths in the data plane as showed in Figure 2. One of them is hardware path - packets are directly forwarded to the output port

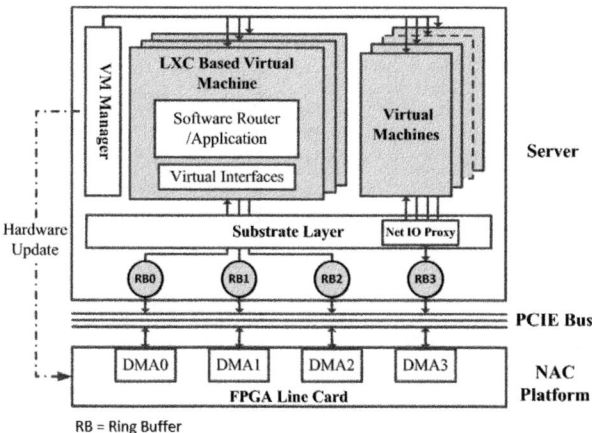

RB = Ring Buffer

Fig. 3. Software architecture of Pearl

in the hardware bypassing the software processing stage. The other is software path - packets are forwarded to the software processing stage and then sent to hardware for output.

In the evaluation of both pathes, we create four VDP instances in hardware data plane and map packets to VDP instances according to their incoming ports. While, the default action rules in these two evaluations are slightly different. In the evaluation of hardware path, packets are forwarded directly from one port to its adjacent one. While, in software path, packets are sent to software and then sent back to hardware and forwarded to adjacent ports of their income ports.

We use TestCenter to generate RFC2544 tests on throughput and latency. All four ports send packets with packet sizes ranging from 64 bytes to 1518 bytes. The testing result is showed in Figure 4.

The hardware path reaches 4Gbps line speed, so does the software path. Whereas raw Linux kernel forwarding is only capable of handling traffic in approximate 500kpps. The latency of hardware path is merely 5 milliseconds to tens of milliseconds, depending on packet size. The latency of software path comes close to raw kernel performance, about 1 to 2 microseconds. The 4 channel DMA engine is the primary contributor to the high throughput of software path. With 4 separated channels, packets can be processed concurrently so that we can benefit more from multi-core processor. What's more, as packet reception is based on ring buffer polling, there is no overhead of context switch due to interrupt. One the other hand, without interrupt, the timing for packet processing becomes uncertain which leads to the latency in software path.

To demonstrate the scalability of our data plane, we set up 4 to 1024 VDP instances in hardware data plane and evaluate the overall throughput. Figure 5 shows the performance of hardware data plane. We can see that the addition of VDP instance does not require additional logic resource in FPGA. Only TCAM

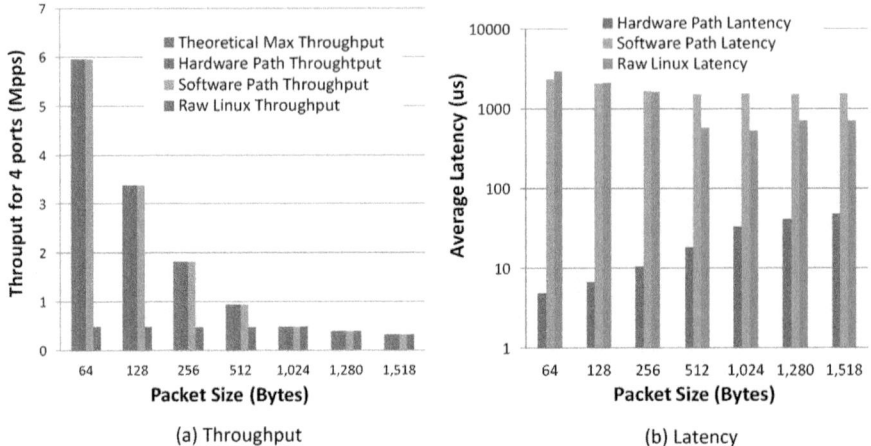

Fig. 4. Throughput and Latency of hardware data plane

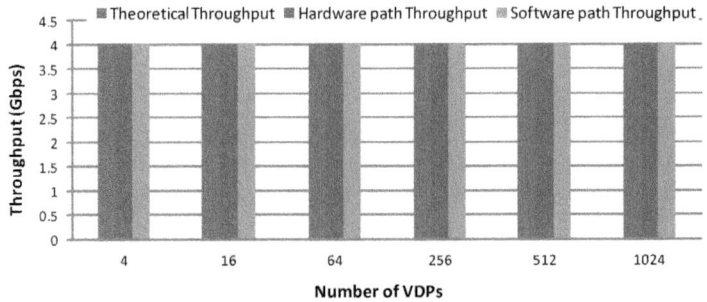

Fig. 5. Scalability of hardware data plane

entries are needed when VDP instance are created. So that the data plane is quite scalable. The overall performance will not be influence by the increase of VDP instances.

6 Conclusion

In this paper, we build a virtual hardware data plane to achieve high performance as well as flexibility and scalability. The TCAM based unified pipeline design facilitates the reconfiguration of performance of VDP instances and provides flexible interface for the configuration of action based processing. The action based packet processing makes the packet processing in each VDP isntance configurable and flexible. The high-performance DMA engine promotes the performance of software and improves the programmability to support customized processing. It manages to support multiple VDP instances which means the hardware data plan is scalable. The evaluation in Section 5 shows that the hardware data plane is low-latency and high-throughput with great scalability. The increase of VDP instances in hardware would not affect the overall performance of the system.

Acknowledgement. This work is supported by National Basic Research Program of China under grant No.2012CB315801, National Natural Science Foundation of China (NSFC) under grant No. 61133015, Projects of International Cooperation and Exchanges NSFC-ANR with grant No.61061130562, Strategic Priority Research Program of the Chinese Academy of Sciences under grant No. XDA06010303, and the Instrument Developing Project of the Chinese Academy of Sciences under grant No. YZ200926.

References

1. Anderson, T., Peterson, L., Shenker, S., Turner, J.: Overcoming the Internet impasse through virtualization. Computer 38, 34–41 (2005)
2. Niebert, N., Khayat, I.E., Baucke, S., Keller, R., Rembarz, R., Sachs, J.: Network virtualization: A viable path towards the future internet. Wireless Personal Communications 45, 511–520 (2008)
3. Tutschku, K., Zinner, T., Nakao, A., Tran-Gia, P.: Network virtualization: Implementation steps towards the future internet. In: Proc. of KiVS-Kommunikation in Verteilten Systemen (KIVS 2009), (KiVS 2009, Kassel) (2009)
4. Rixner, S.: Network virtualization: breaking the performance barrier. Queue 6, 36 (2008)
5. Chowdhury, N.M., Boutaba, R.: A survey of network virtualization. Computer Networks (2009)
6. Chao, H.J., Liu, B.: High performance switches and routers. Wiley-IEEE Press (2007)
7. Anwer, M.B., Motiwala, M., Tariq, M. bin, Feamster, N.: SwitchBlade: a platform for rapid deployment of network protocols on programmable hardware. In: Proceedings of the ACM SIGCOMM 2010 Conference on SIGCOMM, pp. 183–194. ACM, New York (2010)
8. Unnikrishnan, D., Vadlamani, R., Liao, Y., Dwaraki, A., Crenne, J., Gao, L., Tessier, R.: Scalable network virtualization using FPGAs. In: Proceedings of the 18th Annual ACM/SIGDA International Symposium on Field Programmable Gate Arrays, pp. 219–228 (2010)
9. Xie, G., He, P., Guan, H., Li, Z., Xie, Y., Luo, L., Zhang, J., Wang, Y., Salamatian, K.: PEARL: a programmable virtual router platform. IEEE Communications Magazine 49, 71–77 (2011)
10. Jiang, X., Xu, D.: VIOLIN: Virtual Internetworking on Overlay Infrastructure. In: Cao, J., Yang, L.T., Guo, M., Lau, F. (eds.) ISPA 2004. LNCS, vol. 3358, pp. 937–946. Springer, Heidelberg (2004)
11. Bhatia, S., Motiwala, M., Muhlbauer, W., et al.: Trellis: a platform for building flexible, fast virtual networks on commodity hardware. In: Proceedings of the 2008 ACM CoNEXT Conference, vol. 72, pp. 1–6 (2008)
12. Anwer, M.B., Nayak, A., Feamster, N., Liu, L.: Network I/O fairness in virtual machines. In: Proceedings of the Second ACM SIGCOMM Workshop on Virtualized Infrastructure Systems and Architectures, pp. 73–80 (2010)

13. NetFPGA, http://netfpga.org/
14. Anwer, M.B., Feamster, N.: A Fast, Virtualized Data Plane for the NetFPGA. In: NetFPGA Developers Workshop, Stanford, California, pp. 90–94 (2009)
15. Liao, Y., Yin, D., Gao, L.: Network virtualization substrate with parallelized data plane. Computer Communications 34, 1549–1558 (2011)
16. Egi, N., Greenhalgh, A., Handley, M., Hoerdt, M., Huici, F., Mathy, L.: Towards high performance virtual routers on commodity hardware. In: Proceedings of the 2008 ACM CoNEXT Conference, pp. 20:1–20:12. ACM, New York (2008)
17. Yin, D., Unnikrishnan, D., Liao, Y., Gao, L., Tessier, R.: Customizing virtual networks with partial FPGA reconfiguration. ACM SIGCOMM Computer Communication Review 41, 125–132 (2011)
18. McKeown, N., Anderson, T., Balakrishnan, H., Parulkar, G., Peterson, L., Rexford, J., Shenker, S., Turner, J.: OpenFlow: enabling innovation in campus networks. ACM SIGCOMM Computer Communication Review 38, 69–74 (2008)
19. Chowdhury, N.M.M., Boutaba, R.: Network virtualization: state of the art and research challenges. IEEE Communications Magazine 47, 20–26 (2009)
20. Sherwood, R., Gibb, G., Yap, K.K., Appenzeller, G., Casado, M., McKeown, N., Parulkar, G.: FlowVisor: A Network Virtualization Layer (2009)
21. Dong, Y., Dai, J., Huang, Z., Guan, H., Tian, K., Jiang, Y.: Towards high-quality I/O virtualization. In: Proceedings of SYSTOR 2009: The Israeli Experimental Systems Conference, pp. 1–8 (2009)
22. Han, S., Jang, K., Park, K., Moon, S.: PacketShader: a GPU-accelerated software router. In: Proceedings of the ACM SIGCOMM 2010 Conference on SIGCOMM, pp. 195–206. ACM, New York (2010)
23. Spirent TestCenter, http://www.spirent.com

Permutation Routing for Increased Robustness in IP Networks*

Hung Quoc Vo, Olav Lysne, and Amund Kvalbein

Simula Research Laboratory
{vqhung,olav.lysne,amundk}@simula.no

Abstract. We present Permutation Routing as a method for increased robustness in IP networks with traditional hop-by-hop forwarding. Permutation Routing treats routers involved in traffic forwarding as a sequence of resources, and creates permutations of these resources that give several forwarding options. We introduce Permutation Routing as a concept, and use it to create routings where we seek to maximize single link fault coverage. Analogous to the IETF standardized Loop-Free Alternate (LFA), Permutation Routing can easily be implemented for OSPF or IS-IS networks to augment existing ECMP forwarding with additional loop-free forwarding entries for improved load balancing or fault tolerance. Our evaluations show that Permutation Routing can increase single link fault coverage by up to 28% compared to LFA in inferred network topologies.

Keywords: IP networks, Multipath Routing, Resilience, Fault-tolerant.

1 Introduction

The last few years have witnessed the adoption of the Internet as the preferred transport medium for services of critical importance for business and individuals. In particular, an increasing number of time-critical services such as trading systems, remote monitoring and control systems, telephony and video conferencing place strong demands on timely recovery from failures. For these applications, even short outages in the order of a few seconds will cause severe problems or impede the user experience. This has fostered the development of a number of proposals for more robust routing protocols, which are able to continue packet forwarding immediately after a component failure, without the need for a protocol re-convergence. Such solutions add robustness either by changing the routing protocol so that it installs more than one next-hop towards a destination in the forwarding table [1][2][3], or by adding backup next-hops a posteriori to the forwarding entries found by a standard shortest path routing protocol [4][5][6].

---

* Permutation Routing term is used in the literature with slightly different meanings, all related to the act of rearranging network objects (e.g. network devices or network devices and theirs associated radio channels) on which packets are routed from a source to a destination [9] [10].

R. Bestak et al. (Eds.): NETWORKING 2012, Part I, LNCS 7289, pp. 217–231, 2012.

Unfortunately, few of these solutions have seen widespread deployment, due to added complexity or incompatibility with existing routing protocols.

Installing more than one route to a destination has obvious advantages with respect to increased robustness. First, it provides alternative routes that are readily available after a link failure. Second, it provides several paths for load-balancing, which can be used to absorb short-lived spikes in traffic demand and thus avoid congestion. Deployed intradomain routing protocols such as OSPF and IS-IS have traditionally installed only a single shortest path to each destination, or multiple equal-cost shortest paths with the ECMP extension. ECMP allows some degree of multipath routing, but can only give more than one forwarding next-hop to a limited subset of all source-destination (S-D) pairs in a network. Recently, a mechanism called Loop-Free Alternates (LFA) [4] has become available in some routers. LFA allows routers to install alternate forwarding entries that can be used as backup if the primary next-hop fails. LFA improves robustness compared to ECMP, but there will still be many S-D pairs with only a single path (next-hop) available [8].

This paper presents Permutation Routing as a novel and flexible approach for calculating multiple loop-free next-hops in networks with traditional hop-by-hop forwarding. Permutation Routing is based on the observation that routing in any network consists of using a set of resources (links and nodes) *in sequence*. A routing strategy can therefore be expressed as a permutation of the nodes that are involved in traffic forwarding to a destination. Routing will be loop-free as long as traffic can only be forwarded in one direction with respect to the node ordering in this permutation. The main focus in this paper is to use Permutation Routing to create a routing that maximizes the single link fault coverage. Other routing strategies based on permutations of *links* in networks with interface-specific forwarding [3] can also be developed.

In this paper, we provide a simple backtracking algorithm that constructs a permutation of routers for each destination, and a simple forwarding rule that allows us to generate forwarding tables based on the permutations. The properties of the resulting routing are determined by the constraints used at each step in the permutation construction. The input to the construction algorithm is the topology information that is collected by a link state routing protocol, and hence no new control plane signalling is needed with Permutation Routing.

Importantly, Permutation Routing can easily be integrated with existing intradomain routing protocols, and can be used to augment the shortest path routing tables with additional forwarding entries. We show how the constraints in the permutation construction can be designed so that the resulting routing is compatible with normal shortest path routing, while still offering significantly more forwarding options than the existing LFA. With multiple loop-free alternates for a given primary next-hop, OSPF or IS-IS may employ some of them as unequal-cost primary paths and the rest as back-up paths. In the case of multiple primary paths, packets can be distributed evenly among paths or with more intelligent load balancing methods [11][12]. In this paper we focus on finding a set of loop-free paths, and leave the important topic of load-balancing across these paths for future work.

The rest of the paper is organized as follows. Section 2 introduces the notion of Next-Hop Optimal Routing (NHOR), which we will use as our main design objective. Section 3 introduces Permutation Routing and the relationship between a permutation of routers and the corresponding routing graph. Section 4 describes a generic algorithm to construct permutations. Section 5 describes a specific construction algorithm that approximates NHOR, and a modified version that is also compatible with standard shortest path routing. Section 6 shows simulation results evaluating the path diversity and the computational complexity of Permutation Routing. Section 7 reviews related work. Section 8 concludes the paper.

2 Next-Hop Optimal Routing

Before introducing Next-Hop Optimal Routing as our objective function for robust routing, let us introduce some vocabulary that we will use in the rest of this paper. By a *routing*, we refer to the assignment of a set of next-hops for each destination node in each source node. We do not distinguish traffic from different sources, and hence traffic transiting a node is treated in the same way as traffic originated at the node. We require that a routing must be loop-free, and hence a given routing corresponds to a Directed Acyclic Graph (DAG) rooted at each destination node consisting of the links and nodes involved in packet forwarding. With multipath routing, each node may have several outgoing links in the DAG for a destination. A routing where all links are included in each DAG is referred to as a *full connectivity routing*. With a given network topology, many different high or full connectivity routings can normally be constructed. However, they will have different properties with respect to failure recovery and load balancing.

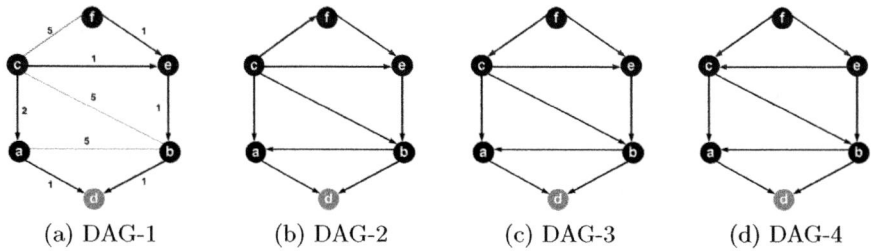

(a) DAG-1   (b) DAG-2   (c) DAG-3   (d) DAG-4

Fig. 1. A network topology with different DAGs

Fig. 1 shows a simple network topology, with 4 different DAGs for the destination node d. In Fig. 1a, DAG-1 is given by shortest path routing with ECMP using the link weights indicated in the figure. Node c can split its traffic over two next-hops, while the other nodes have only a single next-hop towards d. Links (a, b), (c, b) and (c, f) are left idle, and are neither used for backup or load balancing. The DAGs in Fig. 1b, Fig. 1c and Fig. 1d are all full-connectivity routing graphs, where all links can be used for packet forwarding. They differ, however, in their distributions of next-hops. In DAG-2, there are three nodes (a, e and f) that have only a single next-hop towards d. DAG-3 on the other hand, has only two such nodes (a and e). DAG-2 and DAG-3 are both compatible with shortest path routing, because they contain all directed links of DAG-1. DAG-4 is not compatible with shortest path routing: by changing the direction of the link (c, e), the number of nodes with a single next-hop has been reduced to one (the minimum value).

To maximize the single link fault coverage and load-balancing capabilities of a network, it is important that there is more than one next-hop available for as many S-D pairs as possible. This leads us to the following optimization criterion for *Next-Hop Optimal Routing* (NHOR):

Definition 1. *An NHOR is a full-connectivity routing that maximizes number of S-D pairs that have* at least *two next-hops towards a destination.*

As illustrated by the example above, an NHOR will not always be compatible with shortest path routing. For that reason, we also define the *Shortest-Path compatible NHOR* (NHOR-SP):

Definition 2. *An NHOR-SP is a full-connectivity routing that maximizes number of S-D pairs that have* at least *two next-hops while containing the DAG calculated by a shortest path algorithm.*

3   Permutation Routing

We consider a network modeled as a connected graph $G = (V, E)$ where V is a set of nodes and $E \subseteq V \times V$ is the set of links (edges) and its topology. A connected link from node i to node j is denoted by (i, j).

Without loss of generality, we look at the assignment of next-hops for each destination individually. For a destination $d \in V$, let $R_d = (V, E_d)$ be a routing function for packets destined to destination d, where E_d is a set of directed links constructed on E. In R_d, node j is called a *next-hop* of node i if there exists a directed link between node i and node j, denoted by $(i \rightarrow j)$. For a loop-free routing, R_d must be a DAG rooted at destination d.

The routing function R_d contains all valid paths to d, and each path can be considered as a sequence of nodes from a specific source to d. At each node, packet forwarding is the process of sending packets to a next-hop in such a sequence. In the rest of this section, we describe how Permutation Routing can be used as a tool to find such sequences with the goal of realizing NHOR.

Definition 3. *For a given network topology $G = (V, E)$, a permutation P of nodes is an arrangement of all nodes in V into a particular order.*

We write $j < i$ to denote that j occurs before i in P. Our goal is to construct permutations that realize a certain routing strategy.

Definition 4. *A permutation P is a* routing permutation *for R_d if and only if all next-hops of each node occur before it in P: $\forall (i \to j) \in E_d : j < i$.*

According to this definition, the destination node d will always be the first node in a routing permutation for R_d. Nodes further out in the routing permutation will be topologically farther from the destination.

Lemma 1. *Any loop-free routing function R_d can always be represented by a routing permutation in which d is at the left-most position.*

Proof. A loop-free routing function $R_d = (V, E_d)$ is a DAG, rooted at d. Let arrange $i \in V$ into a sequence such that if E_d contains a directed link $(i \to j)$, then j appears before i in that sequence. Such an arrangement can be calculated by a topological sort algorithm [13]. Destination $d \in V$ is the only node that does not have any outgoing link. Following the above ordering, node d, hence, has been placed at the left-most position of the sequence.

In general, there can be more than one routing permutation for one routing function R_d. Starting with a routing permutation P, another routing permutation P' can be generated by swapping *two consecutive* nodes that are not connected by a directed link to each other. For instance, both permutations $\{d\ a\ b\ e\ c\ f\}$ and $\{d\ b\ a\ e\ c\ f\}$ are routing permutations for DAG-1 according to Def. 4.

In the reverse process, routing tables can be generated from a routing permutation, given a *forwarding rule* that defines the relationship between neighboring nodes. For now, we consider a greedy forwarding rule for constructing the routing table, in which all topological neighbors of a node that occur before it in the routing permutation are installed as next-hops. Note that this forwarding rule will result in a full connectivity routing, where all links in the topology are potentially used for traffic forwarding to all destinations. This will maximize the potential for load balancing and failure recovery. More restrictive forwarding rules could also be considered, which would result in a sparser DAG. This can sometimes be beneficial in order to avoid excessively long paths, or to limit the number of next-hops for a particular destination.

With the given forwarding rule, different routing permutations will result in routing functions with different robustness characteristics. Our goal is to find a routing permutation that can realize NHOR. This problem is believed to be NP-hard. In the next section, we present an algorithm that produces routing permutations that approximate NHOR.

4   Generating Routing Permutations

This section introduces a generic method for constructing routing permutations. The construction method is based on a backtracking algorithm, and can be used to construct routing permutations with different optimization objectives. In the next section, we specify the constraints that are used to approximate NHOR.

We reconsider a topology $G = (V, E)$ of N nodes ($|V| = N$), all of which are uniquely identified by a number from 1 to N. Let $P = \{p_1, p_2, \ldots, p_N\}$ be a set of N variables in a *fixed order* from p_1 to p_N, with respective domains

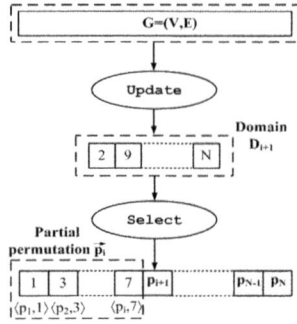

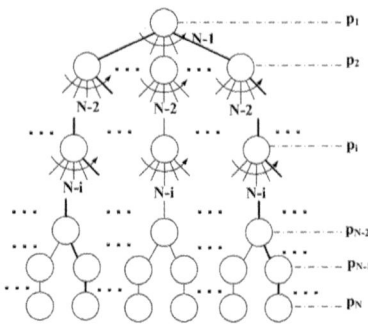

Fig. 2. Basic assignment procedure for variable p_{i+1}

Fig. 3. Search space for routing permutation problem

$D = \{D_1, D_2, \ldots, D_N\}$. We refer to D_i as the *candidate set* for each variable p_i. A candidate set consists of the nodes that can be assigned to variable p_i.

A routing permutation P is constructed by successively assigning a node $u \in D_i$ to each variable $p_i \in P$. Such assignment is said to be valid if it satisfies a specific *constraint function* $C(u)$ which is defined to realize the selected routing objective. Fig. 2 illustrates the basic assignment procedure for variable p_{i+1} in which two key functions Update and Select work as filters to control the assignment. In the figure, each pair $\langle p_i, u_i \rangle$ represents an assignment of the node u_i to variable p_i. The assignment of nodes to a subset of variables $\{p_1, p_2, \ldots, p_i\} \subseteq P$ given by $\{\langle p_1, u_1 \rangle, \ldots, \langle p_i, u_i \rangle\}$ is called *partial routing permutation* with i nodes. For simplicity, we abbreviate it to $\vec{p}_i$.

This basic assignment procedure has been embedded into the well-known backtracking algorithm to obtain the routing permutation P. The algorithm calls function Select (with built-in constraint function $C(u)$) which goes through D_i to find a valid node for the current variable p_i. If Select succeeds in finding a valid assignment, the algorithm calls function Update to generate domain D_{i+1} and proceeds to next variable p_{i+1}. Otherwise, a backtrack step will be executed to revisit the variable p_{i-1}. The algorithm terminates if a routing permutation P of N nodes, also denoted by $\vec{p}_N$, is found or a failure notification returns if all backtracks are examined but no solution is found under $C(u)$.

If the constraint function C allows it, the backtracking algorithm will find *one* routing permutation P among all possible solutions by searching through the *search space* shaped by the number of variables in P and their domains of values. In a naïve implementation, the domain for variable p_{i+1} consists of $(N - i)$ nodes that have not been placed in $\vec{p}_i$. Based on that observation, the search space S of the permutation assignment problem has a form of a tree of depth N rooted at the initial state $\langle p_1, d \rangle$ as illustrated in Fig. 3. Solutions $\vec{p}_N$ are located at its leaves. Two connected states in the search space refer to two instances of p_i and p_{i+1}. Assume that t operations are needed on average to move from state p_i to state p_{i+1}. The complexity in the best case when we do not need any backtrack step (backtrack-free) is $O(t \times N)$. In other extreme, if there is only one solution and we always make the "wrong" choice, the complexity would be $O(t \times N!)$.

The naïve implementation described above results in high computational complexity, and is only feasible for small topologies. Hence, it is important to guide the search to avoid exploring the entire search space. In the next section, we use two main mechanisms to reduce the search space:

1. $C(u)$ should be simple to reduce the computational complexity.
2. Domain D_i can be limited by taking $C(u)$ into account.

5   Next-Hop Optimal Routings

Next, we apply the framework described above to construct two high robustness routings that approximate NHOR and NHOR-SP as defined in Section 2. We call them Approximate NHOR and Approximate NHOR-SP.

5.1   Approximate NHOR (ANHOR)

With Permutation Routing using greedy forwarding, a node in p_i $(i > 2)$ has at least two next-hops if it has at least two topological neighbors that occur before it in the routing permutation. The aim of the ANHOR algorithm is to maximize the number of nodes where this is the case.

The partial routing permutation $\vec{p}_i$ represents a loop-free routing sub-graph towards destination d, denoted by $R_d^i = (V(\vec{p}_i), E_d(\vec{p}_i))$ where $V(\vec{p}_i)$ is the set of i nodes in $\vec{p}_i$ and $E_d(\vec{p}_i)$ is the set of directed edges formed by applying the greedy forwarding rule defined in section 3 on $\vec{p}_i$. To achieve a high robustness routing, the node selected for variable p_{i+1} to form the partial routing permutation $\vec{p}_{i+1}$ should be the node with the maximum number of topological neighbors already placed in $\vec{p}_i$. Correspondingly, the number of directed edges of the routing sub-graph formed by the partial routing permutation $\vec{p}_{i+1}$, resulted from the assignment $\langle p_{i+1}, u \rangle$, must be maximized:

$$|E_d(\vec{p}_{i+1})| = \max_{\forall u \in D_{i+1}} |E_d(\vec{p}_i, \langle p_{i+1}, u \rangle)| \tag{1}$$

For a more efficient implementation, we maintain a counter $c[u]$ for each node u. This counter denotes the number of outgoing links from u to $\vec{p}_i$. In other words, $c[u]$ corresponds to the number of next-hops node u will have if it is selected as the next assignment in the routing permutation. We derive the constraint function $C_{ANHOR}(u)$ to realize the expression (1) as follow:

$$C_{ANHOR}(u) = \begin{cases} \text{True} & \text{if } c[u] = \max_{\forall v \in D_{i+1}} c[v] \\ \text{False} & \text{otherwise} \end{cases}$$

The constraint function $C_{ANHOR}(u)$ implies that the domain D_{i+1} includes all nodes that have *at least* one topological neighbor in $\vec{p}_i$. The domain is, therefore, updated following the recursive relation:

$$D_{i+1} = D_i \cup \{ v \in V \mid (u, v) \in E \} \setminus \{u\} \qquad (2)$$

where u is the node that has been assigned to variable p_i in the i-th assignment.

The computational complexity of ANHOR is the product of the average number of operations to make a move between two successive states and the total number of states visited in the search space.

Proposition 1. *Constraint function $C_{ANHOR}(u)$ gives a backtrack-free algorithm for all connected input topologies.*

Proof. The proof is by contradiction. Assume that the algorithm with the constraint function $C_{ANHOR}(u)$ is not backtrack-free. This means that constraint function returns False for all $u \in D_{i+1}$ at some iteration. That can not happen because D_{i+1} can never be empty in a connected topology before all nodes have been placed in the permutation and all nodes in domain D_{i+1} always have at least one next-hop in $\vec{p}_i$.

Given the backtrack-free property of our algorithm, the complexity of calculating a permutation for *each* destination is $O(|E| + N \times |D|)$, where $|D|$ denotes the average size of the domain. Typically, $|D|$ depends solely on the average node degree of the network topology. In dense topologies, the total complexity of calculating routing permutations for *all* destinations can approach $O(N^3)$. The backtrack-free property also gives a low memory consumption because it does not need to store temporary partial routing permutations.

5.2   Approximate NHOR-SP (ANHOR-SP)

Let $R_d^{SP} = (V, E_d^{SP})$ denote the shortest path tree towards destination d. A routing permutation P whose routing function $R_d = (V, E_d)$ is an ANHOR-SP if R_d satisfies two constraints in following order:

1. R_d contains R_d^{SP}, meaning all routing choices for any node in R_d^{SP} are also valid routing choices for the same node in R_d.
2. ANHOR-SP uses the same selection criterion as ANHOR in order to maximize the number of S-D pairs with at least two next-hops.

The construction of such a routing permutation P is based on the assignment procedure described in Fig. 2. To this end, the shortest path compatibility constraint is implemented in function Update to limit the size of domain D_{i+1} for variable p_{i+1} and the connectivity constraint will be formalized by a constraint function $C_{ANHOR-SP}(u)$ and realized in function Select. Clearly, the constraint function $C_{ANHOR-SP}(u)$ is identical to $C_{ANHOR}(u)$.

The next node in routing permutation P is selected among nodes that have *all* their shortest path next hops towards d already placed in $\vec{p}_i$. Formally, let $R_d^{SP,i} = (V(\vec{p}_i), E_d^{SP}(\vec{p}_i))$ be the shortest path tree of $V(\vec{p}_i)$ nodes and $R_d^{SP,i} \subseteq R_d^{SP}$. The domain D_{i+1} for variable p_{i+1} includes all nodes u such that the assignment $\langle p_{i+1}, u \rangle$, resulting in $\vec{p}_{i+1}$, fulfills:

$$R_d^{SP,i+1} \subseteq R_d^{SP}$$

Let $c_{sp}[v]$ be the number of shortest path next-hops placed in $\vec{p}_i$ and $n_{sp}[v]$ be the total number of shortest path next-hop that can be calculated from R_d^{SP} of node v. The domain D_{i+1} for variable p_{i+1} follows the recursive relation:

$$D_{i+1} = D_i \cup \{ \, v \in V \mid c_{sp}[v] = n_{sp}[v] \, \} \setminus \{u\} \tag{3}$$

where u is the node that has been assigned to variable p_i.

Proposition 2. *Constraint function $C_{ANHOR-SP}(u)$ gives backtrack-free algorithm for all connected input topologies.*

Proof. According to Proposition 1, constraint function $C_{ANHOR-SP}(u)$ always returns True unless D_i is empty. We show here that D_i can never be empty before all nodes have been placed in the permutation. If D_i is empty, there is no node that have all its shortest path descendants in $\vec{p}_i$. In other words, we can follow the shortest path DAG R_d^{SP} from any node that has not been placed and always find a next-hop node that is not placed in $\vec{p}_i$. But this is impossible: since R_d^{SP} is connected and loop-free, any path along the shortest path DAG will eventually reach the destination, which is the first node that was placed in the permutation.

The computational complexity of ANHOR-SP towards one destination includes two parts: the shortest path calculation using Dijkstra's algorithm and routing permutation construction. Due to the property of backtrack-freedom, with sparse topologies the complexity of second part towards *one* destination would be $O(|E| + |E_d^{SP}| + N \times |D|)$ where $|D|$ denotes the average size of the domain. In dense topologies, the total complexity of calculating routing permutations for *all* destinations can approach $O(N^3)$.

With a low computational complexity, ANHOR-SP can be implemented on a per-router basis in OSPF or IS-IS networks. To ensure a consistent permutation routing in the entire network, constraint function $C_{ANHOR-SP}(u)$ must return the same node in each assignment among possible selections. We can break that tie by letting the highest router ID[1] be picked.

6   Performance Evaluation

We evaluate the performance of the proposed algorithms by measuring how well they realize NHOR as defined in Def. 1. Since multipath routing leads to path inflation, we also measure the path length distribution. ANHOR and ANHOR-SP are compared to standard shortest path routing with ECMP, and to LFA. Comparisons to other robust routing methods [3][6] are less relevant, because of their different objectives and implementation costs.

---

[1] The highest router ID means the biggest number numerically. For instance, 192.168.1.2 would be higher than 172.16.3.2, and 172.16.3.2 would also be higher than 172.16.2.1.

6.1   Evaluation Setup

We select six representative network topologies from the Rocketfuel project [19] for our evaluations. The topologies are listed in Table 1 in increasing order of their average node degrees.

Table 1. Network topologies

AS	Name	Nodes	Links	Avg. Degree
1221	Telstra(au)	104	151	2.90
1755	Ebone(eu)	87	161	3.70
3967	Exodus(us)	79	147	3.72
3257	Tiscali(eu)	161	328	4.07
6461	Abovenet(us)	138	372	5.40
1239	Sprint(us)	315	972	6.17

The results for ECMP and LFA depend heavily on the link weight settings used in the topologies. To obtain realistic link weight settings, we run local search heuristic with link load objective function proposed in [15], using a traffic matrix generated by the gravity model [18]. For AS1239, we use unit link weights, because the local search heuristic described in [15] does not scale to a topology of this size. Note that Permutation Routing will work with any link weight settings. We use this approach to show the performance with "typical" link weights.

6.2   Robustness Evaluation

Maximizing Multipath Capability. Fig. 4 shows the fraction of nodes with at least two next-hops with the different routing methods. For reference, Fig. 4 also shows the fraction of nodes in each topology with a node degree larger than 1. Obviously, nodes with a degree of 1 can not have more than 1 next-hop to any destination. We observe that the multipath capability varies strongly between topologies; it is generally higher in more well-connected networks. ANHOR achieves a significant improvement over ECMP and LFA in all networks.

Note that the number of next-hops achieved with ANHOR is independent of link weight settings, while ANHOR-SP is constrained to including the shortest paths in the routing. ANHOR-SP performance is close to ANHOR, and gives a clear improvement over LFA (by up to 28% in AS1239). This shows that Permutation Routing can give a significant gain compared to existing solutions, while being compatible with shortest path routing with realistic link weight settings.

Next-Hop Distribution. Fig. 5 shows the mean and variance for the number of next-hops at each router in our 6 topologies. For increased robustness and load-balancing, it is generally good to have a high mean and a low variance in the number of next-hops. If this variance is high, it means that a few nodes have a high number of next-hops, while others might be left with only one.

Both ANHOR and ANHOR-SP produce full connectivity routings, which means that the mean number of next-hops across all S-D pairs will be equal to half the average node degree in the topology. The mean is somewhat lower for LFA and ECMP, meaning that some links are not used for packet forwarding. The variance, however, is lower with ANHOR than with ANHOR-SP and LFA. This shows how ANHOR achieved a better (more uniform) next-hop distribution than the other routings.

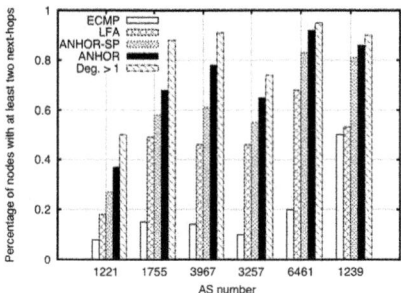
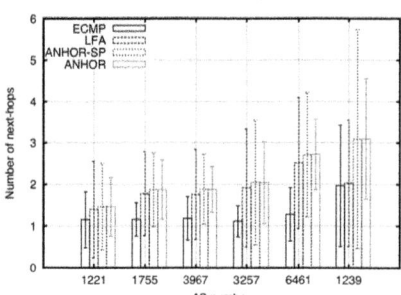

Fig. 4. Fraction of S-D pairs with at least two next-hops

Fig. 5. Means and variances of next-hop distributions

In practice, there are good reasons to limit the number of next-hops that are installed in the forwarding table for a particular destination. Installing more than a few next-hops will not give much benefit with respect to robustness or load-balancing. It will, however, require more fast memory in the forwarding table, and may lead to the unnecessary inclusion of paths that are much longer than the shortest path.

Hence, we look at an approach where the number of next-hops for a particular destination is limited to at most K. We define a *Routing Efficiency* coefficient, which denotes the fraction of bidirectional links that are used for traffic forwarding with a given K.

$$RE = 2 \times |E_d(K)| \, / \, |E| \tag{4}$$

where $|E_d(K)|$ is the number of directed links in the routing DAG when each node can have at most K next-hops and $|E|$ is the number of bidirectional links in the network topology. According to this definition, $0 \leq RE \leq 1$. Note that a high RE is desirable, and corresponds to a low variance in the number of next-hops achieved.

Table 2 shows the RE values for three values of K in the selected topologies. The given value is the average over all S-D pairs. We see that for all routing methods, a higher K gives a higher RE value. ANHOR and ANHOR-SP give the highest RE values, sometimes with a significant improvement over ECMP and LFA even for $K = 2$. The RE values in more well-connected topologies (AS1239) are lower than in sparse topologies. Such topologies contain a high number of nodes with a very high degree (39% nodes has their degrees greater than 15 in AS1239), and a low K will hence exclude many valid (but often unnecessary) paths.

Table 2. Routing Efficiency coefficient

		AS1221	AS1755	AS3967	AS3257	AS6461	AS1239
$K = 2$	ECMP	0.74	0.61	0.61	0.54	0.43	0.49
	LFA	0.80	0.79	0.77	0.77	0.62	0.49
	ANHOR-SP	0.86	0.84	0.85	0.76	0.67	0.58
	ANHOR	0.94	0.90	0.95	0.81	0.81	0.60
$K = 3$	ECMP	0.76	0.62	0.62	0.54	0.46	0.56
	LFA	0.86	0.90	0.87	0.82	0.77	0.57
	ANHOR-SP	0.92	0.95	0.94	0.88	0.85	0.75
	ANHOR	0.98	0.99	1.00	0.95	0.95	0.80
$K = 4$	ECMP	0.77	0.62	0.63	0.54	0.47	0.60
	LFA	0.90	0.94	0.91	0.88	0.85	0.61
	ANHOR-SP	0.96	0.99	0.97	0.93	0.93	0.85
	ANHOR	1.00	1.00	1.00	0.99	0.99	0.92

Path Stretch. High path diversity increases robustness and allows for more load balancing. However, it has a cost in terms of path inflation. Next, we look at the distribution of path lengths. We focus on path lengths in terms of hop counts, since this metric is independent of the link weight settings. Fig. 6 shows the average *path stretch* for $K = 3$ with different routings, where the length of each valid path has been normalized with the shortest path length for that S-D pair. We observe that the superior path diversity in ANHOR and ANHOR-SP comes at the cost of some path inflation, but that the average path lengths are still comparable to those of shortest path routing. The path inflation introduced with multipath routing can be ameliorated with more intelligent load balancing methods [11][12].

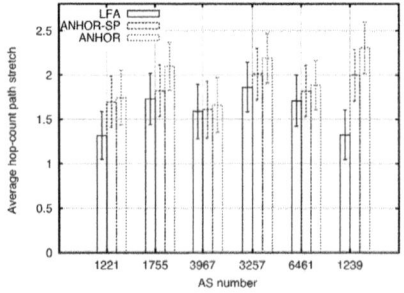

Fig. 6. Average hop-count path stretch with $K = 3$

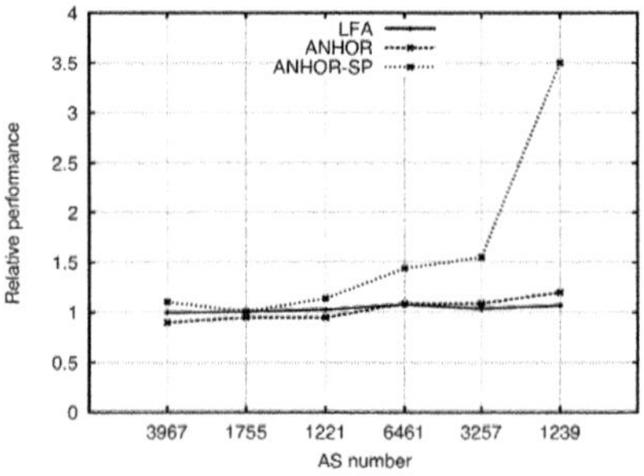

Fig. 7. Relative running time towards all destinations

Running Time and Memory Consumption. The complexity of our proposed algorithms depends on the number of nodes, links and on how efficiently the size of the candidate set can be reduced. The average size of candidate set turns out to be approximately 5 times (AS1755) to 12 times (AS1239) higher than their corresponding average node degrees in our tested topologies.

Fig. 7 shows the relative running time of each routing method to ECMP across six topologies. The AS topologies are listed in an increasing order of number of nodes. The results are achieved with an Intel Core 2 CPU 6300 @ 1.86 GHz machine. ANHOR has a low running time that is comparable to a normal ECMP routing computation. For all destinations, the total time difference is less than 10% for all topologies. As for ANHOR-SP, calculating routing permutations for all destinations take less than four times of ECMP. Across all topologies, the memory consumption never exceeds 6 MB.

7   Related Work

This paper presents Permutation Routing as a method for increased robustness in IP networks with traditional hop-by-hop forwarding. The proposed method can be used to generate routings that give a significant boost in number of nodes that have at at least two forwarding options to a destination. Our proposal shares the same principle with many existing solutions that aim at adding robustness either by changing routing protocols to adopt more than one next-hop towards a destination in the routing table such as DUAL[1], MDVA[2], FIR[3] or adding backup next-hops a posteriori to the forwarding entries found by a standard shortest path routing protocol such as LFA[4], Not-via[5], MRC[6], SPT-DAG[7]. Unlike DUAL and MDVA, Permutation Routing bases its construction solely on readily available topology information and hence no new control plane signalling is required. In addition, Permutation Routing outperforms LFA and does not introduce overhead bits as MRC and SPT-DAG.

Recent solutions focus on centralized routing that can provide added flexibility by improving path diversity that meets different requirements. Examples are O2 [16] and Protection Routing [14]. O2 routing has a similar routing objective to Permutation Routing, but limits itself to finding only two next-hops for each node. Permutation Routing, on the other hand, allows a tunable $K \geq 1$ while still being compatible with traditional link-state routing protocols. In the same category, Protection Routing presents a two-phase heuristic to produce a routing for a given traffic demand in a centralized routing system. In phase 1, the heuristic seeks to minimize number of unprotected nodes towards a destination while minimizing the cost function given in [15]. Although Routing Permutations share the goal of minimizing the number of nodes with only one forwarding option, we prefer to evenly distribute next-hops among nodes rather than performing traffic optimization for a specific traffic demand. We believe that finding a routing that optimizes for a given load function is less important in current Internet where traffic matrix elements vary significantly with time. Instead, we aim at maximizing the available forwarding options for more intelligent load balancing methods such as [11] [12] that are more responsive to traffic variation.

Similar to our method, MARA [17] employed the concept of permutations to generate routings that can maximize alternates. MARA, however, uses a different objective function seeking to maximize the minimum number of next-hops towards a destination. It can easily be shown that in networks without parallel links between nodes, this minimum must always be 1.

8   Conclusion

We have presented Permutation Routing as an approach for calculating more robust routing while being compatible with existing links state routing protocols. The paper proposed a generic algorithm to construct routing permutations. Routings that optimize different objectives can be implemented by modifying the selection criteria that are used in the construction algorithm. The goal of Permutation Routing is to maximize the survivability in a network with traditional hop-by-hop forwarding.

We have evaluated Permutation Routing with simulations on six ISP topologies. The results show that permutation routings outperform existing multipath approaches such as ECMP and LFA in terms of robustness and path diversity. We also showed that the complexity of calculating routing permutations is comparable to that of standard link state routing.

References

1. Garcia-Lunes-Aceves, J.J.: Loop-free routing using diffusing computations. IEEE Trans. on Networking 1(1), 130–141 (1993)
2. Vutukury, S., Garcia-Luna-Aceves, J.J.: MDVA: A Distance-Vector Multipath Routing Protocol. In: IEEE INFOCOM, pp. 557–564 (2001)
3. Nelakuditi, S., Lee, S., Yu, Y., Zhang, Z.-L., Chuah, C.-N.: Fast local rerouting for handling transient link failures. IEEE Trans. on Networking 15, 359–372 (2007)
4. Atlas, A., Zinin, A.: RFC5286: Basic Specification for IP Fast Reroute: Loop-Free Alternates (September 2008)
5. Shand, M., Bryant, S., Previdi, S.: IP Fast Reroute Using Not-via Addresses. Internet-Draft (work in progress, expired in June 2012)
6. Kvalbein, A., Hansen, A.F., Čičic, T., Gjessing, S., Lysne, O.: Multiple routing configurations for fast IP network recovery. IEEE Trans. on Networking 17(2) (2009)
7. Elhourani, T., Ramasubramanian, S., Kvalbein, A.: Enhancing Shortest Path Routing for Resilience and Load Balancing. In: ICC, pp. 1–6 (2011)
8. Francois, P., Bryant, S., Decraene, B., Horneffer, M.: LFA applicability in SP networks. Internet-Draft (work in progress, expired in July 2012)
9. Nakano, K., Olariu, S., Zomaya, A.Y.: Energy-efficient permutation routing in radio networks. IEEE Trans. on Parallel and Distributed Systems 12(6) (2001)
10. Liang, X., Shen, X.: Permutation Routing in All-Optical Product Networks. IEEE Trans. on Circuits and Systems 49(4), 533–538 (2002)
11. Xu, D., Chiang, M., Rexford, J.: DEFT: Distributed Exponentially-Weighted Flow Splitting. In: IEEE INFOCOM, pp. 71–79 (2007)
12. Kvalbein, A., Dovrolis, C., Muthu, C.: Multipath load-adaptive routing: putting the emphasis on robustness and simplicity. In: IEEE ICNP, pp. 203–212 (2009)

13. Cormen, T. H., et al.: Introduction to Algorithms. MIT Press, ISBN 0-262-03293-7
14. Kwong, K.-W., Gao, L., Gurin, R., Zhang, Z.-L.: On the feasibility and efficacy of protection routing in IP networks. In: IEEE INFOCOM, pp. 1543–1556 (2010)
15. Fortz, B., Thorup, M.: Internet traffic engineering by optimizing OSPF weights. In: IEEE INFOCOM, pp. 519–528 (2000)
16. Schollmeier, G., Charzinski, J., Kirstadter, A.: Improving the resilience in IP networks. In: HPSR Workshop, pp. 91–96 (2003)
17. Ohara, Y., Imahori, S., Meter, R.V.: MARA: Maximum Alternative Routing Algorithm. In: IEEE INFOCOM, pp. 298–306 (2009)
18. Nccui, A., Bhattacharyya, S., Taft, N., Diot, C.: IGP link weight assignment for operational Tier-1 backbones. IEEE Trans. on Networking 15, 789–802 (2007)
19. Rocketfuel topology mapping. WWW, http://www.cs.washington.edu

Routing On Demand: Toward the Energy-Aware Traffic Engineering with OSPF

Meng Shen[1,*], Hongying Liu[2], Ke Xu[1], Ning Wang[3], and Yifeng Zhong[1]

[1] Tsinghua National Laboratory for Information Science and Technology
Department of Computer Science, Tsinghua University,
Beijing, 100084, China
{shenmeng,ke.xu,zhongyifeng}@csnet1.cs.tsinghua.edu.cn
[2] School of Mathematics and Systems Science, Beihang University,
Beijing, 100191, China
liuhongying@buaa.edu.cn
[3] Center for Communication Systems Research, University of Surrey,
Surrey, GU2 7XH, United Kingdom
N.Wang@surrey.ac.uk

Abstract. Energy consumption has already become a major challenge to the current Internet. Most researches aim at lowering energy consumption under certain fixed performance constraints. Since trade-offs exist between network performance and energy saving, Internet Service Providers (ISPs) may desire to achieve different Traffic Engineering (TE) goals corresponding to changeable requirements. The major contributions of this paper are twofold: 1) we present an OSPF-based routing mechanism, Routing On Demand (ROD), that considers both performance and energy saving, and 2) we theoretically prove that a set of link weights always exists for each trade-off variant of the TE objective, under which solutions (*i.e.,* routes) derived from ROD can be converted into shortest paths and realized through OSPF. Extensive evaluation results show that ROD can achieve various trade-offs between energy saving and performance in terms of Maximum Link Utilization, while maintaining better packet delay than that of the energy-agnostic TE.

Keywords: Traffic Engineering, OSPF, Energy Saving, Routing.

1 Introduction

Since plenty of bandwidth-intensive applications (*e.g.,* video-on-demand and cloud computing) have come into service, the amount of data being carried on the Internet has dramatically grown [11]. Traditionally, Internet Service Providers

---

* This work is supported by New Generation Broadband Wireless Mobile Communication Network of the National Science and Technology Major Projects (2012ZX03005001), NSFC Project (61170292, 60970104), 973 Project of China (2009CB320501), Program for New Century Excellent Talents in University and EVANS Project (PIRSES-GA-2010-269323).

R. Bestak et al. (Eds.): NETWORKING 2012, Part I, LNCS 7289, pp. 232–246, 2012.

(ISPs) leverage Traffic Engineering (TE) to ensure acceptable network performance. Open Shortest Path First (OSPF) [18] is a commonly used intra-domain routing protocol that can be applied in TE through configuring appropriate link weights. In our previous work [5], we made a change on the traffic splitting scheme in OSPF and proposed an OSPF-based approach to achieve optimal TE.

As bandwidth requirements increase, ISPs need to deploy more and faster network equipments (*e.g.*, routers) to handle these demands, resulting in the growth of energy consumption. Therefore, TE should take both energy efficiency and network performance into consideration rather than merely focusing on the latter. The common idea of the state-of-the-art energy-aware TE approaches [7] [11] is to reduce energy consumption under certain performance constraints. However, we argue that trade-offs exist between energy saving and network performance. Intuitively, an ISP can achieve higher network energy efficiency at the cost of sacrificing partial performance without influencing user experience. Thus, there is a set of solutions rather than a unique solution to solve the energy-aware TE problem, where ISPs can make a flexible adjustment among these solutions.

The main challenge to achieve this goal exists that the conventional TE problem is already difficult due to the elastic nature of network traffic. Therefore, considering energy saving and network performance together will make the problem even harder. This problem should be carefully studied to ensure the network free from frequent oscillations and extra long delays.

This paper presents Routing On Demand (ROD), an OSPF-based routing mechanism that can achieve energy-aware traffic engineering solutions. ROD is formulated based on the Multiple Commodity Flows (MCF) constrains with a weighted objective considering both energy consumption and network performance in terms of Maximum Link Utilization (MLU). In ROD, a specific trade-off requirement can be derived from adjusting a *green factor* in the objective function, where corresponding solutions (*i.e.*, routes) can be converted into shortest paths with respect to a set of link weights. Then we develop algorithms to achieve these link weights, as well as flow splitting ratios based on minor extensions to the Equal-Cost Multiple Paths (ECMPs) function in OSPF that has been proposed in [5], which can be directly configured through OSPF protocol without introducing additional load to the network. The compatibility with existing network protocols eases the deployment of ROD mechanism.

Using real and synthetic network topologies, we evaluate the effectiveness of the ROD mechanism in multiple metrics, including trade-offs between energy saving and performance, link utilization and packet delay. Results show that, through adjusting the green factor, ROD can achieve diverse trade-offs between energy saving and MLU without bias on topology types. Moreover, a promising result for Abilene lies in that a 1% growth of MLU threshold will roughly lead to a 15% decrease in energy consumption. Similar results also hold for other topologies.

The rest of the paper is organized as follows. Section 2 gives an overview of our basic idea. In Section 3, we present the ROD model and the proof of the existence of optimal link weights. The implementation issues are discussed in

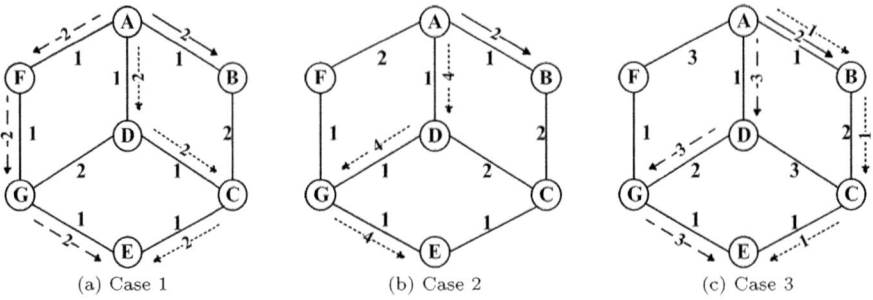

(a) Case 1   (b) Case 2   (c) Case 3

Fig. 1. Optimal Routing in Various Cases

Section 4, which is followed by the evaluation of ROD mechanism in Section 5. We summarize the related work in Section 6 and then conclude the paper in Section 7.

2   Overview of the Basic Idea

In this section, we illustrate through case studying that trade-offs exist between energy saving and network performance.

As shown in Figure 1, each link has a capacity of 5 units. Traffic demands consist of 2 Original-Destination (OD) pairs, namely A→B and A→E, with a bandwidth requirement of 2 and 4 units respectively. For simplicity, we assume that each unused link can be put into sleep mode [8] without any power consumption, while each loaded link consumes the same amount of power. Given traffic demands and network topology, we need to configure the OSPF link weights under which the optimal routes for these demands according to certain constrains and objectives are exactly the OSPF shortest paths.

Through the following three cases with the OSPF link weights beside links, we illustrate the existence of trade-offs between energy saving and network performance in terms of Maximum Link Utilization (MLU) based on OSPF.

Case 1: *Minimizing MLU.* The most widely used objective function of the conventional TE is simply to minimize MLU [14], which aims at reducing the risk of network congestion. The optimal routes are shown in Figure 1 (a), where each path is marked with the same line type and its traffic load. In this case, the optimal routes altogether leverage 7 links with a 2-unit load on each link. Thus the MLU is 40%.

Case 2: *Minimizing energy consumption without MLU constraint.* Based on the energy assumption, this goal is equivalent to minimize the number of loaded links. The optimal routes are presented in Figure 1 (b), where altogether 4 links are involved and the MLU is 80%. Comparing to Case 1, energy consumption of the entire network is reduced approximately by $(7 - 4)/7 = 42.9\%$.

Case 3: *Minimizing energy consumption with certain MLU upper bound.* TE with only energy consideration might increase the risk of network congestion on heavy loaded links when facing traffic bursts. Therefore, an ISP may desire to minimize energy consumption while keeping MLU within certain safety threshold, say 60%. The optimal routes are shown in Figure 1 (c), where the energy saving rate compared with Case 1 is about 14.3% and the MLU is exactly 60%.

We notice that the traffic is not equally split over two ECMPs serving the same OD pair A→E in Case 3. This can be realized by modifying hardware components, which will be discussed in Section 4.

3   ROD Model

In this section, we first present the Routing On Demand (ROD) model to formulate energy-aware TE problem, which extracts the trade-off between energy saving and network performance. Then we prove that the optimal link weights always exist under each considered TE objective function.

3.1   Network Model

We consider a directed network $\mathcal{G} = (\mathcal{N}, \mathcal{E})$, where $\mathcal{N}$ is the set of nodes (i.e., routers) and $\mathcal{E}$ is the set of edges (i.e., links). We assume that all links are directional and each link $(i, j) \in \mathcal{E}$ is attached with its own capacity c_{ij} which is the maximum amount of traffic it can take. Let $\mathcal{M}$ denote the set of source-destination pairs. For each $m \in \mathcal{M}$, a traffic demand of (s_m, t_m) is d_m, which represents the average intensity of traffic volume entering the network at node s_m and leaving from node t_m. Hereafter we use notations N, E and M to denote the cardinalities of sets $\mathcal{N}, \mathcal{E}$ and $\mathcal{M}$, respectively.

A customary way to treat the TE problem is to formulate it based on MCF constraints. We denote the destination node set with $\mathcal{D} = \{t \in \mathcal{N} : \exists \, m \in \mathcal{M} \text{ s.t. } t_m = t\}$. The traffic volume of commodity flow t along edge (i, j) is denoted by f_{ij}^t. The MCF constraints can be formulated as

$$f_{ij} = \sum_{t \in \mathcal{D}} f_{ij}^t \leq c_{ij}, \forall (i, j) \in \mathcal{E} \tag{1a}$$

$$\sum_{i:(s,i)\in\mathcal{E}} f_{si}^t - \sum_{j:(j,s)\in\mathcal{E}} f_{js}^t = d_s^t, \ \forall t \in \mathcal{D}, s \in \mathcal{N} \tag{1b}$$

$$f_{ij}^t \geq 0, \ \forall t \in \mathcal{D}, (i, j) \in \mathcal{E}, \tag{1c}$$

where $d_s^t \geq 0$ in (1b) is the expected traffic entering the network at node s and destined to node t. For $s \neq t$, set $d_s^t = d_m$ if there exists $m \in \mathcal{M}$ such that $s_m = s$ and $t_m = t$, or set $d_s^t = 0$ otherwise. For $s = t$, according to the previous definition of commodity, we set $d_t^t = -\sum_{m:t_m=t} d_m$. Constraint sets (1a) and (1b) represent the capacity and flow conservation constraints, respectively. We say a traffic distribution $\mathbf{f} = (f_{ij}, (i, j) \in \mathcal{E})$ is feasible if there exists $(\mathbf{f}^t, t \in \mathcal{D})$ such that $(\mathbf{f}, \mathbf{f}^t, t \in \mathcal{D})$ satisfies the MCF constraints.

Remark: Here, to formulate the problem in a more compact way, we regard all traffic flows to the same destination as a commodity, *i.e.*, we use the constraints (1b) associated with those traffic flows with their sum for $s = t$, so that we can obtain a formulation with less constraints and variables. It can be easily checked that this formulation is equivalent to the one where each traffic flow is treated as a separate commodity.

Minimizing MLU. If **f** is feasible, the total load and the utilization of link $(i, j) \in \mathcal{E}$ are f_{ij} and f_{ij}/c_{ij}, respectively. The conventional TE problem prefers to minimize MLU to balance traffic distribution over all links, *i.e.*,

$$\text{minimize} \quad \max_{(i,j) \in \mathcal{E}} \frac{f_{ij}}{c_{ij}} \tag{2a}$$

$$\text{subject to } (1a) - (1c). \tag{2b}$$

Minimizing Energy Consumption. If we only consider minimizing energy consumption of the entire network, then the TE problem turns into optimizing traffic distribution under the same MCF constraints in (2) to use as few links as possible, *i.e.*, to minimize $\|\mathbf{f}\|_0$, where $\|\mathbf{f}\|_0$ represents the number of non-zero elements of **f**.

The above formulation, which seeks the vector whose support has the smallest cardinality, is commonly referred to as cardinality minimization and known to be NP-hard. Since the computation time for medium and large-scale networks is a great concern, the cardinality minimization problem is thus usually heuristically solved by minimizing the ℓ_1 norm [6]. Then the TE problem of minimizing energy consumption with the relaxed objective function is

$$\begin{aligned}\text{minimize} \quad &\sum_{(i,j) \in \mathcal{E}} f_{ij} \\ \text{subject to} \quad &(1a) - (1c).\end{aligned} \tag{3}$$

Routing On Demand. We introduce a *green factor* θ to combine the above two models together and propose an integrated ROD model as follows

$$\begin{aligned}\text{minimize} \quad &\theta \max_{(i,j) \in \mathcal{E}} \frac{f_{ij}}{c_{ij}} + \sum_{(i,j) \in \mathcal{E}} f_{ij} \\ \text{subject to} \quad &(1a) - (1c).\end{aligned} \tag{4}$$

3.2   Universal Existence of Optimal Link Weights

We associate the network with an operator, and assume that if the load held by link (i, j) is f_{ij}, then the cost is $\phi(\mathbf{f})$. We also assume that the cost $\phi(\mathbf{f})$ is an continuous *convex* function of **f** $(\mathbf{f} \geq \mathbf{0})$. Without loss of generality, we can apply $\phi(\mathbf{f})$ to represent various objective functions of previous models. The goal of TE thus turns into minimizing $\phi(\mathbf{f})$ over MCF constraints.

For simplicity of presentation, we write the MCF constrains in matrix form. Then the general TE problem is given as TE$(\phi, \mathcal{G}, \mathbf{c}, \mathbf{D})$

$$\text{minimize} \quad \phi(\mathbf{f}) \tag{5a}$$

$$\text{subject to } \mathbf{f} - \sum_{t \in \mathcal{D}} \mathbf{f}^t = \mathbf{0} \tag{5b}$$

$$\mathbf{f} \leq \mathbf{c} \tag{5c}$$

$$\mathbf{A}\mathbf{f}^t = \mathbf{d}^t, \mathbf{f}^t \geq \mathbf{0}, \ \forall t \in \mathcal{D}, \tag{5d}$$

where $\mathbf{A}$, an $N \times E$ node-arc incidence matrix for network $\mathcal{G}$, is introduced to represent the multi-commodity flow constraints (1b). The column corresponding to link (i, j) has a $+1$ entries in row i and a -1 entries in row j.

The partial Lagrangian of $\text{TE}(\phi, \mathcal{G}, \mathbf{c}, \mathbf{D})$ is

$$\mathcal{L}(\mathbf{f}, \mathbf{f}^t, t \in \mathcal{D}; \mathbf{w}) = \phi(\mathbf{f}) - \sum_{(i,j) \in \mathcal{E}} w_{ij} f_{ij} + \sum_{t \in \mathcal{D}} \mathbf{w}^T \mathbf{f}^t$$

From the general theory of constrained convex optimization [1], it follows that $(\mathbf{f}, \mathbf{f}^t, t \in \mathcal{D})$ solves Problem (5) if and only if there exists a Lagrangian multiplier vector $\mathbf{w}$ such that $(\mathbf{f}, \mathbf{f}^t, t \in \mathcal{D})$ solves

$$\begin{aligned} \text{minimize} \quad & \phi(\mathbf{f}) - \sum_{(i,j) \in \mathcal{E}} w_{ij} f_{ij} + \sum_{t \in \mathcal{D}} \mathbf{w}^T \mathbf{f}^t \\ \text{subject to } & \mathbf{f} \leq \mathbf{c}; \quad \mathbf{A}\mathbf{f}^t = \mathbf{d}^t, \mathbf{f}^t \geq 0, \forall t \in \mathcal{D}. \end{aligned} \tag{6}$$

Problem (6) is a separable optimization, since there is no coupling between variable $\mathbf{f}$ and $\mathbf{f}^t$ for all $t \in \mathcal{D}$. Then Problem (5) can be solved through the following distributed method, i.e., $(\mathbf{f}, \mathbf{f}^t, t \in \mathcal{D})$ solves problem (5) if and only if there exists the Lagrangian multiplier vector $\mathbf{w}$ such that $\mathbf{f}$ solves the capacity planning problem $\text{ISP}(\mathbf{w}, \phi, \mathbf{c})$, i.e.,

$$\begin{aligned} \text{minimize} \quad & \phi(\mathbf{f}) - \sum_{(i,j) \in \mathcal{E}} w_{ij} f_{ij} \\ \text{subject to } & \mathbf{f} \leq \mathbf{c} \end{aligned} \tag{7}$$

and for each $t \in \mathcal{D}, \mathbf{f}^t$ solves the minimum cost flow problem $\text{MCF}(\mathbf{w}, \mathbf{d}^t)$

$$\begin{aligned} \text{minimize} \quad & \mathbf{w}^T \mathbf{f}^t \\ \text{subject to } & \mathbf{A}\mathbf{f}^t = \mathbf{d}^t; \quad \mathbf{f}^t \geq 0. \end{aligned} \tag{8}$$

Here we give some engineering interpretations to the separated Problems (7) and (8) for all $t \in \mathcal{D}$. First, the ISP's problem (7) can be interpreted as a capacity planning problem where ISP determines the possible virtual link capacity with each link cost w_{ij}, the total cost $\phi(\mathbf{f})$ generated for flow distribution f_{ij}, and the maximal permission capacity for each link c_{ij} and the objective as the net cost. Then the flows of the t-th class with the same destination t finds a solution that minimizes the total cost under the given link cost w_{ij} without considering the link capacity.

An promising property of the optimal routes is that they can be converted into shortest paths, i.e., the routes for the $t-$th class flow is the shortest path under the link weights w_{ij}. Let $\boldsymbol{\nu}^t$ denote the optimal solution of the dual of Problem (8). By applying the *complimentary slackness* theorem, we have

$$\nu_i^t - \nu_j^t = w_{ij}, \quad \text{if} \quad f_{ij}^t > 0 \tag{9a}$$

$$\leq w_{ij}, \quad \text{if} \quad f_{ij}^t = 0. \tag{9b}$$

Let $p : i_0 i_1 \cdots i_n$ be a possible path of OD pair (s, t), where $i_0 = s$ and $i_n = t$. For example, if $y_p = \min_{k=1,2,\cdots,n} f_{i_{k-1}i_k}^t > 0$, we have $\sum_{(i,j)\in p} w_{ij} = \nu_s^t - \nu_t^t \leq \sum_{(i,j)\in \bar{p}} w_{ij}$ for any other path $\bar{p}$ that connects the same source-destination pair (s, t) under conditions (9a) and (9b).

Hereafter we refer to these as the *optimal link weights* and the *optimal link loads*, which are denoted by $\mathbf{w}^*$ and $\mathbf{f}^*$, respectively. The similar results have been shown by Wang et al. [3]. The major difference existing in our approach is that we present a *close form* of the link weights which are *explicitly* determined by the objective and the optimal traffic distribution. The main results are shown by Theorem 1 and that in Section 3.3.

Let $\mathbf{p}$ be a *subgradient* of a convex function f at $\mathbf{x} \in \text{dom} f$ if

$$f(\mathbf{y}) \geq f(\mathbf{x}) + \mathbf{p}^T(\mathbf{y} - \mathbf{x}), \ \forall \mathbf{y} \in \text{dom} f.$$

Subdifferential of f at $\mathbf{x} \in \text{dom} f$ is the set of all subgradients of f at $\mathbf{x}$ and is denoted by $\partial f(\mathbf{x})$. If f is convex and differentiable at $\mathbf{x}$, then $\partial f(\mathbf{x}) = \{\nabla f(\mathbf{x})\}$.

Theorem 1. *There exists one subgradient $\mathbf{p}$ of ϕ at $\boldsymbol{f}^*$ such that $w_{ij}^* = p_{ij}$ if $f_{ij}^* < c_{ij}$ and $w_{ij}^* \geq p_{ij}$ if $f_{ij}^* = c_{ij}$.*

The proof of Theorem 1 can be conducted based on the convex analysis theory with the ISP's problem (7) and the calculation rule of subdifferential, which is omitted here due to space limitation.

3.3   Optimal Link Weights for Various Cost Functions

In this subsection, we will illustrate that the optimal link weights can be achieved for those cost functions in Section 3.1.

Lemma 1. *Let $f(\mathbf{x}) = \max\{f_1(\mathbf{x}), \cdots, f_m(\mathbf{x})\}$. Define $\mathcal{I}(\mathbf{x}) = \{i : f_i(\mathbf{x}) = f(\mathbf{x})\}$, and the active functions is at $\mathbf{x}$. Then it holds that*

$$\partial f(\mathbf{x}) = \text{conv} \left\{ \cup_{i\in\mathcal{I}(\mathbf{x})} \partial f_i(\mathbf{x}) \right\},$$

For minimizing MLU, i.e., (2), we have $\phi(\mathbf{f}) = \max_{(i,j)\in\mathcal{E}} \frac{f_{ij}}{c_{ij}}$. Let $\mathcal{I}(\mathbf{f}^*) = \{(i,j) \in \mathcal{E} : f_{ij}^*/c_{ij} = \max_{(i,j)\in\mathcal{E}} f_{ij}/c_{ij}\}$, i.e., the set of the link with MLU. By Lemma 1, we have $\mathbf{p} \in \partial\phi(\mathbf{f}^*)$ if and only if $p_{ij} = a_{ij}/c_{ij}$ for $(i,j) \in \mathcal{I}(\mathbf{f}^*)$ and $a_{ij} \geq 0, \sum_{(i,j)\in\mathcal{I}(\mathbf{f}^*)} a_{ij} = 1$, and $p_{ij} = 0$ otherwise.

The results show that the routing minimizing MLU is the shortest path routing for each source-destination pair, where the link weights for the non-bottleneck links are all zeros, and *only* the link weights of the bottleneck links are possibly positive and scale with the inverse of their link capacities.

For minimizing energy consumption, by Theorem 1, routes derived from Model (3) are the shortest paths, where link weights are 1s for unsaturated links (*i.e.*, $f_{ij}^* < c_{ij}$) and greater than 1 for the rest links. Furthermore, any path without including saturated links must be the lowest hop-count path.

For Routing On Demand, in (4), assume the green factor such that the optimal link load $f_{ij}^* < c_{ij}$ for all $(i,j) \in \mathcal{E}$, *i.e.*, all links are unsaturated. By Theorem 1, the routes minimizing Problem (4) are the shortest paths, where the link weight is $\theta a_{ij}/c_{ij} + 1$ for $(i,j) \in \mathcal{I}(\mathbf{f}^*)$ and $a_{ij} \geq 0$, $\sum_{(i,j) \in \mathcal{I}(\mathbf{f}^*)} a_{ij} = 1$, and is 1 otherwise. Furthermore, the path of a source-destination pair must be the shortest hop-count path if it does not cross bottleneck links.

4   Implementation Issues

In this section, we will briefly discuss the implementation issues of ROD mechanism based on the theoretical formulation in Section 3. Similar to the conventional traffic engineering, ROD relies on a centralized controller that is responsible for centrally computing link weights and periodically configuring all routers.

4.1   Collecting and Disseminating Information for ROD

The input information for ROD (*e.g.*, network topology and traffic matrix) can be collected from routers. In OSPF, each router will periodically flood its Link State Advertisements (LSAs) that contain all link state information. In particular, the Traffic Engineering Link State Advertisement (TE-LSA) defined in RFC3630 [17] records link load information. Thus, the latest topology and traffic matrix could be computed based on this information.

Algorithm 1. *Computing the optimal link weights*

Given optimal tolerance ϵ and initial weight $\mathbf{w}(0)$ (such as $w_{ij}(0) = 1), k = 0$;
for the given weight $\mathbf{w}(k)$ **do**
    Solve ISP($\mathbf{w}, \phi, \mathbf{c}$) in (7) to achieve the traffic load $f_{ij}(k)$ for link (i,j);
    Solve MCF($\mathbf{w}, \mathbf{d}^t$) in (8) to achieve the routing variable $\mathbf{f}^t(k)$ for destination t;
    Each link $(i,j) \in \mathcal{E}$ updates its link weight

$$w_{ij}(k+1) = \left(w_{ij}(k) - \gamma_k(\textstyle\sum_{t \in \mathcal{D}} f_{ij}^t(k) - f_{ij}(k))\right)_+;$$

    $k \leftarrow k + 1$;
    Until gap$(\mathbf{w}(k), f(k), \mathbf{f}(k)) < \epsilon$.
  end for

The optimal link weights can be computed through Algorithm 1, which is derived from the dual decomposition of TE($\phi, \mathcal{G}, \mathbf{c}, \mathbf{D}$) defined in (5). Given the link weight $\mathbf{w}$ in each iteration, the ISP's problem in (7) and the MCF problem in (8) will be solved to get their own optimal traffic distribution. Then the dual gap

$$\text{gap}(\mathbf{w}(k), f(k), \mathbf{f}(k)) = \sum_{(i,j)\in\mathcal{E}} w_{ij}(k)(\sum_{t\in\mathcal{D}} f_{ij}^t(k) - f_{ij}(k))$$

is applied as an optimality measure. Here γ_k is the step size and $(z)_+ = \max(0, z)$. If the dual gap is smaller than tolerance ϵ, the iteration will terminate with optimal link weights $\mathbf{w}^*$ and traffic distribution $\mathbf{f}^*$.

These link weights can be directly configured in routes as OSPF link weights. Algorithm 1 is converged because we apply the classic sub-gradient projection method [1] to the dual of $\text{TE}(\phi, \mathcal{G}, \mathbf{c}, \mathbf{D})$ in (5). Since the traffic matrix changes over time, the frequency of periodically computation and configuration may have an impact on network overhead and ROD performance. This is a major issue for our further investigation.

4.2   Data Forwarding under ROD

The packet under ROD mechanism is forwarded hop-by-hop along the destination-based shortest paths, which is the same as in OSPF. The shortest paths are constructed from the optimal link weights derived from Algorithm 1. In case that there are multiple equal-cost shortest paths between the same OD pairs, the flow splitting ratios over these paths can be computed from Algorithm 2. Taking the optimal traffic distribution $\mathbf{f}^*$ derived from Algorithm 1 as input information, Algorithm 2 first topologically sorts all nodes that are involved in data forwarding to a certain destination t, and then calculates the traffic splitting ratio $\tau(s, p, t)$ for each candidate next-hop denoted by p. The computational complexity of Algorithm 2 is $O(N * (N + E))$.

We mentioned the unequal traffic splitting ratios over ECMPs in Section 2. In hardware level, such traffic splitting scheme is usually realized through a hash-based mechanism [9]. Therefore, we can configure the hash table with different weights corresponding to the desired traffic splitting ratios.

Algorithm 2. *Achieving the splitting ratios*

Input the optimal traffic distribution $\mathbf{f}^*$
for each $t \in \mathcal{D}$ **do**
    Define an empty node set $\mathcal{V}$ and an empty link set $\mathcal{L}$
    $\mathcal{V} \leftarrow \mathcal{V} \cup \{t\}$
    for each link $(i, j) \in \mathcal{E}$ **do**
        if $i \notin \mathcal{V}$ and $\mathbf{f}_{ij}^* > 0$ **then**
            $\mathcal{V} \leftarrow \mathcal{V} \cup \{i\}$ and $\mathcal{L} \leftarrow \mathcal{L} \cup \{(i, j)\}$
        end if
    end for
    Do topological sorting on $\mathcal{V}$ to get the sorted node set $\mathcal{V}'$
    for each node $s \in \mathcal{V}'(s \neq t)$ **do**
        $\tau(s, p, t) = f_{sp}^{*t} / \sum_{(s,p)\in\mathcal{L}} f_{sp}^{*t}$
        $\mathcal{V}' \leftarrow \mathcal{V}' \backslash \{s\}$
    end for
end for

Table 1. Topologies for evaluation

Network	Category	Nodes	Links	Flows
Abilene	Backbone	11	28	110
Cernet2	Backbone	20	44	380
Random	Random	50	230	50

Table 2. Power consumption of line-cards

Line-card	Rate/Mbps	Energy/W
1-Port OC3	155.52	60
8-Port OC3	1244.16	100
1-Port OC48	2488.32	140
1-Port OC192	9953.28	174

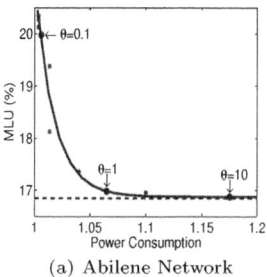

(a) Abilene Network

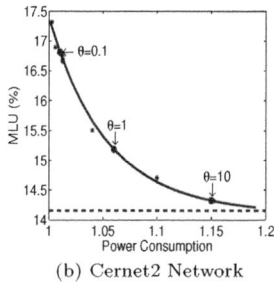

(b) Cernet2 Network

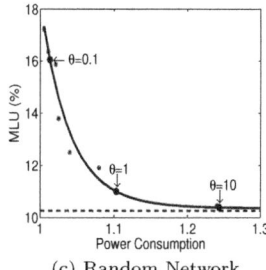

(c) Random Network

Fig. 2. Trade-offs between MLU and Energy Consumption for Different Topologies

5 Evaluation

In this section, we evaluate ROD with real and synthetic networks to show its effectiveness on achieving trade-offs between energy saving and performance.

5.1 Experimental Setup

We resort to ns2 simulator [21] to explore the effectiveness of ROD on small- and large-scale topologies. For comparison, hereafter we refer to mechanisms minimizing MLU and energy consumption as MMLU and MEC, respectively.

Topologies. The topologies used are summarized in Table 1. The router-level topology of Abilene is available at [16]. For simplicity, we consider the two routers at Atlanta as a single node, the entire network thus has 11 nodes altogether. The Cernet2 is the worldwide largest pure IPv6 research network, which has 44 links with 10 Gbps for 8 core links and 2.5 Gbps for the rest. The Random topology is generated by GT-ITM [20], whose capacity is set to 1 Gbps for each link.

Traffic Matrices. For Abilene, we select a subclass of the online available matrices [16], which are measured on March 1st, 2004. The traffic matrix for Cernet2 is generated by the gravity model [15] based on aggregated link load collected from January 10th to 16th, 2010. We generate a sparse traffic matrix for Random through gravity model to achieve a light traffic load scheme.

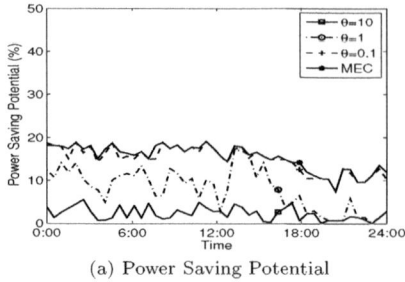
(a) Power Saving Potential

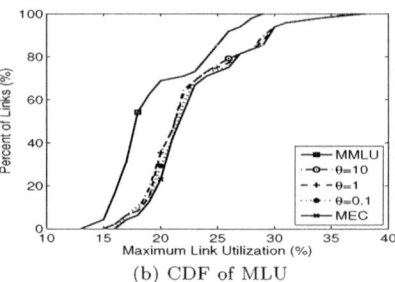
(b) CDF of MLU

Fig. 3. Simulation Results for Abilene on March 1st, 2004

Energy Saving Assumptions. Since line-cards are considered to account for more than 40% of a router's total budget [19], we thus focus on line-cards for energy saving, whose consumption in our evaluation is summarized in TABLE 2. We assume that unloaded line-cards can be automatically put into sleep mode and the time required to enter or wake up from sleep mode is negligible [10].

5.2 Performance versus Energy Saving

In this subsection, we explore trade-offs that exist between performance and energy saving potential in ROD. MLU is usually a major concern for ISPs in TE and thus employed here as an indicator of network performance. For energy saving indicator, we compute the amount of power consumption of line-cards in each simulation.

For each topology, we vary the green factor θ to investigate trade-offs between MLU and power consumption, which are plotted in Figure 2. The x-axis represents power consumption of the entire network and the y-axis represents the MLU. For ease of illustration, we use MEC as a benchmark and normalize the value of energy consumption of ROD. We first asterisk discrete points for a set of selected values of θ and then fit these points to achieve smooth curves.

The result for Abilene in Fig. 2(a) reveals that power consumption varies consistently with θ, whereas MLU decreases as θ increases. We also notice that, if the MLU threshold of Abilene network is relaxed a bit (*e.g.*, from 16.8% to 17.5%), the power consumption will considerably lower down (*e.g.*, from 1.25 to 1.05). Similar results also hold for Cernet2 and Random.

From the three sub-figures, we also notice two stages, 1) keeping θ increasing when it is larger than 10 can slightly bring down MLU but dramatically raises up the energy consumption, which we refer to as ***power-sensitive*** stage; 2) continuing to turn θ down when it is smaller than 0.1 will make a small contribution to energy saving at the cost of a sharp increase of MLU, which we regard as ***performance-sensitive*** stage. As both power- and performance-sensitive stages are undesirable for ISPs, we choose three typical values of θ between these two stages for the following evaluation, namely 0.1, 1 and 10.

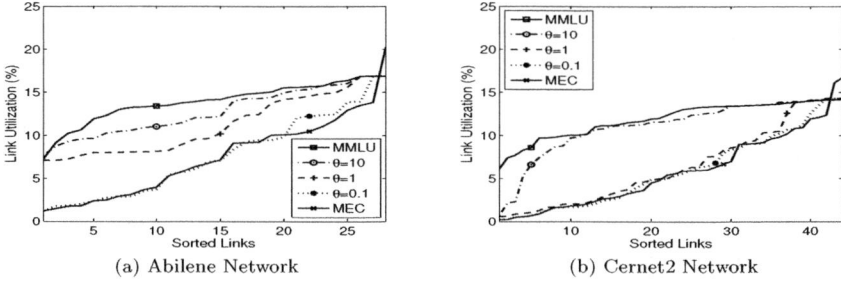

Fig. 4. Sorted Link Utilization for Different Topologies

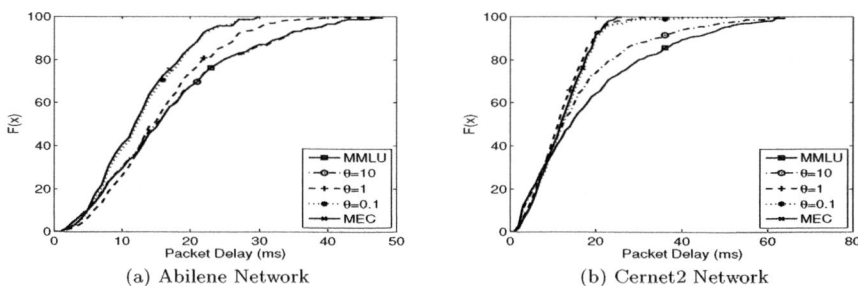

Fig. 5. CDF of Packet Delay for Different Topologies

In Figure 3 (a), we show how the green factor θ affects the energy saving potential of Abilene on March 1st, 2004. For each time interval, the power saving ratio is computed with the total power consumption of MMLU as a benchmark. The average energy saving ratio of ROD with $\theta = 0.1$ is about 15%, which is very close to that of MEC and higher than the other two cases with larger θ. The curve with $\theta = 1$ experiences severe oscillation since it can be easily affected by the changes in traffic matrix when equally considering minimizing MLU and power consumption.

5.3   Link Utilization

In this subsection, we evaluate the impact of ROD on link utilization that is denoted by **f** in Section 3. Using the Abilene traffic matrices collected for 24 hours, we first explore changes of MLU under each routing mechanism. We present the CDF of MLU of Abilene for a whole day in Figure 3 (b). Since the traffic is light, MLU is always under 40% for all cases. The curves for MMLU and MEC mechanisms act as a lower and higher bound, respectively. The three variations of ROD mechanism with different θ have slight difference in terms of MLU over temporal scale.

In Figure 4, we explore the utilization of all links for a single traffic matrix, where the x-axis represents the link indices. For Abilene, the traffic of MMLU is more evenly distributed over all links, since nearly 64.3% of the total 28 links achieve the largest utilization. However, MEC goes to the opposite side and concentrates the traffic on fewer links, leaving the minority of the links with much higher utilization. The curves under ROD mechanism with $\theta = 10$ is similar to that of MMLU, whereas the one with $\theta = 0.1$ is close to MEC, leaving the curve with $\theta = 1$ taking the middle place. The cases for Cernet2 and Random are similar, therefore, we do not present Random result due to space limitation.

5.4   Packet Delay

In this subsection, we evaluate the propagation delay under ROD. Figure 5 shows CDF of packet delay in different routing mechanisms for Abilene and Cernet2 topologies. For Abilene, MMLU exhibits the higher bound of packet delay, whereas MEC acts as the lower bound. The reason lies in that MMLU may result in undesirable overlong paths in terms of hop counts in order to avoid the highly utilized links. For this reason, the reciprocal of green factor in ROD could be considered as a penalty on the unnecessarily long paths. Therefore, the variation of ROD with a smaller θ should have a lower packet delay, which is confirmed through the curves in Figure 5(a). Due to space limitation, we omit the discussion on Cernet2 and the similar results to Random topology.

6   Related Work

Gupta et al. [8] firstly propose an energy saving approach by shifting network interfaces or devices (*e.g.*, routers) into sleep mode during idle periods. Nedevschi et al. [10] investigate two approaches via sleep and rate adaption. The former approach enables line-cards to sleep between packet bursts while the latter can adjust devices to operate at low frequency when traffic load is light. Vasić et al. [11] consider the above two approaches and present EATe to reduce energy consumption at the expense of increasing massage overhead of routers. In contrast, ROD mechanism is compatible with the current OSPF protocol and thus do not introduce unnecessary message cost.

Zhang et al. [7] have recently proposed an MPLS-based intra-domain traffic engineering mechanism, GreenTE, to maximize the number of links that can be put into sleep mode under given performance constrains. However, GreenTE requires excessive management control to achieve its goal because MPLS tunnels should be frequently set up and adjusted according to the elastic traffic. Panarello et al. [12] put forward a trade-off approach for performance and energy saving in access network. However, since devices in backbone networks consume the large majority of energy [13], we thus focus on exploring trade-off that exists in backbone networks.

7   Conclusion

This paper presents Routing On Demand (ROD), which is an OSPF-based routing mechanism that can achieve trade-off solutions to energy-aware TE problem.

We theoretically prove that a set of link weights always exists for each trade-off in energy-aware TE objective, under which solutions (*i.e.*, routes) derived from ROD can be converted into shortest paths and realized through OSPF. Evaluation results show that ROD can achieve different trade-offs between energy saving and performance in terms of Maximum Link Utilization (MLU) while maintaining better packet delay than that of energy-ignore TE. The compatibility of ROD with OSPF eases its deployment. Our future work includes investigating the overhead to realize ROD mechanism as well as the impact of link failures and traffic bursts.

References

1. Bertsekas, D.P.: Nonlinear Programming, 2nd edn. Athena Scientific, Belmont (1999)
2. Awduche, D.: MPLS and Traffic Engineering in IP Networks. IEEE Communication Magazine 37(12), 42–47 (1999)
3. Wang, Z., Wang, Y., Zhang, L.: Internet Traffic Engineering Without Full Mesh Overlaying. In: IEEE INFOCOM, pp. 565–571 (2001)
4. Sridharan, A., Guérin, R., Diot, C.: Achieving Near-Optimal Traffic Engineering Solutions for Current OSPF/IS-IS Networks. IEEE/ACM Transaction on Networking 13(2), 234–247 (2005)
5. Xu, K., Liu, H., Liu, J., Shen, M.: One More Weight is Enough: Toward the Optimal Traffic Engineering with OSPF. In: IEEE ICDCS, pp. 836–846 (2011)
6. Candès, E.J., Tao, T.: Decoding by Linear Programming. IEEE Transactions on Information Theory 51(12), 4203–4215 (2005)
7. Zhang, M., Yi, C., Liu, B., Zhang, B.: GreenTE: Power-Aware Traffic Engineering. In: IEEE ICNP, pp. 21–30 (2010)
8. Gupta, M., Singh, S.: Greening of the Internet. In: ACM SIGCOMM, pp. 19–26 (2003)
9. Cao, Z., Wang, Z., Zegura, E.: Performance of Hashing-Based Schemes for Internet Load Balancing. In: IEEE INFOCOM, pp. 332–341 (2000)
10. Nedevschi, S., Popa, L., Iannaccone, G., Ratnasamy, S., Wetherall, D.: Reducing Network Energy Consumption via Sleeping and Rate-Adaptation. In: USENIX NSDI, pp. 323–336 (2008)
11. Vasić, N., Kostić, D.: Energy-Aware Traffic Engineering. In: The 1st International Conference on Energy-Efficient Computing and Networking, pp. 169-178 (2010)
12. Panarello, C., Lombardo, A., Schembra, G., Chiaraviglio, L., Mellia, M.: Energy Saving and Network Performance: a Trade-off Approach. In: The 1st International Conference on Energy-Efficient Computing and Networking, pp. 41–50 (2010)
13. Tucker, R., Baliga, J., Ayre, R., Hinton, K., Sorin, W.: Energy Consumption in IP Networks. In: ECOC Symposium on Green ICT, p. 1 (2008)
14. Applegate, D., Cohen, E.: Making Intra-domain Routing Robust to Changing and Uncertain Traffic Demands: Understanding Fundamental Tradeoffs. In: ACM SIGCOMM, pp. 313–324 (2003)
15. Roughan, M., Greenberg, A., Kalmanek, C., Rumsewicz, M., Yates, J., Zhang, Y.: Experience in Measuring Backbone Traffic Variability: Models, Metrics, Measurements and Meaning. In: ACM SIGCOMM Internet Measurement Workshop, pp. 91–92 (2002)

16. Yin Zhang's Abilene TM,
    http://www.cs.utexas.edu/~yzhang/research/AbileneTM/
17. Katz, D., Kompella, K., Yeung, D.: RFC 3630: Traffic Engineering (TE) Extensions
    to OSPF Version 2, http://www.ietf.org/rfc/rfc3630.txt
18. Moy, J.: RFC 2328 OSPF Version 2, http://tools.ietf.org/html/rfc2328.txt
19. Power Management for the Cisco 12000 Series Router,
    http://www.cisco.com/en/US/docs/ios/12_0s/feature/guide/12spower.html
20. GT-ITM, http://www.cc.gatech.edu/projects/gtitm/
21. Network Simulator 2, http://www.isi.edu/nsnam/ns/

Minimization of Network Power Consumption with Redundancy Elimination

Frédéric Giroire[1], Joanna Moulierac[1],
Truong Khoa Phan[1], and Frédéric Roudaut[2]

[1] Joint project MASCOTTE, I3S(CNRS-UNS), INRIA, Sophia-Antipolis, France
{frederic.giroire,joanna.moulierac,truong_khoa.phan}@inria.fr
[2] Orange Labs, Sophia Antipolis, France
frederic.roudaut@orange.com

Abstract. Recently, *energy-aware routing* has gained increasing popularity in the networking research community. The idea is that traffic demands are aggregated over a subset of the network links, allowing other links to be turned off to save energy. In this paper, we propose GreenRE - a new energy-aware routing model with the support of the new technique of *data redundancy elimination (RE)*. This technique, enabled within the routers, can identify and remove repeated content from network transfers. Hence, capacity of network links are virtually increased and more traffic demands can be aggregated. Based on our real experiments on Orange Labs platform, we show that performing RE consumes some energy. Thus, while preserving connectivity and QoS, it is important to identify at which routers to enable RE and which links to turn off so that the power consumption of the network is minimized. We model the problem as an Integer Linear Program and propose a greedy heuristic algorithm. Simulations on several network topologies show that GreenRE can gain further 30% of energy savings in comparison with the *traditional energy-aware routing* model.

Keywords: Green networking, Energy-efficient routing, Algorithm.

1 Introduction

Some recent studies [6][12] exhibit that the traffic load of routers only has small influence on their energy consumption. Instead, the dominating factor is the number of active elements on routers such as ports, line cards, base chassis, etc. Therefore, in order to minimize energy consumption, fewer network elements should be used while preserving connectivity and QoS. Intuitively, it is possible to have multiple paths between a pair of source-destination on the network. When traffic load on links is low, we can aggregate the traffic into fewer links so that the other links do not need to carry any traffic. Routers then can turn off idle links (or precisely, two ports on the two routers at the ends of the link) for energy reduction. Although today's routers cannot turn off the ports and bring them back to active state quickly, we believe that this advance will come in near future, especially if it offers big energy savings.

R. Bestak et al. (Eds.): NETWORKING 2012, Part I, LNCS 7289, pp. 247–258, 2012.

In general, link capacity is the main constraint when aggregating traffic flows to a subset of links. In this work, we use an assumption that routers can eliminate redundant data traffic and hence virtually increase capacity of the links. As a result, more traffic flows can be aggregated and more links can be turned off to save energy. Although routers nowadays cannot remove repeated content from network transfers, we notice that there is WAN Optimization Controller (WOC) - a commercial device used in enterprises or small ISPs to eliminate traffic redundancy [5][10][14]. In order to identify the power consumption directly induced by RE, we perform real experiments on the WOCs. Because the principle idea of routers performing RE is similar to that of the WOC (section 2.2), we believe that when a router eliminates traffic redundancy, it also consumes more energy than usual. Therefore, in order to evaluate the global power consumption of the network, we should consider both the number of active links and the number of routers that perform RE. In summary, the contributions of this work are the following:

- We do real experiments to show power consumption of a WOC.
- We define and model the problem using Integer Linear Programing (ILP).
- We propose and evaluate a greedy heuristic algorithm that can be used for large-scale networks.
- By simulation, we present energy savings on real network topologies. In addition, we discuss the impact of GreenRE on route length (or delay also).

The rest of this paper is structured as follows. We summarize related works in Section 2. In Section 3, we model GreenRE using ILP, then propose a greedy heuristic algorithm. Simulation results are presented in Section 4. We have discussion in Section 5 and finally, we conclude the work in Section 6.

2   Related Works

2.1   Reduction of Energy Consumption

Measurement of Energy Consumptions. Several empirical measurements showed that energy consumption of network devices is largely independent of their load [6][12]. In addition, many experiments show that the power consumption depends on the number of active ports on routers [12]. Hence, explicitly turn off unused ports can reduce power consumption of routers.

Energy Minimization. The work on energy consumption of the Internet has been first evoked as a hypothetical working direction in [11]. Recently (2008 - present), researchers have started to massively invest their efforts in this research area [4]. The authors in [7] modeled the problem using ILP and showed how much energy can be saved on different network topologies. In [9], the authors proved that there is no polynomial-time constant factor approximation algorithm for this problem. Then they give theoretical bounds for specific network topologies and present heuristics to find solution close to the optimal one.

2.2   Reduction of Traffic Load

Internet traffic exhibits large amount of redundancy when different users access the same or similar contents. Therefore, several works [1][2][3][10][16][17] have explored how to eliminate traffic redundancy on the network. Spring et al. [17] developed the first system to remove redundant bytes from any traffic flows. Following this approach, several commercial vendors have introduced WAN Optimization Controller (WOC) - a device that can remove duplicate content from network transfers [5][10][14]. WOCs are installed at individual sites of small ISPs or enterprises to offer end-to-end RE between pairs of sites. As shown in Fig. 1, the patterns of previously sent data are stored in database of the WOCs at both sending and receiving sides. The technique used to synchronize the databases at peering WOCs can be found in [10]. Whenever the WOC at the sending side notices the same data pattern coming from the sending hosts, it sends a small signature instead of the original data (called *encoding process*). The receiving WOC then recovers the original data by looking up the signature in its database (called *decoding process*). Because signatures are only a few bytes in size, sending signatures instead of actual data gives significant bandwidth savings.

Recently, the success of WOC deployment has motivated researchers to explore the benefits of deploying RE in routers across the entire Internet [1][2][3][16]. The core techniques used here are similar to those used by the WOC: each router on the network has a local cache to store previously sent data which then is used to encode and decode data packets. Obviously, there are two key challenges that hinder the deployment of RE on routers. First, a significant number of memory accesses and heavy computation are required during various stages of RE. Second, a large amount of memory is required for the local cache at routers. However, Anand et al. [3] have introduced SmartRE which considers these challenges in the design. The authors show that on the desktop 2.4 GHz CPU with 1 GB RAM used for storing caches, the prototype can work at 2.2 Gbps for encoding and at 10 Gbps for decoding packets. Moreover, they believe that higher throughput can be attained if the prototype is implemented in hardware. Hence, the key limitations can be overcome and this technology should come in future's routers, especially if it offers significant bandwidth savings. Several real traffic traces have been collected from many networks such as at 11

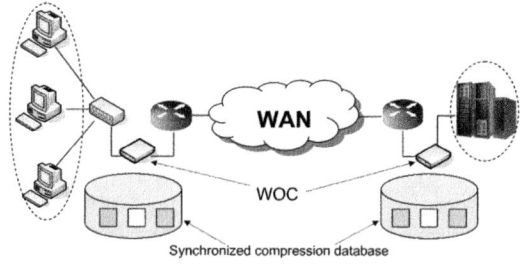

Fig. 1. Reduction of end-to-end link load using WOC

corporate enterprises in US [2], at a large university in US [1] and at 5 sites of a large corporate network in North America [16]. The authors in [1][2][16] conclude that up to 50% of the traffic load can be reduced with RE support.

In this work, we propose GreenRE - the first model of *energy-aware routing* with RE support. We show that RE, which was initially designed for bandwidth savings, is also a potential technique to reduce power consumption of the Internet.

3   Energy-Aware Routing with RE

Traditional energy-aware routing model (without RE) has been presented as a multicommodity integral flow problem with the objective of minimizing the number of links [9]. Our GreenRE model can be modeled in a similar way. However, when performing RE, routers consume some additional energy (section 4.1). Thus both the active links and the RE-routers (*routers that perform RE*) should be considered in the objective function. We model a network topology as an undirected weighted graph $G = (V, E)$, where the weight C_{uv} represents the capacity of an edge $(u, v) \in E$. We represent a set of demands by $D = \{D_{st} > 0; (s, t) \in V \times V, s \neq t\}$ where D_{st} denotes the amount of bandwidth required for a demand from s to t. A feasible routing of the demands is an assignment for each $D_{st} \in D$ a path in G such that the total amount of demands through an edge $(u, v) \in E$ does not exceed the capacity C_{uv}. We present in Fig. 2 a simple example to show the efficiency of using RE-routers in energy-aware routing. The objective is to find a routing solution for the two demands $D_{0,5}$ and $D_{10,15}$ (both with traffic volume requirement of 10) so that it satisfies the link capacity constraint and minimize power consumption of the network. As shown in Fig. 2a, the routing solution without RE-router requires 10 active links, then we can turn off 7 links and save 41% of energy consumption of the network. Meanwhile, when routers 6 and 9 are RE-routers, and assume that 50% of the traffic is redundant, then we have a better solution in which 57% of energy is saved (Fig. 2b). Moreover, if the two demands increase traffic volume to 20, there is no feasible routing solution without RE-router. But it is possible to find a solution if we have RE-routers at 0, 5, 10 and 15.

Energy-aware routing problems are known to be NP-hard [7][9], hence we first define and model the problem using Integer Linear Program (ILP), then we present a greedy heuristic algorithm for large scale networks.

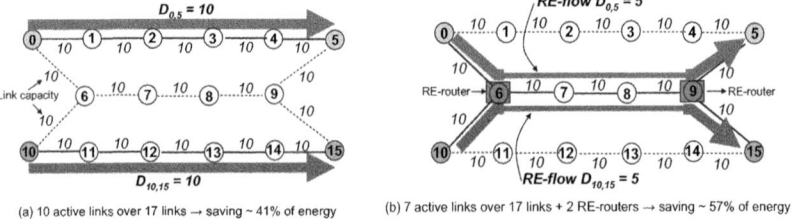

(a) 10 active links over 17 links → saving ~ 41% of energy   (b) 7 active links over 17 links + 2 RE-routers → saving ~ 57% of energy

Fig. 2. Feasible routing solution (a) without RE-router and (b) with RE-routers

3.1   Optimal Solution - Integer Linear Program (ILP)

For each link $(u, v) \in E$, we use a binary variable x_{uv} to determine if the link is used or not. We consider a simplified architecture where each line card on router has one network interface. Therefore, when a link (u, v) is used ($x_{uv} = 1$), the two network interfaces at router u and router v are enabled, and the power consumption of this link (called PL_{uv}) is the sum of power consumption of the two interfaces at router u and router v. We note f_{uv}^{st} be the flow on edge (u, v) corresponding to the demand D_{st} flowing from u to v. We use an assumption that all routers on the network can perform RE, however they can disable RE when unused to save energy. Then, we define a binary variable w_u which is equal to 1 if router u perform RE (called RE-router and it consumes additional PN_u Watts). We also use a redundancy factor γ to represent amount of data that can be eliminated by the RE-routers. We differentiate the usual flow f_{uv}^{st} from the RE-flow fr_{uv}^{st} (the redundant data has been removed), on link (u, v) for a demand (s, t) and $fr_{uv}^{st} = f_{uv}^{st}/\gamma$. When a flow enters a RE-router u, it can happen:

- If the flow is RE-flow, the router u can decode it to usual flow or do nothing (just forward the RE-flow).
- If the flow is usual flow, the router u can perform RE for this flow or just forward it.

The *Objective Function* of the ILP is to find a routing solution that minimizes the total power consumption of the network:

$$\min \left[PL_{uv} \sum_{(uv) \in E} x_{uv} + PN_u \sum_{u \in V} w_u \right]$$

subject to:

"*Flow conservation constraint*": $\forall u, v \in V$, $N(u)$ - neighbors of u

$$\sum_{v \in N(u)} f_{vu}^{st} - \sum_{v \in N(u)} f_{uv}^{st} + \gamma \left(\sum_{v \in N(u)} fr_{vu}^{st} - \sum_{v \in N(u)} fr_{uv}^{st} \right) = \begin{cases} -D_{st} & \text{if } u = s, \\ D_{st} & \text{if } u = t, \\ 0 & \text{otherwise.} \end{cases} \quad (1)$$

"*Capacity constraint*": $\forall (u, v) \in E$, C_{uv} - capacity of link (u, v)

$$\sum_{(s,t) \in D} \left(f_{uv}^{st} + f_{vu}^{st} + fr_{uv}^{st} + fr_{vu}^{st} \right) \leq x_{uv} C_{uv} \quad (2)$$

"*RE-router*": $\forall u, v \in V$,

$$M.w_u \geq \sum_{v \in N(u)} \left(fr_{uv}^{st} - fr_{vu}^{st} \right) \quad (3a)$$

$$M.w_u \geq \sum_{v \in N(u)} \left(fr_{vu}^{st} - fr_{uv}^{st} \right) \quad (3b)$$

Equation (1) states the flow conservation constraint and we differentiate between RE-flows and usual flows. Constraint (2) forces the link load to be smaller than the link capacity. Like some existing works [7][9], we consider a simplified model where link load is the total volume of all the flows passing through this link. In addition, to accommodate traffic bursts, we should limit the maximum utilization over any links in the network. For instance, the capacity using in equation (2) would be set to αC_{uv} where $\alpha = 50\%$. The constraints (3a) and (3b) (where M is a big constant number) are to make sure when the router u disables RE ($w_u = 0$), the difference between the sum of the RE-flows that enter and leave router u is equal to zero.

3.2   Heuristic Solution

The heuristic algorithm, in the first step, tries to find a routing solution for all demands so that it minimizes the number of active links. Because all routers on the network are assumed to be RE-routers, we virtually decrease volume of all the traffic demands to $Dr_{st} = D_{st}/\gamma$. Then for the second step, based on the routing found in the previous step, we try to disable RE on as many routers as possible to save energy.

Algorithm 1. Finding a feasible routing
Input: An undirected weighted graph $G = (V, E)$ where each edge e has a capacity C_e, a residual capacity R_e, an initial metric w_e and a set of demands $Dr_{st} \in D$.

$\forall e \in E, R_e = C_e, w_e = $ total number of demands
Sort the demands in random order
while Dr_{st} *has no assigned route* **do**
  compute the shortest path SP_{st} with the metric (w_e)
  assign the routing SP_{st} to the demand Dr_{st}
  $\forall e \in SP_{st}, R_e = R_e - Dr_{st}, w_e = w_e - 1$
end
return the routing (if it exists) assigned to the demands in D

Starting with the *Algorithm 1*, we compute a feasible routing for the RE-demands. Initially, all links on the network are set up with the same metric w_e which is equal to the total number of demands. We compute the shortest path for each demand with the metric w_e on links. Then, the links that have carried the shortest path is updated with metric: $w_e = w_e - 1$. Using this metric in the shortest path, we implicitly set high priority to reuse links that have already been selected. Then, the *Algorithm 2 - Step 1* is used to remove in priority links that are less loaded. C_e/R_e is used as the load on a link where R_e is the residual capacity on link e when previous demands have been routed.

In *Step 2*, we use the routing solution found in the *Step 1* as the input of the algorithm. Then, we consider all the traffic demands as normal demands without RE. Hence, some links can be congested because the total traffic volume

Algorithm 2. *Input:* An undirected weighted graph $G = (V, E)$ where each edge $e \in E$ has a capacity C_e and a residual capacity R_e

Step 1 - Removing less loaded links:

while *edges can be removed* **do**
  remove the edge e that has not been chosen and has smallest value C_e/R_e
  compute a feasible routing with the Algorithm 1
  if no feasible routing exists, put e back to G
end

return the feasible routing if it exists.

Step 2 - Enabling set of RE-routers:

Consider normal demands (without RE) and routing solution in Step 1

while *network is congested* **do**
  find demands to perform RE first (details later in part (a))
  enable set of RE-routers using end points congestion (details later in part (b))
end

of demands may exceed the link capacity. The heuristic we use in *Step 2* is based on following observations:

(a) Which Demand to Perform RE First? In Fig. 3a, we can see that when performing RE for $D_{0,11}$ on router 1 and 10, the amount of traffic passing all the congested links (links $(1, 3)$, $(8, 9)$ and $(9, 10)$) is also reduced. Then, the heuristic in step 2 decides to perform RE for this flow first. Assume that the redundancy factor $\gamma = 2$, then the links $(8, 9)$ and $(9, 10)$ are still congested. After removing $Dr_{0,11}$, the available capacity of links $(8, 9)$ and $(9, 10)$ are 5 and 10, respectively. Applying the same rule, the next demand to perform RE is $D_{7,11}$. Then, there is no congested link on the network since the link $(9, 10)$ is also released from congestion. Finally, only the routers 1, 8 and 10 are needed to enable RE. In summary, the algorithm will perform RE for the flows that passes through most of the congested links first.

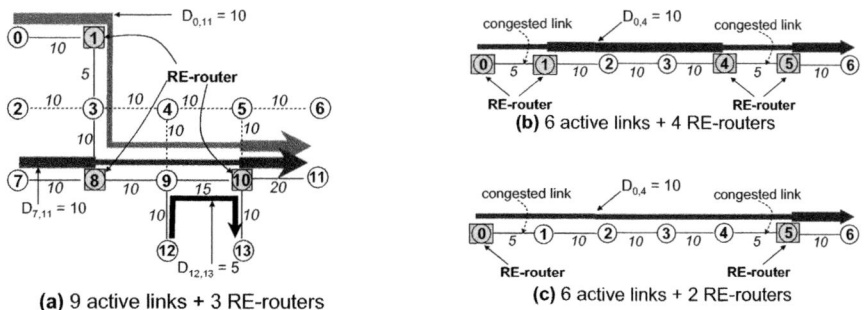

(a) 9 active links + 3 RE-routers

(b) 6 active links + 4 RE-routers

(c) 6 active links + 2 RE-routers

Fig. 3. Congested links and RE-routers

(b) End Points Congestion: in Fig. 3b and Fig. 3c, the demand (0,6) has traffic volume of 10 and the number on links indicates link capacity. Therefore, the two links (0, 1) and (4, 5) are congested. Hence, a naive solution is to enable RE at the two end-point routers of each congested link as shown in Fig. 3b. However, a better solution with less RE-routers should be to enable RE only at the starting (router 0) and ending point (router 5) of all the congested links (Fig. 3c). In summary, the algorithm will look for the longest congested part of the flow to enable RE-routers.

4   Experiments and Simulation Results

4.1   Energy Consumption with WOC

Several results of bandwidth savings using WOC can be found in [10]. We have also performed experiments on the network platform of the project Network Boost at Orange Labs[1]. We installed two WOCs, each at the access links of the two sites (let's call site A and site B). These two sites are connected via a backbone composed of 4 routers. We setup FTP connections for uploading files from site A to site B. As shown in Fig. 4a, power consumption of the WOC is increased (from 26 Watts to 34 Watts) with the number of concurrent FTP sessions. For the next experiment, we keep only one FTP session and let the WOC perform RE for 10 hours in which the size of uploaded files is increased. The results show that the WOC consumes around 30 Watts on average (Fig. 4b). Therefore, for sake of simplicity, we use an average value of power consumption (30 Watts) to represent additional cost for the router to perform RE.

4.2   Simulation Results with GreenRE

We studied ten classical real network topologies extracted from SNDLib [15]. Because CPLEX [8] takes several hours to find an optimal solution even for the

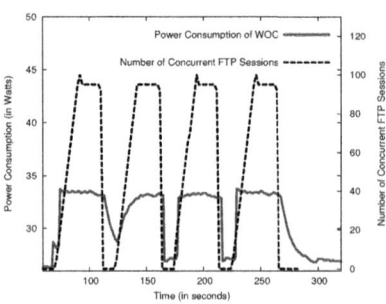

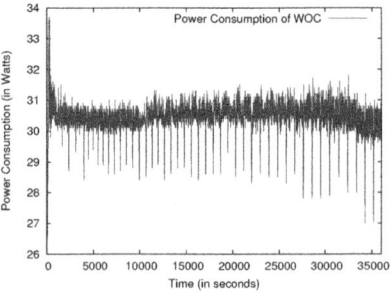

Fig. 4a. With concurrent FTP sessions   Fig. 4b. With increasing size of uploaded files

Fig. 4. Power consumption of the WOC

---

[1] The full figure of the test-bed can be found in [13].

smallest network (Atlanta network), we force CPLEX to stop after *two hours of execution*. According to the results of the works mentioned in Section 2.2, we use a redundancy factor $\gamma = 2$ for all the simulations. We present in Fig. 5 results for Atlanta network topology - optimal solutions without RE-routers (*OPT-Green*) given by [9], ILP results (within two hours) with RE-routers (*ILP-GreenRE*), heuristic with RE-routers (*H-GreenRE*) and shortest path routing with RE-routers (*SP-RE*) given by [1]. For worst-case scenario and for comparison with previous work [9], all links are set up with the same capacity C and the demands are all-to-all (one router has to send traffic to all remaining routers on the network) with the same traffic volume D for each demand. The x-axis in Fig. 5 represents the *capacity/demand ratio* ($\lambda = C/D$). Note that in the simulations, λ has the same value and it represents the traffic load on the network. Large value of λ means that the link capacity is much larger than the required bandwidth of traffic demands or we can say the traffic load on the network is low. The power consumption of a link and of a RE-router used in the simulations are 200 Watts [7] and 30 Watts (section 4.1), respectively. Power consumption of a network with RE-routers are calculated as [(the number of active links)*200+ (the number of RE-routers)*30]. Without considering energy-aware routing and without RE-router, all the links on the network are used. Hence, the power consumption is computed as [(the number of links)*200]. The difference between these two values of power consumption represents how much energy can be saved. As shown in Fig. 5, without RE-router (OPT-Green), there is no feasible routing solution and hence, no energy is saved if $\lambda < 38$. When λ increases, links have more bandwidth to aggregate traffic, hence the solutions with and without RE-router converge to the same amount of energy savings. In general, the heuristic with RE routers works well and approximate to the results of ILP. In addition, we also consider shortest path routing with RE-routers given by [1]. The results show that: without considering energy-aware routing, there is no energy saving even if all the routers have enabled RE. We present in Fig. 6a the number of RE-routers used by the heuristic and by the ILP. When λ increases, it is not

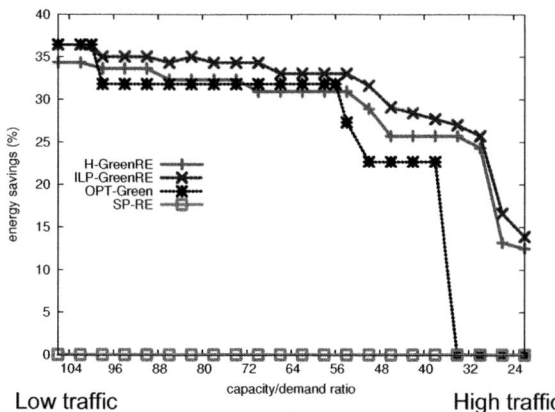

Fig. 5. Simulation results for Atlanta network

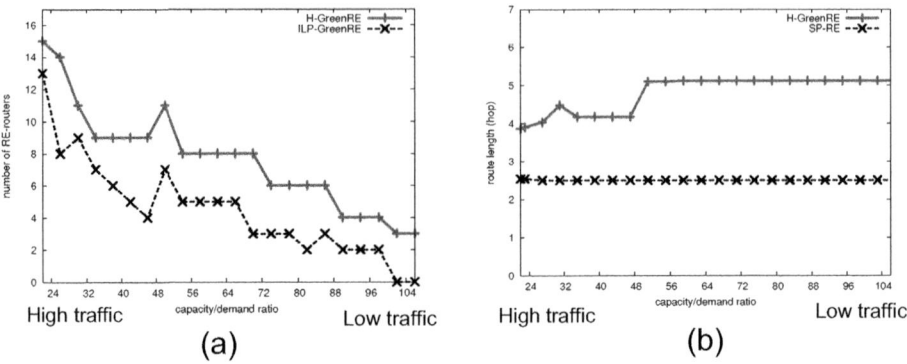

Fig. 6. (a) Number of RE-routers and (b) Average route length

necessary to have many RE-routers because the links have enough capacity for traffic aggregation. As shown in Fig. 6a, in general, the number of RE-routers is reduced when λ increases. However, there are some points where the number of RE-routers increases. It is because at these points, having more RE-routers allows to turn off more links and hence, reducing the global power consumption of the network.

We compare average route length (which also gives an idea of the delay) between the heuristic algorithm (*H-GreenRE*) and the shortest path routing with RE (*SP-RE*) in Fig. 6b. As we can see, when turning off links on the network, we save energy but the traffic demands need to route on longer paths. When λ increases, more traffic demands can be aggregated to a small number of links, resulting in longer routing paths. However, when this ratio is large enough, the routing solutions do not change much, and hence the average route length seems to be unchanged. Because *SP-RE* does not aggregate traffic, the route length is shorter than those in *H-GreenRE*.

We present in Table 1 energy gain for ten classical network topologies extracted from SNDLib [15]. Because these networks are too large to launch ILP, we compare heuristic results of H-GreenRE and H-Green (the heuristic without RE-router given by [9]). We use λ_{min} be the smallest value of capacity/demand ratio that allows to find a feasible route for all the demands without RE (found in [9]). In the simulations, a range of $\lambda = \{\lambda_{min}, 2\lambda_{min}, 3\lambda_{min}\}$ is used to represent high, medium and low traffic load on the network. As shown in Table 1, with RE-routers, it starts to save a large amount of energy (in average 32.77%) even with $\lambda = \lambda_{min}$. Recall that routing with RE-routers is possible even with $\lambda < \lambda_{min}$ meanwhile no feasible solution is found without RE-router. When λ is large enough, it is not necessary to have RE-routers on the network, hence both the solutions (with and without RE-router) converge to almost the same value of gaining in energy consumption.

Table 1. Gain of energy consumption (in %)

Network topologies	$\|V\|$	$\|E\|$	λ_{min}	Traffic volume (capacity/demand ratio λ)					
				with RE-router			without RE-router		
				λ_{min}	$2\lambda_{min}$	$3\lambda_{min}$	λ_{min}	$2\lambda_{min}$	$3\lambda_{min}$
Atlanta	15	22	38	23.6%	30.1%	36.4%	0%	32%	36%
New York	16	49	15	52.2%	61.6%	65.2%	2%	59%	63%
Nobel Germany	17	26	44	23.9%	33.9%	36.8%	0%	35%	39%
France	25	45	67	33.8%	44%	45.4%	0%	42%	44%
Norway	27	51	75	36.2%	45%	47.5%	12%	43%	47%
Nobel EU	28	41	131	27.7%	30.9%	34.2%	12%	32%	34%
Cost266	37	57	175	25.3%	33.6%	35%	3.5%	32%	35%
Giul39	39	86	85	36.5%	48.4%	51.1%	0%	45%	50%
Pioro40	40	89	153	45.3%	53.2%	54.5%	0%	53%	54%
Zib54	54	80	294	23.2%	31.2%	32.7%	0%	30%	33%

5   Discussion

We propose in this paper the GreenRE model with an assumption that all routers on the network can perform RE. However, this model can be extended to support incremental deployment of RE on routers across the Internet. For instance, we have presented in [13] the models that work for the two following cases:

- Only a subset of routers on the network can perform RE. Hence, the model should utilize these RE-routers and find a routing solution so that power consumption of the network is minimized.
- Given a network topology and a maximum number of RE-routers, the model should find where to place these RE-routers and the corresponding routing solution to minimize network power consumption.

6   Conclusion

To the best of our knowledge, GreenRE is the first work which considers RE as a complementary help for energy aware routing. We formally define the model using Integer Linear Programming and propose a greedy heuristic algorithm. The simulations on several network topologies show a significant gain in energy savings. Moreover, in comparison with the traditional energy-aware routing, GreenRE works better especially when the traffic load on the network is high. We also prove by simulations that the heuristic algorithm works well and approximates to the solutions given by the ILP. As part of future work, we will carry on more simulations with real trace redundancy factor on links and real traffic demands. We will also consider the impacts of GreenRE on the network such as network fault tolerance. Moreover, we plan to develop a distributed algorithm for GreenRE.

Acknowledgment. This work has been done in collaboration with Network Boost Project (Orange Labs) and partly funded by the ANR DIMAGREEN. The authors would like to thank Didier Leroy (Orange Labs), Yaning Liu and David Coudert (Project Mascotte) for their advices and support.

References

1. Anand, A., Gupta, A., Akella, A., Seshan, S., Shenker, S.: Packet Caches on Routers: the Implications of Universal Redundant Traffic Elimination. In: Proceedings of ACM SIGCOMM (2008)
2. Anand, A., Muthukrishnan, C., Akella, A., Ramjee, R.: Redundancy in Network Traffic: Findings and Implications. In: Proceedings of ACM SIGMETRICS (2009)
3. Anand, A., Sekar, V., Akella, A.: SmartRE: an Architecture for Coordinated Network-wide Redundancy Elimination. In: Proceedings of ACM SIGCOMM (2009)
4. Bianzino, A.P., Chaudet, C., Rossi, D., Rougier, J.: A Survey of Green Networking Research,
   http://www.cl.cam.ac.uk/teaching/1011/R02/papers/green-survey.pdf
5. BlueCoat: WAN Optimization, http://www.bluecoat.com/
6. Chabarek, J., Sommers, J., Barford, P., Estan, C., Tsiang, D., Wright, S.: Power Awareness in Network Design and Routing. In: Proceedings of IEEE INFOCOM (2008)
7. Chiaraviglio, L., Mellia, M., Neri, F.: Minimizing ISP Network Energy Cost: Formulation and Solutions. IEEE/ACM Transactions on Networking (2011)
8. http://www-01.ibm.com/software/integration/optimization/cplex-optimizer/
9. Giroire, F., Mazauric, D., Moulierac, J., Onfroy, B.: Minimizing Routing Energy Consumption: from Theoretical to Practical Results. In: Proceedings of IEEE/ACM GreenCom (2010)
10. Grevers Jr., T., Christner, J.: Application Acceleration and WAN Optimization Fundamentals. Cisco Press (2007)
11. Gupta, M., Singh, S.: Greening of The Internet. In: Proceedings of ACM SIGCOMM (2003)
12. Mahadevan, P., Sharma, P., Banerjee, S.: A Power Benchmarking Framework for Network Devices. In: Proceedings of IFIP Networking (2009)
13. Phan, T.K.: Minimization of Network Power Consumption with WAN Optimization. Master's thesis, University of Nice Sophia Antipolis (2011),
    www-sop.inria.fr/mascotte/rapports_stages/KhoaPhan_internship-2011.pdf
14. http://www.riverbed.com/solutions/optimize/
15. SNDlib, http://sndlib.zib.de
16. Song, Y., Guo, K., Gao, L.: Redundancy-Aware Routing with Limited Resources. In: Proceeding of IEEE ICCCN (2010)
17. Spring, N.T., Wetherall, D.: A Protocol-Independent Technique for Eliminating Redundant Network Traffic. In: Proceedings of ACM SIGCOMM (2000)

Sign What You Really Care about – Secure BGP AS Paths Efficiently*

Yang Xiang[1,3], Zhiliang Wang[2,3],
Jianping Wu[1,2,3], Xingang Shi[2,3], and Xia Yin[1,3]

[1] Tsinghua National Laboratory for Information Science and Technology (TNList)
[2] Department of Computer Science & Technology, Tsinghua University
[3] Network Research Center, Tsinghua University, Beijing, P.R. China, 100084
{xiangy08,wzl,jianping,shixg,yxia}@csnet1.cs.tsinghua.edu.cn

Abstract. The inter-domain routing protocol, Border Gateway Protocol (BGP), plays a critical role in the reliability of the Internet routing system, but forged routes generated by malicious attacks or mis-configurations may devastate the system. The security problem of BGP has attracted considerable attention, and although several solutions have been proposed, none of them have been widely deployed due to weaknesses such as high computational cost or potential security compromise. This paper proposes Fast Secure BGP (FS-BGP), an efficient mechanism for securing AS paths and preventing prefix hijacking by signing critical AS path segments. We prove that FS-BGP can achieve a similar level of security as S-BGP, but with much higher efficiency. Our experiments use BGP UPDATE data collected from real backbone routers. Compared with S-BGP, FS-BGP only requires a very small cache, and can reduce the cost of signing and verification by orders of magnitude. Indeed, the signing and verification can be accomplished as fast as the most bursty BGP UPDATE arrivals, which implies that FS-BGP will hardly delay the propagation of routing information.

Keywords: Inter-Domain Routing, BGP, Prefix Hijacking, Security.

1 Introduction

As BGP [13] controls the packet forwarding path between Autonomous Systems (ASes), it plays a critical role in the reliability of the Internet. However, routing information received from neighbors can not be validated. Forged routes may cause packets being forwarded along wrong paths. Malicious attacks often use BGP prefix hijacking to drop, intercept or tamper traffic towards specific prefixes. In 2008, US DoD networks were hijacked at least 7 times [21]. Accidental mis-configurations have also resulted in serious routing problems and economic

---

* This work is supported by (1) the National Key Technology R&D Program of China under Grant No. 2008BAH37B03, and (2) the National Basic Research Program of China (973 Program) under Grant No. 2009CB320502.

R. Bestak et al. (Eds.): NETWORKING 2012, Part I, LNCS 7289, pp. 259–273, 2012.

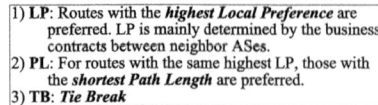

1) **LP**: Routes with the ***highest Local Preference*** are preferred. LP is mainly determined by the business contracts between neighbor ASes.
2) **PL**: For routes with the same highest LP, those with the ***shortest Path Length*** are preferred.
3) **TB**: *Tie Break*

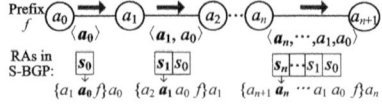

Fig. 1. Decision process in BGP   **Fig. 2.** Route Attestations (RAs) in S-BGP

loss. For instance, in 2008, Pakistan Telecom hijacked YouTube's prefixes and knocked it off the Internet for two hours [14].

Several extensions have been proposed to improve the security of BGP, including S-BGP [9] and many others. However, S-BGP consumes significant resources of computation and storage, and also faces the problem of replay attack. The other solutions either compromise in the security [6,19,7], or bring in more complexity on message size and certification distribution [10].

Towards these unsolved issues, we propose an efficient approach, FS-BGP (Fast Secure BGP), to secure AS paths and prevent prefix hijacking. Through signing critical AS path segments (i.e., adjacent AS triples), FS-BGP can achieve a similar level of security as S-BGP. We evaluate the performance of FS-BGP by BGP UPDATE collected from real backbone routers. Even using a small cache, the signing and verification overhead of FS-BGP account for only 0.56% and 3.9% of that of S-BGP respectively. Indeed, signing and verification can always be accomplished as fast as the most bursty BGP UPDATE arrivals.

This paper is organized as follows. Section 2 introduces backgrounds. Section 3 illustrates our key observation. Section 4 presents the design of FS-BGP and proves its security guarantees. Section 5 reviews the security level of FS-BGP, and section 6 evaluates its performance. Section 7 discusses further extensions including multiple prefixes and complex policies. Finally section 8 concludes.

2   Backgrounds

2.1   BGP and S-BGP

We model the inter-domain routing system as an AS graph, where each AS is denoted by its AS number (ASN). ASes sharing a common edge are called *neighbors*. We are mostly concerned the AS path $p = \langle a_n, \cdots, a_0 \rangle$ embedded in a BGP UPDATE, where the last AS a_0 is the *origin* AS of the path.

BGP is a policy-based routing protocol. An AS only exports a route[1] to a neighbor if it is willing to forward traffic to the corresponding prefix from that neighbor. If an AS receives multiple routes to the same prefix, it chooses and announces the best one according to the decision process as shown in Fig. 1.

In BGP, neither the origin AS nor the AS path is guaranteed to be correct. Secure BGP (S-BGP) [9] is the dominant solution to this problem. S-BGP uses Address Attestations (AAs) for origin authentication, and Route Attestations (RAs) for path authentication. As shown in Fig. 2, an RA is a signature signed

---

[1] We will use route and path interchangeably in this paper.

by an AS to authenticate the existence and position of ASes in the path. We define $\{msg\}a_i$ as the signature on msg generated with AS a_i's private key. In Fig. 2, each AS a_i signs the corresponding path $\langle a_{i+1}, a_i, \cdots, a_0 \rangle$ and the prefix f. The inclusion of the recipient AS a_{i+1} in each signature is necessary to prevent *cut-and-paste* attacks.

S-BGP can protect the network against fabricated routing information, but not against some inside attacks. For instances, a malicious AS can (1) re-announce signed but outdated routes, (2) violate its routing policy and announces routes received from one provider to other providers [5], (3) hijack a prefix through link-cut [3], and (4) selectively drop updates or announce a false withdraw. However, completely securing BGP from inside attacks is difficult. Although S-BGP is not omnipotent, we regard it as currently the most secure scheme with enough capabilities [5], and aim to provide as similar level of security guarantee as S-BGP.

2.2 Related Works

The main concerns about deploying S-BGP in practice include difficulties in maintaining the Public Key Infrastructure (PKI), and the huge computational cost for signing and verifying signatures. Some solutions try to replace PKI but can not guarantee security [12], and some can not provide real-time protection [8,16]. Since PKI has been adopted by the IETF, and regional registries have already started offering related services [15], we believe PKI is an essential part in the BGP security framework, and use it as our basic building block.

On the other hand, since the number of prefixes is much smaller than the number of paths, and most prefix ownerships are quite stable, AAs can be signed and verified out-of-band. Therefore, the dominating barrier for adopting S-BGP is the overhead of processing RAs, that is to authenticate paths.

Toward this direction, there are a bunch of solutions for reducing the overhead of path authentication. SoBGP [19] maintains all authenticated AS edges in a database, but faces the problem of forged paths. IRV [6] builds an authentication server in each AS, but brings the problem of maintaining and inter-connecting these servers, and introduces query latencies. SPV [7] accelerates the signing process by pre-generated one-time signatures based on a root value, but involves a significant amount of state information, and its security can only be guaranteed probabilistically. Signature Amortization [10] uses a bit-vector to indicate the allowed recipients of a route, such that only one signing is needed for all neighbor recipients. However, each AS will need to pre-establish a neighbor list corresponding to the bit vector, and to distribute it to all other ASes.

As we can see, existing methods for alternating S-BGP usually compromise security, or only improve the performance of signing. However, verification happens more frequently, since one signature needs to be verified at multiple places.

Scope of This Paper. Accordingly, it is important to design an efficient method to secure AS paths. Our solution, FS-BGP, builds on the assumption that a PKI is ready for use, and focuses on AS path authentication. For origin authentication, FS-BGP uses the same mechanism as S-BGP.

3   Key Observations

Dilemma of S-BGP. S-BGP's intention of signing every AS path is reasonable, but it is not realistic because of the high computational cost. And more importantly, it's not worth the cost. BGP is a policy-based routing protocol. An AS only exports a route to a neighbor if it is willing to forward traffic from that neighbor. Under a stable AS-level topology, we call a path *available* when the path satisfies the *import and export policies* of all ASes along the path. We further divide all available paths into three categories, according to the decision process of BGP as shown in Fig. 1:

- *Optimal path*: the best path that passes all the three decision steps.
- *Sub-optimal path*: paths with the same Local Preference as the optimal path, but not chosen as the best one.
- *Suppressed path*: paths with lower LP than the optimal path. Expensive paths (i.e., through a provider) are often suppressed by a low LP.

BGP only announces one available path for every prefix each time. However, since failures occur quite frequently in the global routing system, sub-optimal and suppressed path can be easily announced and propagated. For this reason, S-BGP actually authenticates all announced available paths. In extreme cases, S-BGP even authenticates almost *all available paths*. On the other hand, S-BGP can not prevent replay of non-optimal paths. It can only use the expiration-date (up to several days) to roughly control the window exposed to a replay attack.

NBIE. Although complex policies (i.e., route filters [2]) exist, an AS usually does not differentiate those nonadjacent ASes. For example, in Fig. 3, when a_n decides whether routes learned from a_{n-1} can be exported to a_{n+1}, it only considers its relation with the two neighbors (i.e., business partners), but does not consider other ASes along the path (i.e., $a_{n-2}, \cdots, a_0$). We call this phenomenon *Neighbor Based Importing and Exporting* (NBIE).

Because of the dynamic nature of the inter-domain routing system, signing every single path is worthless. We believe that even if a security schema only guarantees that *all authenticated paths are available path*, the protocol also can achieve *a similar level of security as S-BGP*. Inspired by the NBIE observation, we get rid of blindly signing every single path. NBIE abstracts the basic functionality of BGP. According to our measurement using the *whois* database, only a very small portion of routing polices violates the NBIE rule. In deed, our proposal can flexibly support complex routing polices, and we will discuss it in section 7.

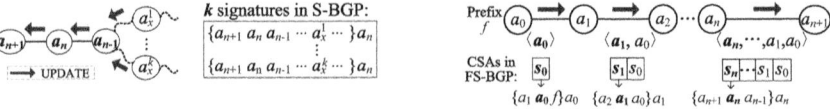

Fig. 3. In S-BGP, a_n signs k paths which share a mutual path segment $\langle a_{n+1}, a_n, a_{n-1}\rangle$

Fig. 4. Critical Segment Attestations (CSAs) in FS-BGP

4   FS-BGP: Fast Secure BGP

4.1   Overview

Following our key observation above, we propose Fast Secure BGP (FS-BGP) to secure AS paths. Given a available path $p=\langle a_{n+1}, a_n, ..., a_0 \rangle$, we define its set of *critical path segments* as $\{c_i, 0 \leq i \leq n\}$, where

$$c_i = \begin{cases} \langle a_1, a_0 \rangle & \text{for } i = 0 \\ \langle a_{i+1}, a_i, a_{i-1} \rangle & \text{for } 0 < i \leq n \end{cases}$$

and we call a_i the *owner* of c_i. Particularly, c_0 is called the *origin* critical path segment owned by a_0. A critical path segment $\langle a_{i+1}, a_i, a_{i-1} \rangle$ actually describes a routing export policy of its owner a_i, and implies that a_i can export all routes imported from a_{i-1} to a_{i+1}.

More specifically, FS-BGP uses Critical Segment Attestations (CSAs) to authenticate AS path. A CSA is simply the signature of the critical path segment signed by its owner. In a path $p=\langle a_{n+1}, a_n, \cdots, a_0 \rangle$, the CSA s_i signed by AS a_i is defined as:

$$s_i = \begin{cases} \{a_1 \ a_0 \ f\}a_0 & \text{for } i = 0 \\ \{a_{i+1} \ a_i \ a_{i-1}\}a_i & \text{for } 0 < i \leq n \end{cases}$$

The inclusion of the prefixes f in s_0 is necessary, because a_0 might be multi-homing and only announces part of its prefixes to a_1.

Fig. 4 and Fig. 2 compare the signatures in FS-BGP and S-BGP, where FS-BGP and S-BGP recursively verify the path segments and path suffixes respectively. It is obvious that the number of distinct critical path segments is far less than the number of distinct paths. As a result, even using a small cache, the number of signing and verification operations in FS-BGP can be greatly reduced. In Fig. 3, a_n needs to sign each of the k paths individually in S-BGP. However, in FS-BGP, all the k different paths can reuse one signature of the common critical segment $\langle a_{n+1}, a_n, a_{n-1} \rangle$.

We argue that, under the NBIE rule, if *every* AS along a path signs the corresponding critical segment it owns, then the path can be authenticated as a available path. We will prove this claim in section 4.2. However, it is possible to forge an available but unannounced path if the security mechanism relies on CSA only, as shown in section 4.3. We will provide an effective solution, Suppressed Path Padding (SPP), to solve this problem in section 4.4.

4.2   Path Authentication in FS-BGP

In this section, we introduce our main argument on CSA based path authentication. We first define some notations as follows. We denote the set of available paths by $\mathcal{P}_A$, and the set of authenticated paths in S-BGP by $\mathcal{P}_S$. $\mathcal{P}_S$ exactly includes actually announced available paths.

We know $\mathcal{P}_S \subset \mathcal{P}_A$, i.e., S-BGP protects a subset of available paths. For FS-BGP, we define the set of all authenticated critical segments as $\mathcal{C}$, and use $\mathcal{P}_{FS}$

to represent the set of paths that can be authenticated using CSAs of $\mathcal{C}$. So $\mathcal{P}_{FS}$ is actually constructed by concatenating those path segments in $\mathcal{C}$. Generally, a constructed path $p_{i-1} = \langle a_i, a_{i-1}, \cdots, a_0 \rangle$ in $\mathcal{P}_{FS}$ must end with an origin critical segment in the form of $c_0 = \langle a_1, a_0 \rangle \in \mathcal{C}$, and can be extended to a longer path p_i by prepending *exactly one* AS a_{i+1} such that $\langle a_{i+1}, a_i, a_{i-1} \rangle \in \mathcal{C}$. Formally, we represent the concatenating procedure by an operator $\odot$ as

$$\langle a_{i+1}, \boldsymbol{a_i}, \boldsymbol{a_{i-1}} \rangle \odot \langle \boldsymbol{a_i}, \boldsymbol{a_{i-1}}, \cdots, a_0 \rangle = \langle a_{i+1}, a_i, a_{i-1}, \cdots, a_0 \rangle$$

We also take for granted that all paths considered here are loop-free, since BGP will drop such paths.

Theorem 1. *Under the NBIE assumption, we have* $\mathcal{P}_S \subset \mathcal{P}_{FS} \subset \mathcal{P}_A$. *That is, paths authenticated by S-BGP can also be authenticated by FS-BGP, and paths authenticated by FS-BGP are guaranteed to be available.*

Proof. The first part, $\mathcal{P}_S \subset \mathcal{P}_{FS}$, is straightforward. We will prove $\mathcal{P}_{FS} \subset \mathcal{P}_A$ by induction. Since the receiver of a path is always included in the signature, AS paths have lengths of at least two.

1) The case for a path $p = \langle a_1, a_0 \rangle \in \mathcal{P}_{FS}$ of length two is trivial.

2) Suppose all paths of length less than $k+1$ that are authenticated by FS-BGP are available. Given a path p_k of length $k+1$ such that $p_k = \langle a_{k+1}, a_k, a_{k-1}, \cdots, a_0 \rangle \in \mathcal{P}_{FS}$, p_k can only be constructed by a path $p_{k-1} = \langle a_k, a_{k-1}, \cdots, a_0 \rangle \in \mathcal{P}_{FS}$ of length k and a critical path segment $\langle a_{k+1}, a_k, a_{k-1} \rangle \in \mathcal{C}$. Due to the induction hypothesis, p_{k-1} is available, then p_k is also available.

By induction, any path that is authenticated by FS-BGP is guaranteed to be available. That is, $\mathcal{P}_{FS} \subset \mathcal{P}_A$.

The implication of Theorem 1 is that, ASes using FS-BGP can still implement the basic BGP functionality in a secure way ($\mathcal{P}_S \subset \mathcal{P}_{FS}$), and can not arbitrarily forge paths, since even if a path is forged, it is still available ($\mathcal{P}_{FS} \subset \mathcal{P}_A$).

4.3   Forged Paths in FS-BGP

This section analyzes the attack faced by FS-BGP, namely, an AS using FS-BGP can construct paths that are not actually announced by others, but avoid CSA based detection. According to Theorem 1, only paths in $\mathcal{P}_{FS} - \mathcal{P}_S$ can be forged and bypass the CSA based verification, and $\mathcal{P}_{FS} - \mathcal{P}_S \subset \mathcal{P}_A - \mathcal{P}_S$. That is, we only need to consider paths in $\mathcal{P}_A - \mathcal{P}_S$, which are sub-optimal and suppressed paths. Such paths can be constructed by concatenating critical segments.

In Fig. 5, p_a and p_b are two authenticated paths received by a_m, and they share a mutual path segment $\langle a_{i+1}, a_i \rangle$. Using critical segments in these two paths[2], a_m can construct a path p_d. According to Theorem 1, p_d is available and can pass the verification, but it may have never been announced before.

We use $p(a, :)$ and $p(:, a)$ to represent the suffix and prefix of the path p, starting from or ending with a respectively. As shown in Fig. 5, consider a

---

[2] Forging a path using critical segments in more than two paths is similar.

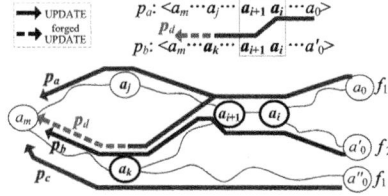

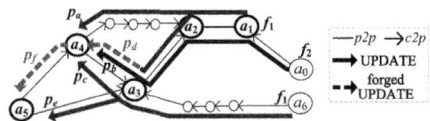

Fig. 5. Manipulator a_m forges path p_d **Fig. 6.** a_4 hijacks a_5's traffic to f_1 by forging the path $p_f = \langle a_5, a_4, a_3 \rangle \odot p_d = \langle a_5, a_4, a_3, a_2, a_1 \rangle$

general case where an intermediate a_k receives two paths $p_d(a_k, :)$ and $p_c(a_k, :)$, both of which can reach prefix f_1. Although a_k ranks $p_d(a_k, :)$ lower than $p_c(a_k, :)$ and does not announce $p_d(a_k, :)$, a_m can still forge the path p_d and propagate it successfully.

Such a forged path could be used for prefix hijacking, as demonstrated in Fig. 6.[3] In this example, a_3 prefers its customer path $p_c(a_3, :)$ to its provider path $p_d(a_3, :)$, so it will not announce $p_d(a_3, :)$ to a_4. However, a_4 receives two paths p_a and p_b with a mutual path segment $\langle a_2, a_1 \rangle$, and it can forge the path p_d for prefix f_1 by concatenating $p_b(, : a_2)$ and $p_a(a_2, :)$. In normal circumstances, a_5 will choose the six-hop path p_e to reach prefix f_1, thus its traffic to f_1 should be forwarded to a_3. However, if a_4 announces the forged five-hop path $p_f = \langle a_5, a_4, a_3 \rangle \odot p_d$ to a_5, a_5 will prefer the shorter path p_f, and forward traffic to a_4 instead of to a_3. As a result, a_4 successfully pollutes the routing information of a_5, and *effectively hijacks* a_5's traffic to prefix f_1.

Although forging paths is possible in FS-BGP, there are still many restrictions on how paths can be forged. First, a path can only be forged by combining non-forged paths which share mutual segments. Second, some part of a forged path must be treated as sub-optimal or suppressed by some AS along the path. Third, forged paths are still available, and can only be used for the right prefixes. Last, forged paths can not be very short.

4.4 Prevent Effective Hijacking

Although there are limitations on forged paths, prefix hijacking is still possible. In this section, we discuss solutions to prevent prefix hijacking. We only concern *effective hijacking*, in a sense that, the recipient of a forged route indeed changes its forwarding path. That is, if when there is no forged route, AS a_m announces p_m but AS a_v does not choose p_m (or a path end with p_m) as its optimal path; and when a_m announces a forged path p_f, a_v changes its optimal path and chooses p_f (or a path end with p_f). In this case, a_v is effectively hijacked by a_m. When a_v receives the forged path constructed by a_m, a_v's decision process will

---

[3] In this paper, we use p2p indicates edge connects two peer ASes, while c2p edge is represented as an arrow from customer AS to provider AS.

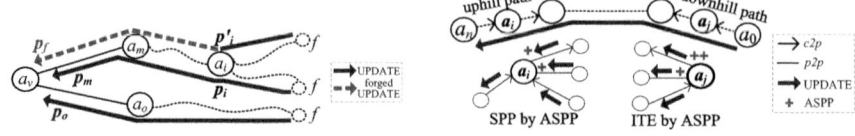

Fig. 7. Manipulator a_m announces a forged path p_f to its neighbor a_v

Fig. 8. ASPP for SPP and ITE

be triggered, as shown in Fig. 1. The necessary condition of an effective hijacking is provided by the following theorem.

Theorem 2. *Under the NBIE assumption, if a forged path is no shorter than the non-forged path BGP should announce, it can not be used for effective hijacking.*

Proof. We first consider the direct recipient of a forged path. As shown in Fig. 7, the manipulator a_m forges a new path $p_f = \langle a_v, a_m, \cdots, p_i' \rangle$ when it wants to hijack traffic from a_v. Notice that p_i' is suppressed by a_i, while p_i is considered as optimal by a_i. The best route a_v originally chooses is $p_o = \langle a_v, a_o, \cdots \rangle$, where $a_o \neq a_m$ in an effective hijacking scenario.

Since both the LP and the rank of TB of p_m and p_f are the same,[4] and the Path Length of p_f is no shorter than that of p_m, p_f is no better than p_m in a_v's decision process. However, the original best route p_o is strictly better than p_m, so a_v will not prefer p_f to p_o. As a result, a_m can not effectively hijack prefix f from its neighbor a_v.

Actually, the analysis above does not depend on whether p_f is forged by a_m, or just part of it was previously forged by another AS and a_m just extends the forged path innocently, so our theorem holds for any circumstance.

We know that only suppressed path can be shorter than the optimal path. Thus, if there is a mechanism to guarantee that all suppressed paths are no shorter than their corresponding optimal paths, the manipulator can no longer effectively hijack a prefix either, according to Theorem 2. This idea can be implemented by using AS Path Pre-pending (ASPP). ASPP is a method to artificially increase the path length by padding multiple local ASNs in the front of an AS path [18]. We believe using ASPP to restrict the length of suppressed path will not bring additional burden to routers, since it is already widely used for In-bound Traffic Engineering (ITE).

We use the example in Figure 6 to explain how ASPP can be applied to FS-BGP. If a_3 intentionally increases the length of p_b by padding itself in the path, and only announces a route $p_b' = \langle a_4, a_3, a_3, a_3, a_2, a_1, a_0 \rangle$, then a_4 can no longer forge a path short enough for effective hijacking. At the same time, when singing its critical segment $\langle a_4, a_3, a_2 \rangle$, a_3 just needs to include the number of its occurrences in the corresponding CSA, i.e., $\{a_4, a_3, 3, a_2\}a_3$.

---

[4] The rank of TB may differ under some rare conditions. However, this is not critical for our theorem, and can be solved by just replacing "no shorter" with "longer".

Algorithm 1. Suppressed Path Padding

Input: local AS a_i, neighbor AS a_{i-1}
Output: k_i: number of times that a_i needs to be added in the paths import from a_{i-1}

1: **if** a_{i-1} has the highest LP **then**
2:     **return** 1
3: $k_i \leftarrow 1$
4: **for all** path p imported from a_{i-1} **do**
5:     $opt(p) \leftarrow$ the optimal path corresponding to p
6:     **if** $PL(p) - PL(opt(p)) > k_i$ **then**
7:        $k_i \leftarrow PL(p) - PL(opt(p))$
8: **return** k_i

We call such a mechanism *Suppressed Path Padding* (SPP), and Algorithm 1 depicts the pseudo code for deciding how many times an AS a_i should pad itself. If a path is imported from a_{i-1} with the highest LP, a_i only appears once (line 1–2). Otherwise, k_i must be large enough such that no suppressed path can be shorter than the corresponding optimal path (line 4–7).

When local preferences are determined by business relationships, paths obey the valley-free rule: a path begins with zero or more *c2p* edges (*uphill path*), followed by zero or one *p2p* edge, and finally ends with zero or more *p2c* edges (*downhill path*) [4]. Fig. 8 compares SPP and ITE, both of which use ASPP. SPP only happens on the uphill path, as shown on the left side. Suppose a_i exports a route p_i which is *imported* from its provider (or peer), a_i need to pad itself in p_i. On the other side, ITE only happens along the downhill path when exporting routes to providers or peers. We can see that, SPP naturally expresses its own interests on neighbors, and has no side effect to ITE. We also note that, although using ASPP on suppressed paths for one prefix may affect routing for another prefix, it is still in the interest of the AS itself, since the AS already uses a lower preference to indicates its preference. As a conclusion, SPP is quite general, natural and easy to implement.

5   Security Level

Table 1 compares the security level of FS-BGP, S-BGP, and soBGP. Ineffective hijacks, such as false withdraw, selective dropping, or longer path announcing, can not effectively hijack a prefix. There are two types of attack, policy violating [5] and link-cut attack [3], which existing security schemes can not defend. As an autonomy organization, AS can completely determine its behaviour, so it is really hard to prevent it from violating its routing policy. For a link-cut attack, it is mainly achieved by wild destroying (i.e., DDoS attack). So defending against this kind of attack should be done through enhancing the robustness of BGP, since this paper aims to secure the AS path, we will not discuss link-cut attack.

We call a path is a *graph path* if it exist in the AS graph, and denote the set of graph paths by $\mathcal{P}_G$. We think that soBGP has a lower level of security

Table 1. Security Level Comparison for FS-BGP, S-BGP, and soBGP

Type of Attack			**FS-BGP**	S-BGP	**FS-BGP (no SPP)**	soBGP
Ineffective hijack			✓	✓	✓	✓
False origin AS			✓	✓	✓	✓
Path not in the AS graph			✓	✓	✓	✓
$\mathcal{P}_G$		Unavailable path	✓	✓	✓	✗
	$\mathcal{P}_A$	Potential path	✓	✓	✓	✗
		Revealed path	✓*	✓	✗	✗
		Outdated path	✓*	✗	✗	✗
Policy violating [5]			✗	✗	✗	✗
Link-cut attack [3]			✗	✗	✗	✗

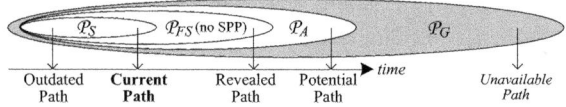

Fig. 9. Categories of available path

compare with FS-BGP and S-BGP, since it can not defend against a cut-and-paste attack. In soBGP, attacker can easily forge an unavailable path through concatenating AS edges in the AS graph. However, we believe that FS-BGP can achieve a similar level of security as S-BGP.

Firstly, we divide all available paths $\mathcal{P}_A$ into four categories, as shown in Fig. 9: (1) outdated path, paths already announced but are temporary down; (2) current path, the recently announced path; (3) revealed path, paths constructed through concatenating authenticated critical segments; and (4) potential path, paths may be announced at some point in the future but can not be constructed even using received critical segments.

As failures occur in the global routing system, available paths are announced one after another. S-BGP actually authenticates outdated path and current path ($\mathcal{P}_S$). It can only use the expiration-date to roughly control the window which is exposed to outdated path replaying. Besides, the expiration-date must be long enough, otherwise there will be a UPDATE surge. According to Theorem 1, the light-weight version of FS-BGP (without SPP) can not defend against outdated path and revealed path attack, but it can defend against potential path attack. We claim that even without SPP, it is very difficult to launch an effective hijacking, since the average path length is very short and keeps on decreasing [20], and forged path can not be very short as it must be constructed by overlapped critical path segments. Armed with SPP, the full version of FS-BGP becomes more secure. It can defend against almost all revealed path and outdated path attack. This is because optimal path always has the longest live-time [11], and no path is shorter than optimal path after using SPP.

As a conclusion, we think that FS-BGP achieves a similar level of security as S-BGP.

6   Performance Evaluation

6.1   Methodology

We use real BGP UPDATE collected by RouterViews [1] to evaluate the performance. We consider S-BGP as the only mechanism that provides enough security guarantee as FS-BGP, and compare their cost on signing and verification. Specifically, we use UPDATEs announced by the busiest monitor (a router in AS7018, owned by AT&T) and all UPDATEs received by the biggest collector (*route-views2*) during the whole month of August 2009 to evaluate how FS-BGP performs on a backbone router.

We assume ECDSA is used for signing and verification, as suggested by the IETF [17], and just count the number of signing/verification operations, since the cost of other operations is negligible compared to cryptographic operations. In the rest of this section, we consider two metrics as follows:

- *INS*: Instantaneous Number of Signings in each second.
- *INV*: Instantaneous Number of Verifications in each second.

Routers can use a cache to effectively improve the performance. Typically there is a limit on the cache size, then the number of cache misses measures the corresponding INS and INV.

6.2   Signing CSA

We use three different cache sizes for signing, i.e., 4K, 16K and 64K. If each signature occupies 256 bits [17], the memory cost will be around several megabytes, and is affordable for a backbone router. Since a signature always includes the corresponding recipient, one UPDATE message must be signed multiple times, once for each recipient. As a rough estimate, we use $m = 32$ as the average number of recipients for each UPDATE.[5]

Fig. 10 depicts the INS of FS-BGP and S-BGP in one month, under a moderate 16K cache size. In most of the time, the INS of FS-BGP is less than 100, while the INS of S-BGP often reaches up to tens of thousands. The maximum peak values of S-BGP reaches 81,920, while the maximum INS of FS-BGP is 12,365, only 15% of that of S-BGP. We also plot the Complementary Cumulative Distribution Function (CCDF) of INS in Fig. 11. In August 2009, there were UPDATE messages announced in 976,043 seconds. Only in 0.28% of that time does FS-BGP need to sign signatures, while the ratio in S-BGP is 44%.

6.3   Verifying CSA

Since there are much more signatures that need to be verified than to be signed, we use larger cache sizes (256K, 512K and 1024K entries) for comparing the verification performance. FS-BGP has another advantage that once a critical

---

[5] As a scaling factor, the actual value of m is not important to our comparison.

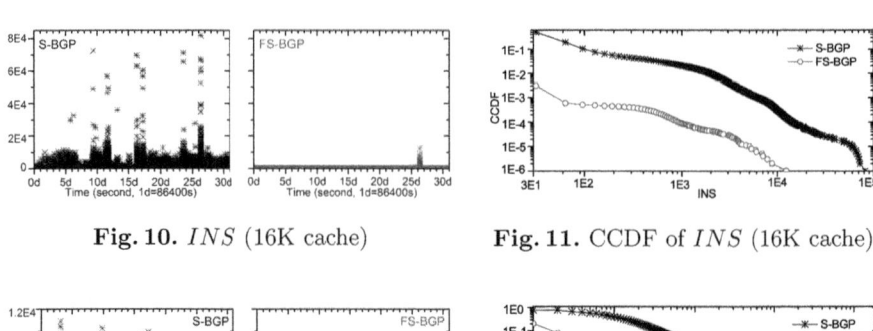

Fig. 10. INS (16K cache)   Fig. 11. CCDF of INS (16K cache)

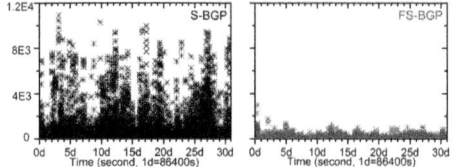
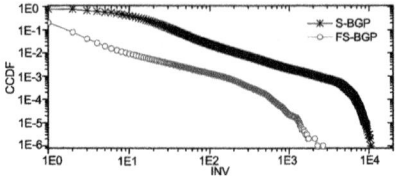

Fig. 12. INV (512K cache)   Fig. 13. CCDF of INV (512K cache)

segment is authenticated, only the path segment needs to be cached, but not the original signature.

Fig. 12 illustrates the INV of FS-BGP and S-BGP while using a moderate cache size of 512K. In most cases, the INV of S-BGP is ten times higher than that of FS-BGP, and sometimes even up to a hundred times. The maximum INV of FS-BGP is 2,919, only 26.7% of that of S-BGP, which is 10,921. The CCDF of INV in Fig. 13 shows that, only in 20% of the time when there are UPDATE messages does FS-BGP need to verify signatures, while S-BGP needs to verify signatures in 78% of the time.

Table 2 numerically compares the computational cost of FS-BGP and S-BGP. Using a cache of 16K entries, FS-BGP only needs to sign 0.56% as many messages as S-BGP. When using a moderate cache of 512K entries, the average verification cost of FS-BGP is only 3.9% of that of S-BGP. A large INS or INV (> 100) will delay the propagation of routing information, but it rarely happens in FS-BGP. In conclusion, FS-BGP performs orders of magnitude better than S-BGP, in both signing and verification. FS-BGP requires a very small cache, and can handle the most bursty BGP UPDATE messages, so it is an efficient and practical solution.

7   Discussions

Multiple Prefixes. As noted before, when signing an origin critical segment, the corresponding prefix f should be included. In practice, an AS may own a large number of prefixes, and it is straightforward to extend the prefix f to a prefix set $\mathcal{F}$. Thus, $s_0 = \{a_1, a_0, k_0, \mathcal{F}\}a_0$, where k_0 is calculated by SPP, and $\mathcal{F}$ is the set of prefixes allowed to be announced to a_1. In practice, most of ASes announce all prefixes to their providers. Under these cases, FS-BGP can omit $\mathcal{F}$ in s_0 to represent no restriction on prefixes.

Table 2. Performance Comparison for FS-BGP (FS) and S-BGP (S)

# Cache	Average INS		$\#(INS > 100)$		# Cache	Average INV		$\#(INV > 100)$	
Entries	FS	FS/S	FS	FS/S	Entries	FS	FS/S	FS	FS/S
4K	0.99	0.55%	507	0.48%	256K	2.35	8.8%	8121	11.3%
16K	0.50	0.56%	457	0.61%	512K	0.88	3.9%	3281	5.50%
64K	0.11	0.37%	133	0.54%	1024K	0.34	1.8%	1128	2.32%
$\infty(>4M)$	0.08	1.70%	43	1.12%	$\infty(>13M)$	0.34	6.7%	1128	11.7%

Complex Policies. Our analysis till now are all based on the NIBE assumption in section 3. However, more complex policies also exist. The Routing Policy Specification Language (RPSL) [2] is commonly used by ASes. There are three kinds of transitive route filters in RPSL: prefix filters, AS path filters, and origin AS filters. We queried all ASes by *whois* and collected 758K import/export expressions in total. Among all these expressions, 1.1% of them use prefix filters. To support prefix filters, FS-BGP can sign the available prefixes together with the critical segment. AS path filters occur in 0.3% of policy expressions, and to support them, an AS can sign the full AS path. Since there is only a very small portion, the influence on computational cost is negligible. About 60% of policy expressions use origin AS filter. To support them, CSAs in FS-BGP can be extended to include the available origin ASes. In most cases, the number of available origin ASes is very small.

Nevertheless, the main purpose of route filters is to protect the routing system against distribution of inaccurate routing information [2]. The use of route filters is mainly due to security considerations rather than policy requirements. We believe that under a security framework (such as FS-BGP), these filters are not needed any more. Even if they do exist, FS-BGP can support them flexibly.

Privacy Concerns. Internet is a commercialized network, an AS may not want to reveal its proprietary information (i.e., customer list) to its competitors. FS-BGP does not require an AS to disclose its proprietary information, since the critical segments are nothing new but already included in BGP UPDATE. FS-BGP does not allow others to obtain the information more easily either, by employing the existing distribution mechanism of BGP. No centralized or public database such as in IRV, soBGP or IRR need to be maintained. In conclusion, we believe FS-BGP preserves the privacy of a business entity.

8   Conclusion and Future Works

This paper introduces an efficient approach, FS-BGP (Fast Secure BGP), to secure AS path and prevent prefix hijacking. Through signing critical AS path segments, FS-BGP guarantees the authentication of all available paths. Through padding suppressed path, FS-BGP prevents almost all replay attacks. We prove that FS-BGP can achieve a similar level of security as S-BGP. In our evaluations based on BGP UPDATE data collected from real backbone routers, FS-BGP performs orders of magnitude better than S-BGP. By using even a very small cache, the signing and verification overhead of FS-BGP account for only 0.56% and 3.9% of that of S-BGP respectively. Indeed, signing and verification can always be accomplished as fast as the most bursty BGP UPDATE arrivals, which implies that FS-BGP will hardly delay the propagation of routing information. In addition, FS-BGP can flexibly support complex routing polices, and can preserve the privacy of an AS.

We plan to design more efficient cache replacement algorithm, and evaluate the influence on convergence time after deploying FS-BGP on a large scale. Besides, we will also investigate the potential to use available paths constructed by critical path segments as backup paths in route protection.

References

1. The routeviews project (2009), http://www.routeviews.org
2. Alaettinoglu, C., Villamizar, C., Gerich, E., Kessens, D., Meyer, D., Bates, T., Karrenberg, D., Terpstra, M.: RFC 2622, routing policy specification language, RPSL (1999), http://tools.ietf.org/html/rfc2622
3. Bellovin, S.M., Gansner, E.R.: Using link cuts to attack Internet routing (2003), http://hdl.handle.net/10022/AC:P:9052
4. Gao, L., Rexford, J.: Stable Internet routing without global coordination. IEEE/ACM Trans. Netw. 9(6), 681–692 (2001)
5. Goldberg, S., Schapira, M., Hummon, P., Rexford, J.: How secure are secure inter-domain routing protocols? In: SIGCOMM (2010)
6. Goodell, G., Aiello, W., Griffin, T., Ioannidis, J., McDaniel, P.D., Rubin, A.D.: Working around BGP: An incremental approach to improving security and accuracy in interdomain routing. In: NDSS (2003)
7. Hu, Y.C., Perrig, A., Sirbu, M.A.: SPV: secure path vector routing for securing BGP. In: SIGCOMM, pp. 179–192 (2004)
8. Karlin, J., Forrest, S., Rexford, J.: Pretty good BGP: Improving BGP by cautiously adopting routes. In: ICNP, pp. 290–299 (2006)
9. Kent, S., Lynn, C., Mikkelson, J., Seo, K.: Secure border gateway protocol (S-BGP). IEEE Journal on Selected Areas in Communications 18, 103–116 (2000)
10. Nicol, D.M., Smith, S.W., Zhao, M.: Evaluation of efficient security for BGP route announcements using parallel simulation. Simulation Modelling Practice and Theory 12(3-4), 187–216 (2004)
11. Oliveira, R., Zhang, B., Pei, D., Izhak-Ratzin, R., Zhang, L.: Quantifying path exploration in the Internet. In: Proc. of the 6th ACM SIGCOMM Internet Measurement Conference (IMC), Rio de Janeriro, Brazil (2006)
12. van Oorschot, P.C., Wan, T., Kranakis, E.: On interdomain routing security and pretty secure BGP (psBGP). ACM Trans. Inf. Syst. Secur. 10(3) (2007)
13. Rekhter, Y., Li, T., Hares, S.: RFC 4271: Border gateway protocol 4 (2006), http://tools.ietf.org/html/rfc4271
14. RIPE: Youtube hijacking: A ripe ncc ris case study (2008), http://www.ripe.net/news/study-youtube-hijacking.html
15. RIPE NCC: Resource certification (2011), http://ripe.net/certification/
16. Subramanian, L., Roth, V., Stoica, I., Shenker, S., Katz, R.H.: Listen and whisper: Security mechanisms for BGP. In: NSDI, pp. 127–140 (2004)

17. Turner, S.: BGP algorithms, key formats, & signature formats (2011),
    http://tools.ietf.org/html/draft-ietf-sidr-bgpsec-algs
18. Wang, J.H., Chiu, D.M., Lui, J.C.S., Chang, R.K.C.: Inter-as inbound traffic engineering via ASPP. Transactions On Network And Service Management 3(1) (2007)
19. White, R.: Architecture and deployment considerations for secure origin BGP (2006), http://tools.ietf.org/html/draft-white-sobgp-architecture
20. Xiang, Y., Yin, X., Wang, Z., Wu, J.: Internet flattening: Monitoring and analysis of inter-domain routing. In: IEEE ICC (2011)
21. Zmijewski, E.: Threats to internet routing and global connectivity (2008),
    http://www.renesys.com/tech/presentations/pdf/20thAnnualFIRST.pdf

Estimating Network Layer Subnet Characteristics via Statistical Sampling

M. Engin Tozal and Kamil Sarac

Department of Computer Science,
The University of Texas at Dallas, Richardson, TX 75080 USA
{engintozal,ksarac}@utdallas.edu

Abstract. Network layer Internet topology consists of a set of routers connected to each other through subnets. Recently, there has been a significant interest in studying topological characteristics of subnets in addition to routers in the Internet. However, given the size of the Internet, constructing complete subnet level topology maps is neither practical nor economical. A viable solution, then, is to sample subnets in the target domain and estimate their global characteristics. In this study, we propose a sampling framework for subnets; derive proper estimators for various subnet characteristics including total number of subnets, subnet prefix length distribution, mean subnet degree, and IP address utilization; and analyze the theoretical and empirical aspects of these estimators.

Keywords: network, topology, sampling.

1 Introduction

At the network layer, Internet topology consists of a set of routers connected via point-to-point or multi-access links or *subnets*. Most of the existing efforts on capturing a *network layer* map of the Internet focus on *router level* maps [15,10,12]. These maps are then studied to understand various topological characteristics of the Internet.

Router level maps typically do *not* consider the nature of the subnets (i.e., point-to-point or multi-access) connecting routers and simply use point-to-point links to represent the connections. On the other hand, subnets are also important building blocks of the Internet topology. An alternative graph representation of the Internet at the network layer may depict subnets as the main entities represented as vertices and consider routers as links connecting those subnets/vertices to each other (see Figure 1). Studying features of *subnet level* maps would improve our understanding of the Internet topology. Among many practical uses of *subnet level* Internet maps are estimating the IP address space utilization in the Internet and developing more representative synthetic Internet graphs.

Despite the benefits of understanding topological properties and evolution of the Internet, drawing complete topology maps turns out to be expensive in terms of time, bandwidth, and computational resources [11,5]. Therefore, a viable alternative for studying Internet topology is to employ statistical sampling and

R. Bestak et al. (Eds.): NETWORKING 2012, Part I, LNCS 7289, pp. 274–288, 2012.

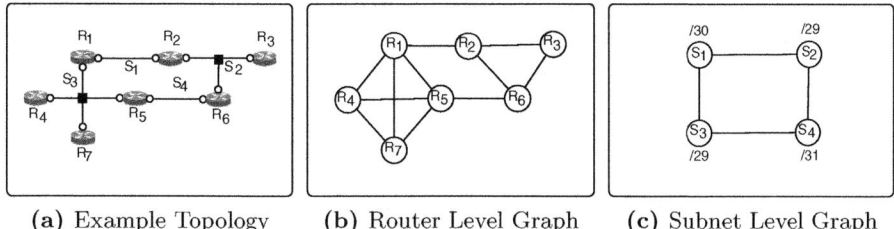

(a) Example Topology   (b) Router Level Graph   (c) Subnet Level Graph

Fig. 1. A network (left) represented as a router (center) and subnet (right) graph

estimate the parameters of the entire population by analyzing a small unbiased sample of the population. This would naturally require to discover individual subnets and define proper estimators to estimate subnet characteristics. In this paper we present a scheme to achieve this for studying subnet level topological characteristics of the Internet.

In general Internet topology sampling suffer from several practical limitations due to the operational characteristics of the Internet [2,8]. As an example, in our context we want to sample subnets randomly which requires a list of subnets to be available a priori. Unfortunately, this information is considered confidential by ISPs and therefore, is not available to us, researchers. On the other hand, through active probing we can identify the list of alive IP addresses utilized in these networks. Thus instead of randomly sampling subnets, we can randomly sample IP addresses. In statistical sampling this phenomenon is referred as the discrepancy between selection units (IP addresses) and observation units (subnets) [13]. The problem induced by this discrepancy is that different subnets may utilize different number of IP addresses. As a result random sampling of IP addresses may not indicate random sampling of subnets. In other words, the subnets hosting more IP addresses are more likely to be sampled compared to the ones accommodating less IP addresses. Finally, unresponsive units, i.e., subnets located behind firewalls, or located within rate limiting ISPs, might introduce estimation error depending on whether they are random or not.

Following the above discussion our sampling frame in this study is IP addresses rather than subnets. Therefore, we develop proper estimators for various subnet features using IP addresses of an ISP network as our sampling frame. Specifically, we develop estimators for total number of subnets, subnet mask distribution, mean subnet degree, and IP address space utilization. Our theoretical and empirical evaluations conducted by collecting population characteristics and samples from six geographically disperse Internet service providers demonstrate that our estimators are fairly accurate and stable with small variances. Note that, the scope of this paper is limited to (i) developing an approach for sampling subnets in the Internet; (ii) defining proper estimators for various subnet characteristics; and (iii) demonstrating that these estimators work well in estimating population parameters.

The rest of the paper is organized as follows. Section 2 discusses the challenges of subnet level Internet topology sampling and develop a step-by-step sampling framework for estimating various features of subnets. Section 3 presents our evaluation results on estimator accuracy. Section 4 discusses the related work and Section 5 concludes the paper.

2   Subnet Sampling

In this section we first introduce `exploreNET`, a subnet inference tool that we used in our sampling process. Then, we discuss the challenges of subnet level topology sampling in the Internet and present how we handle these challenges in our sampling framework. Finally, we define a set of subnet characteristics that we are interested in and explain how we can achieve either unbiased or slightly biased estimators with small variance for these characteristics.

2.1   Subnet Inference with ExploreNET

A subnet is a logically visible sub-section of Internet Protocol network (RFC 950) where the connected hosts can directly communicate with each other. At the network layer, a subnet is independent from its implementation below layer-3 which could be an Ethernet network or an ATM or MPLS based virtual network. For practical purposes we define a subnet S as a set of interfaces (or IP addresses) accommodated by S in addition to a subnet mask (or common IP address prefix length) referring to the IP address range of S.

ExploreNET is a network layer subnet inference tool [16]. Given a target IP address t as input, `exploreNET` uses active probing to discover the subnet S that contains t along with all other alive IP addresses on S. Furthermore, `exploreNET` labels S with its observed subnet mask which corresponds to the minimum of the subnet masks encompassing all observed IP addresses of S. To discover the boundaries of a subnet, `exploreNET` employs a set of heuristics based on hop distance of the subnet being explored, common IP address assignment practices, and routing. Our empirical results show that the proposed subnet inference mechanism achieves 94.9% and 97.3% accuracy rates on Internet2 and GEANT, respectively [14]. Moreover, its success rate is 93% against a ground truth dataset collected by `mrinfo` in the global public Internet [16]. On the other hand, the probing cost of the tool changes between $2|S|$ and $7|S|+7$ depending on the configuration of subnet S and its neighboring subnets, where $|S|$ denotes the size of S. Considering that `exploreNET` is based on a set of heuristics its accuracy rate is subject to change in different network domains and achieving 100% accuracy in every domain is difficult. Since the scope of this paper is limited to statistical subnet sampling, we direct the readers who are interested in the details of subnet discovery to our previous studies [14,16].

A sampling process has two types of errors, namely, *sampling error* and *nonsampling error* [9]. Sampling errors occur because of chance whereas nonsampling errors can be attributed to many sources including inability to obtain information, errors made in processing data, and respondents providing incorrect information. It is impossible to avoid sampling error, hence, the goal would be to minimize the sampling error by defining unbiased estimators with small variance. On the other hand, nonsampling errors could be eliminated by amending data collection and processing phases. In this context, nonsampling error refers to the error introduced by `exploreNET` the subnet inference tool that we used in this study and sampling error refers to the error that occurs in our estimations.

Note that our aim in this study is to define good estimators that minimize the sampling error in estimating various subnet features rather than improving subnet inference to eliminate nonsampling error. On the other hand, the sampling approach and estimators proposed in this study do not necessarily depend on `exploreNET`. Any other future subnet inference tool that collects the subnet accommodating a given IP address should work with our sampling framework.

2.2   Challenges in Subnet Sampling

In this sub-section, we briefly explain the challenges in subnet level Internet topology sampling including the discrepancy between selection and observation units, non-uniform unit sampling, and unresponsive units in sampling.

One challenging issue with sampling is the discrepancy between the selection units and the observation units. Observation units are the main objects that we want to sample and derive their statistical properties, whereas, selection units are the objects in the sampling frame that we can draw from with some probability distribution (preferably uniform). In most cases, selection units and observation units are the same, in some other cases (including Internet topology sampling) however, observation units are not directly available for sampling. To illustrate, we do not have the entire list of the routers or subnets of an ISP to sample from. The only information available to us is the IP address range of ISPs. As a result, the only sampling frame available to us is the list of alive IP addresses.

A consequence of using IP addresses as sampling frame is non-uniform sampling of subnets. In other words we can uniformly sample from the IP address range of an ISP and give the selected IP addresses to `exploreNET` as input but the collected subnets would not be sampled uniformly. Remember that `exploreNET` discovers a subnet as long as one of the IP addresses of the subnet is given as input. As a result subnets with larger degree, i.e., large number of alive IP addresses, appear more frequently in the sampling frame and are observed more frequently compared to the ones having smaller degrees. Put another way, a subnet is drawn with a probability proportional to its degree. Note that we use the term *degree* to denote the number of interfaces of a subnet rather than the number of neighboring subnets of a subnet. Figure 2 shows an example target domain such that large circles depict the subnets and small circles filled in gray denote the IP addresses hosted by the subnets. Considering subnets being sampled through IP addresses, S_3, in Figure 2, is twice as likely to be observed as compared to S_1 because the size of S_1 is half of the size of S_3. As a result, off-the-shelf statistical estimators fail to estimate the characteristics of the subnets due to the unequal drawn probabilities of subnets.

Finally, *unresponsive units* [13] in sampling is another challenge. Most of the tools for collecting topology data in the Internet is based on active probing.

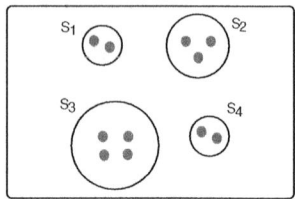

Fig. 2. Subnets hosting more IP addresses are more likely to be sampled compared to the ones with less IP addresses

However, autonomous systems, especially the ones at the Internet edge, are not completely open to active probing because of security and operational concerns. Some portions of their network are completely behind firewalls or they apply rate limiting in the sense that a router remains silent to probe packets if it is busy or drops probe packets if it suspected a security threat. The interior parts (core) of an ISP network provides a reliable, robust, and efficient communication infrastructure to the other networks. Hence, in this paper we study ISP core networks rather than the edge networks. ISP core networks are more compatible with active probing compared to the edge networks. We also introduce artificial delays and multiple probing in our experiments to minimize the rate limiting factor. Assuming that the unresponsive units in the core network are unmethodical our results are not skewed by them.

2.3   Estimating Subnet Characteristics

In this sub-section we present a generic framework based on Hansen-Hurwitz (HH) Estimator [13] for estimating various subnet features. HH-estimator is an unbiased estimator of population total whenever the selection units are drawn with different probabilities and the sampling is done with replacement. Let $\{y_1, y_2, \ldots, y_j, \ldots, y_n\}$ be a sample of n independent observations from a population of size N. Let p_i be the selection probability of the ith unit in the population such that $\sum_{i=1}^{N} p_i = 1$. HH-estimator estimates the population total $\tau_y = \sum_{i=1}^{N} y_i$ and it is defined as follows:

$$\widehat{\tau}_y = \frac{1}{n} \sum_{j=1}^{n} \frac{y_j}{p_j} \tag{1}$$

In Equation (1) dividing observation y_j by its selection probability p_j gives higher weight to the units that are less likely to be selected. An unbiased estimator of population mean is $\widehat{\mu}_y = \widehat{\tau}_y / N$.

In the context of subnet sampling, we define y_i to be the response variable measured on the ith subnet, S_i, of an ISP. Response variables are any characteristic of subnets that we are interested in such as degree, subnet mask, or utilization. Response variables could take discrete values like degree or categorical values like subnet masks. In the former case they are represented as discrete random variables and in the latter case they are represented as indicator random variables. Additionally, the selection probability of a subnet is proportional to its degree, $p_j \propto d_j$, where p_j and d_j are the drawn probability and the degree of subnet S_j, respectively. Then, the drawn probability, p_j, of S_j can be defined in terms of subnet degree as follows:

$$p_j = \frac{d_j}{\sum_{i=1}^{N} d_i} \tag{2}$$

where d_j is the degree of the jth subnet in the sample, d_i is the degree of the ith subnet in the population, and N is the number of subnets in the population. In Equation (2), we neither know the degree, d_i, of each subnet in the population nor the number of subnets, N, in the population in advance. To address this issue, we decided to build a set of alive IP addresses, $F_a = \{a_1, a_2, \ldots, a_k\}$, of the target ISP that we want to estimate its subnet characteristics and utilize the equality between the size of F_a, i.e. $|F_a|$, and the sum of degrees of the observable subnets in the target ISP as $|F_a| = \sum_{i=1}^{N} d_i$. F_a can be obtained as a dataset from LANDER project [5] at USC or be formed in linear probing overhead by pinging each IP address of the target ISP.

Note that, we also use F_a as our sampling frame from which we randomly select n IP addresses with replacement, give each IP address to exploreNET as input and obtain the sample set of n subnets. Rearranging the general estimator given in Equation (1) in the context of subnet sampling results in:

$$\widehat{\tau_y} = \frac{|F_a|}{n} \sum_{j-1}^{n} \frac{y_j}{d_j} \tag{3}$$

Equation (3) does not have any unknown term; sampling frame, F_a, is formed prior to sampling process and response variable, y_j, and degree, d_j, are obtained by our subnet inference tool exploreNET.

As response variables of subnets, y, are independent from each other, sampling distribution of the point estimator $\widehat{\tau_y}$ is approximately normal by Central Limit Theorem [6], i.e., $\widehat{\tau_y} \sim Normal(\tau_y, \sigma_{\widehat{\tau_y}}^2)$ where

$$\sigma_{\widehat{\tau_y}}^2 = \frac{1}{n|F_a|} \sum_{i=1}^{N} d_i \left(|F_a| \frac{y_i}{d_i} - \tau_y \right)^2 \tag{4}$$

Then we can define a confidence interval (CI) with confidence level $(1 - \alpha)100\%$ as $\widehat{\tau_y} \pm z_{\frac{\alpha}{2}} \sqrt{\sigma_{\widehat{\tau_y}}^2}$ such that $z_{\frac{\alpha}{2}}$ is the upper $\alpha/2$ point of the standard normal distribution. Since τ_y is the parameter that we want to estimate and N is usually unknown, an unbiased estimator of the variance of the sampling distribution of $\widehat{\tau_y}$, i.e., $\sigma_{\widehat{\tau_y}}^2$ is defined as follows:

$$s_{\widehat{\tau_y}}^2 = \frac{1}{n-1} \left[\frac{|F_a|^2}{n} \sum_{j=1}^{n} \left(\frac{y_j}{d_j} \right)^2 - \widehat{\tau_y}^2 \right] \tag{5}$$

Here, we can replace $\sigma_{\widehat{\tau_y}}^2$ with $s_{\widehat{\tau_y}}^2$ in confidence interval construction.

2.4   Important Subnet Characteristics

Although we have derived a general formula for estimating population total over any subnet response variable y in Section 2.3, using Equation (3) requires further arrangements and insights for different subnet features that we want to estimate. In this part we define a set of most relevant subnet characteristics including subnet population size, total subnets with a certain subnet mask, average subnet degree, and IP address utilization percentage and show how to deduce their proper point estimators. Note that this set is not complete and one may suggest new subnet characteristics in the future.

Subnet Population Size (N). Subnet Population Size, N, refers to the number of distinct subnets in a particular ISP. Since $\widehat{\tau}_y$ corresponds to the population total over the response variable y, setting $y_j = 1$ in Equation (3) gives us an unbiased estimator, $\widehat{N}$, of total number of subnets, N, in the population as:

$$\widehat{N} = \frac{|F_a|}{n} \sum_{j=1}^{n} \frac{1}{d_j} \tag{6}$$

Total Subnets with Prefix Length /p (τ_p). Observable prefix length (subnet mask) of a subnet refers to the the length of the initial block of bits common to all IP addresses in the subnet. A subnet having prefix length p is said to be a $/p$ subnet. In order to estimate the total number of subnets of $/p$ let the the response variable measured on subnet S_j, i.e., y_j be an indicator random variable such that

$$y_j = \begin{cases} 1 \ \text{if } S_j \text{ is a } /p \text{ subnet} \\ 0 \ \text{otherwise} \end{cases} \tag{7}$$

Then we can estimate the total number of subnets having /p prefix, τ_p, in an ISP as follows:

$$\widehat{\tau}_p = \frac{|F_a|}{n} \sum_{j=1}^{n} \frac{y_j}{d_j} \tag{8}$$

Note that, the variance of the estimator τ_p increases as the prefix gets larger (subnet degree gets smaller). The implication of this raise is wider confidence intervals for larger prefix lengths.

In case we want to estimate the prefix length distribution of the subnets we can resort to the fact that $r_p = \tau_p/N$ where r_p is the ratio of subnets having prefix length $/p$. If N is known then $\widehat{r}_p = \widehat{\tau}_p/N$ would be an unbiased estimator of r_p. On the other hand, if N is unknown $\widehat{r}_p = \widehat{\tau}_p/\widehat{N}$ would be a biased estimator of r_p. Nevertheless our experimental results show that the bias in the latter case is small. One should note that, a slightly biased estimator having a small variance of a population parameter is usually preferable compared to an unbiased estimator having a large variance.

Mean Subnet Degree (μ_d). Degree of subnet S_j, d_j, corresponds to the number IP addresses accommodated by S_j. To calculate the average degree throughout an entire ISP we need to compute the total of all degrees divided by the total number of subnets. An estimator of mean degree, μ_d, is defined as $\widehat{\mu}_d = |F_a|/\widehat{N}$

Note that, although $\widehat{N}$ is an unbiased estimator of N, $\widehat{\mu}_d$ is not an unbiased estimator of μ_d because $E[\widehat{\mu}_d] \neq |F_a|/E[\widehat{N}]$. However, our experimental results show that the bias and variance of $\widehat{\mu}_d$ is extremely small and we get very accurate and stable estimations of μ_d.

IP Address Space Utilization ($\mathcal{U}$). ExploreNET annotates the subnets with their observed prefix lengths while discovering their in-use IP addresses. Prefix length, p, of subnet S_j indicates the observable IP address capacity, c_j, of S_j while its degree d_j indicates the number of IP addresses that have been utilized. Capacity of subnet S_j is defined as

$$
c_j = \begin{cases} 2 & \text{if } S_j \text{ is a /31 subnet} \\ 2^{32-p} - 2 & \text{if } S_j \text{ is a /p subnet} \\ & \text{such that } p \leq 30 \end{cases} \tag{9}
$$

ISP utilization $\mathcal{U}$ is defined as the total number of alive IP addresses divided by the total capacity. Again total number of alive IP addresses corresponds to our sampling frame size, $|F_a|$. On the other hand, we estimate total capacity τ_c as $\widehat{\tau}_c = \sum_{\forall p} c_p \widehat{\tau}_p$ where c_p is the capacity of a subnet having prefix length $/p$ and $\widehat{\tau}_p$ is the estimation of the total number of subnets having prefix length $/p$. Since $E[\widehat{\tau}_c] = \tau_c$, $\widehat{\tau}_c$ is an unbiased estimator of total capacity τ_c. Then we can define ISP utilization as $\widehat{\mathcal{U}} = |F_a|/\widehat{\tau}_c$. Again, $\widehat{\mathcal{U}}$ is not an unbiased estimator of $\mathcal{U}$ but our experimental results demonstrate that it is a very good estimator with small bias and variance.

3   Evaluations

In this section, we evaluate sampling errors introduced by our estimators. For this, we need a sampling frame consisting of alive IP addresses clustered in a number of subnets in a network. Such a network could be generated synthetically; could be collected from the Internet using subnet inference techniques; or could be a genuine network available on its web page or obtained from its network operator. Synthetic networks may not be interesting and genuine topologies are available for a few small sized research networks including Internet2 and GEANT which are not large enough for sampling [1]. Therefore, in our evaluations we decided to work with collected topologies. To the best of our knowledge there is no subnet level topology data available for a large sized network. Hence, we use exploreNET to conduct a *census* of subnets over six geographically dispersed medium sized ISPs [16] including PCCW Global (ISP-1), nLayer (ISP-2), France Telecom (ISP-3), Telecom Italia Sparkle (ISP-4), Interroute (ISP-5), and

MZIMA (ISP-6). The collected population topology via census might posses non-sampling error introduced by the employed subnet inference tool, i.e., collected topology might slightly differ from the underlying genuine topology. In our sampling we use the same subnet inference tool to conduct our survey of subnets, thus, nonsampling error in the census propagates to the survey. Considering that nonsampling error in census and survey cancel each out while calculating sampling error of point estimates, our evaluations in this section are not affected by nonsampling error. On the other hand, using our methodology for a survey in the Internet will produce results that would contain both sampling and nonsampling errors. The magnitude of those nonsampling errors would depend on the accuracy of the subnet inference tool used in sampling (see Section 2.1).

First, we identified the IP address space for each ISP. Then, using an AS relationship dataset provided by CAIDA, we removed the ranges of IP addresses that are assigned to their customer domains. This pre-processing step enables us to focus on the core of ISP networks excluding the topology information of their customer domains which are managed and operated by others. Next, we utilized active probing to identify alive IP addresses which form our sampling frame, F_a. Table 1 shows the sampling frame sizes, $|F_a|$, of our target ISPs.

Table 1. Sampling Frame Sizes, $|F_a|$, for target ISPs

ISP-1	ISP-2	ISP-3	ISP-4	ISP-5	ISP-6	Σ
45,018	54,636	17,170	8,380	21,209	16,453	162,866

When we collected the entire subnet population for six ISPs having $162,866$ IP addresses in total, exploreNET successfully determined a subnet for $155,309$ of them. On the other hand, it failed to explore a proper subnet for $7,557$ of these IP addresses and returned them as /32 subnets. In fact, none of the IP addresses within /29 proximity of any of these $7,557$ IP addresses has responded to the probes sent out by exploreNET.

We then randomly selected 10% of the IP addresses from the sampling frame of each ISP and estimated population parameters including population size, N, prefix length distribution, τ_p, mean subnet degree, μ_d, and IP address space utilization, $\mathcal{U}$.

Note that, correcting the affect of singular IP addresses for which exploreNET failed to return a subnet requires special care in our estimation process. For example, including /32 samples into the estimation causes underestimated mean subnet degree, $\widehat{\mu}_d$. On the other hand, excluding them causes underestimated population size, $\widehat{N}$. Specifically, we include /32 subnets in population size estimation, $\widehat{N}$, because each one represents a subnet and we exclude them from other estimators because we assume that these /32 subnets distributed randomly, i.e., they could be any size subnet in the underlying topology and their omission does not skew the results in any way. The exclusion of /32 subnets is not simply ignoring them in the estimation process but estimating the number of such subnets and subtracting it from the estimated total number of subnets.

Remember that, the scope of this paper is confined to deriving proper estimators for subnet properties in the Internet and validating them, hence, in the rest of this section, we present the estimations, compare them with population parameters, and interpret the discrepancy between them. We plan to conduct another study based on sampled data to analyze and interpret the motifs and patterns of subnet characteristics in the Internet as a future work.

3.1   Subnet Population Size (N)

Remember that population size refers to total number of subnets appearing in an ISP. Table 2 shows the population size, N, and the corresponding point estimate, $\widehat{N}$, for six ISPs. Third column of the table shows sampling error (SE), $\widehat{N} - N$. The values in the table includes /32 singular IP addresses because these IP addresses belong to some subnet that `exploreNET` could not discover.

Table 2. Population subnet sizes and estimations for all ISPs

	N	**Ñ**	**SE**	**E**
ISP-1	4171	4167.53	3.47	247.99
ISP-2	1688	1625.14	62.86	156.38
ISP-3	8809	8791.65	17.35	94.36
ISP-4	3172	3272.01	100.01	116.55
ISP-5	7687	7904.60	217.6	265.44
ISP-6	3119	3180.68	61.68	181.02

Note that the point estimates, $\widehat{N}$, and sampling error, SE, would change for each sample taken from the population. On the other hand, fourth column of the table, E, shows the maximum error of the point estimate for $\widehat{N}$ with 90% confidence level. E is calculated as $z_{0.05} s_{\widehat{N}}$ where $s_{\widehat{N}}$ is the estimated standard deviation of the sampling distribution of $\widehat{N}$ and $z_{0.05}$ is the z-score of the 0.05 right tail probability of standard normal distribution and it demonstrates the fact that 90% of the time the error of the point estimate, $\widehat{N}$, would be less than E. The high deviation of E for some ISPs is because of the variability of the factor $|F_a|/d_j$ in Equation (5) where $y_j = 1$. In other words, HH estimator will have low variance as there is less variability among the values of y_j/p_j with the extreme case y_j/p_j values are exactly proportional in Equation (1) [13].

3.2   Total Subnets with Prefix Length /p (τ_p)

Prefix length (subnet mask) observed by `exploreNET` is the longest prefix length encompassing all alive IP addresses of a given subnet. Consequently, observed prefix length distribution is a natural way of grouping subnet degrees into bins where the range of the bins grow from larger prefix lengths to the smaller. In this part, we evaluate how well the sample prefix length distribution conforms to the population prefix length distribution for six ISPs. Table 3 presents total number of subnets having /p prefix length, τ_p, its estimated value, $\widehat{\tau}_p$, the prefix length probability mass function (p.m.f.), r, and its estimation, $\widehat{r}$, for PCCW Global (ISP-1). In the estimation process we omitted IP addresses that `exploreNET` failed to return a subnet, i.e., /32 subnets. Assuming that these /32 subnets are randomly distributed over the entire population, their exclusion does not affect the accuracy of the estimated prefix length p.m.f, $\hat{r}$.

In our data collection process, we encountered a small number of very large subnets such as the ones having /20, /21, or /22 prefix lengths with at least one thousand IP addresses. A close examination of randomly selected IP addresses from these subnet ranges via DNS name resolution queries revealed the fact that almost all of these large subnets belong to Akamai Technologies, an online content distribution service provider with a global presence in the Internet. An interesting detail in this case is that according to DNS records, those IP addresses belonged to Akamai Technologies while the IP-to-AS mapping dataset that we used mapped those IP addresses to their hosting ASes. Given that the main focus of this work is on building statistical estimations, we leave the close analysis of this type of interesting cases to our future study where we plan to study the data to understand their implications.

Table 3. PCCW Global (ISP-1) prefix length distribution and its estimation

	τ_p	$\hat{\tau}_p$	r_p	$\hat{r}_p$
/20	3	2.94708	0.001266	0.001245
/21	3	2.83139	0.001266	0.001196
/22	7	7.3299	0.002954	0.003096
/23	3	2.62722	0.001266	0.001110
/24	24	24.8725	0.010127	0.010507
/25	25	24.9367	0.010549	0.010534
/26	123	130.435	0.051899	0.055101
/27	152	145.53	0.064135	0.061478
/28	262	284.222	0.110549	0.120066
/29	440	426.242	0.185654	0.180061
/30	899	930.165	0.379325	0.392938
/31	429	385.068	0.181013	0.162668

Although Table 3 demonstrates that the estimated values are close to the population parameters, we need to apply goodness-of-fit test [6] to validate whether the estimated prefix length p.m.f. conforms to the population prefix length p.m.f.. The test statistic for goodness-of-fit test is χ^2 and it is defined as:

$$\chi^2 = \sum_p \frac{(\hat{\tau}_p - E_p)^2}{E_p}, \qquad \forall p \tag{10}$$

where p is prefix length, $\hat{\tau}_p$ is estimated number of subnets having prefix length /p, and $E_p = r_p \sum_{p=0}^{31} \hat{\tau}_p$ is the expected number of subnets having prefix length /p under the population distribution. Null hypothesis, H_0, is the assertion that the estimated and population prefix length distributions are the same, i.e., H_0 : $\{\forall p, \hat{r}_p = r_p\}$. Whereas, alternative hypothesis, H_1, is the opposite of H_0, i.e., H_1 : $\{\exists p, \hat{r}_p \neq r_p\}$.

Computing the χ^2 statistic for PCCW Global results in $X^2 = 8.6591$ with degrees of freedom (*df*) 9 and its related *p-value* is 0.4693. Since *p-value* is greater than the significance level $\alpha = 0.01$ we conclude that there is not enough evidence to reject the null hypothesis, i.e., estimated prefix length distribution conforms to the population prefix length distribution.

Table 4. X^2 scores and *p-values* for all ISPs

	df	X^2	p-value
ISP-1	9	8.659119	0.469317
ISP-2	8	11.53715	0.17308
ISP-3	4	5.513502	0.238545
ISP-4	6	2.1376	0.906618
ISP-5	8	2.02614	0.9802
ISP-6	7	4.97965	0.662447

Instead of tabulating population and estimated prefix length distributions for other five ISPs we present goodness-of-fit test results in Table 4. Since the *p-values* of all ISPs in Table 4 are greater than the significance level $\alpha = 0.01$, we again conclude that there is not enough evidence to reject the null hypothesis asserting that the estimated prefix length distribution conforms to the population prefix length distribution for all ISPs. As a result, prefix length ratio estimator, $\widehat{r}_p$, estimates the population prefix length ratio r_p well in all of our case studies.

3.3 Mean Subnet Degree (μ_d)

Remember that mean subnet degree estimator $\widehat{\mu}_d$ is a biased estimator of the population mean subnet degree μ_d. Even tough the first two columns of Table 5 show that the estimated value $\widehat{\mu}_d$ is pretty close to the population value μ_d, we need to make sure that the maximum error of the point estimator, $\widehat{\mu}_d$, is not much and we can get good estimates of the mean subnet degree most of the time.

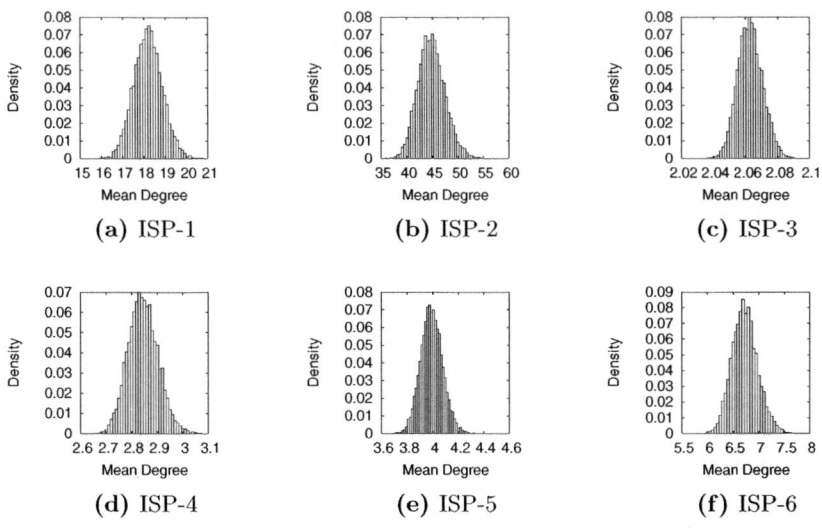

Fig. 3. Mean subnet degree ($\widehat{\mu}_d$) sampling distribution for all ISPs obtained via 10000 instances of Monte Carlo simulation

Table 5. Mean subnet degree figures for all ISPs

	μ_d	$\widehat{\mu}_d$	Min	Max	E	NE
ISP-1	18.235	18.257	15.956	20.975	1.128	1.685
ISP-2	44.904	46.112	36.025	57.12	4.536	1.716
ISP-3	2.063	2.066	2.034	2.095	0.013	1.7
ISP-4	2.849	2.797	2.647	3.079	0.098	1.672
ISP-5	3.997	4.049	3.696	4.39	0.144	1.691
ISP-6	6.742	6.224	5.846	7.986	0.437	1.74

Since variance of the sampling distribution of $\widehat{\mu}_d$ has no closed form solution we resorted to Monte Carlo simulation to determine a confidence interval for μ_d.

We took 10000 independent samples from the population database and estimated mean subnet degree of the sample. Figure 3 shows the histograms demonstrating the approximate sampling distribution of $\widehat{\mu}_d$. The third, *Min*, and fourth, *Max*, columns of Table 5 present the minimum and the maximum of $\widehat{\mu}_d$ point estimations obtained over 10000 samples of the same size (10%), respectively. Although the minimum and maximum estimates fall a little bit off the population mean subnet degree, the bell shape of the distributions suggest that they are outliers. To have a better idea we constructed a confidence interval with confidence level 90% and computed maximum error of the point estimate. Column five, E shows the maximum error that could be obtained 90% of the time while estimating the mean subnet degree. Since each ISP has a different mean subnet degree and variance we normalized the maximum error of estimate for $\widehat{\mu}_d$, E, by dividing it with the estimated standard deviation of $\widehat{\mu}_d$ as shown in the sixth column, *NE*. The *NE* values suggest that for an ISP, 90% of the time the mean subnet degree estimation conveys a maximum error of approximately 1.7 of standard deviation.

Note that, the ISPs having large mean subnet degree are those ones that home very large network layer subnets belonging to Akamai Technologies.

3.4 IP Address Space Utilization ($\mathcal{U}$)

Similar to mean subnet degree, IP address space utilization estimator, $\widehat{\mathcal{U}}$, is a biased estimator of population IP address space utilization, $\mathcal{U}$. The first and second columns of Table 6 presents that the estimated utilization, $\widehat{\mathcal{U}}$, is not far from the population utilization $\mathcal{U}$. However, this is a single instance of sampling and to have a better idea on whether our slightly biased estimator works well most of the time we employed Monte Carlo simulation. Again we repeated sampling for 10000 times from our population database and estimated the value of $\mathcal{U}$ for each sample.

Figure 4 shows the sampling distribution of $\widehat{\mathcal{U}}$ obtained over 10000 samples of the same size (10%) for six ISPs. Third and fourth columns of Table 6 present the minimum and maximum utilization estimates obtained by Monte Carlo simulation. However, the bell shape of the distributions in Figure 4 suggests that the minimum and maximum values are not likely to occur frequently. To have an idea on the error of the estimator we constructed a confidence interval with 90% confidence level. Column six, E, demonstrates the maximum error of $\widehat{\mathcal{U}}$ at

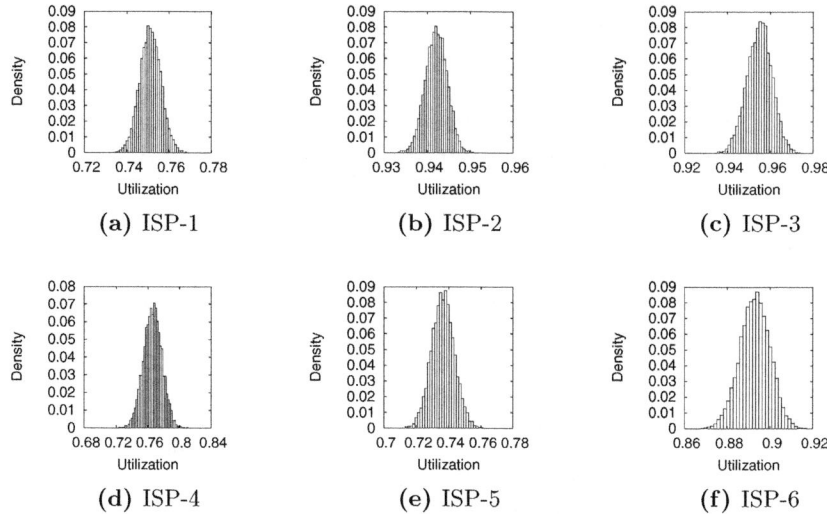

Fig. 4. Utilization $(\widehat{\mathcal{U}})$ sampling distribution for all ISPs obtained via 10000 instances of Monte Carlo simulation

Table 6. Utilization figures for all ISPs

	$\mathcal{U}$	$\widehat{\mathcal{U}}$	Min	Max	E	NE
ISP-1	0.752	0.748	0.734	0.774	0.008	1.645
ISP-2	0.942	0.944	0.933	0.951	0.004	1.654
ISP-3	0.955	0.953	0.929	0.977	0.01	1.691
ISP-4	0.767	0.775	0.722	0.811	0.019	1.653
ISP-5	0.737	0.747	0.709	0.766	0.012	1.656
ISP-6	0.894	0.89	0.865	0.919	0.012	1.67

confidence level 90%. To put in other words, it says that 90% of the time the maximum error of sample utilization, $\widehat{\mathcal{U}}$, is E. Again we normalized the maximum error of estimate for $\widehat{\mathcal{U}}$, E, by dividing it to the standard deviation of the sampling distribution of $\widehat{\mathcal{U}}$. Column NE in Table 6 states that the maximum error of estimate for utilization is about 1.67 of standard deviation for all ISPs.

4 Related Work

Most of the studies in the Internet topology measurement field is based on collecting raw topology data from multiple vantage points by path sampling via `traceroute` or `traceNET`. Then inferring routers and subnets using the raw data [7,12,10]. Early studies in the area used path sampling based topologies to derive conclusions about the topological characteristics of the Internet [3,4]. Later on researchers argued about the validity of those claims by pointing out the limitations of path sampling due to source dependency and routing [8,2].

`TraceNET` [14] infers subnets on an end-to-end path. Although `traceNET` substantially improves existing Internet topology maps by discovering subnet structures, it performs path sampling and hence suffers from above mentioned problems. ExploreNET [16] used in this paper allows us to discover individual subnets rather than subnets on an end-to-end path. In this study, we devised a sampling framework for studying network layer subnet characteristics in the Internet and introduced a set of unbiased and biased with small variance estimators for a set of subnet features. To the best of our knowledge, this is the first study investigating statistical sampling in the domain of subnet inference.

5 Conclusions

Given its large scale, drawing a complete *subnet level* topology of the Internet is neither practical nor economical. Consequently, sampling becomes a viable solution to derive the properties of subnets in the Internet. In this paper, we have proposed a framework for sampling subnets in the Internet along with either unbiased or slightly biased estimators with small variance for various subnet characteristics including total number of subnets, subnet prefix length distribution, mean subnet degree, and subnet IP address range utilization. Our theoretical and empirical evaluations on the estimators show that they work well in estimating population parameters.

References

1. The Internet Topology Zoo, http://www.topology-zoo.org/
2. Achlioptas, D., Clauset, A., Kempe, D., Moore, C.: On the bias of traceroute sampling. In: ACM STOC, Baltimore, MD, USA (May 2005)
3. Faloutsos, M., Faloutsos, P., Faloutsos, C.: On power-law relationships of the internet topology. In: SIGCOMM, New York, NY, USA (October 1999)
4. Govindan, R., Tangmunarunkit, H.: Heuristics for Internet map discovery. In: IEEE INFOCOM, Tel Aviv, Israel (March 2000)
5. Heidemann, J., Govindan, R., Papadopoulos, C., Bartlett, G., Bannister, J.: Census and survey of the visible internet. In: ACM IMC, Vouliag, Greece (October 2008)
6. Irwin Miller, M.M.: John E. Freund's Mathematical Statistics with Applications. Prentice Hall (October 2003)
7. Claffy, K., Monk, T.E., McRobb, D.: Internet Tomography. Nature (January 1999)
8. Lakhina, A., Byers, J., Crovella, M., Xie, P.: Sampling biases in IP topology measurements. In: IEEE INFOCOM, San Francisco, CA, USA (March 2003)
9. Mann, P.S.: Introductory Statistics. Wiley (February 2010)
10. Shavitt, Y., Shir, E.: DIMES: Distributed Internet measurements and simulations, project page, http://www.netdimes.org
11. Sherwood, R., Bender, A., Spring, N.: DisCarte: A disjunctive internet cartographer. In: ACM SIGCOMM, Seattle, WA, USA (August 2008)
12. Spring, N., Mahajan, R., Wetherall, D., Anderson, T.: Measuring ISP topologies with Rocketfuel. IEEE/ACM Transactions On Networking 12(1), 2–16 (2004)
13. Thompson, S.K.: Sampling. Wiley-Interscience (April 2002)
14. Tozal, M.E., Sarac, K.: TraceNET: An internet topology data collector. In: ACM Internet Measurement Conference, Melbourne, Australia (November 2010)
15. Tozal, M.E., Sarac, K.: Palmtree: An ip alias resolution algorithm with linear probing complexity. Computer Communications 34, 658–669 (2011)
16. Tozal, M.E., Sarac, K.: Subnet level network topology mapping. In: IEEE IPCCC, Orlando, FL, USA (November 2011)

Sparsity without the Complexity: Loss Localisation Using Tree Measurements

Vijay Arya[1] and Darryl Veitch[2]

[1] IBM–Research, D4 Mayanta, Bangalore 560045, India
vijay.arya@in.ibm.com
[2] Dept. of EEE, The University of Melbourne, Australia
dveitch@unimelb.edu.au

Abstract. We study network loss tomography based on observing average loss rates over a set of paths forming a tree – a severely under-determined linear problem for the unknown link loss probabilities. We examine in detail the role of sparsity as a regularising principle, pointing out that the problem is technically distinct from others in the compressed sensing literature. While sparsity has been applied in the context of tomography, key questions regarding uniqueness and recovery remain unanswered. Our work exploits the tree structure of path measurements to derive sufficient conditions for sparse solutions to be unique and the condition that ℓ_1 minimization recovers the true underlying solution. We present a fast single-pass linear algorithm for ℓ_1 minimization and prove that a minimum ℓ_1 solution is both unique and sparsest for tree topologies. By considering the placement of lossy links within trees, we show that sparse solutions remain unique more often than is commonly supposed. We prove similar results for a noisy version of the problem.

Keywords: network monitoring, network tomography, loss inference, tree topology, sparsity, ℓ_1 regularization, compressed sensing.

1 Introduction

Network operators and end applications alike would like to localize abnormally lossy links or *loss hotspots*, but how can this be achieved when internal access to the network is limited? Consider a set of nodes instrumented as active probing sources or receivers, generating flows of probes over a set of paths in the network to measure loss. The intersections of these paths result in a set of relations for mutual consistency of the measured path loss probabilities in terms of the constituent link loss probabilities that one would like to recover. This is a network tomography problem, defined over the measurement sub-network traversed by the probes, which can be expressed as a linear system. This system is in general severely under-determined: instead of a unique solution for the link loss rates, an entire family of solutions is consistent with the observed path measurements.

One way to select a particular solution from the family, that is one regularising principle, is *sparsity*: preferring the solution with the smallest number of

R. Bestak et al. (Eds.): NETWORKING 2012, Part I, LNCS 7289, pp. 289–303, 2012.

Table 1. A comparison of problem statements and results for a typical CS problem compared to loss tomography over a tree (with c children under each node)

Sparsity K	Classical CS Results	Tree Loss Tomography Results
Signal x	$\mathbb{R}^n$	link loss vector $\mathbb{R}^{+^n}$
m measurements	$m \ll n$, variable	$m = n(1 - 1/c) + o(1)$, **fixed**
Matrix A	$m \times n$, $a_{ij} \sim \mathcal{N}(0, 1/m)$	$m \times n$ binary matrix representing the tree measurement topology
Uniqueness	every $2K$ columns of A independent	Every branch node has at least 2 lossless incident links ($K \leq m$)
Efficient Recovery	$m = O(K \log(n/k))$, RIP conditions on A	Every branch node has at least 1 lossless child link ($K \leq m$)

lossy links. Sparsity is in keeping with Occam's razor which seeks the simplest explanation to a given set of observations, and is a natural fit to the assumption that hotspots are rare. It also sits well with an operational need to provide a short list of potential hotspots worthy of closer attention. This paper examines in detail the role of sparsity in network loss tomography in the important special case of a tree-like measurement topology or *measurement matrix A*. Trees have been considered in a few prior works treating sparsity in loss tomography (see section 2), but mostly in an implicit sense, as a default special case of general networks. In this paper we show that the structure of trees can be exploited to allow an essentially complete picture to be obtained including conditions for uniqueness of sparse solutions, the relationship of sparse to ℓ_1 solutions, and the condition that minimum ℓ_1 recovers the true underlying solution.

Tree measurement topologies arise in several practical contexts – for instance unicast path measurements between a web server and its clients form a tree. Thus information on locations of hotspots could be used to direct clients to replica servers or have the hotspots resolved in cooperation with the concerned ISP [1]. Moreover, since any general measurement infrastructure can be configured for use as a tree, our results are relevant in practice. Furthermore, loss rates inferred from multiple intersecting trees over the same infrastructure can also be used to quickly obtain important partial information about general measurement topologies/matrices. Exploiting tree solutions and insights to gain purchase on the more general problem is a new direction and the subject of ongoing work. An early result in this direction is Lemma 1 in section 4.3.

There are three main differences between our problem setting and that of compressed sensing (CS), which has in recent years exploited sparsity for signal recovery (see Table 1). First, network tomography deals with positive quantities such as link loss probabilities and delays, whereas CS generally treats real valued signals. Second, in CS one inquires after the nature of the measurement matrix A, and its size (number of observations m) needed to recover a solution of given sparsity K uniquely. In network tomography both the nature and size of A are highly constrained by the availability of measurement nodes, and the lack of control over the packet routing between them. In the case of a tree, adding a new 'measurement' is non-trivial as it implies installing an active probe receiver in a new location. Finally, the measurement matrices studied in CS are designed to satisfy strong technical conditions such as the restricted isometry property (RIP [2, 3]). In contrast, for a tree the measurement matrix A is given rather than designed, and has a specific dependence structure. Studying sparse solutions over trees is quite different from 'traditional CS' approaches, however, the key questions of interest are the same: **hardness, uniqueness, and recovery**.

Compared to prior CS work, our task is nontrivial in that we do not benefit from properties such as RIP and must develop fresh techniques, but easier in that the structure of trees is simple and powerful. As we see below, the net result is that the tree context is more tractable, enabling detailed and complete solutions with desirable properties. Our main contributions include:

Hardness (Complexity of Computing a Sparse Solution): Recovering the sparsest or min ℓ_0 (pseudo) norm solution of under-determined systems is in general NP-hard [4]. For a tree we show that the sparsest solution(s) may be characterised precisely and found with a fast linear time algorithm.

Uniqueness (Conditions for Sparsest Solution to Be Unique): Provided the number of lossy links at every internal/branch node with node degree g is at most $g - 2$, the sparsest solution is unique. This result takes into account the locations of loss within the tree, and shows that uniqueness may hold for much higher values of sparsity K than the worst case analysis typically used in CS would suggest. When solutions of a given sparsity are not unique, the alternative solutions can be precisely localised and characterised.

Recovery (Conditions That Min ℓ_1 Solution is the True Solution): For the general problem, the min ℓ_1 norm solution does not always have the min ℓ_0 norm and need not even be unique. For the tree problem, we show that min ℓ_1 solution always has the min ℓ_0 norm and is unique. Provided every internal/branch node has at least one lossless child link, the min ℓ_1 recovers the true underlying solution. We define the 'UpSparse' algorithm, a fast single-pass linear-time algorithm which outputs the min ℓ_1 solution. For the general problem, the min ℓ_1 solution is recovered through a linear program (cubic complexity).

Since in practice only a finite number of probes can be sent, the measured path loss probabilities can only be known approximately. We formulate and study a 'noisy' version of the problem that addresses this key practical concern. As before, we exploit the nature of the tree, and characterise the minimal norm solutions and present fast algorithms to recover them. The noisy problem is briefly discussed in section 4.4 with results in the technical report [5].

We begin by discussing related work in section 2. Section 3 presents the general solution for the hotspot localisation problem in trees. Section 4 characterises the sparse and min ℓ_1 solutions and shows how to find them. Section 5 compares our algorithms with CS optimization techniques. Section 6 presents experimental results where we explore the relationship between the sparse and true solutions.

2   Related Work

Several problems [1, 6–11], all related to inferring link parameters using either unicast or multicast path measurements, have been studied under the purview of Network Tomography. Whereas multicast measurements utilise observations at the per-probe level, the unicast tomography problem works with average observations of paths and reduces to an under-determined linear system in terms of unknown link parameters. However the most common approach to solve an underdetermined system, namely choosing the min ℓ_2 norm solution, may not be suitable when the link quantity is concentrated at particular locations. For example, for loss inference, it tends to spread the loss over all links in the network.

More recent work borrows techniques of recovering sparse solutions to under-determined linear systems from compressed sensing (CS) [2, 3, 12]. In CS, a fixed but randomly generated (generally Gaussian) matrix A is used to 'measure' an unknown signal $\boldsymbol{x}$ as $\boldsymbol{y}_{m\times 1} = A_{m\times n}\boldsymbol{x}_{n\times 1}$; $m \ll n$ so that x is underdetermined. CS results show that a minimally sparse $\boldsymbol{x}$ can be recovered with high probability (i.e. for most A) using ℓ_1 minimization [13]. As outlined above, network loss tomography is quite different: $\boldsymbol{y}$ represents the path observations, $\boldsymbol{x}$ the unknown link parameters, and A, which determines which paths traverse each link, cannot be chosen freely and has unknown properties in general.

Despite these differences, ℓ_1 minimization has been used as a black box to recover sparse solutions in tomography. In [9], Bayesian experimental design is used to determine the set of paths to measure in a network and a variant of ℓ_1 minimization is used to infer link parameters. In [10], variance in path measurements across multiple measurement intervals is used to identify a prior, and an ℓ_1 minimization formulation from [9] is used to find a sparse solution close to it. In [1], path measurements between a server and its clients are used to recover link loss rates by using sampling, Bayesian inference, and a variant of ℓ_1 minimization. None of these works provide insight into the nature of the sparse or ℓ_1 solutions, how they interact, or their uniqueness, the central focii of our work. In [6], locations of 'bad' network links is inferred from path measurements forming a tree. For this, each path is classified as 'good' (0) or 'bad' (1) and the smallest set of bad tree links consistent with the binary path observations is recovered. In section 6, we see that this two-step approach fails to recover the true locations of hotspots more often than our approach that directly recovers a minimally sparse link loss solution. In addition, we recover both the locations and loss rates of lossy links. Given a measurement matrix, [14] uses expander graphs [15] to determine conditions for recovering unique sparse solutions in networks. However, for trees expander graphs do not bring any additional insight.

The work most closely related to our own is [16], which answers some of the key questions for CS over graphs. The key difference is that we work with trees instead of general networks. The simpler tree topology enables far greater insight into the sparse and ℓ_1 solutions, and allows explicit solutions and fast algorithms to be defined. In [16] the authors determine the number of random measure-ments over underlying network paths needed to uniquely recover sparse link so-lutions. Random measurements however are difficult to justify in the tomography context. Conversely, for a given measurement matrix they provide upper bounds on the number of lossy links consistent with uniqueness of the sparsest solution. These bounds are quite restrictive for trees. For example for any ternary tree, irrespective of its size, the largest allowed number of lossy links is 2, and for a binary tree the price of uniqueness guarantee is that only a single link may be lossy. In section 6, we see that for a ternary tree with 25 links, even when 4 links are lossy, the sparsest solution is still unique for 95% of feasible link loss vectors, and the proportion grows with tree size. In section 4.3, we also show how the recovery of a K-sparse vector relates to the degree of the measurement graph.

3   The Tree Hotspot Problem and Solution

In this section we describe how we model the loss process over a tree, and how to formulate the resulting problem as a linear system. We then solve the system formally, and make some preliminary observations.

Tree Model. Let $\mathcal{T} = (V, L)$ denote the logical tree consisting of a set of nodes V and links L. Let $\mathcal{O} \in V$ denote the root node, $R \subset V$ be the set of leaf nodes, and $I = V \setminus \{\mathcal{O} \cup R\}$ the set of internal nodes. A link is an ordered pair $(j, k) \in \{V \times V\}$ representing a logical link (one or more physical links) from node j to node k. For each node except the root there is a unique node $j = f(k)$, the father of k, such that $(j, k) \in L$. The set of children of a node k is denoted by $c(k)$, thus $c(k) = \{j \in V : (j, k) \in L\}$. All nodes have at least two children, except the root (just one) and the leaves (none). The depth of a node is the number of links in the (unique) path of ancestors leading to the root. By level l of a tree we mean the set of nodes of depth l, with the root being of depth zero. We denote the *height* of the tree (the depth of the deepest leaves) by H. The *top link* is the unique link adjacent to the root node.

For convenience, we refer to link $(f(k), k)$ simply as link k, and similarly, we also use I to refer to the set of *internal links* corresponding to the internal nodes, R to refer to the leaf links as well as nodes, and so on. Let n denote the number of links in the tree, $m = |R|$ the number of leaves, and $d = n - m$ the number of internal links. From each leaf there is a unique path to the root, so m is also the number of paths. Clearly $n \geq m + 1$. It is convenient to label nodes/links as follows: First, the leaf nodes are labelled by $k = 1, 2 \ldots m$ from left to right. Then beginning with the child of the root the counting continues in a *preorder* traversal of the internal nodes of the tree (recursively: node, left subtree right subtree). With this convention, the labels of leaf nodes can double as convenient path labels. In other words, path j terminates at leaf node j, with paths labelled as $j = 1, 2 \ldots m$ from left to right. Examples are given in the figures.

The topology of tree is captured by the $m \times n$ measurement matrix A, where the entry $A_{jk} = 1$ if link k forms part of the path j, 0 otherwise. Row j of the matrix gives the links in path j, and column k gives the paths that cross link k.

Modelling Link Loss. The marginal probability of loss on link k is given by $b_k \in [0, 1]$, and we denote the $(n \times 1)$ vector of loss probabilities over all links by **b**. We assume stationarity so that **b** is constant. As in all prior work, spatial independence is assumed, i.e., all link loss processes are mutually independent. It follows that the path loss probability is easily expressed via the product of the link passage probabilities: $p_j = 1 - \prod_{k:A_{jk}=1}(1 - b_k)$ where the product is over the links on path j. We assume that we have access, through measurements based on a large number of probes, to the exact path loss probability vector, $\mathbf{p} = [p_1, p_2, \ldots, p_m]^T$.

3.1   System Solution

Define the *addloss function* as $\mathcal{L}(b) = -\log(1 - b)$, $b \in [0, 1)$. We write $x_k = \mathcal{L}(b_k)$ and $y_j = \mathcal{L}(p_j)$. Since $\mathcal{L}$ is a monotonically increasing function, mapping $[0, 1)$ to $[0, \infty)$, the link loss vector **b** is replaced by the equivalent link *addloss vector* $\boldsymbol{x} = [x_1, x_2, \ldots, x_n]^T$, and similarly **p** is replaced by $\boldsymbol{y} = [y_1, y_2, \ldots, y_m]^T$. The relation $p_j = 1 - \prod_{k:A_{jk}=1}(1 - b_k)$ is now $y_j = \sum_{k:A_{jk}=1} x_k$, and the relationship between path and link loss takes the linear form

$$\boldsymbol{y} = A\boldsymbol{x}, \quad x_k \geq 0, y_j \geq 0. \qquad (1)$$

The term 'addloss' is justified by the additive nature of link addloss, together with the fact that values of x_k and y_j can still be interpreted directly as loss for many purposes. In particular 0 addloss implies zero loss. Since we use addloss exclusively in this paper, we will use 'loss' as a shorthand for addloss.

$$A = \begin{bmatrix} 1 & 0 & 0 & | & 1 & 1 \\ 0 & 1 & 0 & | & 1 & 1 \\ 0 & 0 & 1 & | & 1 & 0 \end{bmatrix}$$
$$= \begin{bmatrix} I_m & | & A_I \end{bmatrix}$$

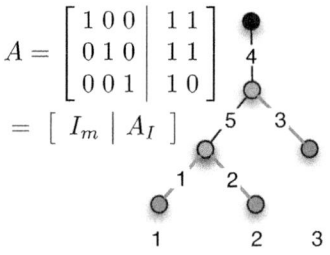

Consider the tree in figure 1 together with its measurement matrix A, where the vertical divider separates the m columns corresponding to receiver links $\boldsymbol{x}_R = [x_1, x_2, \ldots, x_m]^T$ on the left, from those of the internal links $\boldsymbol{x}_I = [x_{m+1}, \ldots, x_n]^T$ on the right. The $(m \times m)$ identity matrix I_m in the left appears because

Fig. 1. A tree with 3 leaves (and paths) and 2 internal links, with its measurement matrix A. Receiver links correspond to the identity matrix I_m.

each leaf link belongs to just one path (and because of our link and path naming conventions). This is true in general for any tree, and we may partition any measurement matrix into I_m and the $(m \times d)$ matrix A_I that shows how internal links contribute to paths. We can now rewrite (1) as

$$\boldsymbol{y} = A\boldsymbol{x} = \begin{bmatrix} I_m & A_I \end{bmatrix}\begin{bmatrix} \boldsymbol{x}_R \\ \boldsymbol{x}_I \end{bmatrix} = \boldsymbol{x}_R + A_I\boldsymbol{x}_I \Rightarrow \boldsymbol{x}(\boldsymbol{x}_I; \boldsymbol{y}) = \begin{bmatrix} \boldsymbol{x}_R \\ \boldsymbol{x}_I \end{bmatrix} = \begin{bmatrix} \boldsymbol{y} - A_I\, \boldsymbol{x}_I \\ \boldsymbol{x}_I \end{bmatrix} \quad (2)$$

It is clear that A has full rank m, as its column rank is clearly at least m due to the embedded identity matrix, and row and column rank are equal. The general solution will therefore have $d = n - m$ free parameters. Since there are also d internal links, it is convenient to select $\boldsymbol{x}_I$ to span this space, in terms of which a formal solution can be immediately written as above where the path observation vector $\boldsymbol{y}$ appears as a parameter.

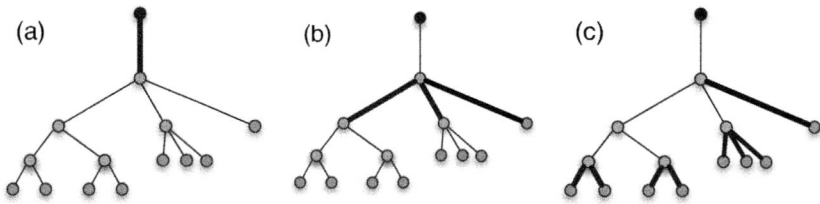

Fig. 2. Example of ambiguity in the locations of lossy links (bold), each of loss x. In each case, receiver vector $\boldsymbol{y} = [x, \ldots, x]^T$, but the no. of lossy links varies - 1, 3, and 8.

The choice of $\boldsymbol{x}_I$ as the independent variables has the advantage of making $\boldsymbol{y}$ appear in a simple way in the solution. For e.g., setting $\boldsymbol{x} = 0$, it is clear that $\boldsymbol{y} = \boldsymbol{0}$ is the corresponding observation. Setting $\boldsymbol{x}_I = 0$, we see that $\boldsymbol{y} = \boldsymbol{x}_R$, which also reveals that any set of (non-negative) observations is possible in general. We call $\boldsymbol{x}_I = 0, \boldsymbol{x}_R = \boldsymbol{y}$, the *receiver solution*.

4   Regularizing the Solution Using Sparsity

Equation (2) is a d-dimensional family of loss solutions all equally consistent with a single observed $\boldsymbol{y}$. From the point of view of an observer whose end goal is to identify a unique set of candidate loss hotspots, this represents a significant and problematic *ambiguity*. Our main regularising principle is that of *sparseness*, i.e. minimizing the number K of lossy links which are consistent with any observed $\boldsymbol{y}$. A smaller number is preferred because it is more likely under a priori assumption that loss is rare, and focusses attention on a smaller number of candidate problem links, which has practical advantages. If in fact solutions are sparse, i.e. given a bound on the sparsity K, we wish to determine the conditions under which a sparsest solution is the unique solution consistent with observed $\boldsymbol{y}$.

Consider figure 2 which gives three (of several) possible solutions consistent with the observations $\boldsymbol{y} = [x, \ldots, x]^T$, $x > 0$. Figure 2(a) shows the solution $\boldsymbol{x}_I = [x, 0, \ldots, 0], \boldsymbol{x}_R = 0$ where only the top link is lossy while figure 2(c) shows the receiver solution $\boldsymbol{x}_I = 0, \boldsymbol{x}_R = \boldsymbol{y}$. If sparsity $K \leq 1$, then 2(a) is the unique sparsest solution consistent with $\boldsymbol{y}$.

Finding the sparsest solution is equivalent to minimizing the ℓ_0 (pseudo) norm of $\boldsymbol{x}$: $K = \|\boldsymbol{x}\|_0 = \sum_{i=1}^n |x_i|^0 = \sum_{i:x_i>0} 1$. Results from CS have shown that minimizing with respect to ℓ_1 norm often identifies solutions with min ℓ_0 norm but at a lower computational cost. We therefore also explore ℓ_1 below, both in its own right, and as a secondary principle which can be used to further reduce ambiguity. Whilst a priori information on the likely locations of lossy links within the topology may be available in some contexts, this is not always so. In this paper we treat the case where there is no such information, corresponding informally to a uniform prior over all links.

4.1   Local Regularisation

It is useful to understand ambiguity in a *local complex*. This is a 'building block' consisting of an internal/branch node and its adjacent links. We will explore it using the two level tree of figure 3(a), where link and path labels have been dropped in favor of their loss values. The general solution corresponding to the observed $\boldsymbol{y}$ is $\boldsymbol{x}' = [y_1 - x, y_2 - x, \ldots, y_m - x, x]^T$, parameterised by $x \in [0, y_{\min}]$, where $y_{\min} = \min_j y_j$ is the smallest path loss. We examine sparsity in the local complex as a function of the parameter $x \in [0, y_{\min}]$.

$y_{\min} = 0$: the family collapses to a unique solution, $x = 0$.

$y_{\min} > 0$: x is not uniquely determined by $\boldsymbol{y}$ and we speak of an *ambiguous complex*. At $x = 0$ the internal link is lossless and the child link losses are maximized. We call this the *downstate*, and it has sparsity $K = m$. As x increases over the range $x \in (0, y_{\min})$ loss is 'pulled up' equally and in parallel from each child link to the internal link. In these *mixed states* all links are lossy and sparsity is $K = m + 1$, the largest possible. This upward transfer of loss ceases at the *upstate* when $x = y_{\min}$, where suddenly the loss on all $m_{\min}$ links sharing the minimum value $y_{\min}$ becomes zero, and K drops to $K = m - m_{\min} + 1$.

   To summarise, requiring sparsity excludes the mixed states, singling out the downstate ($x = 0$) and upstate ($x = y_{\min}$). If $y_{\min}$ is not unique ($m_{\min} > 1$) then the upstate solution is the unique sparsest solution. For example when moving from figure 2(b) to figure 2(a), K drops locally around the top internal node from 3 to 1 as the complex moves from the downstate to the upstate. If instead $m_{\min} = 1$, then both the upstate and downstate have sparsity $K = m$, so ambiguity, though greatly reduced, remains.

   If instead of sparsity we consider ℓ_1 the conclusions are similar but ambiguity vanishes. The ℓ_1 norm of the local complex illustrated in figure 3(a) is given by $\|\boldsymbol{x}'\|_1 = \sum_{k=1}^{n} |x_k| = \sum_{j=1}^{m} y_j - x(m - 1)$. It is clearly minimized by setting x as large as possible, namely $x = y_{\min}$, i.e. the upstate. Hence for the ℓ_1 norm upstate is always both optimal and unique, whereas for ℓ_0 it is always optimal but not always unique. A simple example with $\boldsymbol{y} = [2, 3, 4]$ is shown in Figure 3(b).

4.2   Global Regularisation

The local ambiguity above can occur centered on any internal node in the tree. A natural question is whether global effects may resolve local ambiguities, or alternatively result in new forms of global ambiguities. Consider the scenario of figure 3(c) where the original sparsity is K. The lower complex is initially in a downstate (left) with $m_{\min} = 1$ and the upper complex is in the upstate. After the lower complex moves to the upstate, the sparsity remains at K (middle), however the upper complex is no longer minimally sparse. Let $x_{\min}$ denote the minimum loss of the child links in the upper complex **after** the move, and $m'_{\min}$ its associated multiplicity. Since $x_{\min} > 0$ and $m'_{\min} = 2 > 1$ sparsity can be reduced to $K + 1 - m'_{\min} = K - 1$ by moving the upper complex to the upstate.

   There are two important observations to make from the above example. First, choosing the upstate always achieves minimal local sparsity. Second, choosing the

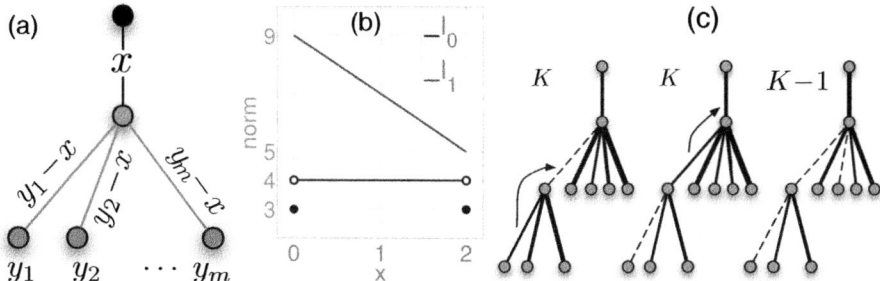

Fig. 3. (a) Ambiguity at a local complex, which could be centered on any internal node, illustrated by a simple tree. Links and paths are annotated with their loss values, obeying the general solution parameterized by x. Loss can be moved up or down subject to $x \in [0, y_{min}]$. (b) ℓ_0 and ℓ_1 norms of solutions shown for a 3-receiver complex with observations $y = [2, 3, 4]$. Loss can be moved up or down subject to $x \in [0, y_{min} = 2]$. Both downstream and upstate solutions at $x = 0$ and $x = 2$ have the equal sparsity $K = m = 3$, but upstate achieves minimal ℓ_1. (c) An example of the coupling of two local complexes resulting in lower global ambiguity (link thickness proportion to loss, dashed links lossless). Left: initial configuration: lower complex in downstate, upper in upstate. Middle: system remains K-sparse when the lower complex changes state, but upper complex is no longer minimally sparse. Right: moving the upper complex into the upstate yields lower sparsity.

upstate locally may also enable sparser states to be found in complexes higher in the tree. Together these motivate the following algorithm which defines a global solution based on the systematic exploitation of local sparsity with a preference for the upstate solution in case of local non uniqueness. Recall that local complexes are centered on internal nodes at levels $l = 1, 2 \dots H - 1$.

UpSparse Algorithm
Begin with an arbitrary feasible solution x. For all ambiguous local complexes at the deepest level select the up sparse state. Move up to the next level and repeat. Terminate at level 1.

Function $UpSparse(\boldsymbol{x})$	**Function** $PutInUpState(i, \boldsymbol{x})$
% arbitrary initial solution $\boldsymbol{x}$	% Put node i in upstate
1: **for** $\ell = H - 1$ downto 1 **do**	1: $\delta \leftarrow \min_{j \in c(i)} \{x_j\}$
2: **for all** nodes i at level ℓ **do**	2: $x_i \leftarrow x_i + \delta$
3: $\boldsymbol{x} \leftarrow PutInUpState(i, \boldsymbol{x})$	3: **for all** nodes $j \in c(i)$ **do**
4: **end for**	4: $x_j \leftarrow x_j - \delta$
5: **end for**	5: **end for**
6: **return** $\boldsymbol{x}$ % *UpSparse* Solution	6: **return** $\boldsymbol{x}$ % Link i & children updated

---

We call the state of the loss vector tree after the application of the algorithm the *UpSparse solution*. An example of initial solution is the receiver solution $\boldsymbol{x}_I = 0, \boldsymbol{x}_R = \boldsymbol{y}$. Reading figure 2 from right to left provides an example of the algorithm in action. We now give its main properties. These are proved in [5].

UpSparse Properties

Let $\boldsymbol{x}^*$ be the output of UpSparse with input $\boldsymbol{x}$, then

Property 1. Each local complex in $\boldsymbol{x}^*$ is in upstate (by construction).

Property 2. *UpSparse uniqueness*: For fixed $\boldsymbol{y}$ and any feasible input $\boldsymbol{x}$, UpSparse outputs the same $\boldsymbol{x}^*(\boldsymbol{y})$, defined $\forall i \in V\backslash\{\mathcal{O}\}$ by:

$$x_i^* = \gamma_i \text{ if } f(i) = \mathcal{O}, \text{ else } x_i^* = \gamma_i - \gamma_{f(i)}, \text{ where } \gamma_i = \min \{ \boldsymbol{y}_{R(i)} \} \qquad (3)$$

$\boldsymbol{y}_{R(i)}$ denotes the observations in subtree rooted at node i.

Property 3. *UpSparse solution has minimal ℓ_0 and ℓ_1 norms*: For any solution $\boldsymbol{x}$, $\|\boldsymbol{x}^*\|_0 \leq \|\boldsymbol{x}\|_0$ and $\|\boldsymbol{x}^*\|_1 \leq \|\boldsymbol{x}\|_1$.

Property 4. $\|\boldsymbol{x}^*\|_0 \leq m = |R|$.

Property 5. $\min \ell_1$ *uniqueness*: UpSparse solution $\boldsymbol{x}^*$ is the unique solution with $\min \ell_1$ norm.

Property 6. $\min \ell_0$ *uniqueness*: If each local complex in $\boldsymbol{x}^*$ is uniquely sparse, then $\boldsymbol{x}^*$ is the unique sparsest solution, otherwise not.
Corollary: When the number of lossy links at internal node i with degree g_i is at most $g_i - 2$ $\forall i$, then the sparsest solution is unique. This degree condition ensures that no internal node can be moved from upstate without increasing sparsity, and hence $\boldsymbol{x}^* = UpSparse(\boldsymbol{x}) = \boldsymbol{x}$.

Property 7. $\min \ell_1/UpSparse$ *solution = true solution?*: If there exists at least one lossless child link at every internal node, UpSparse solution = true solution.

4.3 A Sufficiency Condition for Non-uniqueness in Graphs

Consider a measurement matrix A that defines a graph (instead of a tree) using paths from an underlying general network. Each column vector in A corresponds to a link in the graph. For any branch or internal node, let g^{in} (resp. g^{out}) denote its in-degree (resp. out-degree), that is the number of links covered by its incoming (resp. outgoing) paths. Then:

Lemma 1. *Let $K \geq \max\{g^{in}, g^{out}\}$ for some branch node. Then there exists a K-sparse non-negative vector $\boldsymbol{x}$ that is not the unique sparsest solution to $\boldsymbol{y}(\boldsymbol{x}) = A\boldsymbol{x}$.* (The proof is given in [5])

Lemma 1 gives a worst case bound relating the degree of networks and their amenability to sparse recovery. If networks have small degree, then sparsity minimization can fail even for small K if lossy links are concentrated at a branch point. For a binary tree, the bound for non-uniqueness given by lemma 1 is $K \geq 2$ since all branch nodes have $g^{in} = 1$ and $g^{out} = 2$. However property 6 above shows that sparsity minimization will still succeed for a binary tree for higher K provided the number of lossy links at each branch point is < 2.

4.4   The Noisy Problem

We briefly describe the noisy problem for completeness with details in [5]. The model defined in section 3 is based on knowing the mean loss observed at each receiver exactly. This assumption may fail in a number of ways, the most important of which is that, in practice, loss is estimated based on a finite number of observations, resulting in receivers seeing only an estimate $\widehat{y}$ of the true path loss observation vector $\boldsymbol{y}$. We model this by associating a confidence interval $[y_j^\ell, y_j^u]$ for each receiver j with any $\boldsymbol{y} \in [\boldsymbol{y}^\ell, \boldsymbol{y}^u]$ (i.e. $\forall j, y_j \in [y_j^\ell, y_j^u]$) an equally possible observation vector. As before, we characterise the min ℓ_0 and min ℓ_1 solutions locally and globally within the tree. We observe that, unlike the noiseless case, the min ℓ_1 solution is no longer unique and need not always have the min ℓ_0 norm. We define **UpSparse**$^+$, a fast single-pass algorithm, which retrieves the min ℓ_1 solution out of all solutions with min ℓ_0.

5   Comparisons with CS Algorithms

Noiseless	Noisy
$\min\limits_{x} \|\boldsymbol{x}\|_0 : A\boldsymbol{x} = \boldsymbol{y}, x_i > 0 \; \forall i \qquad (4)$ $\min\limits_{x} \|\boldsymbol{x}\|_1 : A\boldsymbol{x} = \boldsymbol{y}, x_i > 0 \; \forall i \qquad (5)$	$\min\limits_{x} \frac{1}{2}\|\boldsymbol{y} - A\boldsymbol{x}\|_2^2 + \lambda\|\boldsymbol{x}\|_1 \qquad (6)$ $\min\limits_{x} \|\boldsymbol{y} - A\boldsymbol{x}\|_1 + \lambda\|\boldsymbol{x}\|_1 \qquad (7)$ $\min\limits_{x} \|\boldsymbol{y} - A\boldsymbol{x}\|_1 + \lambda\|\boldsymbol{x} - \boldsymbol{\mu}\|_1 \qquad (8)$

We now compare UpSparse to other optimisation methods from CS, of higher computational complexity, which can also find min ℓ_0/ℓ_1 solutions in trees.

In the noiseless case, the optimal sparsity optimisation problem is (4) which is non-convex. In general such problems are NP-hard. In contrast, UpSparse provides a low cost single pass algorithm, which achieves global optimality through only $O(n)$ local operations, the number of links in the tree. In the ℓ_1 case, the optimisation problem is (5) which is not only convex but a linear program. Known as the basis pursuit formulation [13], it is used as a substitute for (4). Although its solution is straightforward using linear programming, UpSparse offers a low cost direct alternative which fully exploits the underlying tree topology.

The noisy problem has been tackled using the approaches (6-8). Eq.(6) is the basis pursuit denoising formulation and the unconstrained Lasso [17] formulation. Eq. (7) is used as an alternative to (6) as it is a linear program. It is used in [9]. Eq. (8) is proposed in [9] and used in [10] to choose a solution close to a prior $\boldsymbol{\mu}$. Compared to these approaches which require the introduction of a penalty term which is traded off with the ℓ_1 norm (used to approach ℓ_0) as well as a tradeoff parameter, UpSparse$^+$ uses ℓ_0 directly, supplemented by ℓ_1 when needed. A relevant special feature here of the tree problem is that **any** noisy observation $\widehat{y}$ has a feasible solution (for e.g. the receiver solution). There is no need for regularisation in the sense of finding the closest feasible solution to $\widehat{y}$.

6   Experiments

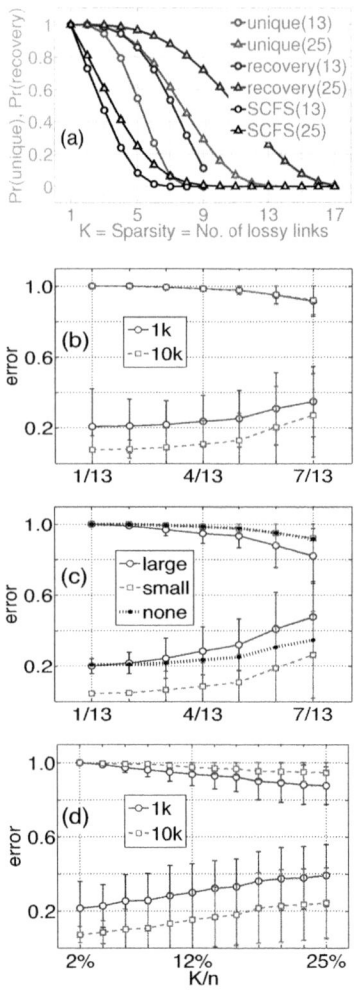

Fig. 4. Top to bottom: (a) Probabilities of uniqueness and recovery of true solutions. (b-d): Effect of measurement noise for trees as a function of K/n(sparsity/number of links). Top curves show the success rate e_0 and bottom curves show relative error e_2 (b, c): Results for UpSparse and UpSparse$^+$ solutions respectively for a ternary tree with $n = 13$ links, and (d): Results for realistic trees from AT&T network.

Uniqueness of $\min \ell_0$ *and effectiveness of* ℓ_1 *recovery in trees:* Figure 4(a) plots different probabilities of interest as a function of sparsity K by exhaustively looking at all $\binom{n}{K}$ links in trees. These are shown using two ternary trees with $n = 13$ ($m = 9$ leaves) and $n = 25$ ($m = 17$ leaves) links for $1 \leq K \leq m$. The red curves show the probability that the sparsest solution is unique. The blue curves show the probability that the $\min \ell_1$ solution is the true underlying solution. Property 6 of section 4.2 gives the condition for uniqueness of sparsest solution and 7 gives the condition that $\min \ell_1$ solution = true solution. For $K \leq 2$, $\min \ell_0$ is guaranteed to be unique since for all internal nodes, the node degree $g = 4$ (3 children, 1 parent) and $K \leq g - 2$ holds. For $K > 2$, the probabilities gradually decay with increasing number of lossy links. For $K > 2$, a vector of sparsity K need not be unique in general (lemma 1). It is unique however if corollary of property 6 is satisfied. We see that ℓ_1 minimization can recover the true solution even when the sparsest solution is not unique. Since ℓ_1 picks one of the sparsest solutions that is in upstate, when the true solution is the upstate solution, ℓ_1 minimization recovers it. It is clear that UpSparse or any ℓ_1 minimization algorithm effectively recovers the true loss hotspots in a tree, provided that only few of them exist. Next, we compare UpSparse to SCFS (smallest consistent failure set) algorithm [6], which recovers the sparsest link binary vector (each link either good/lossless or bad/lossy) given the path binary vector (each path either good or bad). The black curves show the probability that SCFS recovers the true link binary vector *i.e.* the locations of all bad/lossy links. We see that SCFS has a lower success rate than UpSparse (blue curves) even though UpSparse recovers both the locations as well as loss rates of all lossy links. The binary approach yields higher ambiguity than the loss approach. For example, at a branch node, if one of the child links and the parent link are both bad/lossy, SCFS will report only the parent as bad. However UpSparse will report both links as lossy.

Effect of measurement noise: Next we conduct probing experiments where instead of the true path probabilities only an estimate is available, based on a fixed number of probes. We compute the minimal norm solutions and plot errors as a function of both sparsity and increasing number of probes. We assume a scenario where we have no prior knowledge of where the lossy links may lie. Thus we pick K links uniformly at random. For each of these, loss is set uniformly at random from $1 - 10\%$. For the remaining links, loss is set to 0. Using $\boldsymbol{x}$, we simulate the passage of probes on each path and derive the noisy observation $\widehat{\boldsymbol{y}}$ and its confidence intervals $\widehat{\boldsymbol{y}}^{\ell}$, and $\widehat{\boldsymbol{y}}^{u}$. These are used by UpSparse and UpSparse$^+$ to yield $\widehat{\boldsymbol{x}}$. Finally we recover the true loss estimates $\widehat{\mathbf{b}} = \mathcal{L}^{-1}(\widehat{\boldsymbol{x}})$.

We compute two quantities: (*i*) relative ℓ_2 error $e_2 = \|\mathbf{b} - \widehat{\mathbf{b}}\|_2 / \|\mathbf{b}\|_2$ and (*ii*) Success rate e_0 which is the number of common lossy links between $\mathbf{b}$ and $\widehat{\mathbf{b}}$ normalised by $\|\mathbf{b}\|_0$. e_0 attempts to measure the effectiveness of UpSparse in identifying the correct locations of lossy links.

1) UpSparse Solutions: Figure 4(b) shows benchmark results using a ternary tree with $n = 13$ links and height 3. The top curves show the success rate e_0 and bottom ones show the error e_2 averaged over 100 repetitions for 1k and 10k probes when UpSparse is given the noisy $\widehat{\boldsymbol{y}}$ in each repetition. We see that even with 1k probes (blue curve), UpSparse identifies the correct locations of majority of the lossy links. As probes increase, $\widehat{\boldsymbol{y}}$ approaches $\boldsymbol{y}$ and e_2 decreases.

2) UpSparse$^+$ Solutions: Figure 4(c) shows results for UpSparse$^+$ using 1k probes for intervals of small and large sizes. In each repetition, UpSparse$^+$ is given intervals $\widehat{\boldsymbol{y}}^{\ell}$, and $\widehat{\boldsymbol{y}}^{u}$ that contain the true $\boldsymbol{y}$. When interval sizes are large, UpSparse$^+$ can find solutions which are even sparser than $\boldsymbol{x}$ resulting in higher error. As intervals get narrower, UpSparse$^+$ gives better results. The curve in the centre shows results when the actual noisy observation is used by UpSparse.

3) Large Realistic trees: Finally 4(d) shows results for real tree topologies cut out from the publicly available router level map of the AT&T network obtained by Rocketfuel [18], with about 48 links on average per tree. The figure shows results for solutions computed using UpSparse$^+$ for 1k and 10k probes when $\widehat{\boldsymbol{y}}^{\ell}$, and $\widehat{\boldsymbol{y}}^{u}$ are t-distributed intervals for 90% confidence, centered around $\widehat{\boldsymbol{y}}$. We see that e_2 increases as expected as large trees will need more probes to get accurate path probabilities. However success rate e_0 remains high implying that minimal norm solutions identify the correct locations of majority of the lossy links.

7   Conclusion

The sparsity principle has been mostly used as a black box in loss tomography without detailed characterization of conditions under which minimal norm solutions are unique or recover the true underlying solution. These conditions are important in practice as network operators wish to know when sparsity could be used to accurately localise hotspots using few monitoring points.

In this work, we study the problem of loss hotspot localization in tree topologies (e.g. server-based measurements) using the principle of sparsity. We derive explicit solutions and fast algorithms for both min ℓ_0 and min ℓ_1 norms that give deep insight into the nature of sparsity in trees. We provide conditions under which minimal norm solutions are unique and when they recover the true underlying solution. We show that when lossy links are well separated, sparse solutions remain unique in many cases. We conduct experiments to measure the ability of the minimally sparse solution to approach the actual sparse solutions in practice. We see that minimal norm solutions can identify the locations of most lossy links, however as their number increases it becomes much harder to identify true loss rates. We also observe that minimally sparse link loss solutions can localize hotspots better than minimally sparse link binary solutions used in prior work.

Future work will extend our work to graphs, study recovery conditions for the binary performance problem, and test our results with real measurements.

References

1. Padmanabhan, V.N., Qiu, L., Wang, H.J.: Server-based inference of internet link lossiness. In: IEEE INFOCOM, pp. 145–155 (March 2003)
2. Candes, E., Tao, T.: Decoding by linear programming. IEEE Transactions on Information Theory 51(12), 4203–4215 (2005)
3. Candés, E.J., Romberg, J., Tao, T.: Robust uncertainty principles: exact signal reconstruction from highly incomplete frequency information. IEEE Transactions on Information Theory 52(2), 489–509 (2006)
4. Natarajan, B.K.: Sparse approximate solutions to linear systems. SIAM Journal on Computing 24(2), 227–234 (1995)
5. Arya, V., Veitch, D.: Sparsity without the complexity: Loss localisation using tree measurements. CoRR abs/1108.1377 (2011), http://arxiv.org/abs/1108.1377
6. Duffield, N.G.: Network tomography of binary network performance characteristics. IEEE Transactions on Information Theory 52(12), 5373–5388 (2006)
7. Chen, Y., Bindel, D., Song, H.H., Katz, R.H.: An algebraic approach to practical and scalable overlay network monitoring. In: ACM SIGCOMM, pp. 55–66 (August 2004)
8. Chua, D., Kolaczyk, E., Crovella, M.: Network Krigging. IEEE Journal on Selected Areas in Communications 24(12), 2263–2272 (2006)
9. Song, H.H., Qiu, L., Zhang, Y.: Netquest: a flexible framework for large-scale network measurement. IEEE/ACM Trans. Netw. 17(1), 106–119 (2009)
10. Ghita, D., Nguyen, H., Kurant, M., Argyraki, K., Thiran, P.: Netscope: practical network loss tomography. In: IEEE INFOCOM, pp. 1262–1270 (March 2010)
11. Adams, A., Bu, T., Caceres, T., Duffield, N.G., Friedman, T., Horowitz, J., Lo Presti, F., Moon, S.B., Paxson, V., Towsley, D.: The use of End-to-end Multicast Measurements for Characterising Internal Network Behavior. IEEE Communications Magazine 38(5), 152–158 (2000)
12. Donoho, D.L.: Compressed sensing. IEEE Transactions on Information Theory 52(4), 1289–1306 (2006)
13. Chen, S.S., Donoho, D.L., Saunders, M.A.: Atomic decomposition by basis pursuit. SIAM Rev 43(1), 129–159 (2001)

14. Firooz, M.H., Roy, S.: Network tomography via compressed sensing. In: IEEE GLOBECOM, pp. 1–5 (December 2010)
15. Indyk, P.: Sparse Recovery Using Sparse Random Matrices. In: López-Ortiz, A. (ed.) LATIN 2010. LNCS, vol. 6034, pp. 157–157. Springer, Heidelberg (2010)
16. Xu, W., Mallada, E., Tang, A.: Compressive sensing over graphs. In: IEEE INFO-COM, pp. 2087–2095 (March 2011)
17. Tibshirani, R.: Regression shrinkage and selection via the lasso. Journal of the Royal Statistical Society Series B Methodological 58, 267–288 (1996)
18. Spring, N., Mahajan, R., Wetherall, D., Anderson, T.: Measuring ISP topologies with Rocketfuel. IEEE/ACM Trans. Netw. 12(1), 2–16 (2004)

Efficient and Secure Decentralized Network Size Estimation

Nathan Evans[1], Bartlomiej Polot[2], and Christian Grothoff[2]

[1] Symantec Research Labs,
nathan_evans@symantec.com
[2] Free Secure Network Systems Group,
Network Architectures and Services,
Technische Universität München,
{bart,grothoff}@net.in.tum.de

Abstract. The size of a Peer-to-Peer (P2P) network is an important parameter for performance tuning of P2P routing algorithms. This paper introduces and evaluates a new efficient method for participants in an unstructured P2P network to establish the size of the overall network. The presented method is highly efficient, propagating information about the current size of the network to all participants using $O(|E|)$ operations where $|E|$ is the number of edges in the network. Afterwards, all nodes have the same network size estimate, which can be made arbitrarily accurate by averaging results from multiple rounds of the protocol. Security measures are included which make it prohibitively expensive for a typical active participating adversary to significantly manipulate the estimates. This paper includes experimental results that demonstrate the viability, efficiency and accuracy of the protocol.

Keywords: Peer-to-Peer, protocol design, network security.

1   Introduction

Individual peers participating in unstructured networks, such as Peer-to-Peer (P2P) networks, ad-hoc wireless networks and sensor networks, can benefit from knowing the size (total number of participants) of the network. Peers in unstructured P2P networks which know the network size can make intelligent decisions with respect to content replication, message routing and forwarding and the overall cost of operations. Additionally, nodes in a sensor network can use such data to gauge the health of the overall network, calculate on/off time to save energy, selectively route messages, and generate alerts.

This paper describes the design, implementation and experimental results of a protocol that provides all peers in a structured or unstructured P2P network with an accurate estimate of the total number of peers in the network. The primary motivation for our work are the requirements of various P2P routing protocols [4,6,12]; these protocols explicitly require a network size estimate to tune parameters responsible for routing, path selection and content replication.

R. Bestak et al. (Eds.): NETWORKING 2012, Part I, LNCS 7289, pp. 304–317, 2012.

The focus of our design is to provide security in the context of an open and completely decentralized network architecture. While it would be possible to strengthen the security of our design with trusted centralized services — for example by preventing a Sybil-attack with a centralized registration requirement — our design does not require a centralized authority and is intended to provide security in the presence of actively participating adversaries. A key difference to existing proposals is that in our design there are no peers with special roles in the process; this eliminates the possibility of malicious peers abusing such roles.

A central goal of our design — which is generally not satisfied by many other network size estimation algorithms [1,11,14] — is that all peers are supposed to participate in calculating a network size estimate at roughly the same time and obtain the same result. This is achieved by a controlled flood of the network with the size estimation information, costing $O(|E|)$ messages per round. In practice, the constant factor is typically between one and two; in other words, the algorithm can be expected to only generate $|E|$ messages per round.

The basic idea behind our algorithm is to flood the network with the identity of the peer whose identity is closest to a particular key T. Each peer's identity is generated when the peer starts the first time. The key T is not chosen by any peer but instead is generated from the start time S of the current round. Despite using time, we specifically do not assume that the clocks in the network are closely synchronized; our protocol ensures that in the worst case individual peers with significant clock skew only cause a bounded amount of additional network traffic.

Our protocol considers many other important networking issues as well. The protocol is very efficient as it requires only $O(1)$ state per peer and does not require peers to establish new connections (we assume the network graph is connected). The amount of work required by each node is based only on the number of edges of the node, so the load between peers is typically balanced. Given that our protocol floods the network with size estimation information, peers randomly delay messages to avoid spikes in network traffic. Finally, our design handles network churn well, and allows the system designer to trade-off computational efficiency for security and bandwidth for accuracy.

The remainder of the paper is structured as follows. In Section 2, we present related work on methods for network size estimation for P2P networks. Section 3 then presents our protocol. In Section 4 we analyze the security aspects of the protocol. Experimental results obtained using large-scale emulation are given in Section 5.

2   Related Work

Algorithms for estimating the size of a P2P network can be categorized into algorithms for structured overlays, which typically exploit statistical properties of an existing routing table from a DHT, and algorithms for unstructured overlays, which make no assumptions about the structure of the underlying network.

2.1   Network Size Estimation for Structured Overlays

Structured overlays construct routing tables at each peer according to particular rules that enable efficient routing of messages to the peer with the "closest" identifier with respect to a given key [15,18]. In these structured overlays, the distance to the nearest neighbors in the routing table can be used as a first network size estimate as it correlates with the network size [17].

As node identifiers are often not perfectly uniform, searching the structured overlay for the closest node to various randomly selected keys can be used to get accurate network size estimates [17]. Given a DHT routing algorithm with a typical cost of $O(\log n)$, network size estimation for all nodes using this method would be $O(n \log n)$. When compared to the method presented in this paper, a key disadvantage of existing methods for structured networks is that they rely on the security of the underlying routing algorithm; actively participating malicious nodes have thus the potential to significantly skew the network size estimate. Furthermore, for any of the structured methods that we are aware of, different nodes will virtually always compute somewhat different network size estimates.

2.2   Network Size Estimation for Unstructured Overlays

Several algorithms for unstructured overlays are based on sampling. Examples include Sample & Collide, Random Tour and Hops Sampling. In Sample & Collide [14], each peer starts bounded random walks to sample random peers in the network and uses the collision information and the birthday paradox to estimate the size of the network. In Random Tours [14], a message tours the network until it reaches the initiator; the size estimate is then computed based on a counter in the message that was incremented by each peer on the tour. Hops Sampling [11] works by flooding the network with a message containing a hop count. Peers report back to the initiator with a probability inverse to the hop count they received. The network size estimate is then the sum over all distances of the number of replies received for a particular distance divided by the reply-probability for that distance.

As described, these methods generate results for just one peer in the network, resulting in a high amount of bandwidth used overall (assuming each peer requires an estimate). Also, different peers will have potentially significantly different network size estimates. While these approaches do not assume a particular network structure for routing, they do still make implicit assumptions about the structure of the overlay topology and may significantly underestimate the network size if the overlay topology happens to have a structure that is unfavorable to the algorithm. For example, a circular topology would result in a network size estimate of n^2 for Sample & Collide.

Other algorithms, such as Gossip-based Aggregation [8], achieve a somewhat more uniform estimate for all participating nodes at the cost of sensitivity to node failures. Gossip-based Aggregation starts with one peer setting a local state to 1 while all other peers set their local state to 0. Peers continuously connect to randomly selected peers, and exchange states in pairs. Each peer then replaces its state with the average of both values. After a predefined number of iterations, all peers are supposed to end up with a value close to $1/n$ where n is the size of the network. A method that addresses the problem of who should set the state to 1 has been proposed [20], but only works in certain structured networks and retains other shortcomings of this approach, including high vulnerability to malicious peers.

A special case is the method proposed in [1] which attempts to produce a network size estimate using only "local" information. The idea behind this algorithm is to observe the number of new neighbors discovered in a breadth-first search of the network and estimate the network size based on the growth of this function. The authors claim to obtain accurate results with a breadth-first search of depth three, which makes this a "local" method. However, the way they constructed the topologies for their experiments does not seem to properly model the structure of actual networks. We were unable to reproduce their results on other network topologies.

The accuracy and performance of various size estimation methods for unstructured networks are compared in [16] using simulation. The authors identify the Sample & Collide method as the strongest algorithm and state that it requires about 50 million messages in a random-graph topology of 100,000 nodes for an accuracy of $\pm 4\%$. It should be noted that this is the overhead for an individual node to obtain an estimate; if each of the 100,000 nodes were to run the Sample & Collide protocol, it would take 5 *trillion* messages to achieve this degree of accuracy.

None of these papers mention concrete implementations or discuss security concerns. Furthermore, all of them are clearly vulnerable to malicious participants. For example, in the case of sampling-based algorithms, malicious participants can manipulate walks that pass through them (allowing virtually unbounded manipulation of the network size estimates) or achieve a significant multiplier $(O(\sqrt{n}))$ to their network bandwidth in a denial-of-service attack by continuously initiating size estimation requests. Similarly, an active adversary can manipulate the exchanged values in gossip-based methods to change the size estimate in any direction.

3   Our Approach

We generate node identifiers by hashing the public key of the respective node. Node identifiers for benign nodes should therefore be statistically equivalent to random numbers from a uniform distribution. Furthermore, nodes are able to cryptographically sign messages using their respective private key.

Similar to the network size estimation algorithms for structured overlays, our network size estimation approach is based on the largest number of leading overlapping bits between any node identifier and a random key:

Theorem 1. *Let $\bar{p}$ be the expected maximum number of leading overlapping bits between all n random node identifiers in the network and a random key. Then the network size n is approximately $2^{\bar{p}-0.332747}$.*

Proof. Let X be the random variable for all n identifiers and let X_i be the number of overlapping bits for an individual random node identifier i.

The probability that a single random node identifier i overlaps with at least α bits with a random key is

$$P(X_i \geq \alpha) = 2^{-\alpha}. \tag{1}$$

Then, the probability that a single random node identifier overlaps with less than α bits with a random key is

$$P(X_i < \alpha) = 1 - 2^{-\alpha}. \tag{2}$$

The probability that the maximum number of leading overlapping bits for all n random nodes is strictly less than α is

$$P_n(X < \alpha) :=: P\left(\bigwedge_i X_i < \alpha\right) = (P(X_i < \alpha))^n = \left(1 - 2^{-\alpha}\right)^n. \tag{3}$$

Then $E_n(X)$, the expected maximum number of leading overlapping bits between n random node identifiers in the network is:

$$
\begin{aligned}
E_n(X) :=: \sum_{\alpha=0}^{\infty} \alpha \cdot P_n(X = \alpha) &= \sum_{\alpha=1}^{\infty} P_n(X \geq \alpha) \\
&= \sum_{\alpha=1}^{\infty} (1 - P_n(X < \alpha)) = \sum_{\alpha=1}^{\infty} \left(1 - \left(1 - 2^{-\alpha}\right)^n\right) \\
&= \sum_{\alpha=1}^{\log_2 n} \left(1 - \left(1 - 2^{-\alpha}\right)^n\right) + \sum_{\alpha=\log_2 n+1}^{\infty} \left(1 - \left(1 - 2^{-\alpha}\right)^n\right)
\end{aligned}
$$

Suppose n is sufficiently large such that we can use $\lim_{n\to\infty}(1 - \frac{x}{n})^n = e^{-x}$. By substituting $\beta := \alpha - \log_2 n$ and $\gamma := \log_2 n - \alpha$ we then get:

$$
\begin{aligned}
E_n(X) &= \log_2 n - \sum_{\gamma=0}^{\log_2 n-1} \left(1 - 2^{\gamma-\log_2 n}\right)^n + \sum_{\beta=1}^{\infty} \left(1 - \left(1 - 2^{-(\beta+\log_2 n)}\right)^n\right) \\
&= \log_2 n - \sum_{\gamma=0}^{\log_2 n-1} \left(1 - \frac{2^{\gamma}}{n}\right)^n + \sum_{\beta=1}^{\infty} \left(1 - \left(1 - \frac{2^{-\beta}}{n}\right)^n\right) \\
&\approx \log_2 n - \sum_{\gamma=0}^{\log_2 n-1} e^{-2^{\gamma}} + \sum_{\beta=1}^{\infty} \left(1 - e^{2^{-\beta}}\right) \\
&\approx \log_2 n - 0.521865 + 0.854613 = \log_2 n + 0.332747
\end{aligned}
$$

Thus, for sufficiently large values of n,

$$E_n(X) \approx \log_2 n + 0.332747. \tag{4}$$

$\square$

Given Theorem 1, the key remaining challenge is thus to efficiently and securely find a closest node identifier (with distance measured in terms of leading overlapping bits) to a random key in an unstructured network.

In our design, all nodes in the network periodically participate in a global network size estimation operation at a frequency of f. Each round results in all peers learning a discrete approximation p (the number of overlapping leading bits for a particular random key) for $\bar{p}$ (the theoretically expected number of overlapping leading bits). The specific frequency f is chosen based on the expected level of network churn and the desired accuracy. f is a design parameter and fixed in the implementation. The results from the last k iterations are averaged locally by each peer to obtain an approximation $\tilde{p}$ for $\bar{p}$. A standard deviation can also be computed if an estimate for the error of the size estimate is desired. Furthermore, the current p value is used by the protocol as a parameter to (slightly) improve the performance for the next round. We will refer to the number of overlapping leading bits from the previous round as p'.

3.1   Generating a Random "Key"

Given a frequency f, the random target key T for each round is generated by hashing the start time S, which is the absolute UTC time at times that are zero modulo f. For example, if $f = 1h$, then a fresh key could be generated every hour by hashing "DD-MM-YYYY HH:00:00". Using this method, all peers will generate exactly the same key at (roughly) the same time. Generating the key this way has the advantage that it will be known to all peers without communication and that malicious participants cannot influence the process. However, it should be noted that while the keys satisfy the statistical properties of being random and uniform, it is trivial to compute them in advance.

Our method requires all peers to calculate the current key T at the respective start time S. The network size estimation protocol's goal is to communicate to all peers an identity I_T of a peer with the largest proximity p with respect to T. More specifically, all peers are supposed to learn one of the closest peer identities I_T between time S and time $S + f$. Given I_T, each peer can then calculate p, the average $\tilde{p}$ of the p-values from the last k rounds and finally the current network size estimate $2^{\tilde{p}-0.332747}$.

Note that p is a discrete value representing the number of leading matching bits between the key T and a peer's identity. As such, it is quite likely that many peers have identities with the same number of leading matching bits and hence the same proximity p. Our protocol deliberately ignores all bits after the first mismatch to improve performance; if multiple peers have the same proximity score, it does not matter which of these equivalent identities is propagated as they will all ultimately result in the same proximity estimate p.

3.2   Starting the Flood

Our protocol essentially floods the network with the identity of a closest peer I_T. If only the identities of closest peers are propagated, this operation would create less than $2|E|$ messages (up to two per edge in the network). The challenge is to avoid creating significantly more than $2|E|$ messages, which is difficult since in an unstructured network a peer with the closest identity I_T cannot be certain that there is no other peer that is closer to T. We address this problem by delaying the flood based on proximity.

First, each peer evaluates its own proximity x with respect to T. How close the peer is to T is then used to determine how soon the peer will initiate the flood of the network with a message claiming that he is the closest peer. We use the previous network size estimate as a guide to time the release. Specifically, given a proximity of x leading overlapping bits and a network size estimate p' from the previous round, we use the following function to determine the time when a peer starts to flood the network:

$$r(x) := S + \frac{f}{2} - \frac{f}{\pi} \cdot \arctan\left(x - p'\right) \qquad (5)$$

Using this function, if the peers proximity x is equal to the proximity of the last iteration, the peer floods the network at time $S + \frac{f}{2}$. If the peer's proximity is 0 bits, the peer floods the network close to time $S + f$, which is at the end of the time interval for the current round; if the peer's proximity is significantly higher than p', the peer floods at the beginning of the round, which is close to time S. It should be noted that since $\lim_{x \to p'} \frac{\partial r}{\partial x}(x) = 1$, Equation 5 maximizes the difference between release times for nodes with proximities that are close to the previous number of overlapping bits p' and as such minimizes the chance of two peers releasing floods for different network size estimates around the same time — assuming the network size estimate did not change significantly.

3.3   Processing the Flood

Peers that receive the resulting network size estimation messages first perform a series of validation steps before continuing to forward the message. First, each peer checks if a notification from a closer peer has already been received for round S. Messages with proximity scores equal to the currently known best score for the current round are simply discarded. Messages with lower proximity scores should only occur if there is significant clock skew, and are answered immediately with a message indicating the higher proximity score. If the message contains a higher proximity than what was previously known for the current round, the peer checks if the proximity p of the given message justifies receiving it at the current time. If not, further processing is delayed until the local peer's time is past $r(p)$. Finally, before forwarding and further processing, the format of the message is validated (this is discussed in more detail in Section 3.5).

Assuming the message validates, the peer then proceeds to forward it to all of its neighbors. For each neighbour, the message is forwarded with a peer-specific random delay. If a peer receives a message with an equivalent proximity score during the delay, the transmission is canceled. As a consequence, the delay helps to both avoid an explosion of messages on the network in a tiny amount of time, and to improve the chances of traversing each edge in only one direction per link (as it decreases the chances of equivalent messages being sent in both directions at the same time).

The permissible delay L is calculated using the time difference between $r(p)$ and $r(p-1)$ divided by an estimate of the network's diameter. The network diameter D is estimated using the maximum of the hop counters in the network size estimation messages from the previous k iterations. For each neighbor, the peer then applies a delay chosen uniformly at random from the interval $[0, L)$ where L is defined as

$$L := \frac{r(p-1) - r(p)}{D}. \tag{6}$$

As a result, each peer in the network is expected to receive a proximity notification with proximity p before any peer with notifications for proximity $p-1$ would even begin to flood the network. Naturally, there are various causes that could increase the number of messages above $|E|$ in a real-world network; for example, different system times between peers, high network latencies, and peers with unusually high distances to I_T can increase the total number of messages. However, all benign peers that form a connected component are guaranteed to eventually receive I_T. Furthermore, given that the number of bits in the key is a small constant, the total number of messages can never exceed $O(|E|)$ per round.

3.4 Joining the Network

A peer that is freshly joining the network lacks results from previous rounds for network size estimation. In order to bootstrap the protocol, each peer starts with a network size estimate based on its own key in relation to the key T from the previous round, and a network diameter estimate of one. Whenever a connection between two peers is established, they exchange the network size estimation result from the previous round (and the current round if their local time is past $f(p)$). As a result, all nodes can always be expected to use the same value for p' in Equation (5).

3.5 Proof of Work

The presented design is vulnerable to an adversary that creates fake identities (Sybil attack [2]). Such an adversary could create identities that are "close" to the respective key for each time $S + \mathbb{Z}r$. By flooding the network with the respective messages at the right time, the adversary can then make the network appear to be larger than it is.

Our design defends against this attack by requiring a proof of work [19] for the identity of the peer as part of the network size estimation message. Specifically, we require the originator to produce a value with a W-bit hash collision with the peer's identifier, and a cryptographic signature to demonstrate that the identifier was derived from a valid public-private key pair.

The complete message format for the network size estimation messages is described in our technical report [7].

4   Security Analysis

For our security analysis, we assume that an active adversary is participating in the P2P network. The adversary is allowed to control a certain percentage of colluding malicious nodes in the network. Individual malicious and benign nodes are assumed to have the same amount of computational resources; all nodes are assumed to have sufficient bandwidth to participate in the protocol in the absence of an attack.

We can imagine three different high-level goals an adversary may pursue with an attack. First, an adversary may try to cause nodes to significantly underestimate the size of the network. Second, an adversary may try to cause nodes to significantly overestimate the size of the network. Finally, an adversary may want to use the protocol for a denial-of-service attack where the P2P network uses significantly more traffic for network size estimation, possibly causing other components of the system to be left with insufficient bandwidth.

The best method for an adversary to cause peers to underestimate the size of the network is to not participate in the protocol. If the adversary controls $X\%$ of the network, that will cause the protocol to underestimate the size of the network by $X\%$. Furthermore, if the adversary is able to control an ϵ-separator of the network graph [9,10,13] (removing an ϵ-separator from a graph reduces the size of the largest remaining connected component to ϵn), then the overall network size estimate would be reduced to less than ϵn for all nodes in the network. Given that in all of these cases the network size estimate would correspond to the size of the network after the removal of the adversaries' nodes, this attack is not particularly disruptive in relation to the strength of the adversary. Thus, an adversary cannot make the network appear significantly smaller than it is.

If the adversary wants to make the network look larger by M nodes, it needs to first compute (and store) M public-private key pairs. Then, at every time interval f, the adversary needs to compute collisions costing an additional $O(2^W)$ to generate the required W-bit collision. Actually joining the network with M "fake" peers is not required. If W is chosen so large that the adversary cannot solve the problem at frequency f, it is still possible for the adversary to cause an increase in the network size estimate by solving the problem every $c \cdot f$ (for an appropriate choice of c based on the adversaries computational resources), which would still affect the computed medium-term averages computed for subsequent intervals. Using such an attack, an adversary can make the network appear significantly larger than it is, as long as the adversary has access to sufficient computational resources.

Finally, for a denial-of-service attack, an adversary would first generate additional identities and generally perform the same steps as for increasing the estimated network size. Now, suppose the adversary has created m identities that are closer to the current key than the closest actual peer in the network by $1 \ldots m$-bits respectively. Then, just after the identity of the peer that is actually closest to the key has been broadcast to the network, the adversary can cause m additional broadcasts by transmitting its m "fake" identities in order of increasing proximity to the key. Each time, the network will presume that a closer peer was "late" with its transmission (for example, due to clock-skew or network latency) and broadcast the update. If the network is already of size n, the expected one-time cost for the adversary to create m such identities is $O(n2^m)$; the attack then requires an additional $O(m2^W)$ operations for the hash collisions at frequency f. Therefore, if we neglect the high one-time cost of computing identities, the adversary can cause $|E|$ traffic on the network at the cost of $O(2^W)$ computations.

Analytical Worst-Case Analysis

The following scenario describes the theoretical worst case in terms of bandwidth consumption by the protocol. Without loss of generality, suppose a 512-bit hash function is being used. Then, the worst-case network would for $\mu \in [1, 512]$ have exactly one peer with μ matching bits with the target key T (in each round); all other peers in the network would have zero matching bits. The peers that do have matching bits should be connected to the main network via a long chain (with larger distances for peers with larger μ), causing the network diameter D to be large. (As a result, the algorithm will calculate $L \approx 0$.) Peers with μ matching bits should furthermore have (or pretend to have) a late system time that causes them to transmit effectively at time $S + \mu\epsilon$; all other peers have fast clocks that cause them to accept any message at any time, causing 512 network-wide floods per round. Furthermore, in the worst case, network delays in the main network would be so large that each of the 512 messages would traverse each link in both

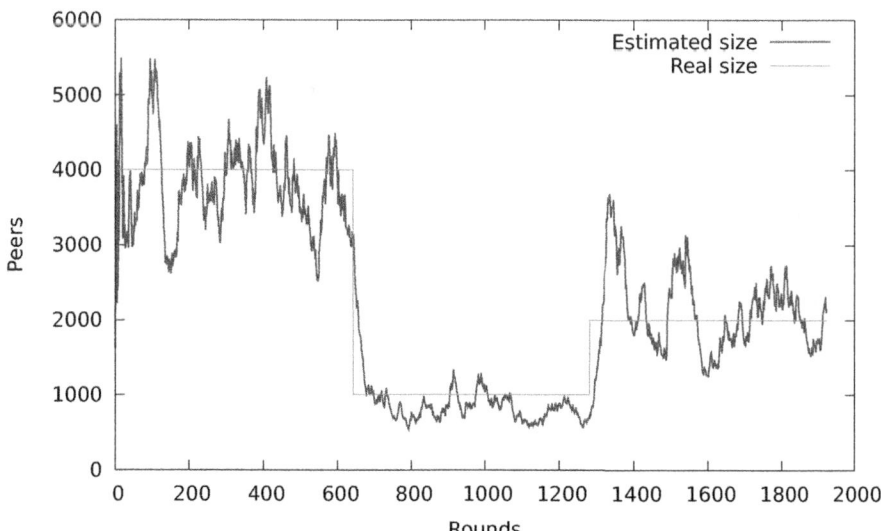

Fig. 1. Network size estimates as actually observed by a node in relation to the (changing) total network size over time for a random graph topology. All nodes arrive at the same estimate, except for nodes that recently joined the network.

directions, creating a total of $1024 \cdot |E| \in O(|E|)$ transmissions. Note that this scenario covers the worst-case and includes an adversary with infinite computing power and full control over the network topology.

5   Experimental Results

We have implemented the presented protocol in the GNUnet P2P framework[1], and evaluated the behavior of the proposed protocol using large-scale emulation [5].

To evaluate the network size estimation quality, we show the network size estimate based on the average of the previous 64 rounds. Figure 1 shows the evolution of the network size estimate for a random graph topology [3,5] with a minimum node degree of 5 and an average node degree of 10 for the 4,000 node network. It should be noted that the shape of the network topology has no impact on the size estimate. The experiment was started with an initial network size of 4,000 nodes for 640 rounds. Then, we decreased the network size to 1,000 nodes for 640 rounds (the remaining peers stayed connected) and finally increased it to 2,000 nodes for another 640 rounds.

The number of rounds used to calculate the result has an impact on the precision of the estimate. The trade-off between more measurements and the resulting precision is plotted in Figure 2. Precision is measured as $|\tilde{p} - \overline{p}|$. Averaging over

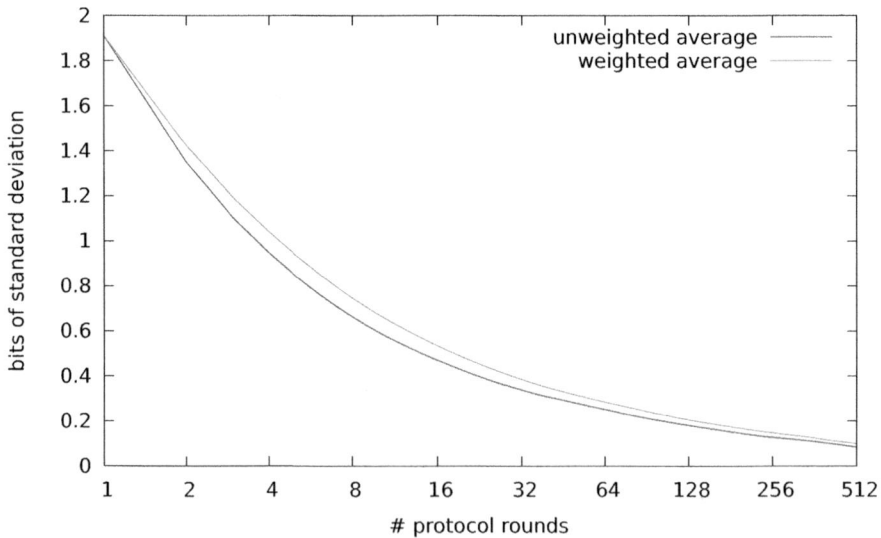

Fig. 2. Trade-off between the precision of the network size estimate vs the number of rounds used to calculate $\tilde{p}$. Naturally, this plot assumes that the network size does not change during the measurement.

---

[1] https://gnunet.org/svn/gnunet/src/nse/

four rounds gives results with a standard deviation of one. As the network size is calculated as $2^{\bar{p}-0.332747}$ (Theorem 1), a standard deviation of one means that the network size estimate is in an interval between half and double the actual network size 68% of the time and between a quarter and four times the actual network size 95% of the time. The 64 rounds we used for Figure 1 correspond to a standard deviation of under 0.3. This means that 95% of the time the network size estimate is accurate up to a factor of ≈ 1.5.

In the proof for Theorem 1 we made an approximation that is valid if "n is sufficiently large". However, what constitutes a sufficiently large n in practice is not obvious. Figure 3 shows the results of a simulation that determined $\tilde{p}$ from 50,000 rounds for networks of size $n \in [1, 2^{24}]$. The difference $\tilde{p} - \log_2 n$ quickly converges to the constant calculated in Theorem 1 (0.332747). It should be noted that even with 50,000 rounds the values for $\tilde{p}$ still exhibit some visible fluctuation. Figure 3 shows that for a reasonable number of rounds of measurement ($\leq$ 50,000), the "sufficiently large values of n" are values larger than 2^5.

Additional experimental results, such as an analysis of the impact of the network topology and clock skew on bandwidth consumption, can be found in our technical report [7]. The key result from those experiments is that the protocol

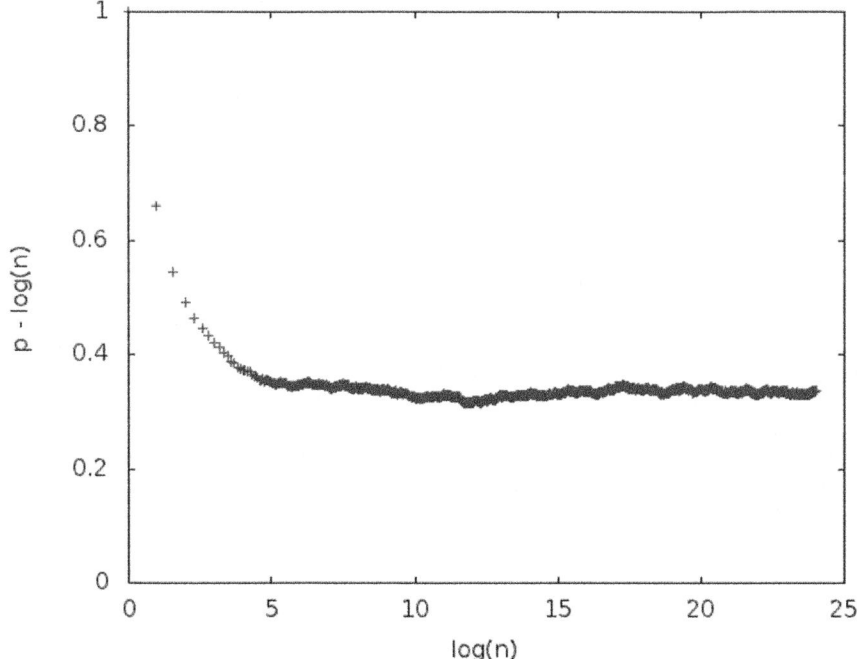

Fig. 3. Differences observed between $\log_2 n$ and the average observed value for p over 50,000 iterations in relation to the network size. The average difference was $0.33 \approx 1/3$. Since peer-to-peer networks smaller than $2^5 = 32$ peers are not really relevant for network size estimation techniques, we use a uniform correction of $1/3$ to compensate for the observed difference when estimating n from $\bar{p}$.

works for any topology, tolerates clock skew and under realistic conditions uses fewer than $2|E|$ messages per round.

6   Conclusion

We have presented the first protocol for securely and efficiently estimating the size of a P2P network. Our protocol combines proximity to a deterministic sequence of (pseudo-)random values, staggered triggering of messages and a proof-of-work component. The scheme works for structured and unstructured networks, is inexpensive in terms of bandwidth, perfectly distributed imposing equal requirements in terms of computation and bandwidth on all nodes, and is quite accurate even for networks under churn. The protocol is secure against adversaries trying to make the network appear smaller and makes it computationally expensive (based on a parameter W) to make the network appear larger or to flood the network with unwarranted traffic.

Acknowledgments. This work was funded by the Deutsche Forschungsgemeinschaft (DFG) under ENP GR 3688/1-1. We thank Mikhail Atallah for his help proving Theorem 1 and Christopher Wolf for an insightful discussion on an earlier draft of this paper.

References

1. Bustos-Jimenez, J., Bersano, N., Schaeffer, S.E., Piquer, J.M., Iosup, A., Ciuffoletti, A.: Estimating the size of peer-to-peer networks using lambert's w function. In: Gorlatch, S., Fragopoulou, P., Priol, T. (eds.) Grid Computing, pp. 61–72. Springer, US (2008)
2. Douceur, J.R.: The Sybil Attack. In: Druschel, P., Kaashoek, M.F., Rowstron, A. (eds.) IPTPS 2002. LNCS, vol. 2429, pp. 251–260. Springer, Heidelberg (2002)
3. Erdős, P., Rényi, A.: On random graphs. I. Publ. Math. Debrecen 6, 290–297 (1959)
4. Eugster, P.T., Guerraoui, R., Handurukande, S.B., Kouznetsov, P., Kermarrec, A.-M.: Lightweight probabilistic broadcast. ACM Trans. Comput. Syst. 21, 341–374 (2003)
5. Evans, N., Grothoff, C.: Beyond simulation: Large-scale distributed emulation of p2p protocols. In: 4th Workshop on Cyber Security Experimentation and Test (CSET 2011). USENIX Association (2011)
6. Evans, N., Grothoff, C.: R5n: Randomized recursive routing for restricted-route networks. In: 5th International Conference on Network and System Security. IEEE, Milan (2011)
7. Evans, N., Grothoff, C., Polot, B.: Efficient and secure decentralized network size estimation. Technical report, Technische Universität München (2011)
8. Jelasity, M., Montresor, A., Babaoglu, O.: Gossip-based aggregation in large dynamic networks. ACM Trans. Comput. Syst. 23, 219–252 (2005)
9. Kernighan, B.W., Lin, S.: An Efficient Heuristic Procedure for Partitioning Graphs. The Bell System Technical Journal 49(1), 291–307 (1970)
10. Kleinberg, J., Sandler, M., Slivkins, A.: Network failure detection and graph connectivity. SIAM J. Comput. 38, 1330–1346 (2008)

11. Kostoulas, D., Psaltoulis, D., Gupta, I., Birman, K., Demers, A.: Decentralized schemes for size estimation in large and dynamic groups. In: Proceedings of the Fourth IEEE International Symposium on Network Computing and Applications, pp. 41–48. IEEE Computer Society, Washington, DC (2005)

12. Malkhi, D., Naor, M., Ratajczak, D.: Viceroy: a scalable and dynamic emulation of the butterfly. In: PODC 2002: Proceedings of the Twenty-First Annual Symposium on Principles of Distributed Computing, pp. 183–192. ACM, New York (2002)

13. Marx, D.: Parameterized graph separation problems. Theor. Comput. Sci. 351, 394–406 (2006)

14. Massoulié, L., Le Merrer, E., Kermarrec, A.-M., Ganesh, A.: Peer counting and sampling in overlay networks: random walk methods. In: Proceedings of the Twenty-Fifth Annual ACM Symposium on Principles of Distributed Computing, PODC 2006, pp. 123–132. ACM, New York (2006)

15. Maymounkov, P., Mazières, D.: Kademlia: A Peer-to-Peer Information System Based on the XOR Metric. In: Druschel, P., Kaashoek, M.F., Rowstron, A. (eds.) IPTPS 2002. LNCS, vol. 2429, pp. 53–65. Springer, Heidelberg (2002)

16. Le Merrer, E., Kermarrec, A.-M., Massouli, L.: Peer to peer size estimation in large and dynamic networks: A comparative study. In: 15th IEEE International Symposium on High Performance Distributed Computing 2006, pp. 7–17 (2006)

17. Polot, B.: Adapting blackhat approaches to increase the resilience of whitehat application scenarios. Master's thesis, Technische Universität München (2010)

18. Rowstron, A., Druschel, P.: Pastry: Scalable, Decentralized Object Location, and Routing for Large-Scale Peer-to-Peer Systems. In: Guerraoui, R. (ed.) Middleware 2001. LNCS, vol. 2218, pp. 329–350. Springer, Heidelberg (2001)

19. Serjantov, A., Lewis, S.: Puzzles in p2p systems. In: 8th CaberNet Radicals Workshop, Corsica (2003)

20. Shafaat, T.M., Ghodsi, A., Haridi, S.: A Practical Approach to Network Size Estimation for Structured Overlays. In: Hummel, K.A., Sterbenz, J.P.G. (eds.) IWSOS 2008. LNCS, vol. 5343, pp. 71–83. Springer, Heidelberg (2008)

A Panoramic View of 3G Data/Control-Plane Traffic: Mobile Device Perspective

Xiuqiang He[1], Patrick P.C. Lee[2], Lujia Pan[1], Cheng He[1], and John C.S. Lui[2]

[1] Noah's Ark Lab, Huawei Research, China
[2] Dept of Computer Science & Engineering,
The Chinese University of Hong Kong, Hong Kong
{hexiuqiang,panlujia,hecheng}@huawei.com,
{pclee,cslui}@cse.cuhk.edu.hk

Abstract. Users can access the Internet via 3G/4G cellular data networks using various types of user devices (e.g., smartphones, tablets, datacards). We conduct a detailed measurement study on the impact of different device types on the data/control-plane performance of a commercial, city-wide 3G cellular data network in China. We present a methodology that correlates different data/control-plane datasets collected at different points in the network core, and identify more than 60K devices of different types per day on average. For the devices we identify, we investigate how their commonly used Internet applications and internal heartbeat mechanisms lead to distinct data/control-plane behaviors. For example, we observe that datacard devices contribute a large volume of IP traffic in the data plane, while smartphones introduce significant resource overhead in the signaling control plane. Our measurement study provides insights for network operators to strategize pricing and resource allocation for the data/control planes of their cellular data networks with regard to the market penetrations of various device types.

Keywords: mobile device traffic, data/control-plane, 3G networks, measurement.

1 Introduction

With the wide deployment of 3G/4G wide-area cellular data networks, we have witnessed the tremendous growth of mobile Internet access worldwide. There are many types of devices that enable mobile Internet access, such as smartphones, tablet computers, or datacards attached to laptops/PCs. There have been studies (e.g., [1]) in forecasting the explosive growth of mobile data traffic from these devices. Compared to traditional wireline networks, cellular networks not only have relatively limited data-plane link capacity, but also have higher control-plane overhead that increases the loading of the network core [16] and power consumption of mobile devices [19]. From the perspectives of network operators, it is necessary to understand how the traffic patterns of *different* device types influence the performance of cellular data networks in both data and control planes. Understanding this influence can shed light on how to provision network resources, and if necessary, how to provide differentiated pricing or priority services across different device types or applications.

R. Bestak et al. (Eds.): NETWORKING 2012, Part I, LNCS 7289, pp. 318–330, 2012.

In this paper, we conduct an in-depth measurement study on the impact of different device types on the data/control-plane performance of a commercial 3G UMTS network deployed in a metropolitan city in China. We analyze several large-scale datasets, including IP packets with complete payload, UMTS-compliant signaling messages, and radio resource control (RRC) logs. We identify more than 60K devices of different types per day on average, as well as different IP applications. The extensive scales of our datasets allow us to conduct comprehensive analysis. In summary, the contributions of this paper are two-fold:

- We first present a methodology that correlates all these datasets, so that we can study the interactions of data and control traffic of *each* device type. We also characterize the IP applications commonly used by different device types. Note that the formats of our datasets follow the standard 3G specifications. Thus, our methodology is also applicable for general 3G networks, and will be useful for network operators of interest to carry out network planning by collecting the same datasets within their own managed networks.
- We conduct an extensive measurement study mainly based on a 24-hour span of data/control-plane traces. We make the following observations. Even when there are only 7.5% of datacard devices, they contribute disproportionately about 46% of IP data traffic; for smartphone or tablet devices, they contribute less data traffic than datacard devices, but consume significantly high radio resource usage due to the frequent RRC connection setups/releases. Furthermore, we identify that iPhone/iPad devices have some internal heartbeat mechanisms that trigger a substantial number of RRC connection setups/releases, even though they generate minimal data traffic. Our study shows the importance of how mobile devices interact with the control plane and what kind of behavior can be expected.

The remainder of the paper proceeds as follows. Section 2 reviews related work on 3G traffic measurement. Section 3 describes how we correlate different datasets collected at different points of a 3G network. Section 4 presents our findings from the datasets. Finally, Section 5 summarizes the lessons learned from our study.

2   Related Work

There have been measurement studies focusing on in-network data traffic collected inside 3G networks. Kilpi and Lassila [14] analyze the round-trip times of TCP flow data collected at a GPRS/UMTS network. Ridoux *et al.* [22] collect data from a CDMA2000 network, and compare the similarities and differences with wireline data traffic. The DARWIN+ group [3] collects IP packets at a GPRS/UMTS network, and analyzes various issues such as TCP performance [21] and traffic anomalies [2]. The above studies mainly analyze the data-plane performance of 3G networks.

Some recent studies analyze the control-plane performance of 3G networks. Lee *et al.* [16] study the signaling overhead from the security perspective. They show that if an attacker generates data traffic following specific patterns, then a small amount of data traffic can sufficiently trigger heavy signaling overhead in the network core. On the other hand, they only analyze synthetic traces. Qian *et al.* [19] infer and analyze the RRC state transitions of user sessions using data-plane traces, and mainly focus on power consumptions of mobiles. Paul *et al.* [18] collect data packet headers and various signaling messages in a national 3G network, and analyze their temporal and spatial variations. Our work differs from [18] in two aspects. First, our work focuses on distinguishing different device types and analyzing each of their data/control-plane performance. Also, our datasets contain full IP payload information that enables us to accurately identify more user applications and infer their impact to the network.

Several studies analyze the data traffic behavior of different types of handheld devices. Maier *et al.* [17] analyze the mobile data traffic of handheld devices inside a residential DSL network, and Gember *et al.* [8] compare handheld and non-handheld devices in a campus WiFi network. In the context of 3G cellular data networks, Falaki *et al.* [6,7] study the smartphone traffic and identify the differences of user behavior based on the traces collected from a number of individual devices. Huang *et al.* [10] propose a tool called 3GTest, which runs on thousands of smartphones over a wide geographic coverage. Each 3GTest-enabled device generates probe traffic to measure the network performance. Instead of collecting data from smartphone devices, Xu *et al.* [23] analyze the IP data traffic collected inside a tier-1 network core, and study the usage behavior of a variety of smartphone applications from the operator's perspective. Note that the above studies mainly focus on the data-plane performance. Our work complements the prior studies by also analyzing the control-plane performance. Specifically, we play the role of a network operator and look into the interactions of both data-plane and control-plane traffic traces collected within a 3G network.

3   Datasets and Preprocessing

We first describe several datasets that are used for our measurement study of a 3G UMTS network. We then explain how we correlate these datasets to obtain the information needed for our analysis. Finally, we address the limitations of our datasets.

3.1   Datasets

Data/Control Traffic. We capture traffic in both data and control planes from a commercial, city-wide 3G UMTS network in China. Fig. 1 shows the simplified topology of the 3G network we consider. The data path between a *user equipment (UE)* and the Internet traverses a *base station (NodeB)*, a *Radio Network Controller (RNC)*, and the *core network (CN)*, all of which are defined in the 3GPP UMTS standard. Here, we focus on packet-switched domain of the CN, which comprises the *Serving GPRS Support Node (SGSN)* and the *Gateway GPRS Support Node (GGSN)*. In a high level, the CN can be viewed as the central point that relays all UE-Internet data traffic as well as UE-UE data traffic. A 3G data network typically has a hierarchical architecture, i.e., multiple UEs communicate with a NodeB, multiple NodeBs communicate with an RNC, and multiple RNCs communicate with the CN.

We capture traffic in the integrated interface (a.k.a. *Iu-PS*) that connects 16 RNCs and the CN. We collect a 7-day span of traces from November 25, 2010 to December 1, 2010. The traces have a total size of around 13TB with about 27.6 billion packets and

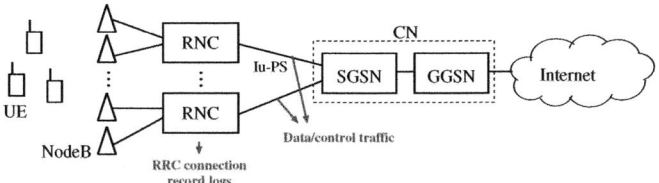

Fig. 1. A simplified architectural view of a 3G UMTS network

Table 1. Summary of categories of applications

Category	Description
Web browsing	HTTP-based web browsing activities
IM	instant messaging applications such as Yahoo Messenger, MSN, Tencent QQ, etc.
File access	audio/video files, software download/updates, FTP, etc.
Streaming	online video services such as YouTube, PPLive, QQLive, Tudou, etc.
Email	email services such as IMAP, POP3, SMTP, etc.
P2P	file sharing via P2P software such as BitTorrent, Thunder, etc.
Network Admin	network protocols such as DHCP, DNS, MDNS, NetBios, NTP, etc.
Tunneling	tunneling protocols such as Socks4, Socks5, SSL
Others	other successfully identified applications

383 million 5-tuple flows, including IP data packets with headers and full payloads and the *Radio Access Network Application Part (RANAP)* signaling messages. RANAP is the UMTS signaling protocol for the radio connection between a UE and the CN.

As shown in Section 4.1, we note that the data traffic patterns are very similar across each day over the 7-day span. Thus, in this paper, we focus on the 24-hour span of traces collected on November 28, 2010. The 24-hour traces have a total size of around 1.9TB with about 4.10 billion packets and 56.5 million 5-tuple flows.

Given that our captured data packets contain full payloads, we apply a deep packet inspection (DPI) module to identify the application protocols. The DPI module has been available as part of a commercial product [11]. It maintains a database of signatures of hundreds of applications, and maps each 5-tuple flow to an IP application based on the pre-defined payload signatures. Table 1 summarizes the top application categories that we focus on in our analysis.

RRC Connection Record Logs. A UE sends (receives) data to (from) the Internet via a *Radio Resource Control (RRC)* connection. An RNC keeps the information for every RRC connection that we refer to as an *RRC record*. We collect the log files of the RRC records from all the 16 RNCs in the network (see Fig. 1). The log files span the same period as our data/control traffic trace, and account for a total of 168 million records with total size around 28GB from November 25, 2010 to December 1, 2010. From the RRC records, we then extract the useful fields including: (i) *International Mobile Subscriber Identity (IMSI)*, the unique identification associated with each GSM- or UMTS-based UE, (ii) *type allocation code (TAC)*, the code that uniquely identifies the UE device type and is composed of the initial 8-digit portion of the 15-digit *International Mobile Equipment Identity (IMEI)* code, (iii) RNC-ID, which uniquely identifies an RNC, and

Table 2. Summary of device types

Device type	Description
iPhone	smartphones developed by Apple
Symbian	smartphones using Symbian as the operating system
Windows mobile	smartphones using Windows mobile as the operating system
Blackberry	e-mail and smartphone devices developed by RIM
Android	smartphones using Android as the operating system
Bada	smartphones developed by Samsung
Linux	smartphones using Linux as the operating system
Feature phone	modern low-end mobile phones
iPad	tablet computers developed by Apple
Datacard	laptops or PCs using datacards to access 3G services

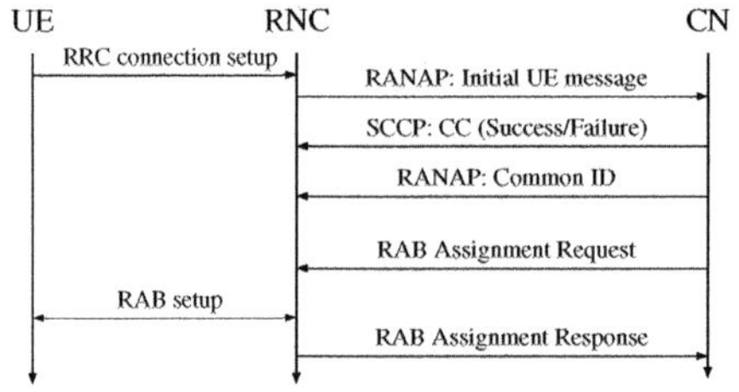

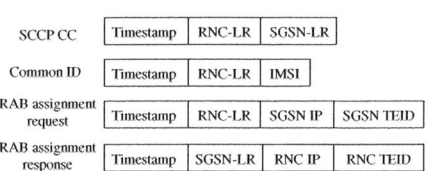

Fig. 2. Signaling message flow of the RRC connection setup

Fig. 3. Signaling message format (LR denotes local reference)

(iv) setup/release times of the RRC connection, which enable us to keep track of the connection duration.

Device Type Mappings. For each TAC, we identify the corresponding device type, including the hardware model and operating system of a UE. We then establish the device type mappings. Such mappings can be obtained from different public sources such as [13,20]. Table 2 shows the device types considered in our analysis.

3.2 Correlating Datasets

We now describe how we correlate the datasets that cover both data-plane and control-plane information. Our correlation seeks to identify the control-plane behaviors of the corresponding data-plane packets generated by different device types and different IP applications.

Extracting Signaling Messages. We first extract the RANAP signaling messages that are later used for correlation with RRC records. Fig. 2 depicts the signaling message flow of the RRC connection setup. When the UE wants to send or receive data, the RNC first makes a *Signaling Connection Control Part (SCCP)* connection request with the CN, which replies a *SCCP Connection Confirm (SCCP CC)* message. SCCP can be viewed as the transport protocol for RANAP. The CN also replies a *Common ID* message, which contains the IMSI of the UE. The RNC associates the IMSI with the

RRC connection for the UE and keeps this association throughout the RRC connection. Note that a single RRC connection may consist of one or multiple *Radio Access Bearers (RABs)* assigned for data communications. Each RAB assignment defines the actual radio resources for transmitting data traffic between the UE and the CN. The CN sends a RAB Assignment Request message to the RNC, which then executes the RAB setup protocol with the UE. Finally, the RNC replies a RAB Assignment Response message to the CN. From this onwards, the UE can send or receive data.

Based on the above signaling message flow, we can identify four signaling messages that are important for our correlation: SCCP CC, Common ID, RAB Assignment Request, and RAB Assignment Response. Fig. 3 summarizes the fields of the signaling messages that we use for our correlation.

Correlating Data/Control Traffic and RRC Record Logs. The goal of our correlation is to identify the data/control traffic for each RRC connection. One major challenge is that the data/control traffic and the record logs are obtained at different capture points that have *loosely synchronized system clocks* with around 60-150 second difference. Thus, we need to allow tolerance of such capture time differences in our correlation.

The basic idea of our correlation process is similar to the relational join operation by correlating the common fields in different datasets. We elaborate the details as follows:

1. We join the four signaling messages that have identical RNC-LR and SGSN-LR values and are seen within 15 seconds. RNC-LR and SGSN-LR are the local references of the RNC and SGSN associated with the RRC connection, respectively.

2. We join the output record in Step 1 with the RRC record that has the same IMSI field and is logged within 150 second difference. The time difference accommodates the imperfect synchronization among different captured points.

3. We then join the output record in Step 2 with the data packets. The correlation is based on the *Tunnel Endpoint Identifier (TEID)* field, which identifies the data communication tunnel. Each correlated data packet must have the same RNC IP, SGSN IP, and TEID fields as in the RRC record. Also, its timestamp must either be within the duration of the RRC connection, or differ from the setup/release times by no more than 150 seconds. If a data packet matches more than one RRC record, then we correlate it with the RRC record that has the closest setup/release time with the packet timestamp.

4. Finally, we obtain the device information of each packet with the device type mappings based on the TAC field.

The final output is that each data packet is associated with its corresponding RRC record and the device type. We then parse the data packets and perform the analysis.

Implementation. Our correlation is implemented as a MapReduce [4] program and run on a Hadoop [9] platform. This allows us to *parallelize* the correlation analysis. We utilize the *repartitioned sort-merge join* provided by Hadoop to realize the correlation. Our Hadoop platform consists of one namenode and seven datanodes that are connected by Gigabit switches. The platform has 112 CPU cores and 112GB memory in total, and it takes around 10 hours to complete the correlation process for one day of traces.

3.3 Limitations

One limitation of our work is that our datasets were collected in November 2010 (about a year ago from the time of this writing). Given the dramatic expansion of the smartphone market worldwide [12], we expect that there is a dramatic growth of the volume of data/control-plane traffic in 3G networks as of today. Also, there have been continuing version upgrades for smartphone operating systems (e.g., in Android and iPhone/iPad), and such upgrades may change the underlying data transmission behaviors. Nevertheless, since our correlation methodology is based on the standard 3G specifications, it remains applicable for today's 3G networks in general, provided that the same datasets are available. One important future work is to validate the findings of our measurement study with the latest datasets.

Another limitation is that we cannot validate the accuracy of the DPI tool used in our flow-based application identification, given that the tool is part of a commercial product and is closed-source. Nevertheless, we expect that the identification process is sufficiently accurate for our analysis due to the commercial nature of the tool. Some of our findings (see Section 4) also conform to our intuition. Note that our flow-based application identification can also be substituted by other approaches, such as port-based analysis [6] or dynamic protocol detection [5].

4 Measurement Results

4.1 Overview

We first give a general overview of our captured data traffic. After pre-processing, we identify more than 80% of traffic (out of ~13TB) that can be successfully associated with the corresponding signaling traffic. Fig. 4 plots the IP traffic volume (per minute) over time in our 7-day traces, and Fig. 5 shows the number of devices identified on each day. We observe that the distributions of each day are fairly stable. In particular, the distributions do not indicate significant differences across weekdays and weekends.

In the following analysis, we only focus on the 1-day traces on November 28, 2010. Given the regularity of the traffic, we expect that our observations are consistent for other days of traffic.

Fig. 6 illustrates the number of devices for different types listed in Table 2. There are 64,398 devices in total, in which iPhone leads all devices with a proportion of 32%.

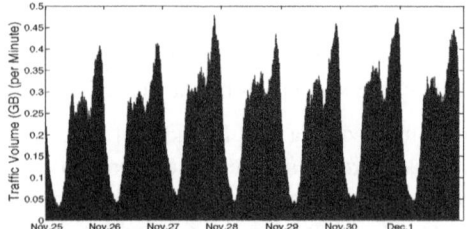

Fig. 4. Total traffic volume (per minute) over one week

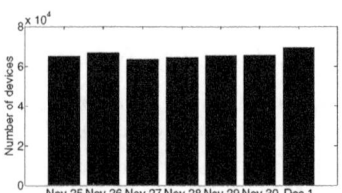

Fig. 5. Number of devices per day over one week

Mobile hand-held phones (including smartphones or low-end feature phones) dominate the majority of devices, while iPad and Datacard only account for 7.9% and 7.5% of devices, respectively. We point that over 99% of the devices can be successfully identified with the corresponding device types.

Fig. 7 shows the total traffic volume for each traffic type. Datacard devices contribute 46% of the total traffic, although they only account for 7.5% of devices. We will explore possible reasons for this phenomenon later in Section 4.2.

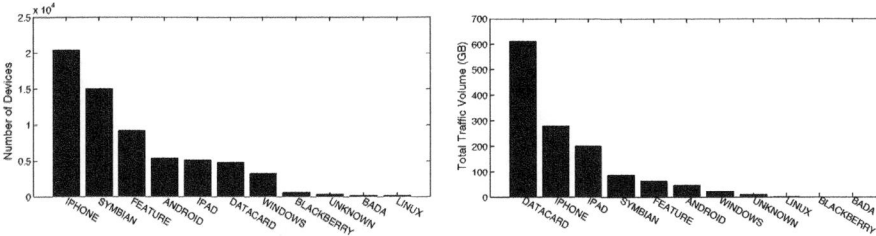

Fig. 6. Number of devices for each device type   **Fig. 7.** Total traffic volume of each device type

We now evaluate the control-plane performance of different device types, based on our RRC record logs. Fig. 8 shows the average number of RRC connections per device for each device type. As shown, iPhone triggers the most RRC connections per device (237 times), followed by iPad (174 times). Even datacard devices dominate the largest portion of IP traffic, each datacard device triggers a relatively small number of RRC connections (68 times) compared to smartphones (e.g., Android, Windows Mobile). Note that each RRC connection setup/release triggers a number of signaling messages between a UE and an RNC, and such signaling messages consume radio resources of the cellular network [16]. With the frequent RRC connection setups/releases, we can see that iPhone/iPad devices can bring large signaling overhead of an RNC and consume substantial radio resources of the whole cellular network.

Fig. 9 shows the average RRC connection duration per device for each device type. Note that iPhone has the smallest duration among all (30 seconds). Since iPhone triggers more RRC connections and each RRC connection has a short duration, it implies that iPhone devices inject a high intensity of signaling workload into the network. In contrast, Datacard has the longest RRC connection duration, so we expect that Datacard devices tend to run long-lived data transfer applications that keep an RRC connection active for a long time.

We thus far observe that different device types have distinct data/control-plane performance. In the following subsections, we explore the possible reasons behind.

4.2 Application Characteristics

Let us examine the application usage patterns of different device types, and explore how the traffic compositions of the device types lead to the distinct data/control-plane behaviors that we observe. We first analyze the overall traffic distributions of different IP applications listed in Table 1. Fig. 10 shows the traffic volume distribution of different IP applications, where web browsing (38%), streaming (21%), P2P (10%) and

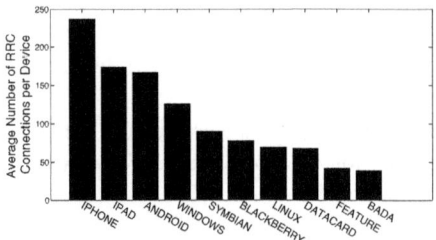

Fig. 8. Average number of RRC connections per device

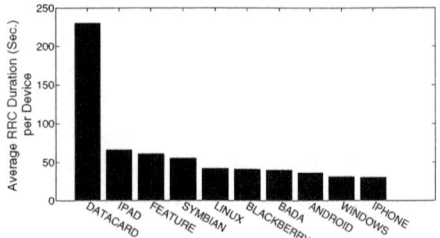

Fig. 9. Average RRC duration per device

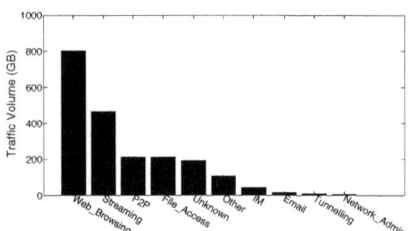

Fig. 10. Traffic volume of applications

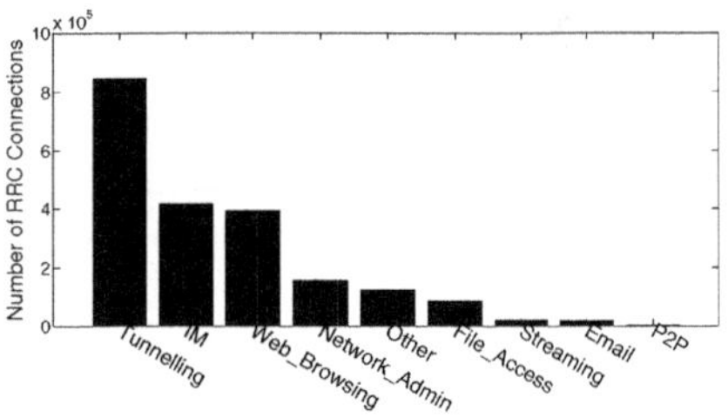

Fig. 11. Total number of RRC connections of applications

file access (10%) are ranked the top four. IM and Email applications contribute 2% and 1% of the total traffic, respectively. The traffic that cannot be identified, i.e., labeled as "Unknown" in Fig. 10, only contributes 9% of the traffic. In other words, over 90% of total traffic can be successfully identified by our DPI tool.

Fig. 11 shows the total number of RRC connections triggered by each application. In this work, we define that an application triggers an RRC connection if the first data packet transmitted in the RRC connection belongs to the application. We see that Tunneling triggers the most RRC connections among all applications. The reason is that a tunneling session is first established for providing a secure path for any other application protocols, or supporting the roaming service. IM triggers 21% of RRC connections, while generating only 2% of the total traffic (Fig. 10). On the other hand, P2P triggers only 0.1% of the total number of RRC connections, while contributing 11% of the total traffic. This is an important observation because it clearly shows that different applications have *inherently different* data/control-plane behaviors.

Let us investigate the application usage patterns of different device types. Fig. 12 illustrates the total traffic volume of applications for each device type, in which we only look at the top device types that contribute the most traffic. We make two key observations. First, datacard devices contribute 85% and 48% of all P2P and streaming traffic, respectively. Since most datacards are attached to static terminals such as PCs or laptops, these static terminals tend to run long-lived applications. Thus, they contribute a large volume of data-plane traffic even the population of datacard devices is small. Also, P2P and streaming applications trigger very few but long-lived RRC connections, and this validates our previous observations in Figs. 8 and 9. Second, web browsing, streaming, and file access are the top three applications that account for the most traffic

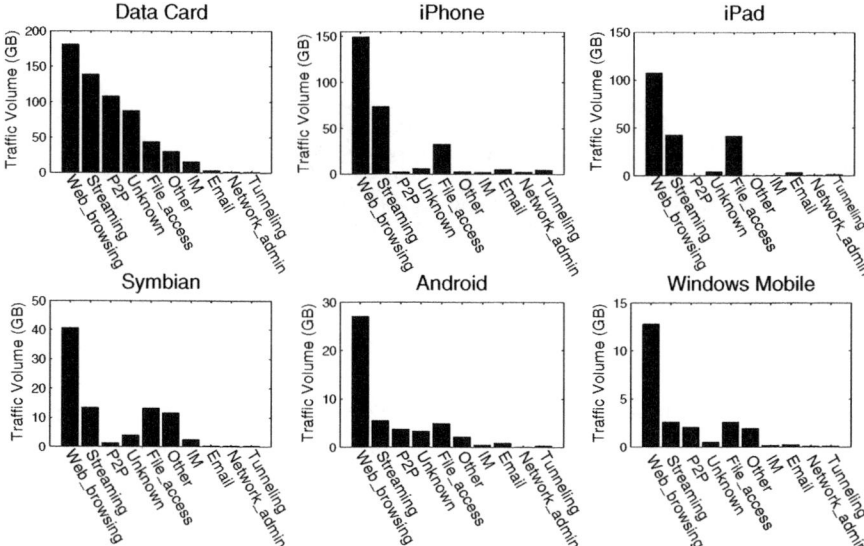

Fig. 12. Traffic volume distribution of applications for each device type (note that the y-axis scale is different for each device type for clear presentation)

volume on smartphones (i.e., iPhone, Android, Symbian, Windows Mobile), and they altogether contribute over 80% of the smartphone traffic.

4.3 Active Devices

We now explore the behaviors of *active* devices and see their impact on the data/control-plane performance. We say that a device is active within time $[t_1, t_2]$ if it starts and releases an RRC connection at times t_1 and t_2, respectively. Figs. 13 and 14 plot the distributions of the traffic volume and the number of active devices (per minute) for different device types over a 24-hour period (on Nov. 28), respectively. As expected, Fig. 13 shows that the traffic volume peak appears in day time, and the traffic trough appears at late night. While the number of active devices in Fig. 14 shows a similar pattern, the peak-trough difference is significantly different. We find that the traffic volume at the trough is 93% less than the peak traffic volume in Fig. 13, while the number of active devices at the trough is only 52% less than the peak value in Fig. 14.

After further investigation, we find that iPhone and iPad have a very similar pattern that is distinct from other device types, i.e., the numbers of active devices of iPhone and iPad remain quite stable during the 24-hour period. On the other hand, other device types exhibit obvious troughs and peaks in the number of active devices, with the troughs also appearing at late night.

We find that the internal heartbeat mechanism of iPhone/iPad keeps the devices active, even there is no data traffic initiated by user applications. To demonstrate, Fig. 15 shows the probability density of the inter-arrival times of RRC connection setups of iPhone. The inter-arrival times of RRC connections of iPhone occur more often at two values: 64 seconds (18.1%) and 589 seconds (4.9%). Similar observations are also made

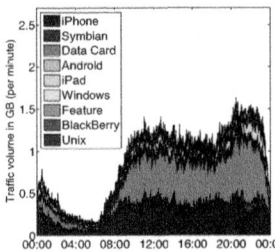
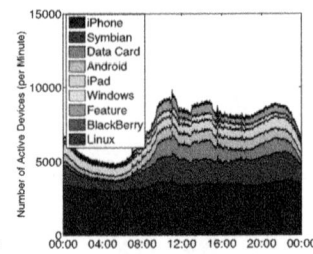
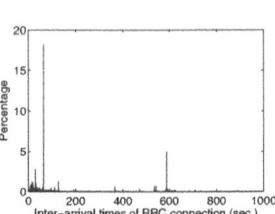

Fig. 13. Traffic volume (per minute) distribution

Fig. 14. Number of active devices (per minute)

Fig. 15. Density of interarrival times of RRC connection setups of iPhone

for iPad. We investigate and find that iPhone/iPad devices are "always-on" and generate heartbeat packets around every 60 seconds, and such packets are sent to Apple servers. Each heartbeat packet triggers an RRC connection, which will be released if there is no data traffic for an idle period that is less than 60 seconds. On the other hand, we do not observe any explicit heartbeat patterns in Android (not shown in the figure), possibly because there are many variants of Android devices and their heartbeat designs are different. In short, the heartbeat packets that we observe in the iPhone/iPad devices introduce many RRC connection setups/releases, which could lead to high signaling overhead in the network core.

Note that there are a few upgrades of iOS (Apple's mobile operating system) in 2011. In iOS 4.2 (which was launched in late November 2010), a technology that enables mobile devices and the network core to collaboratively minimize network congestion and mobile battery consumption [15]. This may change the heartbeat behavior that we observe in our datasets. One future work is to validate such a heartbeat behavior with the latest datasets, based on the same correlation methodology as described in Section 3.2.

5   Summary, Implications, and Conclusions

We now summarize the key observations and potential challenges through the analysis of our massive data traces.

- In the data plane, datacard devices contribute almost 50% of the total traffic, while accounting for only 7% of the device population. Network operators may devise special resource allocation schemes for datacard users. Also, iPhone/iPad devices altogether account for around 40% of devices, and contribute nearly 40% of the total traffic due to their large market shares.
- Our application-based analysis clearly shows that different user groups (distinguished by device types) have distinct user behaviors with specific preferences of choosing applications. For example, in all Apple devices (iPhone and iPad), the dominant applications are web browsing, streaming, and file access, and they altogether contribute more than 90% of iPhone/iPad traffic. From the network operators' perspectives, there are challenging issues of providing QoS guarantees and QoE improvement for different user groups.

– In the signaling control plane, among all device types, iPhone has the highest average number of RRC connections per device. Meanwhile, iPhone/iPad devices have a very similar heartbeat behavior that triggers significantly more RRC connections than any other device types. Recall that each RRC connection setup/release involve a number of signaling messages exchanged between a UE and an RNC [16]. These results reveal that specific device types can increase the signaling overhead to the network.

This paper analyzes the data/control-plane performance of different device types using massive traces collected from the core of a 3G UMTS city-wide network. We describe a correlation methodology that studies the interactions of different data/control-plane datasets, and make several key observations from our analysis that could be of interest to network operators and researchers. Our measurement study motivates the importance of characterizing the data/control-plane workloads of a 3G network and designing adequate strategies for network planning, resource allocation, and pricing. Since our measurement study is based on the datasets that are compliant with the 3G standard, our analysis methodology can be generalized for other 3G operational networks.

Acknowledgment. The work of Patrick P. C. Lee is supported in part by grant GRF CUHK413711 from the Research Grant Council of Hong Kong.

References

1. Cisco. Cisco Visual Networking Index: Global Mobile Data Traffic Forecast Update, 2010–2015 (February 2011)
2. D'Alconzo, A., Coluccia, A., Ricciato, F., Romirer-Maierhofer, P.: A Distribution-Based Approach to Anomaly Detection for 3G Mobile Networks. In: IEEE Globecom (2009)
3. DARWIN, http://userver.ftw.at/~ricciato/darwin/
4. Dean, J., Ghemawat, S.: Mapreduce: simplified data processing on large clusters. Commun. ACM 51, 107–113 (2008)
5. Dreger, H., Feldmann, A., Mai, M., Paxson, V., Sommer, R.: Dynamic Application-Layer Protocol Analysis for Network Intrusion Detection. In: Proc. of USENIX Security Symp. (2006)
6. Falaki, H., Lymberopoulos, D., Mahajan, R., Kandula, S., Estrin, D.: A First Look at Traffic on Smartphones. In: Proc. of ACM IMC (November 2010)
7. Falaki, H., Mahajan, R., Kandula, S., Lymberopoulos, D., Govindan, R., Estrin, D.: Diversity in Smartphone Usage. In: Proc. of ACM MobiSys (June 2010)
8. Gember, A., Anand, A., Akella, A.: A Comparative Study of Handheld and Non-handheld Traffic in Campus Wi-Fi Networks. In: Spring, N., Riley, G.F. (eds.) PAM 2011. LNCS, vol. 6579, pp. 173–183. Springer, Heidelberg (2011)
9. Hadoop, http://hadoop.apache.org/
10. Huang, J., Xu, Q., Tiwana, B., Mao, Z.M., Zhang, M., Bahl, P.: Anatomizing Application Performance Differences on Smartphones. In: Proc. of ACM MobiSys (2010)
11. Huawei. MSCG hierarchical DPI solution (2011),
    http://www.huawei.com/products/datacomm/catalog.do?id=1219
12. IDC. Worldwide Smartphone Market Expected to Grow 55% in 2011 and Approach Shipments of One Billion in 2015, According to IDC (June 2011),
    http://www.idc.com/getdoc.jsp?containerId=prUS22871611

13. IMEI lookup, `http://imei-number.com/imei-lookup/`
14. Kilpi, J., Lassila, P.E.: Micro- and Macroscopic Analysis of RTT Variability in GPRS and UMTS Networks. In: Boavida, F., Plagemann, T., Stiller, B., Westphal, C., Monteiro, E. (eds.) NETWORKING 2006. LNCS, vol. 3976, pp. 1176–1181. Springer, Heidelberg (2006)
15. Kingsley-Hughes, A.: iOS 4.2 Supports Network Controlled Fast Dormancy (December 2010), `http://www.zdnet.com/blog/hardware/ios-42-supports-network-controlled-fast-dormancy/10586`
16. Lee, P.P.C., Bu, T., Woo, T.: On the detection of signaling DoS attacks on 3G/WiMax wireless networks. Computer Networks 53(15), 2601–2616 (2009)
17. Maier, G., Schneider, F., Feldmann, A.: A First Look at Mobile Hand-Held Device Traffic. In: Krishnamurthy, A., Plattner, B. (eds.) PAM 2010. LNCS, vol. 6032, pp. 161–170. Springer, Heidelberg (2010)
18. Paul, U., Subramanian, A.P., Buddhikot, M.M., Das, S.R.: Understanding Traffic Dynamics in Cellular Data Networks. In: Proc. of IEEE INFOCOM (2011)
19. Qian, F., Wang, Z., Gerber, A., Mao, Z.M., Sen, S., Spatscheck, O.: Characterizing Radio Resource Allocation for 3G Networks. In: Proc. of ACM IMC (2010)
20. radiorraiders, `http://www.radioraiders.com/gsm-IMEI-TAC.php/`
21. Ricciato, F., Vacirca, F., Karner, M.: Bottleneck detection in UMTS via TCP passive monitoring: a real case. In: Proc. of ACM CoNEXT (October 2005)
22. Ridoux, J., Nucci, A., Veitch, D.: Seeing the difference in IP traffic: wireless versus wireline. In: Proc. of IEEE INFOCOM (2006)
23. Xu, Q., Erman, J., Gerber, A., Mao, Z.M., Pang, J., Venkataraman, S.: Identifying Diverse Usage Behaviors of Smartphone Apps. In: Proc. of ACM IMC (November 2011)

Towards a Robust Framework of Network Coordinate Systems

Linpeng Tang[1,2], Zhiyong Shen[2,*], Qunyang Lin[2], and Junqing Xie[2]

[1] Shanghai Jiao Tong University
chnttlp@gmail.com
[2] HP Labs, China
{peng.tang,zhiyongs,qun-yang.lin,jun-qing.xie}@hp.com

Abstract. Network Coordinate System (NCS) is an efficient and scalable mechanism to predict latency between any two network hosts based on historical measurements. Most NCS models, such as metric space embedding based, like Vivaldi, and matrix factorization based, like DMF and Phoenix, use squared error measure in training which suffers from the erroneous records, i.e. the records with large noise. To overcome this drawback, we introduce an elegant error measure, the Huber norm to network latency prediction. The Huber norm shows its robustness to the large data noise while remaining efficiency of optimization. Based on that, we upgrade the traditional NCS models into more robust versions, namely Robust Vivaldi model and Robust Matrix Factorization model. We conduct extensive experiments to compare the proposed models with traditional ones and the results show that our approaches significantly increase the accuracy of network latency prediction.

Keywords: Network Coordinate Systems, Robust Error Measure, Metric Space Embedding, Matrix Factorization.

1 Introduction

The prediction of network latency, e.g. Round-Trip Time(RTT), has been a hot research topic over the last few years. Predicting network latencies between two network hosts without involving physical measurements may benefit many networking applications, like geometric routing[14], large scale online game systems [4], locality-aware data center selection[6], and P2P file sharing.

One way to predict network latency is via a Network Coordinate System (NCS)[9,17,15,8]. NCS tries to assign each network host some coordinates representing its virtual location in the network and the latency between two hosts is computed using a *prediction function* over the coordinates. To learn the coordinates for each host, we only need to measure a small fraction of all the pairs of the hosts and adjust the coordinates in order to minimize the *prediction error* on these records. Since the coordinates of one host are just a constant number of

---

* Corresponding author. 5F, Block A, SP Tower, No. 1, ZhongGuanCun East Road, Haidian District Beijing 100084, P.R. China

R. Bestak et al. (Eds.): NETWORKING 2012, Part I, LNCS 7289, pp. 331–343, 2012.

values, from Valiant's learning theory [18] we could learn the coordinates with just $O(N)$ measurements and predict the N^2 latencies between all host pairs, which makes the system scalable.

NCS approaches could be categorized into two classes: metric space embedding models and matrix factorization models. One of the most representative metric space embedding based approach is Vivaldi [9], which maps the network hosts into a Euclidean space and approximates network latencies with the Euclidean distances between the image points. In matrix factorization models, such as Phoenix [8] and DMF [15], the pairwise latency matrix is approximated by the factorization of two low-rank matrices, one for the source hosts and the other for the destination hosts. Most of these traditional NCS's use the ℓ^2-norm[1] as the measure of the prediction error, due to its simplicity in model learning. However, the complexity of the Internet may cause extremely large noise in the measurement of the latency between hosts, and the quadratic ℓ^2-norm shows high sensitivity to such erroneous data records. In this paper, we address the issue of large data noise based on an elaborately designed error measure—the Huber norm, which is more robust for the erroneous data records while remaining simplicity of learning the model. Based on the Huber norm, we propose a robust framework of NCS and further increase the accuracy of network latency prediction. The key contributions of this paper are threefold:

1. To the best of our knowledge, we are the first to introduce and study the Huber norm for network latency prediction, based on which we upgrade the traditional Vivaldi and matrix factorization models to more robust versions.
2. We derive two algorithms to learn the robust models—stochastic gradient descent and alternative damped-Newton method. The former has simple implementation and runs faster in terms of CPU time, while the latter takes fewer iterations for convergence.
3. We conduct extensive experiments to demonstrate that our robust models significantly outperform the traditional ones. Specifically, we find the robust version of Vivaldi shows much higher accuracy than other methods when neighbors of one host are fewer than 32 and may be the preferred model in many applications.

1.1   Preliminaries

Before we introduce the NCS approaches, let's formally define the problem of network coordinating. Assume that we have N network hosts. We may use an $N \times N$ matrix Y to denote the pairwise latency matrix, and let W denote the indices set of known entries, i.e., $Y_{i,j}$ is known(measured) for $(i,j) \in W$ and generally $|W| \ll N \times N$. For each entry in the latency matrix Y, whether or not known, we may predict its value using $\hat{Y}_{i,j} = pred_{\mathcal{C}}(i,j)$, where $pred_{\mathcal{C}}(i,j)$ is a universal representation of the prediction function and the prediction is made based on the virtual coordinates assigned to the hosts, noted as $\mathcal{C}$. The problem is now how to assign and adjust the coordinates for each network host in order to make

---

[1] The ℓ^2-norm of an error vector $\mathbf{e} = (e_1, e_2, ..., e_n)$ is defined as $\|e\|_2 = \sqrt{\sum_{i=1}^{n}(e_i)^2}$.

the predicted latency matrix $\hat{Y}$ as *close* to the matrix of true measurements Y as possible. We use the prediction error measure function $\phi(\hat{Y}, Y)$ to characterize the closeness between the $\hat{Y}$ and Y. Then in a NCS, we generally aim to solve the following optimization problem

$$C^* = \operatorname*{argmin}_{\mathcal{C}} \phi(\hat{Y}, Y) \tag{1}$$

Both the prediction function $pred_{\mathcal{C}}(i,j)$ and the error measure function $\phi(\hat{Y}, Y)$ should be specified for a NCS.

2   Related Work

2.1   Metric Space Embedding Models

In metric space embedding models, each host is associated with a position in a metric space, such as Euclidean space [9] and hyperbolic space [16], etc. The latency between two hosts is then estimated as the distance between their positions in the space. Vivaldi [9] is the most representative metric space embedding based approach, whose variants have been employed in real-world systems, such as Htrae [4]. In this model, the network hosts are embedded into a D-dimensional Euclidean space. Assume that two nodes i, j have embedded coordinates $X_i = (X_{i,1}, ..., X_{i,D})$ and $X_j = (X_{j,1}, ..., X_{j,D})$. We may calculate the Euclidean distance between i, j as $\|X_i - X_j\|_2$.

In Vivaldi, the authors also define quantities of *heights* $H = (h_1, h_2, ..., h_N)$ $(h_i \geq 0)$ for each host to characterize the host's intrinsic contribution to the latency, say the latency from a host to the nearest ISP. The prediction function is then defined as

$$pred_{X,H}(i,j) := \|X_i - X_j\|_2 + h_i + h_j, \tag{2}$$

so we have $\hat{Y}_{i,j} = \|X_i - X_j\|_2 + h_i + h_j$. The coordinates are learned by optimization algorithms so as to minimize the summation of prediction errors over the known entries in Y. Formally, in Vivaldi, we achieve the coordinates X and H by solving the the following optimization problem with an ℓ^2-norm objective function

$$X^*, H^* = \operatorname*{argmin}_{X,H} \phi(\hat{Y}, Y) = \operatorname*{argmin}_{X,H} \left\| \hat{Y} - Y \right\|_2 \tag{3}$$

$$= \operatorname*{argmin}_{X,H} \sum_{(i,j) \in W} [\|X_i - X_j\|_2 + h_i + h_j - Y_{i,j}]^2 \tag{4}$$

The above problem could be solved using *stochastic gradient descent* [5] and due to the space limitation, we refer to [9] for details.

2.2   Matrix Factorization Models

The representative matrix factorization NCS approaches include IDES [17], Phoenix [8] and DMF [15] etc. In this kind of models, we generally seek the following approximation for the latency matrix Y.

$$Y \approx UV^T$$

The two factor matrices U, V represent the out-coordinates and the in-coordinates of the hosts. Specifically, the i-th row of U is the out-coordinate of host i and the jth row of V is the in-coordinate of node j. Let $\hat{Y} = UV^T$, so $\hat{Y}_{i,j} = \sum_{k=1}^{K} U_{i,k} V_{j,k}$. The prediction function here is specified as

$$pred_{U,V}(i,j) := \sum_{k=1}^{K} U_{i,k} V_{j,k} = (UV^T)_{i,j}. \tag{5}$$

Similar to Vivaldi, we may find U and V by minimizing the prediction error $\phi(\hat{Y}, Y)$ over the known entries, and formally we have

$$U^*, V^* = \operatorname*{argmin}_{U,V} \phi(\hat{Y}, Y) = \operatorname*{argmin}_{U,V} \left\| \hat{Y} - Y \right\|_2 \tag{6}$$

$$= \operatorname*{argmin}_{U,V} \sum_{(i,j) \in W} [(UV^T)_{i,j} - Y_{i,j}]^2 \tag{7}$$

From (6) and (3), we may see that traditional NCS's mostly use l^2-norm in their objective function. This choice of prediction error measure has its easiness for optimization. In the next subsection, we'll discuss whether this is a proper choice for network latency prediction.

2.3   Issue of Data Noise

We must also, however, recognize the inherent complexity of network latencies. Because of the existence of inefficient routing rule, network congestions, malicious attack etc., many of data records we get and use to build the NCS may contain extremely large noise. If the NCS tries to fit these data records, it may deviate from the true distance model behind the whole network. Many people have recognized this phenomenon, and try to overcome it by introducing a "faith" factor on the host or the records. For example, Vivaldi introduced a weight in $[0, 1]$ on each host to indicate how reliable its coordinates are. Similarly, Phoenix introduced a weight on each record indicating how reliable it is. Hosts or records with smaller weights will have less impact on the NCS. While such weight based approach can improve the stability and accuracy of NCS, it is very heuristic and often involves several parameters in the model that make tuning difficult.

We attack the issue of data noise from another direction by leveraging a noise robust error measure—the Huber norm, based on which we propose the robust versions of Vivaldi and matrix factorization models as described in the next section.

3 Algorithm Design

3.1 Huber Norm

A simple choice of robust measure is the ℓ^1-norm[2], which has shown considerably less sensitivity to large measurement errors than ℓ^2-norm measures[13]. However, ℓ^1-norm is not continuously differentiable so its numerical minimization is difficult. In [12], Huber presents an elegant error-measure defined as (8) and illustrated in Figure 1.

$$M_\epsilon(r) = \begin{cases} \frac{r^2}{2\epsilon}, & |r| \le \epsilon \\ |r| - \frac{\epsilon}{2}, & |r| \ge \epsilon \end{cases} \tag{8}$$

The Huber norm is a combination of ℓ^1/ℓ^2 norm (see in Figure 1). The tradeoff of ℓ^1 or ℓ^2 is controlled by the parameter ϵ : for small errors $|r| \le \epsilon$, it assumes the ℓ^2 norm, while for large errors $|r| > \epsilon$, it assumes the ℓ^1 norm. We'll empirically investigate the role of parameter ϵ in our models in Section 4.3. Compared to ℓ^2 norm, the Huber norm is more robust since large errors only has a linear impact. On the other hand, it is easier to optimize than ℓ^1 norm in some sense because it is continuously differentiable.

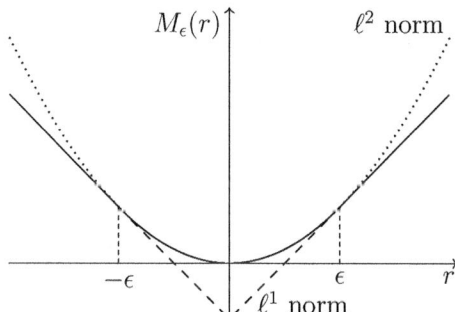

Fig. 1. The Huber norm

By leveraging the Huber norm, the learning objective of NCS (1) becomes

$$C^* = \operatorname*{argmin}_{C} \sum_{(i,j) \in W} M_\epsilon(pred_C(i,j) - Y_{i,j}) \tag{9}$$

In the next two subsections we will describe how we specify and solve (9) for the robust versions of Vivaldi and matrix factorization models, respectively.

3.2 Robust Vivaldi

Substituting (2) into (9), we may train the model to learn the coordinates in the Euclidean space X and the heights H via optimizing

$$X^*, H^* = \operatorname*{argmin}_{X,H} \sum_{(i,j) \in W} M_\epsilon[\|X_i - X_j\|_2 + h_i + h_j - Y_{i,j}] \tag{10}$$

---

[2] The ℓ^1-norm of an error vector $\mathbf{e} = (e_1, e_2, ..., e_n)$ is defined as $\|e\|_1 = \sum_{i=1}^{n} |e_i|$.

We leverage two alternative methods, stochastic gradient descent and alternative damped-Newton to solve the above problem.

Stochastic Gradient Descent: One way to solve the problem is by stochastic gradient descent [5]. For each measurement record $Y_{i,j}$, we do the following update on the coordinates X_i and heights h_i:

$$F_{i,j} = M_\epsilon(\|X_i - X_j\| + h_i + h_j - Y_{i,j}) \tag{11}$$

$$X_{i,d} \leftarrow X_{i,d} - \delta \frac{\partial}{\partial X_{i,d}} F_{i,j} \quad (d = 1 \cdots D) \tag{12}$$

$$h_i \leftarrow \max(0, h_i - \delta \frac{\partial}{\partial h_i} F_{i,j}) \tag{13}$$

where δ is the learning rate. In each round, we first randomly shuffle the training data, and then do the updates as (11), (12), (13).

Alternative Damped-Newton: We then utilize another optimization scheme–alternative damped-Newton method [7].

Instead of considering one sample at a time, alternative damped-Newton method considers all neighbors[3] of one host at the same time. Specifically, for one host i, fixing the coordinates of other hosts, it tries to optimize the sum of prediction errors from i to all its neighbors:

$$F_i = \sum_{j:(i,j)\in W} M_\epsilon(\|X_i - X_j\|_2 + h_i + h_j - Y_{i,j}) \tag{14}$$

To achieve this efficiently, we first consider Newton's method. Let $\mathcal{G} = \nabla_{X_i} F_i$ be the gradient, $\mathcal{H} = \mathcal{H}_{X_i} F_i$ be the Hessian. Then Newton's Method updates $X_i \leftarrow X_i - \mathcal{G}\backslash\mathcal{H}$[4]. However, since $M_\epsilon''(x) = 0$ for $|x| < \epsilon$, $\mathcal{H}$ can often be singular and $\mathcal{G}\backslash\mathcal{H}$ undefined. To overcome this problem, we use Damped-Newton Method. Specifically, we introduce the *damping factor* λ and modify the update rule to:

$$X_i \leftarrow X_i - \mathcal{G}\backslash(\mathcal{H} + \lambda I) \tag{15}$$

Note that when λ is small, the update rule approximates a Newton update, but when λ grows larger, the update becomes similar to a gradient descent with learning rate $1/\lambda$. We use an adaptive scheme to choose a proper λ—starting with $\lambda = 1$, try $\lambda \leftarrow 2\lambda$ until F_i decreases or λ becomes too large, and this finishes one update.

3.3   Robust Matrix Factorization

By instituting (5) into (9), we get the objective of our Robust Matrix Factorization model.

$$U^*, V^* = \underset{U,V}{\operatorname{argmin}} \sum_{(i,j)\in W} M_\epsilon[(UV^T)_{i,j} - Y_{i,j}] \tag{16}$$

---

[3] The neighbors of one host i are the hosts j s.t. $Y_{i,j}$ is known, noted as $j : (i,j) \in W$.
[4] $X = A\backslash B$ is the solution of $A \times X = B$.

Both stochastic gradient descent and alternative damped-Newton method can be applied in similar ways. Note that we only give the update rules for the out-coordinates U and the rules for V are symmetric.

Stochastic Gradient Descent: Let $F_{i,j}$ denote the prediction error of record $Y_{i,j}$ (17). We may minimize it using gradient descent on U_i.

$$F_{i,j} = M_\epsilon(\sum_{d=1}^{D} U_{i,d}V_{j,d} - Y_{i,j}) \tag{17}$$

$$U_{i,d} \leftarrow U_{i,d} - \delta\frac{\partial}{\partial U_{i,d}}F_{i,j} \qquad (d = 1 \cdots D) \tag{18}$$

Alternative Damped-Newton: Let F_i denote the sum of prediction error from i to all its neighbors (19). We try to minimize F_i by updating U_i with alternative damped-Newton method (15). Again, assuming $\mathcal{G} = \nabla_{U_i}F_i$ and $\mathcal{H} = \mathcal{H}_{U_i}F_i$, we have

$$F_i = \sum_{j:(i,j)\in W} M_\epsilon(\sum_{d=1}^{D} U_{i,d}V_{j,d} - Y_{ij}) \tag{19}$$

$$U_i \leftarrow U_i - \mathcal{G}\backslash(\mathcal{H} + \lambda I) \tag{20}$$

3.4 Discussion

As can be seen, stochastic gradient descent and alternative damped-Newton actually work in two fashions. The former considers one measurement record at a time while the latter update the coordinate of one host based on the current coordinates of all its neighbors. Intuitively, methods that consider the coordinates of all neighbors of one node converges faster, i.e., requiring fewer iterations to reach a optimal value. This property is especially desirable when each node maintain its own coordinates and communicates with its neighbors to adjust the coordinates. Faster convergence, in this case, means that it takes less time for the dynamic system to become stable.

However, stochastic gradient descent also has its merits. Since it has no need to compute the Hessian, and do the matrix inversion, the computational cost of one iteration is extremely cheap. Therefore, although it may take more iterations, it is still considerably faster than alternative damped-Newton method when running on one computer. In Section 4.4, we'll further investigate the convergence property of these two methods.

4 Experiments

In this section we empirically compare performance of our models: Robust Vivaldi and Robust MF, with traditional models: Vivaldi, DMF, IDES and Phoenix. We use three large public data sets for our experiments: Meridian [2], P2PSim

[3], Harvard [1], which are all collected with King method[11]. The data sets are preprocessed by removing some hosts to guarantee that all the hosts have at least 32 neighbors. In addition, we are predicting the RTT between hosts, which should be symmetric (not considering factors such as network congestion), so for the original latency matrix Y, we transform it to $(Y + Y^T)/2$. After these preprocessing steps, the three data sets have 2,500, 1740 and 1818 hosts and average latency of 91.8ms, 75.8ms, 85.7ms, respectively.

4.1   Overview of the Experiments

We have three parameters to be specified in the experiments: K − the number of measured neighbors of a host, D − the dimension of the coordinates and ϵ − the tradeoff parameter between ℓ^1 and ℓ^2 norm in the Huber norm (8). Note that in all experiments, the K neighbors of any host are chosen randomly. In the experiments, we consider the cases when $K = 16, 32, 64$ and the choice of D are chosen by experiments to let the algorithms to achieve best accuracy. We found that Robust-Vivaldi always works best when $D = 2$. (This coincides the finding of [9] that increasing the dimension can't significantly improve the prediction accuracy in Vivaldi model.) The matrix factorization methods should have a slightly lower dimension when K are small in order to avoid overfitting, but a higher dimension when K are large in order to improve accuracy. So we set $D = 5$ when $K = 16$, $D = 6$ when $K = 32$ and $D = 7$ when $K = 64$. In Robust Vivaldi and Robust-MF, we set $\epsilon = 7$ and in Section 4.3 we will conduct further investigation on the choice of ϵ.

In Section 4.2, we give the results on the distribution of absolute error. We then used other two evaluation criteria for later experiments, Mean Absolute Error (MAE) and Rooted Mean Squared Error (RMSE), defined as follows

$$\text{MAE} = \underset{(i,j)\in W^*}{\text{average}} \left(\left| \hat{Y}_{i,j} - Y_{i,j} \right| \right) \quad \text{RMSE} = \sqrt{\underset{(i,j)\in W^*}{\text{average}} \left[(\hat{Y}_{i,j} - Y_{i,j})^2 \right]} \quad (21)$$

where, W^* denotes the indices of a held out test set. For both of the two criteria, smaller values mean better performance.

4.2   Comparison on Prediction Accuracy

In this subsection, we compare the prediction accuracy of our algorithms with Phoenix, DMF, IDES and Vivaldi. The comparison results of error distributions are given in Figure 2(a)-(i) and we also specify the comparison of the 80-percentile error in Figure 2(j). From these results, we see that robust methods significantly outperform their non-robust counterparts. This observation verifies the robustness of the Huber norm in the application of network latency prediction.

It is also remarkable that although Vivaldi does poorly in all our test cases (which coincides with the conclusion of Phoenix [8] and DMF [15] that matrix factorization supersedes (non-robust) Vivaldi), Robust Vivaldi achieves much

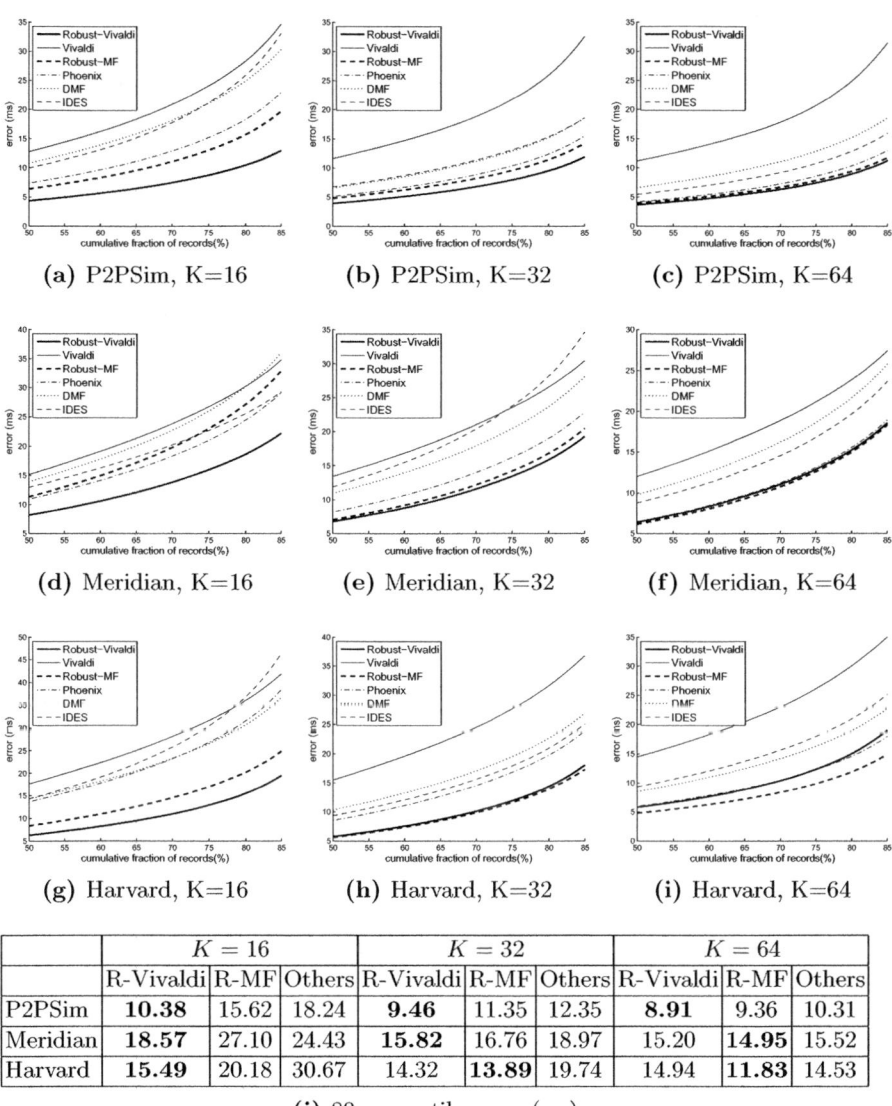

	$K = 16$			$K = 32$			$K = 64$		
	R-Vivaldi	R-MF	Others	R-Vivaldi	R-MF	Others	R-Vivaldi	R-MF	Others
P2PSim	**10.38**	15.62	18.24	**9.46**	11.35	12.35	**8.91**	9.36	10.31
Meridian	**18.57**	27.10	24.43	**15.82**	16.76	18.97	15.20	**14.95**	15.52
Harvard	**15.49**	20.18	30.67	14.32	**13.89**	19.74	14.94	**11.83**	14.53

(j) 80-percentile error (ms)

Fig. 2. Results of error distribution shown in (a)-(i). The x-axis is the cumulative fraction of records (in percentiles), and the y-axis is the corresponding prediction error (in milliseconds). In Table (j), we compare the 80-percentile error, where *R-Vivaldi* stands for Robust Vivaldi, *R-MF* stands for Robust MF and *Others* stands for the best result of IDES, Phoenix, DMF and Vivaldi in each test case.

higher accuracy than other methods when K is small. When $K = 16$, it supersedes all other methods by 25%–30% in terms of 80-percentile error in all 3 data sets (see in Figure 2(j)). When K is smaller, the advantage is even more obvious. So Robust Vivaldi may be the preferred model in many applications.

The advantage of Robust Vivaldi may profit from that the number of parameters utilized by Vivaldi like models (three for each host) is much smaller than that of matrix factorization based models ($2D$ for each host). According to the principle of Occam's razor, when there is no sufficient training data, a model with fewer parameters faces lower risk of the issue of *overfitting* and may perform better. On the other hand, performance of the methods based on matrix factorization improves gradually as K increases, taking lead when $K = 64$. When we have sufficient training data, matrix factorization may capture more information with more parameters and achieve higher prediction accuracy. In summary, Vivaldi like model is preferred when we have fewer than 32 neighbors for each host, but when we really know much information about the system (say, an online P2P game system), matrix factorization based methods has the potential to provide better accuracy.

4.3   The Choice of ϵ

In Figure 3, we vary ϵ, the threshold between ℓ^1 norm and ℓ^2 norm, and give the results of both MAE and RMSE on the P2PSim data set. According to the

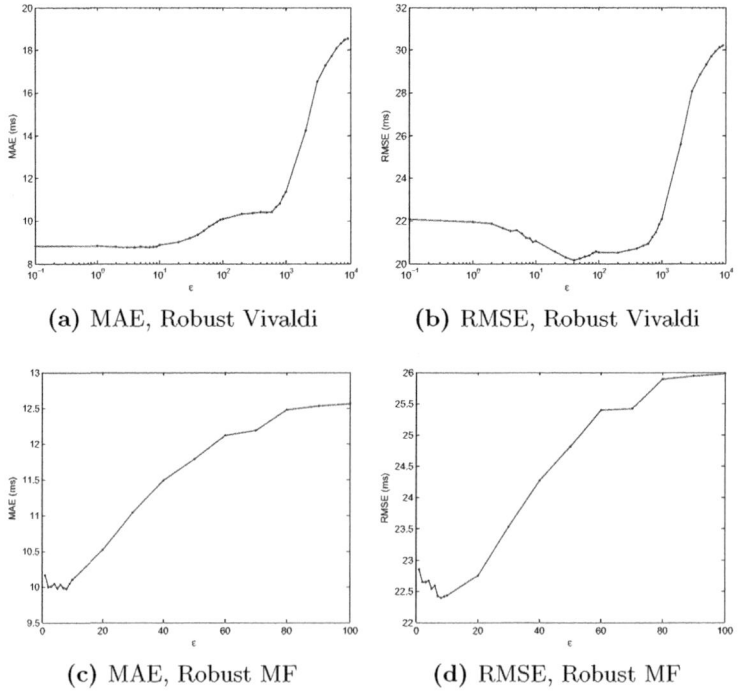

(a) MAE, Robust Vivaldi   (b) RMSE, Robust Vivaldi

(c) MAE, Robust MF   (d) RMSE, Robust MF

Fig. 3. Effect of ϵ on the prediction error on the P2PSim data-set, $K = 32$

definition in (21), MAE and RMSE actually corresponds to the ℓ^1 and ℓ^2 error measures respectively. So one would expect that if we minimize the ℓ^1 error (set $\epsilon = 0$), MAE would be minimized; and if we use the ℓ^2 error (set a large ϵ), RMSE would be minimized.

Interestingly, this is not the case (see in Figure 3). Neither MAE is best optimized at $\epsilon = 0$, or is RMSE optimized when we increasing ϵ. For Robust Vivaldi, RMSE is best optimized when $\epsilon = 40$. For Robust MF, RMSE is best optimized when $\epsilon = 7$. As ϵ grows beyond this threshold, the performance deteriorates greatly. On the other hand, when ϵ is relatively small ($\epsilon < 10$), the MAE remains stable for Robust Vivaldi. For Robust MF, MAE drops a little when ϵ increases as long as $\epsilon < 7$.

Theoretically speaking, ϵ is the threshold between small errors and large errors. As a rule of thumb, we recommend to use the 80-percentile error as ϵ, so a value between 5–15 is recommended (choose a percentile error as ϵ was recommended in [10]).

4.4   Convergence Analysis

In this section we investigate the convergence of the two optimization schemes, stochastic gradient descent and alternative damped-Newton. Figure 4 shows how the MAE criteria of the two algorithms (on p2psim, $K = 32$) evolve as the number of iterations increases on both Robust Vivaldi and Robust MF models. Eventually the two methods converge to almost the same value. Apparently, alternative damped-Newton converges much faster, reaching a stationary point within 20 iterations. Stochastic gradient descent, on the other hand, taking more (about 80) iterations for convergence. However, due to its efficiency in each round, it still runs much faster than alternative damped-Newton. On a typical 2.4GHz machine, Stochastic Gradient Descent takes less than 10 seconds to compute a coordinate for any data sets we have used while Alternative Newton can take about 2 minutes. These empirical results coincide the discussion in Section 3.4.

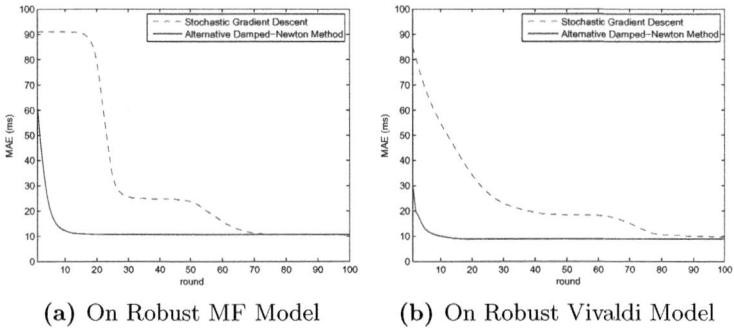

(a) On Robust MF Model   (b) On Robust Vivaldi Model

Fig. 4. Convergence investigation

5   Conclusion and Future Works

In this paper, we argue that the ℓ^2-norm widely used in traditional NCS approaches is not robust for the noisy network latency data. We therefore introduce an elegant error measure, the Huber norm to network latency prediction, based on which we upgrade the traditional Vivaldi and matrix factorization NCS approaches into more robust versions. We conduct extensive experiments and verify the robustness of the upgraded models. According to the results, we recommend Robust Vivaldi as the first choice. However, when there are sufficient historical data, Robust MF is also considerable. We also provide two different learning methods for the models: the alternative damped-Newton method which takes fewer iterations to converge, and the stochastic gradient descent method which is slower in convergence but faster in terms of CPU time. As our future work, we hope to find factors that could capture the noise and further boost the accuracy of NCS.

References

1. Harvard data set, http://www.eecs.harvard.edu/~syrah/nc/king/lats.n8.gz
2. Meridian data set, http://www.cs.cornell.edu/People/egs/meridian/data.php
3. P2psim data set, http://pdos.csail.mit.edu/p2psim/kingdata/
4. Agarwal, S., Lorch, J.: Matchmaking for online games and other latency-sensitive p2p systems. ACM SIGCOMM Computer Communication Review 39(4), 315–326 (2009)
5. Barto, A.G., Anandan, P.: Pattern-recognizing stochastic learning automata. IEEE Transactions on Systems, Man, & Cybernetics (1985)
6. Brin, S., Page, L.: The anatomy of a large-scale hypertextual web search engine. Computer networks and ISDN systems 30(1-7), 107–117 (1998)
7. Buchanan, A.M., Fitzgibbon, A.W.: Damped newton algorithms for matrix factorization with missing data. In: IEEE Computer Society Conference on Computer Vision and Pattern Recognition, CVPR 2005, vol. 2, pp. 316–322. IEEE (2005)
8. Chen, Y., Wang, X., Song, X., Lua, E., Shi, C., Zhao, X., Deng, B., Li, X.: Phoenix: Towards an Accurate, Practical and Decentralized Network Coordinate System. In: Fratta, L., Schulzrinne, H., Takahashi, Y., Spaniol, O. (eds.) NETWORKING 2009. LNCS, vol. 5550, pp. 313–325. Springer, Heidelberg (2009)
9. Dabek, F., Cox, R., Kaashoek, F., Morris, R.: Vivaldi: A decentralized network coordinate system. In: ACM SIGCOMM Computer Communication Review, vol. 34, pp. 15–26. ACM (2004)
10. Guitton, A., Symes, W.W.: Robust inversion of seismic data using the huber norm. Geophysics 68(4), 1310 (2003)
11. Gummadi, K., Saroiu, S., Gribble, S.: King: Estimating latency between arbitrary internet end hosts. In: Proceedings of the 2nd ACM SIGCOMM Workshop on Internet Measurment, pp. 5–18. ACM (2002)
12. Huber, P.J.: Robust regression: asymptotics, conjectures and monte carlo. The Annals of Statistics 1(5), 799–821 (1973)
13. Ke, Q., Kanade, T.: Robust l1 norm factorization in the presence of outliers and missing data by alternative convex programming. In: IEEE Computer Society Conference on Computer Vision and Pattern Recognition, CVPR 2005, vol. 1, pp. 739–746. IEEE (2005)

14. Ledlie, J., Pietzuch, P., Mitzenmacher, M., Seltzer, M.: Wired geometric routing. In: Proc. of IPTPS, Citeseer (2007)
15. Liao, Y., Geurts, P., Leduc, G.: Network Distance Prediction Based on Decentralized Matrix Factorization. In: Crovella, M., Feeney, L.M., Rubenstein, D., Raghavan, S.V. (eds.) NETWORKING 2010. LNCS, vol. 6091, pp. 15–26. Springer, Heidelberg (2010)
16. Lumezanu, C., Spring, N.: Playing vivaldi in hyperbolic space (2006)
17. Mao, Y., Saul, L.K., Smith, J.M.: Ides: An internet distance estimation service for large networks. IEEE Journal on Selected Areas in Communications 24(12), 2273–2284 (2006)
18. Valiant, L.G.: A theory of the learnable. Communications of the ACM 27(11), 1134–1142 (1984)

BSense: A Flexible and Open-Source Broadband Mapping Framework

Giacomo Bernardi[1], Damon Fenacci[1],
Mahesh K. Marina[1], and Dimitrios P. Pezaros[2]

[1] School of Informatics, The University of Edinburgh, UK
[2] School of Computing Science, University of Glasgow, UK

Abstract. We present, BSense, a flexible broadband mapping system for broadband coverage and quality assessment of broadband connections in a given geographic region. For coverage related analysis, it relies on data that is either obtained from ISPs or generated based on technology models and information about infrastructure sites. Broadband quality assessment in BSense is via host-based measurements using our multi-platform and technology-adaptive software client which periodically runs as a background process on users' computers. The host-based software measurement approach employed in BSense is not only cost-effective but is also flexible and reduces measurement bias. BSense also incorporates a flexible broadband quality index for summarizing the collective effect of various underlying attributes such as download/upload speeds and latency. BSense system has been implemented based on open-source software components. The usefulness of the BSense system is demonstrated using two real world case studies, one on identifying notspots in Scotland and the other on broadband quality assessment in a rural part of Scotland through pilot deployment.

Keywords: Broadband mapping, performance measurement, broadband quality index, broadband access technologies.

1 Introduction

Broadband mapping is the process of assessing broadband coverage, quality and market for a given geographical region (e.g., country, province, city). Broadband *coverage* assessment is aimed at identifying *"notspots"*, i.e., locations not serviced by any broadband access technology. For areas that are covered, assessing broadband *quality* in those areas is more interesting. Quality is measured using a set of performance metrics such as download/upload speeds, latency, jitter and packet loss rate. Several technology-specific and network provisioning factors affect quality in practice (e.g., line length, number of concurrent users, contention ratio, backhaul capacity). *Choice* and *cost* associated with broadband subscription are additional aspects that are of interest for broadband mapping, especially to consumers and regulators. In order to determine the amount of choice that a consumer has, one needs to find out various access technologies and Internet Service Providers (ISPs) available at the consumer's location. Greater choice usually

R. Bestak et al. (Eds.): NETWORKING 2012, Part I, LNCS 7289, pp. 344–357, 2012.

also implies lower cost (per Mbps) for the consumer. Moreover, choice and cost both tend to depend on the coverage and quality aspects — poor broadband coverage or quality in a region correlates well with lack of choice and/or higher costs for consumers in that region. Note that all these four aspects – coverage, quality, choice and cost – vary with time though timescales of change may differ widely between them.

Interest in broadband mapping has been growing recently in tune with increase in consumer awareness and recognition by governments on the importance of high-speed Internet access for all citizens. Different countries have launched national broadband mapping programs (e.g., [1,2,3,4]) to quantify the existing state of broadband delivery and to track the progress towards achieving targets set forth in national broadband plans, especially in view of the on-going debate on the role of public funding and regulation in enabling fast and universal broadband access. Beyond these government-initiated efforts, other broadband mapping examples include [5,6]. *Despite these various efforts, we identify the lack of an open and flexible broadband mapping framework, key to effective and consistent mapping exercises.* Open specification of assessment methodology and metrics is important for audit, whereas the use of open-source software results in lower cost implementations which in turn enable broadband mapping efforts in developing regions. Flexibility is also important to accommodate the wide range of broadband access technologies and to factor in latest advances and best practice in terms of measurement techniques. Moreover, these efforts take one of two different approaches we refer to as *model based* (e.g., [7,8,9]) and *measurement based* (e.g., [10,11,5,12,6,13,14]) even though *neither is enough* — model based approaches fail to capture the discrepancy between expected and actual broadband quality experienced by consumers, whereas measurement based approaches are clearly not useful for identifying notspots. We further discuss these different broadband mapping approaches in §2.

In this paper, we present an open and flexible broadband mapping framework called BSense (Fig. 1(a)) along with its implementation based on open-source software components (§3).

— BSense incorporates both model based and measurement based approaches keeping in mind the observation that each is useful for a different purpose. For coverage related analysis, it relies on data that is either obtained from ISPs or generated based on technology models and information about infrastructure sites. Broadband quality assessment in BSense is based on host-based continuous measurement stream obtained using our multi-platform and technology-adaptive software client, which periodically runs as a background process on users' computers. As elaborated in §2, *the measurement paradigm employed in BSense results in a lower-cost and flexible alternative to [10,11], and reduces measurement bias compared to [5,12,14].*

— *Unique to BSense is a flexible broadband quality index* for summarizing the collective effect of various underlying attributes such as download/upload speeds and latency. Specifically, we propose to separately model user preference concerning various performance attributes through a specific

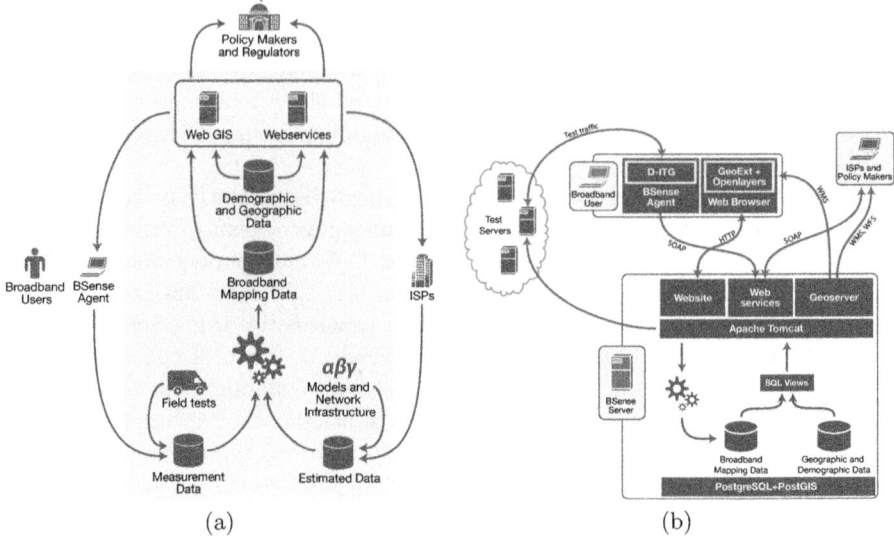

Fig. 1. (a) BSense broadband mapping framework illustrated. (b) BSense software architecture.

instantiation from a flexible family of utility functions and then combine them to produce an overall index by leveraging the general framework of multi-attribute utility theory [15].

To demonstrate the usefulness of the BSense system, we use two real world case studies — one on identifying notspots in Scotland and the other on broadband quality assessment in a rural part of Scotland through pilot deployment involving 60 real users over a three month period (§4). Also, evaluation of our measurement methodology shows that it compares favorably to the hardware based measurement approach [10] as well as with a recently proposed sophisticated measurement technique [16]; see [17] for these results. We should, however, clarify that our focus is not on measurement techniques. Though we use D-ITG [18], an open-source measurement technique we have access to, in our current BSense implementation, our design is flexible enough to allow the use of other measurement techniques such as [16,19]. *We regard the BSense framework, its realization using open-source components and the two real-world case studies demonstrating its utility as our key contributions in this paper.*

2   Related Work

Broadband mapping approaches can be broadly classified into two categories: (1) *model based* and (2) *measurement based*. With the model based approach, broadband coverage and speeds are estimated based on theoretical or empirically derived models of access technologies, knowledge of network infrastructure (e.g., locations of phone exchanges, mobile network base stations) and configurations (e.g., contention ratio, radio parameters). For ADSL, see [7,8] for examples of such models and their use in estimating broadband coverage. Similar approaches can be followed to estimate 3G mobile broadband coverage [9]. Such data is inherently optimistic as it does not consider various practical impediments (e.g., line quality, contention). Nevertheless, it is useful for coverage analysis in the absence of any measurement data. Note that this is the approach followed in [1,2,3] using the data obtained from ISPs.

Measurement based approaches involve actual measurement of broadband connections, a necessity for assessing broadband quality in a region. None of the existing measurement approaches we are aware of rely on measurement data from ISPs. Measurement based approaches can be further divided into *hardware-based* and *software-based* approaches.

The hardware-based measurement approach (also called gateway-based approach) involves deploying a customized hardware box that directly connects to home broadband router for a representative *sample* of users, and using the gathered statistics across all such boxes to estimate statistics for the whole *population*. It is pioneered by SamKnows [10] for UK Ofcom and US FCC sponsored broadband speed studies, and is also considered recently in the academic community [11]. This approach can be expensive than an equivalent software based measurement approach. Moreover, it is also limited in terms of flexibility — unviable for mobile/wireless broadband quality assessment and for analysis at a fine-grained geographic granularity (e.g., city) when planned at national level.

Software-based measurement approaches broadly come in three varieties:

- *Web based*: Using web-based speed tests such as [5,12] can only gather sporadic and geographically non-uniform measurement data and also suffer from measurement biases (e.g., users taking speed tests may have poor broadband connections or may not belong to a representative sample).
- *Host daemon based*: This approach relies on a measurement agent running in the background on user computers and can overcome the limitations of the web-based approach. However, existing systems following this approach [13,6] lack openness and flexibility expected from a broadband mapping system.
- *Consumer and ISP independent*: Dischinger et al. [14] present an interesting approach that does not require cooperation from either the consumers or the ISPs. This approach relies instead on certain specific but standard functionality from routers (e.g., responding with TCP RST packets upon receiving unsolicited ACKs). Such functionality may be disabled due to security concerns. If a particular ISP does not support this functionality on all its broadband routers, then that ISP is effectively ignored by this approach, introducing a measurement bias and thus undesirable from a broadband mapping perspective.

BSense uses a model based approach as well as a host daemon based software measurement approach, each for a different use case.

3   BSense Broadband Mapping Framework

As shown in Fig. 1(a), BSense framework views broadband mapping as a cooperative exercise involving various stakeholders — consumers, ISPs, policy makers and regulators. In [17], we elaborate on the incentives that can be provided to different stakeholders to continually contribute to the exercise. BSense brings together different types of data. Estimated broadband data from ISPs is fed into the BSense database via webservice API calls. Broadband users (consumers) are the key source of continuous measurement data for the mapping system. This is enabled by a lightweight software agent termed *BSense Agent* that runs in the background on a user computer and periodically communicates with BSense *Test Servers* to measure technical attributes of user's broadband characteristics such as download speed, upload speed and latency. Digital geographic data from country-specific sources and demographic data from population census are additionally used as layers underneath estimated or measured broadband statistics to generate broadband coverage or quality maps, respectively.

3.1   Design and Implementation

Fig. 1(b) depicts the BSense software architecture, which is described in the rest of this section.

BSense Server. The BSense server side includes a web server for hosting a public website that users can access for registering and downloading the BSense agent software and subsequently to retrieve their broadband connection statistics. The web server also supports a set of web service API calls over SOAP for interaction between the BSense system and various stakeholders. The current API consists of the following calls: BroadbandTestRecord, AddPackage, EditPackage, DeletePackage, AddEstimatedData, EditEstimatedData, DeleteEstimatedData and LookUpMappingData. The purpose of these calls is evident from their names; for further elaboration, see [17]. These API calls are handled by a server side component that enforces security and access control, validating the input and checking whether an API call is made by a party with the required permissions.

In addition to SOAP based web services, BSense provides external access via the Open Geospatial Consortium's standard WMS (Web Map Service) and WFS (Web Feature Service) to obtain raster and vector geo-referenced images, respectively, of a geographical area of interest; most open-source and commercial GIS software products can directly use WMS and WFS services. BSense also provides an in-built web application based on WMS, developed using open-source GeoExt (http://www.geoext.org/) and OpenLayers (http://openlayers.org/) frameworks, to further ease access to broadband maps and their visualization.

Coming to data management, our implementation uses the open source PostgreSQL (http://www.postgresql.org/) database management system augmented with the PostGIS (http://postgis.refractions.net/) extensions to handle spatial data. See [17] for further details (database schema, etc.). Access to the BSense database via WMS and WFS is enabled by the well known open-source GeoServer (http://www.geoserver.org), a Java software that allows users to view and edit geospatial data. Besides broadband mapping data, BSense allows the storage of additional geospatial layers with geographic and demographic data that are useful when generating maps.

BSense Agent. Each broadband user participating in the BSense based mapping exercise runs a software agent (BSense Agent) that facilitates continuous and cost-effective measurement of the user's broadband connection. As such, the agent is a key element of the BSense framework for gleaning the quality of broadband provisioning in a given region. Given the diversity of operating system (OS) platforms used by consumers in the real world, the agent should function on different commonly used platforms to avoid measurement bias. The BSense agent was designed explicitly keeping in mind this requirement for multi-platform support. Specifically, it has been developed using Qt (http://qt.nokia.com), an open-source and cross-platform application/UI framework; Qt is available for Windows, Linux and Mac OS X.

We first give a high-level overview of the measurement process. A participating broadband user would download the agent from a public website like the one we developed (http://broadbandforall.net) and install it on the user's home computer. The agent runs in the background and periodically wakes up to perform a measurement test of the user's broadband connection[1]. Each measurement test consists of the following sequence of steps (Fig. 2(a)):

1. The agent queries BSense server to get the details of the measurement test to be performed.
2. BSense server replies with an "experiment definition" (elaborated below) as well as details for a test server to be used (e.g., IP address, port number).
3. BSense server also simultaneously notifies the test server about the impending measurement test from the user's agent.
4. The agent interprets and follows the experiment definition received, generating the traffic flow requested and/or receiving the incoming traffic to/from the specified test server.
5. Upon test completion, the agent summarizes the traffic traces from the test and uploads it to the BSense server.

Since the effectiveness of the mapping framework improves with a larger number of participating users, the mapping system should be scalable and robust to server failures. In the case of BSense server, this can be achieved through the use of a server farm (the current approach) or by installing the BSense server on a cloud-based hosting. The use of multiple test servers as in our current design also contributes towards scalability and fault tolerance. As regards to the location of

---

[1] The time interval between measurements (i.e., measurement frequency) is a customizable parameter whose setting is a tradeoff between gathering fine-grained measurement samples over time and measurement overhead. The value is set to 15 minutes in the current implementation.

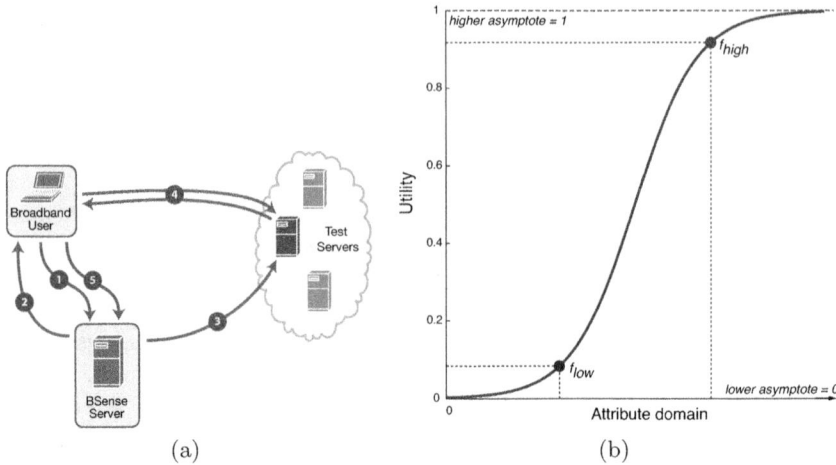

Fig. 2. (a) The sequence of steps making up a measurement test initiated by an instance of the BSense agent. (b) The sigmoid utility function can be seen as a transfer function between a given attribute f and the perceived utility associated with specific values of f. In fact, the function shown is a modified sigmoid function given in equation (1) to realize zero utility when the value of f is zero.

test servers, we advocate their deployment at neutral Internet exchange points (IXPs) (e.g., the ones listed at http://www.euro-ix.net/) to avoid introducing bias against users of some ISPs. In the current implementation, however, test servers are co-located with the BSense server farm.

BSense agent could in principle use any multi-platform network performance measurement tool. In our implementation, we choose a widely used traffic generator called D-ITG [18] with BSense Agent acting as a wrapper application. While D-ITG can be seen just as a placeholder in the current implementation, it also has several attractive features that have influenced our choice such as the following: open source; can be made to work on different platforms and behind most common types of NATs with minimal effort; and provides a high degree of flexibility when it comes to traffic generation.

We now briefly describe the experiment specified to the agent every time it is about to do a measurement test. An experiment is defined as a set of traffic session specifications with each session potentially consisting of multiple concurrent or partially overlapping flows. Specifically, each experiment in our context is a sequence of three traffic sessions: initial ping-like UDP session with short packets to measure latency, jitter and packet loss rate, followed by an upstream traffic session and then a downstream traffic session. See [17] for detailed parameter settings. Note that in a host-based software measurement approach like ours, it is possible to have in-home wireless network use and other active use of the broadband connection from other devices within the home concurrently with broadband connection measurement traffic. However, the problem of cross-traffic on the network path is unavoidable with end-to-end performance measurement even with hardware-based measurement approach. We believe it is best dealt with via statistical filtering.

3.2 Broadband Quality Index

As noted at the outset, coverage and quality are the two key aspects of interest for broadband mapping. While broadband coverage in a particular location can be quantified as a binary variable, the same is not true for broadband quality as the latter is dependent on several underlying technical attributes such as download speed, upload speed and round-trip latency. Due to the lack of standard ways to summarize the collective impact of those several attributes, the focus is often solely on download speeds even though it is widely recognized that upload speeds and latency also have to be considered at the very least.

Defining an *index* is a common approach to deal with problems of the above nature. Only work we are aware of that tries to address the issue of developing a broadband quality index is surprisingly a sociological study [20] relying on expert surveys to determine the relative importance of various technical attributes. More importantly, the work in [20] only provides a very specific approach to defining the broadband quality index while we are interested in defining a more flexible and general framework.

Our main idea in addressing this issue is to model each attribute impacting broadband quality as a *utility function* and then draw upon the *multi-attribute utility theory (MAUT)* [15] to define the *broadband quality index (BQI)* as a composite function of utility function values for the individual attributes.

In our model, denote $F = [f_1, ..., f_n]$ for the set of network attributes (features) to be included in the BQI, covering important attributes characterizing a broadband connection. For the sake of concreteness, we focus on three key performance attributes in this paper: download speed (d, in Mbps), upload speed (u, in Mbps) and round-trip latency (l, in milliseconds).

We first aim to identify a suitable family of *realistic* single-attribute utility (SAU) functions for modeling user preferences about individual attributes, and then consider their composition into a multi-attribute function. For the attributes under consideration (i.e., download speed, upload speed and latency), we observe that sigmoid functions, whose graphs are "S-shaped" curves (see Fig. 2(b)), better reflect user satisfaction. This is because improvement in utility from improving any of these attributes beyond a point is marginal. Equally, when these attributes are below a certain threshold (for speeds) and above a certain threshold (for latency), the change in utility is again marginal. In between these extremes, the improvement in utility with improvement for any of these attributes is noticeable and substantial.

Thus, we define a set of utility functions $u_f(f)$, each defined on a given attribute $f \in F$ as:

$$u_f(f) = \frac{e^{a+bf} - e^a}{1 + e^{a+bf}} \qquad (1)$$

where parameters a and b determine the nature of each utility function curve.

A pragmatic approach to specify a utility function for each attribute would be to have the BSense administrator pick two strategic values of each attribute f and provide their corresponding utility u values:

$$(f^o, u^o) \text{ and } (f^*, u^*) \quad \text{s. t.} \quad u^o = u_f(f^o) < u^* = u_f(f^*) \tag{2}$$

These two points can be carefully picked so that they represent the utility of low-end and high-end broadband connections (e.g., with $u^o = 0.2$ and $u^* = 0.8$ or $u^o = 0.1$ and $u^* = 0.9$). Intuitively, the lower knee in the curve represent the value of the attribute which is deemed as insufficient, and the upper knee describes the attribute value which is good enough for the service. As a consequence, "poor" broadband connections that are only able to offer attribute values (e.g., download speeds) below the lower point offer only marginal utility to the users. Similarly, the incremental utility above the upper threshold is only marginal. The low and high values can, for example, be based on current policies and regulations (e.g., the "Universal Service Obligation" to set the bottom bar that ISPs have to provide and the policy maker must enforce) or current state-of-the-art (e.g., the fastest commercially available service to set the top bar).

The parameters a and b for each of our SAU functions can then be derived from their two specified corresponding 'knobs' by solving the following pair of equations obtained by substituting values from (2) in equation (1).

To provide a quality index for each broadband connection, we first need to generate a summary statistic from all the measurements that have been gathered by BSense for that connection. In the current implementation, we use the median values of download speed, upload speed and latency. These values along with their respective a and b parameter values as input to equation (1) determine the utilities of the broadband connection in question with respect to each of those three attributes.

Multi-attribute utility theory assists us in combining the various SAU functions in a single equation, whose form depends upon the particular independence conditions fulfilled by the different SAU functions. For simplicity, we assume mutual additive independence in this paper. Then the resulting multi-attribute utility function can be represented as:

$$u(f) = \sum_{f \in F} k_f u_f \quad \text{s.t.} \quad \sum_{f \in F} k_f = 1 \quad \text{and} \quad k_f > 0, \quad \forall f \in F \tag{3}$$

A further simplifying assumption would be to have all scaling constants (weights) k_f to be equal. This is reasonable given that our main purpose is to demonstrate the value of multi-attribute utility theory and utility functions in providing a flexible framework for defining broadband quality index.

4   Case Studies

4.1   Broadband Coverage Analysis for Scotland

In this case study we show the benefit of BSense for understanding broadband coverage, using Scotland as the setting. Such studies would rely upon estimated

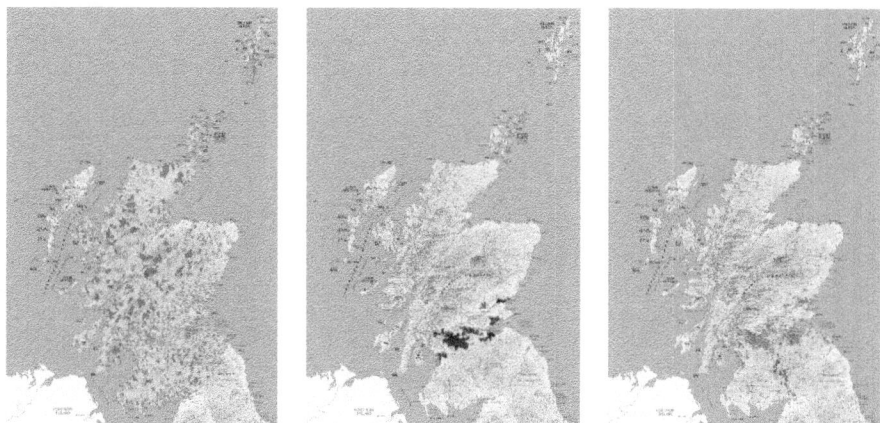

Fig. 3. BSense generated broadband map for various access technologies and ISPs based on their estimated data: (a) ADSL — BT Wholesale; (b) Cable — Virgin and Smallworld; (c) 3G mobile broadband — Orange. In each of these maps, postcode areas with the corresponding service are colored with darker colors indicating faster expected speeds or better mobile coverage.

coverage and speed data from ISPs whenever available. For this study, we mimicked the way ISPs would contribute to the BSense mapping system by trawling through the public websites of different ISPs to determine whether an ISP covers a particular postcode and if so, the estimated download speeds from the ISP's viewpoint, for each of the 152,000 postcodes in Scotland. This information is then fed into the BSense estimated database via the web service API calls.

Fig. 3 shows the broadband coverage in Scotland for different access technologies based on the estimated data from ISPs collected as described above. For the 3G mobile broadband case, we show data for only one network operator for clarity but the coverage for other mobile network operators is similar. From these maps, we observe that ADSL is the dominant access technology with cable and mobile confined mainly to population centers in the central belt and north east.

Now focusing on ADSL alone, we estimate the notspots in Scotland with respect to a threshold download speed. Specifically, a postcode area is considered to be a notspot if ADSL service with estimated download speed above the specified threshold cannot be supported within that area. This may be because residences in the postcode area are too far away from their nearest phone exchanges, for example. We consider three different threshold values (512Kbps, 2Mbps, 8Mbps). Resulting notspot maps produced using BSense are shown in Fig. 4. It can be clearly seen that most postcode areas outside of the central belt of Scotland (with the two main cities of Edinburgh and Glasgow and having the largest population concentration) become notspots as the threshold is increased. While it is true that satellite based broadband covers virtually the whole of Scotland, the large round-trip latencies associated with the satellite technology (as shown using measurements in the next case study) makes it less attractive.

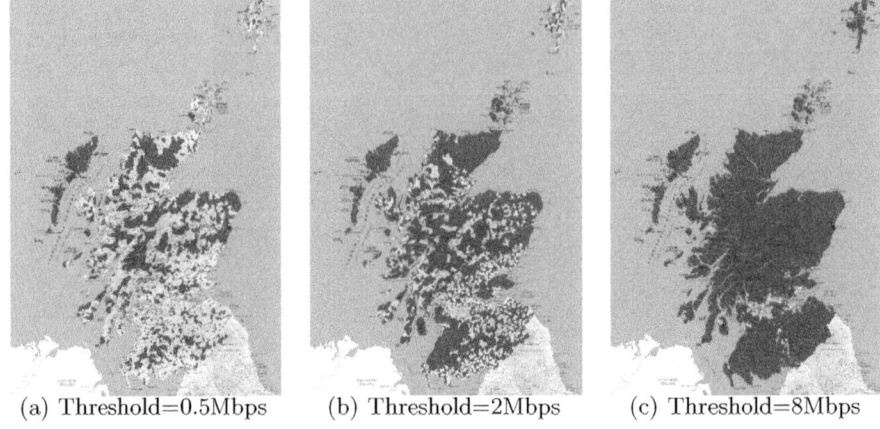

(a) Threshold=0.5Mbps   (b) Threshold=2Mbps   (c) Threshold=8Mbps

Fig. 4. BSense generated map of notspots in Scotland that lack an ADSL broadband service supporting download speed greater than the indicated threshold. Notspot postcode areas are shaded in red.

4.2 Broadband Quality Measurement: A Pilot Study

In this case study, we assess the broadband quality in a rural part of Scotland. Specifically, we focus on the area around the Isle of Skye located in the northwest of Scotland. We also consider the neighboring archipelago of the 'Small Isles' and the mainland rural areas of Glenelg and Knoydart peninsulas. This region is quite diverse in terms of demographics, terrain and broadband service provisioning, making it a well suited region for our broadband performance measurement study. It has a population of around 10 thousand people spread across a handful of small towns, several small villages and scores of isolated dwellers in the farming lands. Several different access technologies used for broadband provisioning. In total, 15 phone exchanges are located in the area. Although every resident has access to a landline, broadband connection types vary. A few phone exchanges are enabled for ADSL2/ADSL2+, which is available only from the telecom incumbent (BT). Other exchanges offer ADSL (8Mbps download speed) service, and a few are enabled only for "Exchange Activate" (512Kbps) ADSL service. There are no FTTH deployments in the area, and cable and 3G coverage are non-existent. Due to a recent broadband reach initiative from the Scottish government, some of the users in rural and remote areas in previously notspot areas in Scotland, including those in our study area, now connect via subsidized yet relatively expensive satellite connections. In addition, residents in a small part of this area connect via Tegola, an experimental/community long distance WiFi network we have deployed four years ago [21].

Through publicity of our pilot broadband quality assessment initiative via email, local press and word of mouth, we managed to find 60 volunteers in the area who were willing to install and run our BSense agent software. Half of these users are connected to the Internet via ADSL lines to different exchanges and differing line lengths, whereas 18 users connected via our Tegola network; remaining volunteers used satellite connections. Over a 3-month period, we measured the broadband connections of each of the volunteer users, keeping track of median values of download/upload speeds and latency measurements for each user. We collected around 40,000 measurements in total. To study the broadband quality index across users and access technologies, we used the following parameter settings for the individual utility functions (see Section 3.2), all reasonable given the type of broadband connections in the study area:

- *Download speeds.* Low-end: 2Mbps with a utility of 0.1. High-end: 24Mbps with a utility of 0.9.
- *Upload speeds.* Low-end: 1Mbps with a utility of 0.1. High-end: 5Mbps with a utility of 0.9.
- *Latency speeds.* Low-end: 200ms with a utility of 0.1. High-end: 20ms with a utility of 0.9.

Fig. 5(a) shows the results. The top graphs show the utility function values for each of the three performance attributes. Each data point in the plots corresponds to a user with the color of the data points indicating the access technology used. Clearly and as expected, satellite users have poor utility values and are clustered together at the worst extreme. Wireless users on the Tegola network, on the other hand, not only experience high speeds exceeding 20Mbps but also are subject to greater variability in speeds because of the shared nature of access. ADSL users also exhibit greater variability in speeds like wireless users but because of different reasons — due to differences in broadband capabilities of the associated phone exchanges and differences in line lengths; most ADSL users fall in between satellite users and wireless users in this area. The bottom graph

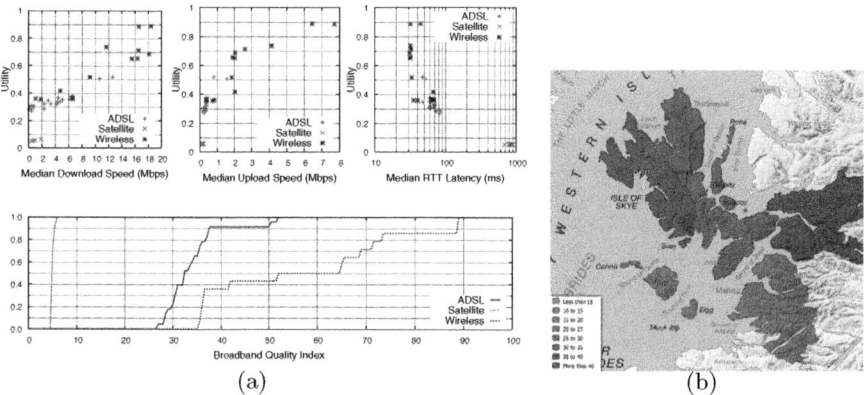

(a)   (b)

Fig. 5. (a) Pilot study results. Top graphs showing the utilities for different attributes, different users and different access technologies. Bottom graph summarizes the broadband quality index values for users of different access technologies as CDFs. (b) Map generated using the results from the pilot study aggregating users belonging to a ward together and colors representing different ranges for broadband quality index values.

shows the combined effect of the three attributes. For ease of interpretation, we scale up each user's index value to a percentage value between 0% and 100%. Results for different access technologies shown as CDFs follow directly from the top graphs given our choice for the multi-attribute utility function (an equal weighted sum of individual attribute utilities). When index values for different users are geographically rendered on a map, however, the result is quite revealing (see Fig. 5(b)). Here coloring is done at the ward level — all users belong to a ward (an area with about 50 residents and 20 households) are aggregated together. We observe that remote parts of Knoydart and the Small Isles (colored in red) fare poorly, whereas adjacent ward above Knoydart has the best index as a result of high-speed wireless connections from the Tegola network. Wards on the Isle of Skye have intermediate index values as it mainly consists of ADSL users.

5   Conclusions

In this paper, we have developed a flexible framework for broadband mapping called BSense that incorporates both model based broadband coverage data and broadband performance measurement data from users. BSense framework also incorporates a flexible specification of broadband quality index based on utility functions and multi-attribute utility theory. We implemented BSense using open-source tools and use it demonstrate the value of BSense approach for broadband coverage and quality assessment with two real-world case studies. Our future work will focus on enhanced measurement techniques that are robust to various sources of variability. We would like to also enhance the broadband quality index by considering additional attributes, variability of each attribute and relationships among various attributes. Finally, we would like to extend BSense for mobile broadband mapping.

References

1. US National Broadband Map, http://broadbandmap.gov/
2. UK Fixed Broadband Map, http://maps.ofcom.org.uk/broadband/index.html
3. Broadband Map for Germany, http://www.breitbandatlas.de
4. Ofcom. UK Fixed Broadband Speeds (March 2011),
   http://stakeholders.ofcom.org.uk/market-data-research/
   telecoms-research/broadband-speeds/speeds-nov-dec-2010/
5. Ookla Speedtest, http://www.speedtest.net/
6. Grenouille, http://www.grenouille.com/
7. Grubesic, T.H.: Spatial Data Constraints: Implications for Measuring Broadband. Telecommunications Policy 32(7) (August 2008)
8. Ofcom. Assessment of the Theoretical Limits of Copper in the Last Mile (July 2008), http://www.ofcom.org.uk/research/~technology/research/
   emer_tech/copper/
9. Ofcom. 3G Coverage Maps (July 2009), http://www.ofcom.org.uk/radiocomms/
   ifi/~licensing/classes/broadband/cellular/3g/maps/3gmaps/

10. SamKnows, `http://www.samknows.com/broadband/`
11. Sundaresan, S., et al.: Broadband Internet Performance: A View From the Gateway. In: Proc. ACM SIGCOMM (2011)
12. Network Diagnostic Tool (NDT), `http://www.measurementlab.net/measurement-lab-tools`
13. Isposure, `http://www.isposure.com/`
14. Dischinger, M., Haeberlen, A., Gummadi, K., Saroiu, S.: Characterizing Residential Broadband Networks. In: Proc. Internet Measurement Conference, IMC (2007)
15. Russell, S., Norvig, P.: Artificial Intelligence: A Modern Approach, ch.16. Pearson (2010)
16. Kanuparthy, P., Dovrolis, C.: ShaperProbe: End-to-end Detection of ISP Traffic Shaping using Active Methods. In: Proc. Internet Measurement Conference, IMC (2011)
17. Bernardi, G., Fenacci, D., Marina, M.K., Pezaros, D.P.: BSense: A Flexible and Open-Source Broadband Mapping Framework. The University of Edinburgh, Informatics Research Report EDI-INF-RR-1408 (December 2011), `http://homepages.inf.ed.ac.uk/mmarina/papers/EDI-INF-RR-1408.pdf`
18. Botta, A., Dainotti, A., Pescapè, A.: Multi-Protocol and Multi-Platform Traffic Generation and Measurement. In: Proc. IEEE INFOCOM (2007); Demo
19. Kreibich, C., Weaver, N., Nechaev, B., Paxson, V.: Netalyzr: Illuminating The Edge Network. In: Proc. Internet Measurement Conference, IMC (2010)
20. Vicente, M., Gil de Bernabé, F.: Assessing the Broadband Gap: From the Penetration Divide to the Quality Divide. Technological Forecasting & Social Change 77(5) (June 2010)
21. Bernardi, G., Buneman, P., Marina, M.K.: Tegola Tiered Mesh Network Testbed in Rural Scotland. In: Proc. ACM MobiCom Workshop on Wireless Networks and Systems for Developing Regions (WiNS-DR 2008) (September 2008)

Validity of Router Responses
for IP Aliases Resolution

Santiago Garcia-Jimenez, Eduardo Magaña, Mikel Izal, and Daniel Morató

Public University of Navarre, Campus Arrosadia, 31006 Pamplona, Spain
santiago.garcia@unavarra.es

Abstract. In order to obtain close-to-reality Internet maps, IP aliases resolution allows identifying IP addresses belonging to the same router. Mainly, active probing is used for IP aliases resolution following direct and indirect schemes. Also, different types of probe packets are used (ICMP, UDP, etc.) focusing on different header fields and characteristics of IP and higher layers. Responsiveness of routers is different not only in the number of response packets received, but also in the validity of those packets to be used in IP aliases identification. Therefore, specific behavior of routers generating those response packets can decide the success or failure of specific IP aliases resolution methods. In this paper, an in-depth analysis of router behaviors is provided considering not only router responsiveness, but also the validity of those responses to be used in IP aliases resolution. Our results show that although responsiveness is better for indirect probing, direct probing with ICMP Echo probe packets and IPID-based behavior provide the best identification ratio for IP aliases resolution.

Keywords: IP aliases resolution, router responsiveness, active probing, direct and indirect probing.

1 Introduction

Several attempts have been put forward over the last decade to obtain an Internet map, like ARK [1], iPlane [2], Skitter [3] and DIMES [4]. They are mostly based on the traceroute tool, launched periodically between a high number of vantage points (controllable nodes that generate probe packets). This generates a graph composed by nodes (IP addresses) and links between nodes (adjacencies obtained from traceroute paths).

A special case of Internet map is the one in which the nodes in the graph are routers instead of IP addresses. This is a router-level Internet map where the graph links represent the connectivity between interfaces of different routers. Those router-level Internet maps are useful in network simulation, P2P protocol optimization, improvement of routing protocols, geolocation of IP addresses and many other applications.

This topology information is not provided by almost any Internet carrier or Autonomous System. The reasons are related to security and reluctance to share

R. Bestak et al. (Eds.): NETWORKING 2012, Part I, LNCS 7289, pp. 358–369, 2012.

network information with competing ISPs. This means that router-level Internet maps have to be inferred from passive or active monitoring schemes. In passive monitoring, traffic at specific network points is captured and analyzed looking for some specific information. In active monitoring, probe packets are sent to the network infrastructure, and the responses are analyzed to discover network characteristics. In topology discovery, mainly active monitoring is used because it allows discovering remote networks from a limited number of vantage points.

The traceroute tool discovers the IP addresses of the routers in the path to the target IP address. When probing a high number of target IP addresses from different vantage points, an approximation of an Internet map can be provided. To construct a router-level Internet map, IP addresses of the same router have to be aggregated. Those IP addresses are called *IP aliases* and the process of aggregation is called *IP aliases resolution* [5]. Therefore, two phases have to be performed: IP addresses discovery and IP aliases resolution. Some projects like Rocketfuel [5] and, more recently, MIDAR [6] perform IP aliases resolution at large scale.

IP aliases resolution techniques are also based mainly on active probing and, therefore, it is important to select the right type of probe packet and measurement procedure to obtain the highest number of responses possible. The percentage of response packets over the number of probe packets is called *responsiveness* [7]. However, the responses, depending on their characteristics, could be useless to perform IP aliases resolution and, therefore, the responsiveness indicator is not enough to determine the final results from the IP aliases resolution procedure. Router responsiveness to active probing was analyzed in [7] and an evaluation of performance for several IP aliases resolution schemes is available in [8]. However, there are not studies about the *validity* of this responsiveness: this means the ratio of responses that really are useful to apply IP aliases resolution techniques. Validity of responses will depend on the great variety of router implementations and configurations in Internet. This paper focuses on this router behavior and it will identify those active monitoring schemes that provide the best ratios of valid responses in IP aliases resolution. In fact, it will be shown that strategies with more responsiveness will not always provide the best identification ratios in IP aliases resolution.

The rest of the paper is organized as follows. Section 2 presents different schemes of IP aliases resolution techniques in the state of the art that will be evaluated in the paper. In section 3, the network scenario used in the evaluation is presented. Router behaviors related to IP aliases resolution are presented in section 4. Section 5 presents the evaluation of IP aliases resolution based on the types of routers behaviors. Finally, conclusions are presented.

2   Related Work

There are different possibilities of probing schemes that can be used by IP aliases resolution techniques. These schemes can be classified depending on multiple aspects: the need of probe traffic, the directiveness of the measurement, the type of probe packets and the type of router behavior.

First, depending on the necessity of sending probe packets, IP aliases resolutions schemes can be classified in active and inference-based. Active probing techniques are based on sending probing packets to the routers and analyzing the responses. They provide the best performance in IP aliases resolution [9], but they are intrusive and they need to be controlled in order not to get confused with network scanning or attacks. Inference techniques are the other possibility. They try to deduce aliases information by analyzing data extracted from traceroute paths or from out-of-band measurements such as checking similarities of DNS names in router interfaces. We will focus in active probing techniques.

Second, looking at the measurement directiveness, indirect methods send the probe packet to different IP addresses than the target IP addresses of the aliases resolution technique. Traceroute tool is an example of indirect method for network discovery. In direct methods, the probe packet is addressed at the target IP interface of the aliases resolution technique. Sending ICMP Echo Request packets is an example of direct method. As stated in [7], router responsiveness is greater in indirect methods than in direct ones. However, the validity of those responses to be used in IP aliases techniques was not evaluated in that work. We will perform an analysis of validity of probe responses in section 4.

Third, several types of probe packets are used in IP aliases resolution techniques: UDP, TCP, ICMP Echo Request and ICMP Timestamp Request. As stated in [7], ICMP Echo Request provides the best responsiveness results. The validity of those responses is analyzed in section 4.

Finally, IP aliases resolution techniques are based on different peculiarities of router behaviors. Routers fill up some fields of response packets following specific patterns that can be used to identify aliases. The main behaviors used in IP aliases resolution techniques are IPID-based, Timestamp-based and Source IP-based.

The IPID is the identifier field in the IP header. This IPID is originally used in the procedures of fragmentation and reassembly of IP packets. Typical TCP/IP implementations use a counter which is incremented by one for each packet generated (not forwarded) by the router, independently of destination, protocol or service. Therefore, several packets received from the same router and near in time will have close IPID values, following an incremental pattern. Probing different IP addresses of the same router simultaneously, an incremental sequence of IPIDs is obtained. This behavior was used first by the Ally technique [5], with 3 UDP probe packets being sent and allowing an IPID offset of 200 IPIDs between the first and the third response in order to consider both IP addresses to belong to the same router.

Timestamp-based behavior uses the prespecified timestamp option in the IP header that allows selecting up to four IP addresses and receiving the timestamps from those IP addresses. Typical implementations provide milliseconds timestamps that allow checking if two IP addresses are aliases (they will have the same timestamp). It was used for the first time in the Prespecified Timestamp technique with direct ICMP probe packets [10].

Source IP-based behavior uses special probe packets to generate ICMP Error response packets from the target routers. Probe packets are usually UDP packets sent to a random port at the target IP address. The corresponding router answers with an ICMP Error Port Unreachable packet whose IP address can be different from the destination IP address of the probe packet. In fact, the source IP address of the response is usually chosen from the interface with shortest path to the destination. Therefore, probing different IP addresses from the same vantage point, they will be aliases if the response packets have the same source IP address. This behavior was used for the first time in the Mercator technique [11], using UDP probe packets.

Some of the most commonly used techniques for IP aliases resolution, besides those previously described, are presented below. TraceNet [12] uses direct/indirect probing, source-IP based behavior and ICMP/UDP probe packets. It infers the subnetworks, and it tries to obtain aliases at the same time that the traceroute is performed. It is based on distance to provoke ICMP Error TTL exceeded responses.

Palmtree [13] uses direct probing, source-IP based behavior and ICMP/UDP probe packets sent to inferred /30 and /31 subnetworks. Those probe packets have bounded TTL in order to obtain ICMP Error TTL exceeded responses with the desired source IP addresses.

In [14], Ally-based techniques are proposed extending the types of probe packets (ICMP, TCP) and the number of probe packets compared with the standard Ally. The rate at which probe packets are generated is also controlled with 0.3 secs inter-packet delay.

Radargun [15] uses direct probing, IPID-based behavior and UDP/TCP probe packets to apply a velocity modeling to characterize IPID evolution per router. It allows to check for IPID evolution in thousands of IP addresses simultaneously. Also, Midar [6] proposal argues to identify aliases with an improved variation of IPID-based behavior, but the current version (September 2011) is limited to 200 IP addresses and its identification results are not as good as expected.

Focusing in active probing, the following sections analyze the validity of the responses obtained as a funtion of the directiveness of the measurement, the type of probe packets and the type of router behavior. Finally, a performance comparison for previous IP aliases resolution techniques will be presented.

3   Network Scenario

In order to compare the IP aliases resolution techniques over the same network scenario, specific requirements are needed for this scenario. Some techniques, like TraceNet, need a large set of vantage points to perform indirect probing between them. We have used vantage points belonging to the Planetlab measurement infrastructure [16]: 25 Planetlab nodes have been used as vantage points to perform IP aliases resolution for IP addresses of the routers in between. Those IP addresses have been discovered using paris-traceroutes [17] between each pair of vantage points, resulting in 2037 different IP addresses discovered, some of them belonging to the same router (aliases).

As the underlying topology is unknown, the quality of IP aliases resolution techniques can not be checked for the existence of false positives and false negatives. There are some NRENs (National Research and Educational Networks) that provide public information about their network topologies, but they are small and they are composed by similar router behaviors (same manufacturer and even router models in many cases). Some examples of those networks are Geant [18], Canet4 [19] and GlobalNOC [20], but they do not provide more than 500 IP addresses at most. Besides, we do not have enough nodes in the border of those networks to be used as vantage points (needed for indirect methods).

Planetlab provides a bigger topology with a great variability of router behaviors as different Internet service providers are traversed. Networks are not only academic because several Planetlab nodes are connected to commercial Internet trunks or these commercial Internet trunks are traversed in the interconnection. However, the main reason to use Planetlab has been the necessity of having distributed vantage points around the network topology that would enable to perform indirect probing. We did not have access to similar vantage points for the above-mentioned NRENs.

IP addresses of the Planetlab topology between vantage points have been obtained by indirect probing (paris-traceroute) and there is no knowledge of real IP addresses. Therefore, direct probing will be performed over those IP addresses, and responsiveness in that case will correspond to the subset of IP addresses that are the intersection between direct probing responsiveness and indirect probing responsiveness. As our analysis focuses on the validity of the responses, the set of IP addresses will not imply any limitation.

4   Analysis of Router Behaviors

Unresponsiveness, as stated in [7], can be due to several reasons. The main one is the configuration of routers to ignore or filter certain types of probe packets, mainly for security reasons but also in order to avoid extra processing load. Rate limiting of ICMP responses at the target router is another reason for not receiving response packets. This rate limiting can depend on the internal router congestion and be applied in order to reduce the impact of this low priority traffic on the router. Finally, the routers can have private or duplicated public IP addresses and, therefore, not be reachable from the public Internet.

Besides router responsiveness, finding the expected header fields with the right content in returned packets is imperative to apply specific IP aliases resolution techniques. The different alternatives that can be found in router behavior are explained in the following subsections depending on the IP aliases resolution technique: IPID-based, Timestamp-based and SourceIP-based. Not all responses to packet probes will be useful for IP aliases resolution. Those useful will be called valid responses.

4.1   IPID-Based Router Behaviors

In IPID-based techniques, the routers are expected to increase their internal IPID counter for each IP packet they generated. The probe packets sent to some routers will originate response packets with IPID fields following an incremental sequence, useful for IPID-based techniques. This behavior is called *Incremental*, but several others have been detected in Internet routers. In *Zero* behavior, the IPID field is always filled up with zero value. In *Random* behavior, the IPID field is filled up with a random value for each packet. Finally, in *Copy* behavior, the IPID field is a copy of the IPID field in the probe packet received by the router.

All these four behaviors are present in direct probing, but only Incremental, Zero and Random behaviors have been found in indirect probing. Only the Incremental behavior can provide positive aliases in IPID-based techniques. The other behaviors can be used only to identify negative aliases because different behaviors can not be present simultaneously in the same router depending on the network interface or the network path followed by the probe request/response packets.

Experimental measurements have been performed over the 2037 IP addresses in the Planetlab scenario. Series of 20 probe packets of different types (ICMP Echo/Tstamp, UDP, TCP) have been sent to each IP address (direct probing) or to IP addresses in the border nodes (indirect probing), and responses have been analyzed looking for the IPID behavior. In indirect probing, TTL-limited probes are used to scan intermediate IP addresses in the path to each target IP address. In tables 1 and 2, percentages for each type of IPID behavior are presented in indirect and direct probing cases respectively. They show the percentage of each IPID behavior obtained in responses for different types of probe packets. As IP addresses have been obtained from indirect probing (paris-traceroutes), responsiveness is total for indirect probing and partial for direct probing as expected. However, validity of the responses is quite different as only Incremental behavior is useful to proceed with positive IP aliases resolution. In general, Incremental behavior appears in a bigger percentage in direct probing. Specifically, for ICMP Echo probes, the responses following Incremental behavior in the experiments are 35.87 % in indirect probing and 48.40% in direct probing. With UDP probes, indirect probing provides better results, but with TCP probes it is direct probing that gives the best results (33.08% compared to 26.53% for the indirect alternative). The column called valid responsiveness in tables 1 and 2 represents the percentage of responses by pair of IP addresses that contributes with positive or negative aliases identification results. It includes responses in which both IP addresses are incremental (positive or negative aliases) and responses in which each IP address has different behavior (negative aliases). In indirect probing with ICMP Echo, with 100% responsiveness, only 61.65% is useful responsiveness. The valid responsiveness increases to 70.51% in direct probing keeping the same type of probe packet (ICMP Echo).

UDP probes provide almost a negligible percentage of responses in direct probing. This makes unusable that kind of probe packets for direct probing. This happens because routers are usually configured to not respond with ICMP

Table 1. IPID-based behaviors in indirect probing

Type of probe packet	Zero (%)	Incremental (%)	Random (%)	Copy (%)	Unresponsive (%)	Valid responsiveness(%)
ICMP Echo	37.94	35.87	26.17	0	0	61.65
UDP	41.23	20.77	37.98	0	0	53.09
TCP	41.21	26.53	32.24	0	0	57.17

Table 2. IPID-based behaviors in direct probing

Type of probe packet	Zero (%)	Incremental (%)	Random (%)	Copy (%)	Unresponsive (%)	Valid responsiveness(%)
ICMP Echo	0	48.40	13.59	35.00	2.99	70.51
ICMP Tstamp	0	25.92	6.67	16.54	50.85	18.74
UDP	0.78	0.04	0.29	6.23	92.63	0
TCP	3.04	33.08	63.81	0.04	0	55.26

Error port unreachable packets. On the other hand, IPID-based aliases resolution methods will obtain better identification results using ICMP Echo as probe packets, for both direct and indirect probing.

Another interesting finding is that router responsiveness is different depending on whether the router is in an access or core network (close or not to the network border). In figure 1, response ratio of routers located at different hop distances from the vantage points is plotted. In the left-one figure, responsiveness for ICMP Echo probes depends clearly on the distance from the vantage point, being more responsiveness access routers compared to core routers. However, the important parameter to IP aliases resolution is the valid responsiveness, that is related to the incremental behavior plotted in the right-one figure. In ICMP Echo responses, incremental behavior (related with valid responsiveness) is reduced for access routers and greater for core routers. The differentiation in valid responsiveness for access and core routers will need a specific future work.

4.2 Timestamp-Based Router Behaviors

In timestamp-based techniques, IP prespecified timestamp option is used in probe packets. Those timestamps can be accounted in milliseconds since midnight UTC (standard) but if the time is not available in milliseconds or cannot be provided with respect to midnight UTC, then any time may be inserted as timestamp (non-standard).

Timestamp-based techniques need routers that fill up the timestamp for their interfaces if they are requested. However, several behaviors are obtained in response to those probe packets:

- *N-tstamp*: the router is able to fill up N timestamps in the IP option, with $4 \geq N \geq 1$
- *Always*: the router always fills up the timestamps even for IP addresses not belonging to it. This behavior is undesirable.

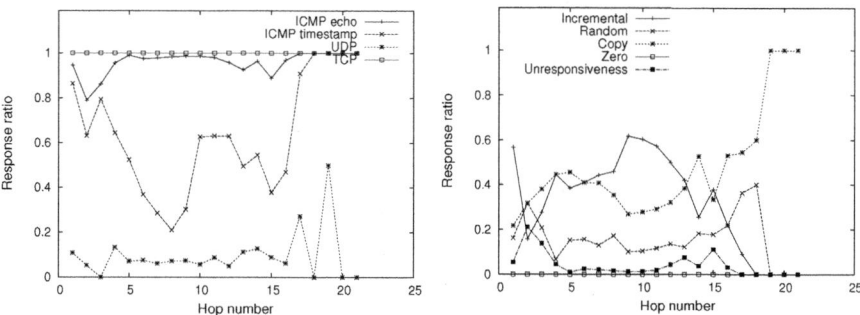

Fig. 1. Per-hop response ratio for each type of probe packet (left) and ICMP Echo behaviors (right) in direct probing

- *None*: the router does not answer with the IP prespecified timestamp option enabled.

Timestamp-based techniques can be applied only if at least 2 timestamps belonging to the target router are filled up (valid responsiveness). Therefore, only N-tstamp behaviors with $4 \geq N \geq 2$ can be used for Timestamp-based techniques.

The 2037 IP addresses in the Planetlab scenario have been checked for valid responsiveness with timestamp-based router behaviors. Table 3 presents the results obtained in responses to probes with IP prespecified timestamp option in direct and indirect probing. Percentages of responsive and unresponsive routers are shown. Again, for this technique, direct probing provides better values of valid responsiveness: 42.87% compared to 21.10% for indirect probing.

Table 3. Timestamp-based behaviors in direct and indirect probing

Type of probe packet	1-tstamp (%)	2-tstamp (%)	3-tstamp (%)	4-tstamp (%)	Always (%)	None (%)	Unresp. (%)	Valid resp.(%)
ICMP Echo (direct)	9.51	6.08	0.09	36.70	1.47	0.09	45.94	42.87
ICMP Echo (indirect)	8.00	0.19	0.0	20.91	0.00	0.09	71.13	21.10

4.3   SourceIP-Based Router Behaviors

Source IP address in response packets is the base of IP aliases resolution techniques such as Mercator. In this case, there are two expected behaviors:

- *Same-interface*: source IP address of response packet matches always target IP address of probe packet.
- *Different-interface*: source IP address of response packet does not always match target IP address of probe packet.

In this case, only UDP probes make sense because they produce ICMP Error responses. Those ICMP Error response packets can be generated from a different interface (and therefore IP address) than the incoming probe packet. In ICMP or TCP, the response packets are always generated from the same IP address that was the target IP address of the probe packet. Also, direct probing is the only way to perform this type of IP aliases resolution.

In the case of UDP probe packets, the behavior needed to apply IP aliases resolution is Different-interface behavior and, therefore, it will be considered to identify the valid responsiveness. If a response packet is received from a different source IP address than the original target IP address of the probe packet, both IP addresses are considered aliases.

Experimental results have been performed over the Planetlab scenario, with series of 20 packets sent to each one of the 2037 IP addresses in direct probing. In table 4, percentages of occurrences for each SourceIP-based behavior are presented. As stated in [14], very low values of responsiveness are present in this method with direct UDP probing. Besides, valid responsiveness has also very low values: 7.26% of routers answer with Different-interface behavior usable to apply this technique for IP aliases resolution.

Table 4. SourceIP-based behavior in direct probing

Type of probe packet	Same-interface (%)	Different-interface (%)	Unresponsive (%)	Valid responsiveness(%)
UDP	0.09	7.26	92.63	7.26

5   Behaviors Applied to IP Aliases Resolution

The identification ratio for each type of IP aliases resolution scheme depends on the responsiveness and valid responsiveness ratios presented in previous sections. Table 5 shows the identification ratios obtained using the most frequent IP aliases resolution techniques for the Planetlab scenario described in section 3. The experiments were run using the original software provided by the creators of each technique (Palmtree, Tracenet, Radargun, Ally-based) or with custom software where the original software was not available (Mercator, Ally, Prespecified timestamps). All software and data files used in this paper are available online at [21].

Table 5 shows the percentage of positives, negatives, error and unknown identifications over the total number of pairs of IP addresses presented. "Positives" indicate the pairs of IP addresses identified as aliases by each technique. "Negatives" identify those not aliases. "Errors" are those pairs of IP addresses with some error in the technique like not responding with the desired header field (they did not provide any information at all). "Unknown" are those pairs of IP addresses that have not provided enough information to identify the aliasing (they provided some but not enough information). Take note that the percentage of false positives and false negatives can not be provided because the real network topology is unknown (those ratios can be found in studies like [14]).

The column called Identified is the sum of positives and negatives, indicating the total pairs of IP addresses identified as being aliases or not. This is the main column in order to compare the different techniques. Also, a column with the number of resulting nodes in the network graph after applying the technique is shown. The column called "X-based" indicates the type of technique: IPID-based, Timestamp-based and SourceIP-based. It will indicate whether a technique will be affected by some router behavior or another. Finally, there is a column indicating if the specific technique uses direct or indirect probing.

As expected, IPID-based techniques provide better identification results, mainly Radargun and Ally-based techniques. Both use ICMP Echo with direct probing that provided the best valid responsiveness in previous sections. In fact, valid responsiveness reviewed in previous sections is the most important factor in determining results of IP aliases resolution techniques. However, results are not as good as expected by the valid responsiveness in Prespecified Timestamps technique. The reason is that the number of negatives is very low. To check for negative aliases both routers whose IP addresses are being checked must be in the same path from the vantage point. This is not feasible with a reduced number of vantage points compared with the number of IP addresses to check for aliases that would be the most common case.

The technique called "All" is a merge of the results coming from all the techniques, representing the expected results if all methods could be used simultaneously to verify IP aliases in a certain network topology. It is very expensive in terms of time and amount of probing traffic, but it provides the best results in the identification.

Table 5. IP aliases resolution results for more important methods

Technique	Positives (%)	Negatives (%)	Identified (%)	Error (%)	Unknown (%)	Resulting nodes	X-based	Direct/ Indirect
Mercator	0.00	0.00	0.00	0.00	99.99	2029	SourceIP	Direct
Palmtree	0.03	-	0.03	99.97	-	1343	SourceIP	Direct
Tracenet	0.10	-	0.10	99.9	-	857	SourceIP	Indirect
Ally	0.00	0.04	0.04	99.96	0.00	2025	IPID	Direct
Radargun	0.11	20.27	20.39	79.49	0.11	1625	IPID	Direct
Ally-based (6 packets)	0.07	19.72	19.80	7.65	72.55	1212	IPID	Direct
Ally-based (20 packets)	0.12	62.66	62.79	0.33	36.85	1129	IPID	Direct
Prespecified timestamps	0.06	0.24	0.31	99.68	-	1523	Timestamp	Indirect
All	0.34	73.85	74.19	0.03	25.77	492	All	Both

6 Conclusions

This paper has analyzed the impact of different router behaviors in answering to probing schemes for IP aliases resolution techniques. Schemes that provide a high percentage of response are not enough. It has been discussed how important is the quality of the responses. Only part of the responses can be used in an IP aliases process, and this subset comprises the so-called valid responsiveness.

Although routers are more responsive to indirect probing, valid responsiveness is greater in direct probing. Therefore, direct probing provides better results in IP aliases resolution. Besides, the type of probe packet is very important. If possible, ICMP Echo probe packets should be used as they provide the best results in valid responsiveness, reaching ratios of almost 70% of valid responses. Indirect probing makes sense to be used in network topology discovery. Direct probing will be the best alternative in IP aliases resolution.

The percentage of identification in IP aliases resolution follows the same criteria as with valid responsiveness. In this case, techniques with direct probing and ICMP Echo probe packets provide the best identification results. For example, Ally-based techniques reach almost 62% of identification with respect to the total number of pairs of IP addresses in the network scenario.

In order to propose new IP aliases resolution techniques, as a rule of design, it is recommended to consider a direct probing scheme combined with ICMP Echo probe packets in order to get the best ratios of valid responsiveness in IPID-behavior schemes that provide the best identification ratios.

References

1. CAIDA. ARK, Archipelago Measurement Infrastructure (2002),
   http://www.caida.org/projects/ark/
2. Madhyastha, H.V., Isdal, T., Piatek, M., Dixon, C., Anderson, T., Krishnamurthy, A., Venkataramani, A.: iPlane: An information Plane for Distributed Services. In: 7th USENIX Symposium on Operating Systems Desing and Implementation, Seattle, WA, pp. 367–380 (November 2006)
3. McRobb, D., Claffy, K., Monk, T.: Skitter: CAIDA's macroscopic Internet topology discovery and tracking tool (1999),
   http://www.caida.org/tools/measurement/skitter/
4. Shavitt, Y., Shir, E.: DIMES: Let the Internet Measure Itself. ACM SIGCOMM Computer Communication Review 35(5), 71–74 (2005)
5. Spring, N., Mahajan, R., Wetherall, D.: Measuring ISP topologies with Rocketfuel. In: Proc. ACM SIGCOMM, Pittsburgh, pp. 133–145 (August 2002)
6. Keys, K., Hyun, Y., Luckie, M., Claffy, K.: Internet-Scale IPv4 Alias Resolution with MIDAR: System Architecture. Technical report, Cooperative Association for Internet Data Analysis (CAIDA) (May 2011)
7. Gunes, M.H., Sarac, K.: Analyzing Router Responsiveness to Active Measurement Probes. In: Moon, S.B., Teixeira, R., Uhlig, S. (eds.) PAM 2009. LNCS, vol. 5448, pp. 23–32. Springer, Heidelberg (2009)
8. Keys, K.: Internet-Scale IP Alias Resolution Techniques. ACM SIGCOMM Computer Communication Review (CCR) 40(1), 50–55 (2010)
9. Gunes, M.H., Sarac, K.: Resolving IP aliases in building traceroute-based Internet maps. IEEE/ACM Transactions on Networking 17, 1738–1751 (2009)

10. Sherry, J., Katz-Bassett, E., Pimenova, M., Madhyastha, H.V., Anderson, T., Krishnamurthy, A.: Resolving ip aliases with prespecified timestamps. In: Proceedings of the 10th Annual Conference on Internet Measurement, IMC 2010, pp. 172–178. ACM, New York (2010)
11. Govindan, R., Tangmunarunkit, H.: Heuristics for internet map discovery. In: Proc. IEEE INFOCOM (March 2000)
12. Kamil Sarac, Engin Tozal, M.: Tracenet: An internet topology data collector. In: Internet Measurement Conference IMC, pp. 356–368 (November 2010)
13. Kamil Sarac, Engin Tozal, M.: Palmtree: An ip alias resolution algorithm with linear probing complexity. Computer Communications 34(5), 658–669 (2011)
14. Garcia-Jimenez, S., Magaña, E., Morató, D., Izal, M.: On the performance and improvement of alias resolution methods for Internet core networks. Annals of Telecommunications 66, 31–43 (2011)
15. Bender, A., Sherwood, R., Spring, N.: Fixing Ally's Growing Pains with Velocity Modeling. In (IMC 2008) 8th ACM SIGCOMM Conference on Internet Measurement, pp. 337–342. ACM, New York (2008)
16. Chun, B., Culler, D., Roscoe, T., Bavier, A., Peterson, L., Wawrzoniak, M., Bowman, M.: Planetlab: An overlay testbed for broad-coverage services. ACM SIGCOMM Computer Communications Review 33, 3–12 (2003)
17. Augustin, B., Cuvellier, X., Orgogozo, B., Viget, F., Timur Friedman, M.L., Magnien, C., Teixeira, R.: Avoiding traceroute anomalies with paris traceroute. In: 6th ACM SIGCOMM, Rio de Janeiro, Brazil, pp. 153–158 (October 2006)
18. Geant official site, http://www.geant.net/pages/home.aspx
19. Canet4 looking glass web tool, http://dooka.canet4.net/lg/lg.php
20. Globalnoc looking glass tool, http://routerproxy.grnoc.iu.edu/
21. Garcia-Jimenez, S., et al.: Tools and data sets used in this paper, http://www.tlm.unavarra.es/~santi/research/paper11.html

Semantic Exploration of DNS

Samuel Marchal, Jérôme François, Cynthia Wagner, and Thomas Engel

SnT - University of Luxembourg, Luxembourg
`firstname.lastame@uni.lu`

Abstract. The DNS structure discloses useful information about the organization and the operation of an enterprise network, which can be used for designing attacks as well as monitoring domains supporting malicious activities. Thus, this paper introduces a new method for exploring the DNS domains. Although our previous work described a tool to generate existing DNS names accurately in order to probe a domain automatically, the approach is extended by leveraging semantic analysis of domain names. In particular, the semantic distributional similarity and relatedness of sub-domains are considered as well as sequential patterns. The evaluation shows that the discovery is highly improved while the overhead remains low, comparing with non semantic DNS probing tools including ours and others.

1 Introduction

DNS (Domain Name System) [15] is critical for the well functioning of Internet as it is mainly used for locating a host in the Internet based on a human readable name. Service availability is improved by dynamic reallocation to another machine without changing the DNS name. However, this mechanism is also employed by attackers to improve the robustness and the efficiency of the attacks [18]. Hence, DNS has recently gained interest from the security community and especially the naming scheme for discovering malware hosting domains [18].

This paper focuses on DNS probing, *i.e.* guessing domains that are in use. This is an alternative to IP address scanning, which is fastidious and quite visible whereas DNS requests go through intermediate DNS servers, which hide the attackers. An attacker commonly uses dictionaries to probe existing domain names and aims to discover the networking organization, as well as potential vulnerable hosts. A common example is to check the hostnames of common services like FTP (File Transfer Protocol) or SSH (Secure Shell) by probing domains such as `ftp.example.com`. In [10], the authors show that DNS scanning allow to identify potentially vulnerable IPv6 addresses quicker than with a classical random IP scanning due to the large address space, they apply this technique to ease the spread of worms in IPv6 Internet.

Thus, penetration testing and security assessment are based on an initial recon by discovering subdomains and hosts. With a proper DNS configuration, this cannot be gathered directly and therefore requires brute-forcing. In this paper, the DNS brute forcing tool is semantically extended since we have observed that human based names usually follow semantic schemes. This includes the word semantic as well as numerical semantics (series of numbers) of DNS names.

R. Bestak et al. (Eds.): NETWORKING 2012, Part I, LNCS 7289, pp. 370–384, 2012.
© IFIP International Federation for Information Processing 2012

The paper is organized as follows. Section 2 presents DNS. An overview of the semantic exploration system and SDBF (Smart DNS Brute-Forcer) [21] is given in section 3. Semantic extensions are covered in section 4. Our approach is assessed in section 5. Related work is presented in section 6 and conclusions are drawn in section 7.

2   DNS Background

To keep the paper self-contained, this is a short overview of DNS, but the reader may read [16,15,17] for further explanation.

The main objective of DNS is to provide a map between human readable and remainable names to IP addresses. The organization of DNS is hierarchical with a root server at the top and dedicated authoritative servers for each subdomain. Assuming the domain name `www.uni.lu`, `lu` is the top level domain (TLD) which is the parent of all `.lu` subdomains (second level domain) including `uni.lu`. The third level domain is `www.uni.lu`. When a user needs the IP address of `www.uni.lu`, the first step is to query a recursive DNS server, usually maintained by his operator. This server is responsible to find the host by iteratively querying the authoritative servers of the subdomains. So, it starts by asking a root server which replies back with the DNS server in charge of the `lu` domain. The recursive DNS server of the client can also contact it to know, which server is in charge of `uni.lu`. Finally, when `uni.lu` is queried, it returns the IP address of `www.uni.lu`, which is then forwarded to the client by the recursive server.

DNS messages are mainly composed of Resource Records (RRs), which refers to different types of resolution. For the most common one the type is A or AAAA for getting the IPv4 or IPv6 address corresponding to a domain name. PTR are defined to enable inverse resolution (IP address to name).

A DNS name uses a dotted format to separate several components, *i.e.* a sequence of labels. In this paper, $label_i$ refers to the ith component, starting from the right. Thus, the top level domain is defined by $label_0$. For example, `www.uni.lu` has three labels: $label_0 = lu$, $label_1 = uni$, $label_2 = www$. Even if a recent extension allows non-ASCII characters [7], this paper doesn't consider them, as most of domains are still composed only of ASCII characters.

3   Exploration of DNS

3.1   System Overview

Our approach aims to automatically discover DNS names, and in particular, some subdomains of a domain by generating labels. Assuming a domain d, most of the current techniques rely on testing labels sequentially, l, stored in a dictionary (www, ns, ftp, smtp, etc. but also atlanta, boston, host, etc.) by concatenating the label l with the domain d to form a new subdomain $l.d$. In this paper, our prior tool SDBF [21] is used to generate new names after a learning stage. Samples are required to learn how valid labels of domain names look like. They are collected through a passive DNS platform [22], which consists in monitoring

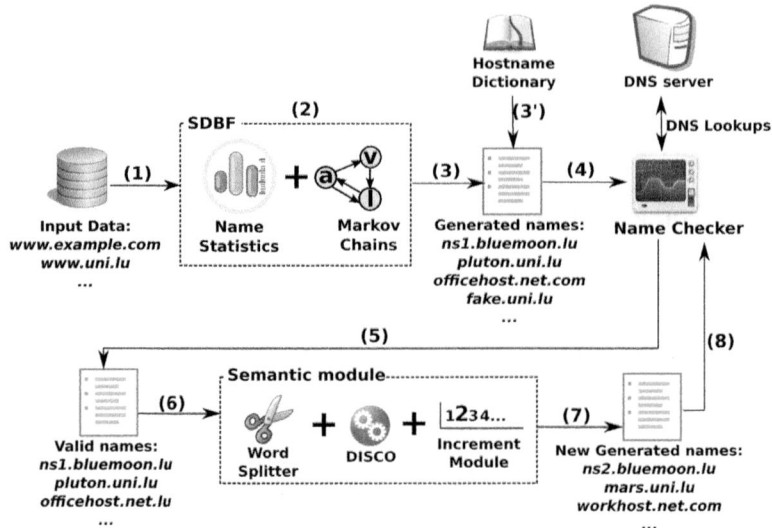

Fig. 1. System overview

and storing requests sent and replies received by a recursive DNS server. In our case, only valid names *i.e.* with valid DNS replies, are stored in a database.

Two key ideas have emerged from observations we made during our personal experience, as well as by mining the passive DNS database:

- subdomains of the same domain are semantically related, in particular end hosts. For example, using planets, cities, countries, or character names of cartoons is a frequent habit of network administrators,
- a domain may present sequential patterns. For example, enumeration is a standard naming convention within a company or a university that leads to hostnames like room1-pc1, room1-pc2, room2-pc1, ns1, ns2, etc.

As shown in figure 1, the two main steps for discovering the DNS names are:

- the construction of an initial list of names using SDBF [21] (2) or a dictionary based-approach (3')
- the extension of that list by exploiting the semantics of names (5)-(8)

3.2   SDBF

Features. The main features in SDBF are based on linguistic parameters. We assume an input list (1) of DNS names $N = \{n_1, ..., n_P\}$, a set of DNS label levels $L = \{l_1, ..., l_S\}$, a set of used characters $C = \{c_1, ..., c_M\}$ and a set of n-grams, $G_x = \{x_1, ..., x_T\}$.

The **statistical** features include: $\#wlen_n$ - the number of DNS names with n labels, $\#wlen_{i,j}$ - the number of labels of the ith level (with $i \in L$) having j characters, $\#firstchar_{i,j}$ - the number of labels at the ith level (with $i \in L$)

starting with character $j \in C$ and $\#ngram_{i,j,k}$ - the number of times that a character $j \in C$ is succeeded by $k \in C$ at the ith level with $i \in L$.

These features are transformed into distributions as follows ((2) in figure 1):

- the distribution for domain lengths (in label levels):

$$distwlen(X = j) = \frac{\#wlen_j}{\sum_k \#wlen_k} \tag{1}$$

- the distribution of the lengths of labels (in number of characters) for a given level l:

$$distwlen_l(X = j) = \frac{\#wlen_{l,j}}{\sum_k \#wlen_{l,k}} \tag{2}$$

- the distribution of the first characters of labels for a given level l:

$$distfirstchar_l(X = j) = \frac{\#firstchar_{l,j}}{\sum_k \#firstchar_{l,k}} \tag{3}$$

- the N-gram distribution for a certain level l and a character c is given by:

$$ngram_{l,c}(X = i) = \frac{\#ngram_{l,c,i}}{\sum_k ngram_{l,c,k}} \tag{4}$$

N-gram Model: N-grams [14] are successive character sequences of length $n \in \mathbb{N}$, extracted from a string. For example, an n-gram with $n = 2$ is called *bigram*. Consider the following DNS name, test.uni.lu, bigrams can be: te, es, st, un, ni, lu... For generating the names of labels, the different estimated distributions are applied to a Markov chain. A Markov chain is defined for each label level, l, as a set of states $S=\{s_1, s_2,...,s_r\}$ representing the characters that have been observed at this level. The probability of the transition between the two nodes representing by the two characters c_i and c_j is equivalent to $ngram_{l,c_i}(X = c_j)$. By applying k steps, this model allows to generate a label of k characters.

An example for the n-gram model Markov chain is given in figure 2. This means, the probability that a character 'u' is followed by character 'n' is 0.4 and the probability that an 'i' is followed by another 'i' is only 0.2.

Name Generation: Once the system is trained, SDBF can generate new names to probe, by first defining how long the new name should be in terms of number of labels. To achieve this, a random number following the distribution of number of labels, $(distwlen)$, is generated. As SDBF is designed to be highly customizable, this value can also be set by the user. The same process is applied to determine the length of labels in characters for each label l to generate: $distwlen_l$. Again, the user can set the value. Finally, for a label with a length k, the first character will be generated following the distribution of the first characters corresponding to the label level, $distfirstchar_l$, and the remaining $k-1$ characters are generated by applying the Markov Chain.

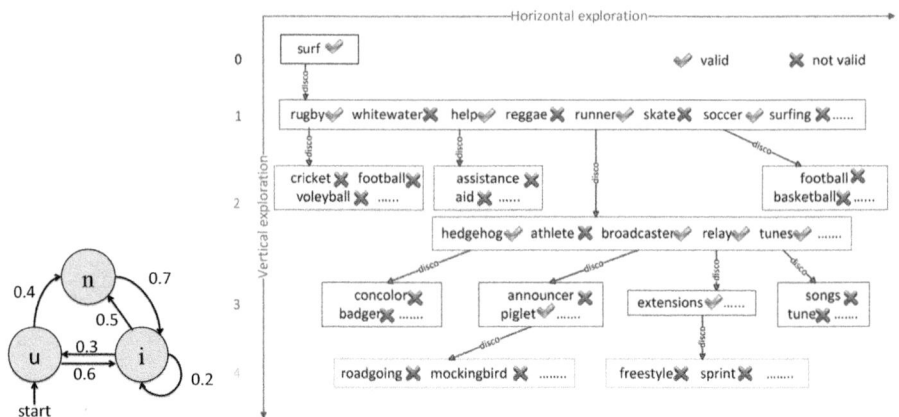

Fig. 2.   Markov   **Fig. 3.** Example semantic exploration from surf.apple.com
chain for n-grams

As the Markov chain is limited to a fixed set of transition, some transitions
are not possible. For instance, if the bigram "sn" was never observed, the word
"snt" could not be generated. To strengthen the discovery, we consider that a
transition between any two pairs of characters is possible with a probility ϵ. For
this, the other tansitions probabilities are slightly decreased to keep the sum of
probabilities for outgoing transitions equal to one.

Because it is common usage to scan a domain, the user can set fixed parts
for a domain. For example, the objective may be fixed to discover all domains
following the *.uni.lu or ns.*.lu or www.*.*.

Once names are generated, (3) in figure 1, their existence is checked by the
name checker (4), which makes a DNS query. This formally corresponds to a
function $valid(D)$ returning the valid domains of a set D.

4   Semantic Extension

As illustrated in figure 1, the semantic module takes as input a list of names,
where the validity has been checked (5). The goal is to extend this list of discovered names by analyzing individual labels. There are two modules, DISCO
and the incremental module that can be used individually or combined together,
whereas the splitter module is an optional preprocessing step.

4.1   Similar Names

The first semantic extension aims to discover names that are similar or related.
These are distinct notions[4]. Similarity refers to words having a close meaning
(for example, notebook and laptop). Semantic relatedness refers to words sharing the same semantic field like mars and venus, which are different planets.

As claimed by Kilgraff et. al [11], these usual notions impliy a manual analysis to establish relationships between words that limits its applicability and its extension to further language or semantic domain.

In this paper, we leverage DISCO (extracting DIStributionally related words using CO-occurrences) [12], which is based on an efficient and accurate method for approximating automatically (based on leanring samples) these two notions within one metric, called the similarity afterwards. DISCO considers the distance between two words within a window by defining $||w, r, w'||$, the number of times the word w' occur after r words after the word w, where $-3 \leq r \leq 3$. For example, table 1 represents windows centered on services. The window is moving along all the database samples to compute the counting that is transformed into frequencies, $i.e.$ $f(w, r, w')$, by dividing by the total number of counted co-occurences for any $||w, r, w'||$.

Intuitively, two words w_1 and w_2 are considered similar, if both of them have many co-occurrences with the same words, in particular, if the positions the latter regarding w_1 and w_2 are similar. DISCO uses the following definition, initially proposed in [13]:

$$sim(w_1, w_2) = \frac{\sum_{(r,w) \in T(w_1) \cap T(w_2)} I(w_1, r, w) + I(w2, r, w)}{\sum_{(r,w) \in T(w_1)} I(w_1, r, w) + \sum_{(r,w) \in T(w_2)} I(w_2, r, w)} \quad (5)$$

where $I(w, r, w')$ is the mutual information between w and w' [9] and $T(w)$ all the pairs (r, w') where $I(w, r, w')$ is positive.

Assuming a domain d including the label l, the objective is to find similar labels l'. The exploration goes into two directions. The first one is the horizontal exploration, which may be adjusted by lim_h. This corresponds to select the most lim_h similar words from DISCO. This result is set into a new set of labels $Expl_H(l, lim_h)$ which are tested by the Name Checker (figure 1) by concatenating with unmodified labels (other levels). By this, a new set is obtained, denoted by $Valid(Expl_H(l, lim_h))$.

The second exploration examines the vertical dimension by looking for additional similar names starting from this new set. The limit of the vertical

Table 1. Example of co-occurrence counting (2 windows centered on services)

position	-3	-2	-1	0	+1	+2	+3
sample 1	a	client	uses	services	of	the	platform
sample 2	the	platform	provides	services	to	the	client

$||services, -3, a|| = 1$	$||services, -3, the|| = 1$
$||services, -2, client|| = 1$	$||services, -2, platform|| = 1$
$||services, -1, uses|| = 1$	$||services, -1, provides|| = 1$
$||services, 1, of|| = 1$	$||services, 1, to|| = 1$
$||\mathbf{services, 2, the}|| = \mathbf{2}$	$||services, 3, client|| = 1$
$||services, 3, platform|| = 1$	

exploration is set by lim_v and is defined by repeating the previous process lim_v times with new discovered valid names:

$$Expl_V(l, lim_v) = \begin{cases} \emptyset & \text{if } lim_v = 0 \\ \bigcup_{l' \in Expl_H(l,lim_h)} Valid(Expl_H(l', lim_H)) & \text{if } lim_v = 1 \\ \bigcup_{l' \in Expl_V(l,lim_v - 1)} Valid(Expl_H(l', lim_H)) & \text{otherwise} \end{cases}$$
(6)

In order to reduce the search space only validated labels are considered for further extensions, as noticed by the use of $Valid$ in the equation (6). The vertical exploration stops once no new correct labels are found. So, lim_v does not need to be manually set, which improves the easy use of our tool.

The vertical exploration is actually recursive and highlighted in figure 1 by the loop (5)-(6)-(7)-(8). Figure 3 represents a subset of a real probing by starting from the label surf, the horizontal exploration reveals unsuccessful (surfing, skate) and successful (rugby, soccer...) labels. Then, the vertical exploration entails a horizontal extension for each of the latter.

4.2   Incremental Discovery

In many cases, machines and services are replicated and/or respect a systematic naming scheme as for example pc1, pc2, etc. Assuming that one of them has been discovered, the others can be generated by finding out the numerical components and using the following heuristic: test all possible values (including $\oslash$) for each individual digit. This limits the exploration to a number of the same or of smaller power of ten (0 to 9 will in the previous example). Preliminary experiments have shown that increasing the search range to bigger numbers does not improve the results while, the overhead highly increases.

4.3   Splitter

Labels of DNS names can be composed of several words like linuxserver or linux-server. Applying DISCO on such names cannot provide any results since it performs over single words. Therefore, the labels have to be divided automatically in advance. Using a list of separating characters, as for instance "-" is too restricted and our tools refer to the word segmentation method described in [20]. The process is recursive by successively dividing the label in 2 parts, to find the best combination, *i.e.* with the maximum probability, of the first word and the remaining part. Therefore, a label l is divided in 2 parts for each position i and the probability is computed:

$$P(l, i) = P_{word}(pre(l, i))P(post(l, i))$$
(7)

where $pre(l, i)$ returns the substring of l composed of the first i characters and $sub(l, i)$ of the remaining part. $P_{word}(w)$ returns the probability of having the word W equivalent to its frequency in a database of text samples.

Additionally, the splitter modules can also discover the incremental part of a domain. A label like `computer23` is split as `computer` and 23 which is helpful for the first step of the incremental process (see previous subsection). This may also detect non numerical increments, as observed in our database (`servera`, `serverb`, etc.), which can be incremented afterwards using ASCII codes.

5 Evaluation

5.1 Methodology

Assuming a domain d, dictionary based techniques probe by iterating over a set of labels, l, to form the hostname $l.d$. In the current evaluation, SDBF is configured similarly and two dictionary-based tools are also tested: Fierce[1] and DNSenum[2]. Both are included in Backtrack [1], a Linux distribution designed for digital forensics and penetration testing. The dictionary from Fierce includes only 1 895 words, whereas the one from DNSenum includes 266 930 entries. Hence, SDBF was configured to generate as many labels as DNSenum. The reader should refer to [21] for an evaluation of these tools without semantic extension. The main result is that SDBF and Fierce provide the best results, but all of them are complementary, *i.e.* they do not find the same names.

Based on the discovered names, new ones are probed using the semantic extensions with one of the following strategies:

- Similar names (DISCO)
- Similar names (DISCO) + Splitter
- Similar names (DISCO) + Splitter + Incremental discovery

Except if mentioned, the last one is applied. The original databases provided with the semantic tools [12,20], like Wikipedia[3], are used to train them.

The targeted domains in our experiment are extracted from the top 50 websites ranked by Alexa (*www.alexa.com*), where only 19 domains have been selected such as `google.com`, `ebay.com`, `baidu.com`... This selection discards domains performing wildcarding *i.e.* these domains will always respond positively to DNS requests regardless of the query. Furthermore, similar domains with different TLD have also been discarded, since hostname results are similar in this case. For example, `google` has no less than twelve domain names with different TLDs in the top 50 Alexa. We also choose five *popular* domains from Luxembourg including the one of our university (`uni.lu`) which is probed from internal network. All these domains are presented in figure 5.

5.2 Main Metrics

In our experimental evaluation we consider $Init_i$ with $i \in \{SDBF, DNSenum, Fierce\}$, the initial list of discovered domains for each tool and we also define:

---

[1] http://ha.ckers.org/fierce/
[2] http://code.google.com/p/dnsenum/
[3] http://www.wikipedia.org

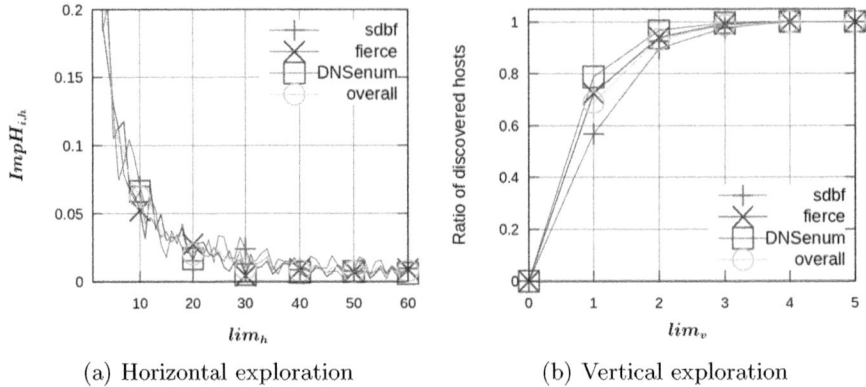

(a) Horizontal exploration   (b) Vertical exploration

Fig. 4. Vertical and horizontal depth analysis (average overall domain)

$$Init_{overall} = Init_{SDBF} \cup Init_{DNSenum} \cup Init_{Fierce} \qquad (8)$$

For the evaluation, we present New_i with $i \in \{SDBF, DNSenum, Fierce, overall\}$ the set of new discovered domains thanks to every initial dataset $Init_i$. Assuming $|S|$ as the cardinality of a set S, the improvement is defined as:

$$\%Imp_i = \frac{|New_i|}{|Init_i|}, i \in \{SDBF, DNSenum, Fierce, overall\} \qquad (9)$$

It represents the percentage of new discovered names regarding to the initial dataset. A significant value of $\%Imp_i$ shows that our method is able to find new hostnames which previous methods have not found, even when they are combined.

5.3   Exploration Parameters

Horizontal Search: The horizontal search may be configured by adjusting lim_h, which limits the exploration to the top lim_h similar words, as noticed in section 4. On the one hand, we can assume that the more words we have and test, the more hostnames we find. On the other hand, each DNS request is expensive in time and this may lead to the detection of the DNS probe. Figure 4(a) represents the evolution of the hostname discovery regarding lim_h, which varies between 1 and 200. The plotted metric, $ImpH_{i,h}$ represents the proportion of new discovered names when $lim_h = h$ compared to $lim_h - 1$. Assuming $\%Imp_{i,h}$, the value of $\%Imp_i$ when $lim_h = h$, we define:

$$ImpH_{i,h} = \begin{cases} \%Imp_{i,h} & \text{if } h = 1 \\ \%Imp_{i,h} - \%Imp_{i,h-1} & otherwise \end{cases} \qquad (10)$$

Figure 4(a) shows that having an exploration limit higher than 40 words does not significantly improve the results. That is why we set lim_h to 40 but, in case

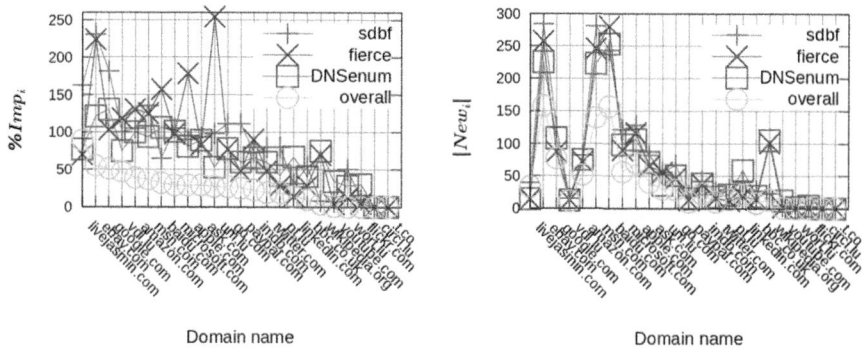

(a) Percentage ($\%Imp_i$) of newly discovered hostnames

(b) Number ($|New_i|$) of newly discovered
hostnames

Fig. 5. Efficiency of semantic exploration

a deep domain investigation is required, by increasing it, it can still discover
new names, as the curves are still positive. Besides, performances are equivalent,
whatever the initial tool is.

Vertical Search: As our probing method is based on previous discovered hostnames, we can launch it over new hostnames, gathered by the different process
iterations. This number of performed probes is called the vertical depth and
fixed through lim_v. The process also stops once no new generated names are
valid (see section 4). In our case, this leads to a maximal number of 5 iterations.
Figure 4(b) represents the ratio of discovered names compared to the maximum
($lim_V = 5$). Between 55 and 80 % of the domain names are found in the first
iteration and more than 95 % before the fourth one, so we can reasonably limit
the probe to three iterations.

5.4 Gain Evaluation

Figure 5 shows the result of our probe, made on 24 domain names using DISCO
with the previously tuned parameters. Regarding the individual improvements,
in many cases the number of discovered hostnames is doubled ($\%imp > 100$)
or even more. For instance with the original dataset from SDBF, the number
of names related to domains, as go.com, msn.com or google.com, is increased
by more than 100 %, moreover for ebay.com, we reach an improvement of more
than 200 % for both, SDBF- and Fierce-based intialization. Similar results can
be observed for DNSenum and the mean improvement over the 24 domains is
between 84% and 102% as shown in Table 2.

Furthermore, this tool provides a real solution to discover new hostnames
that existing solutions are unable to find, even if all the three other tools are
combined (overall in table 2). For instance, a global improvement of 55% for
ebay.com, 51% for google.com or 30% on the overall domains set is observed.

Table 2. Probing results – top 10 and over all domains

Domains	SDBF			Fierce			DNSenum			Overall		
	$\|Init\|$	$\|New\|$	$\%Imp$	$\|Init\|$	$\|New\|$	$\%Imp$	$\|Init\|$	$\|New\|$	$\%Imp$	$\|Init\|$	$\|New\|$	$\%Imp$
livejasmin.com	24	39	162	20	14	70	18	14	77	37	33	89
ebay.com	123	284	230	115	257	223	185	225	121	284	158	55
google.com	69	125	181	84	87	103	83	108	130	149	77	51
vdl.lu	15	15	100	11	13	118	16	12	75	23	11	47
amazon.com	78	82	105	55	72	130	75	75	100	132	52	39
msn.com	207	281	135	196	246	125	236	223	94	372	140	37
baidu.com	369	243	65	178	280	157	238	253	106	478	157	32
microsoft.com	115	121	105	91	90	98	97	98	101	189	56	29
apple.com	141	128	90	65	116	178	130	106	81	241	70	29
ask.com	88	82	93	78	65	83	79	71	89	135	40	29
all domains	2057	1739	84	1520	1558	102	1788	1565	87	3170	954	30

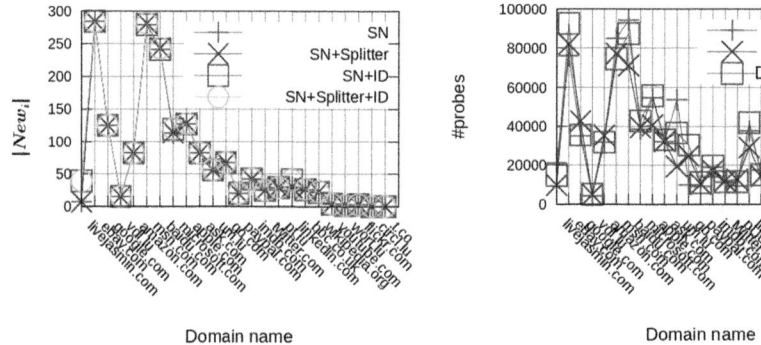

Fig. 6. Efficiency of the different semantic extension when initialized with SDBF

Fig. 7. Number of probes per domain to discover $|New_i|$

This proves the usefulness and accuracy of semantic exploration as the most common hostnames have already been discovered by one of the initial tools (SDBF, Fierce or DNSenum). From a domain name such as mars.pt.lu, merkur.pt.lu and jupiter.pt.lu have been found or from kangaroo.apple.com, we discover camel.apple.com, porcupine.apple.com and piglet.apple.com. Our first assumption deduced from observations that hostnames are attributed by human and by this, a semantic relation exists between hostnames, proves correct.

5.5   Strategy Evaluation

As introduced in section 5.1, different strategies are tested by combing DISCO (SN - *similar names*), the splitter and the incremental modules. Figure 6 shows the efficiency of each startegy initialized with SDBF. We clearly see that *Similar Names* leads to discover the main part of new DNS names, as curves of other strategies mainly coincide with the one from *Similar Names*.

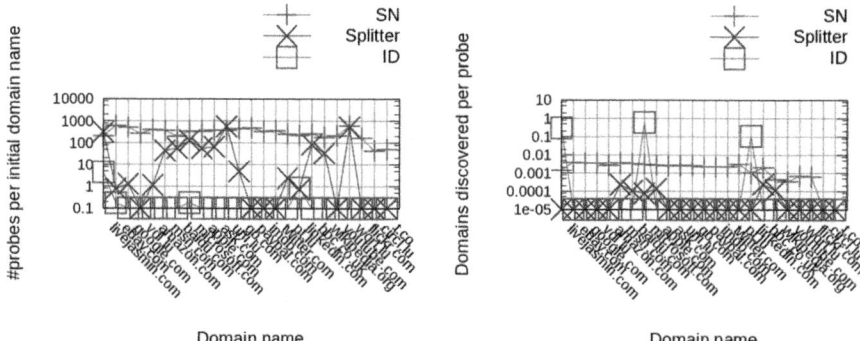

(a) Number of probes made per domain in the initial dataset

(b) Number of newly discovered hostnames per probe

Fig. 8. Ratio of probes due to each individual module

The second observation is that *Splitter* provides few signs of improvement to *Similar Names* and *Incremental Discovery* (ID) brings some results, especially for the domain livejasmin.com. In fact through this method, the hostname news10. livejasmin.com leads to discover 31 new hosts (newsX with $X \in \{1; 9\} \cup \{11; 32\}$).

Therefore, the strategy has to be carefully chosen. For fast probing of many domains, only the DISCO based extension should be used, but if the objective is to probe deeply one domain, all of them have to be combined, since each of them may improve the results.

5.6 Overhead

The overhead is defined as the number of additional DNS requests (*#probes*). As previously mentioned, SDBF and DNSenum require more than 250 000 DNS probes to produce their results. In Figure 7, we can observe that our method always needs to perform less than 100 000 DNS requests, but this discovery is based on a list established by a prior tool. The biggest probes are made for the biggest initial datasets (ebay.com, msn.com, baidu.com) but, half of the domains require less than 20 000 probes. Figure 8(a) shows that the *Similar names* module has a quite steady ratio of probes per initial name (between 200 and 500 requests). The efficiency of this module, as we can see in figure 8(b), is also steady, it discovers around 1 domain name for 200 probes. Other modules perform less requests than the previous one, as we can see in figure 8(a), but figure 8(b) shows that applying *Splitter* is less efficient than *Similar names*, whereas *Incremental discovery* needs to perform very few probes to discover new domains.

These results show that our method is far less expensive than initial ones (at least 4 times for SDBF or DNSenum) for approximatively discovering the same number of domain names (section 5.4). As a basis the *Similar names* module should be used, which provides the steadiest results although, the efficiency of the other tools is dependent of the targeted domain.

6   Related Work

In DNS research, major works deal with the detection of DNS attacks as for example, fast-flux, spamming, anomalies in DNS traces,... and present mots various defensive measures for these threats. Statistical evaluation is used in [2], respectively whitelists and classifiers are referred to, to detect anomalous patterns in RR data for rervealing poisoning attacks. The authors in [3] describe a large-scale passive DNS tool, where features are used to detect anomalies, as for example euclidean distances between entries to identify changes in the lifetimes of domains, etc. In [18], suspicious flux networks are detected by passively capturing DNS traffic. The data evaluation is based on the Jaccard index, similar to [8]. To classify the services, the authors refer to supervised learning, where the C4.5 algorithm is used to separate malicious flux and benign services. In [19], the authors perform analysis and visualization of DNS traffic in different modes, off-line, near-real-time and real-time by combining aggregation to clustering. In [6], the authors show that regular expressions improve filtering capabilities for malicious domain detection and provide more accurate results than black-lists.

In this paper, a more semantic approach is used to explore domains in the Net. Natural language processing (NLP) techniques emerged in the research areas of forensics and security. In [5], an automatic domain name generator is constructed by combining different NLP techniques, as for example by using a syllable to construct new passwords or usernames. A major difference to this work is, in [5] full words are generated. By using different statistical tools, as Kulback-Leibler divergence or Levenshtein edit distances, domain names related to botnets can be detected [24]. In the same context of generating new passwords is the work presented in [23]. Here, a new approach relying on probabilistic context-free grammar is used to generate rules in order to crack passwords.

7   Conclusion

In this paper, DNS brute forcing tools are enhanced by using semantics, *i.e.* the average improvement is higher than 80%. When combined with SDBF, the tool only needs a passive DNS database and a set of text samples like Wikipedia. Hence, it may easily be applied and it can be continuously reinforced since the previous databases are continuously evolving. Depending on the context, this paper has assessed the benefit of different strategies, as well as the implied overhead. Future work will deal with distributed probing.

References

1. Backtrack linux - penetration testing distribution (accessed on 08/22/11), www.backtrack-linux.org
2. Antonakakis, M., Dagon, D., Luo, X., Perdisci, R., Lee, W., Bellmor, J.: A Centralized Monitoring Infrastructure for Improving DNS Security. In: Jha, S., Sommer, R., Kreibich, C. (eds.) RAID 2010. LNCS, vol. 6307, pp. 18–37. Springer, Heidelberg (2010)
3. Bilge, L., Kirda, E., Kruegel, C., Balduzzi, M.: Exposure: Finding malicious domains using passive dns analysis. In: Network and Distributed System Security Symposium, NDSS (2011)
4. Budanitsky, A., Hirst, G.: Evaluating wordnet-based measures of lexical semantic relatedness. Comput. Linguist. 32 (March 2006)
5. Crawford, H., Aycock, J.: Kwyjibo: automatic domain name generation. Software Practice and Experience 38, 1561–1567 (2008)
6. Dagon, D., Lee, W.: Global internet monitoring using passive dns. In: Proceedings of the 2009 Cybersecurity Applications & Technology Conference for Homeland Security, pp. 163–168. IEEE Computer Society, Washington, DC (2009)
7. Faltstrom, P., Hoffman, P., Costello, A.: Internationalizing Domain Names in Applications (IDNA). RFC 3490 (Proposed Standard) (March 2003), http://www.ietf.org/rfc/rfc3490.txt, obsoleted by RFCs 5890, 5891
8. Hao, S., Feamster, N., Pandrangi, R.: An internet wide view into DNS lookup patterns. Tech. rep., School of Computer Science, Georgia Tech (June 2010)
9. Hindle, D.: Noun classification from predicate-argument structures. In: 28th Annual Meeting on Association for Computational Linguistics, ACL. Association for Computational Linguistics (1990)
10. Kamra, A., Feng, H., Misra, V., Keromytis, A.: The effect of dns delays on worm propagation in an ipv6 internet. In: Proceedings of IEEE Infocom. IEEE, Miami (2005)
11. Kilgarriff, A.: Thesauruses for natural language processing. In: Natural Language Processing and Knowledge Engineering (October 2003)
12. Kolb, P.: Experiments on the difference between semantic similarity and relatedness. In: 17th Nordic Conference of Computational Linguistics NODALIDA. Northern European Association for Language Technology (2009)
13. Lin, D.: Automatic retrieval and clustering of similar words. In: 17th International Conference on Computational Linguistics - COLING. Association for Computational Linguistics (1998)
14. Manning, C.D., Schütze, H.: Foundations of Statistical Natural Language Processing. MIT Press, Cambridge (1999)
15. Mockapetris, P.: Rfc 1035: Domain names - implementation and specification
16. Mockapetris, P.: Rfc 1034: Domain names - concepts and facilities (1987)
17. Mockapetris, P., Dunlap, K.: Development of the domain name system. In: Proceedings of the 1988 ACM SIGCOMM, pp. 123–133. IEEE Computer Society, Stanford (1988)
18. Perdisci, R., Corona, I., Dagon, D., Lee, W.: Detecting malicious flux service networks through passive analysis of recursive dns traces. In: Proceedings of ACSAC 2009, pp. 311–320 (2009)
19. Plonka, D., Barford, P.: Context-aware clustering of dns query traffic. In: Proceedings of the 8th ACM SIGCOMM Conference on Internet Measurement, IMC 2008, pp. 217–230. ACM, New York (2008)

20. Segaran, T., Hammerbacher, J.: Beautiful Data: The Stories Behind Elegant Data Solutions, ch. 14. O'Reilly Media (2009), http://norvig.com/ngrams/

21. Wagner, C., François, J., State, R., Engel, T., Dulaunoy, A., Wagener, G.: Sdbf: Smart dns brute-forcer. In: To Appear in IEEE/IFIP Network Operations and Management Symposium - NOMS, Miniconference. IEEE Computer Society (2012)

22. Weimer, F.: Passive DNS replication. In: Conference on Computer Security Incident Handling (2005)

23. Weir, M., Aggarwal, S., Medeiros, B.D., Glodek, B.: Password cracking using probabilistic context-free grammars. In: Symposium on Security and Privacy. IEEE

24. Yadav, S., Reddy, A.K.K., Reddy, A.N., Ranjan, S.: Detecting algorithmically generated malicious domain names. In: Proceedings of the 10th Annual Conference on Internet Measurement, IMC 2010, pp. 48–61. ACM, New York (2010)

On the Dynamics of Locators in LISP

Damien Saucez[1] and Benoit Donnet[2]

[1] INRIA, Sophia Antipolis, France
[2] Université de Liège, Liège, Belgium

Abstract. In the Internet, IP addresses play the dual role of identifying the hosts and locating them on the topology. This design choice limits the way a network can control its traffic and causes scalability issues. To overcome this limitation, the Locator/Identifier Separation Protocol (LISP) has been introduced. In LISP, the addresses used to identify end hosts (i.e., identifiers) are independent of the addresses used to locate them (i.e., locators). LISP maps identifiers into a list of locators and provides a mean to transport the packets with the appropriate locator. A key feature of this separation is that several locators can be associated to a given identifier, leading to more control for an end-site on the path selection to reach a given destination.

In this paper, we show that the choice of the locator can have an impact on the performance and the reliability of the communication in a LISP environment. To this aim, we build a mapping between identifiers and locators as if LISP were deployed today. In addition, we extensively collect delay data between locators and demonstrate that the locator selection for a given identifier prefix impacts the performance of the LISP path in 25% of the cases. Finally, we measure the locators availability over time and demonstrate that it remains quite stable.

1 Introduction

During the last decade, the Internet has strongly evolved. Its natural growth combined with factors such as multihoming and interdomain traffic engineering has lead to an increase of the BGP routing tables [1, 2] and the BGP churn [3].

To cope with these problems, several solutions have been proposed. Most of them assume two different types of addresses: *identifiers* and *locators*. An identifier is used on an host to identify a connection endpoint while a locator refers to a node attachment point in the Internet topology. The proposals are divided in two categories: those attaching locators directly to hosts (HIP, SHIM6,... [4]) and those attaching locators to routers (see for instance LISP [5, 6]). It is worth to notice that, in the current Internet, an host address is at the same time its identifier and its locator.

A key advantage of the addresses separation is to offer the possibility of associating several locators to a given identifier. This implies the availability of multiple paths between two identifiers and, as shown by several studies, those paths often offer different characteristics in terms of delay, jitter, bandwidth, etc [7–10].

R. Bestak et al. (Eds.): NETWORKING 2012, Part I, LNCS 7289, pp. 385–396, 2012.

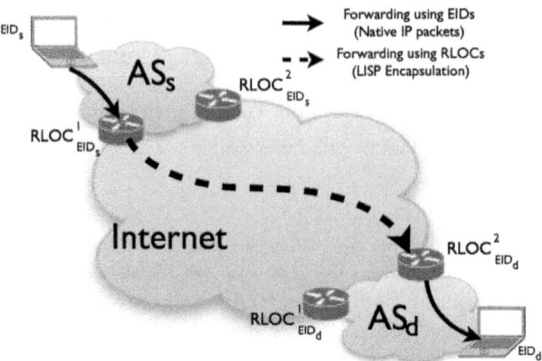

Fig. 1. LISP general behavior

In this paper, we quantify the effects of controlling paths between locators in the Locator/Identifier Separation Protocol (LISP). A key feature of LISP is the *mapping* between locators and identifiers. As LISP is not yet globally deployed[1], we cannot rely on an existing solution for this mapping. Therefore, in this paper, we propose our own mapping as if LISP were deployed today. We generate our mapping based on BGP prefixes and Archipelago dataset [11]. Our mapping considers more than 14,000 locators associated to more than 1,800 identifiers.

We next evaluate the performance of each locator by extensively measuring the delay to each locator from a point located in Belgium. Based on this dataset, we demonstrate that the locator selection, for a given identifier, has an impact on the communication performance in 25% of the cases. We further evaluate the selected locators availability over time and show that it remains quite stable.

The remainder of this paper is organized as follows: Sec. 2 introduces how LISP works; Sec. 3 explains how we build a mapping between locators and identifiers as if LISP were deployed today and how we collected our dataset; Sec. 4 discusses the impact of the locator selection on the delay; Sec. 5 discusses reachability issues; finally, Sec. 6 concludes this paper by summarizing its main contributions.

2   LISP Background

The *Locator/Identifier Separation Protocol* (LISP) [5, 6] has been introduced to separate the identifier and the locator roles of IP addresses. LISP considers two independent address spaces: the *Routing Locator Space* (RLOC), that is globally routable, and the *Endpoint Identifier Space* (EID), that is only locally routable. With LISP, routers in the core of Internet are handling RLOCs like today and routes are maintained so that packets can be forwarded between any router.

---

[1] A LISP testbed is available. See http://www.lisp4.net/

On the contrary, endpoints (typically hosts) receive EIDs. As an EID is only locally routable, it means that routers in the core of the Internet do not maintain routes towards EIDs.

LISP is used when packets are exchanged between two endpoints located in different sites. First, the source endpoint generates a regular IP packet to the destination endpoint and forwards it. When the packet arrives at the edge of the source endpoint site, it is processed by a LISP router called *Ingress Tunnel Router* (ITR). As the destination is an EID, it cannot be forwarded on the Internet. The ITR must first query the *mapping system* [6, 12, 13], a global database associating EIDs to RLOCs.[2] Note that several RLOCs might be associated to a given EID. Once the ITR has received the mapping for the destination EID, it selects one RLOC and encapsulates the original message in a LISP packet. The source address of the LISP packet is the ITR RLOC and the destination address is the selected RLOC. As the source and destination addresses of this LISP packet are RLOCs, it can be forwarded over the Internet. The router with the destination RLOC is at the destination EID site and is called *Egress Tunnel Router* (ETR). ETRs are responsible of decapsulating LISP packets so that the original packet can be forwarded to its final EID destination. This behavior is illustrated in Fig. 1.

The mapping system is a key element of LISP. A mapping associates an EID prefix to a list of $< RLOC, priority, weight >$ tuples. The priority and weight associated to each RLOC help the ITR in preferring an RLOC for reaching a given EID. RLOCs with the lowest priority are always preferred. When RLOCs share the same lowest priority, the weight is used for load balancing among themselves. One of the key feature of LISP mapping systems is that the mappings can be specific to each ITR. For instance, it is possible to return a different mapping to a site attacking you than the mapping returned to valid sources of traffic. Such incoming traffic engineering is not possible today with BGP.

The flexibility and control offered by LISP come at the cost of increasing the reachability detection complexity. Indeed, identifiers are not directly routable and have to be mapped to locators, themselves being assigned to the border routers. Therefore, it is no longer possible for the routing layer to move traffic to another border router in case of a router failure. The ITRs are thus in charge of checking the reachability of the selected ETR. To test the reachability, the ITRs can periodically probe the ETRs or monitor the traffic. Unfortunately, both solution provide poor detection times.

Reachability tests can be made in an active or passive way [5]. Active testing consists of sending probes to the ETR to ensure that it is reachable. This technique is implemented with the *RLOC Probing Algorithm* but does not scale for short time failure detection. On the contrary, passive reachability tests can be used when a short time scale detection is required. This is done by tagging a LISP packet with a nonce and the targeted RLOC has to tag the packets back to the source RLOC with the same nonce. The locator is considered as unreachable

---

[2] To avoid requesting for mapping all the time, ITRs enable a mapping cache.

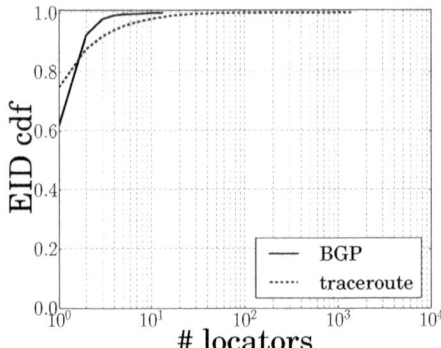

Fig. 2. Comparison between the number of locators per mapping as discovered with BGP AS paths and traceroutes analysis

if the nonce is not echoed within a given time frame. This technique is called the *Echo Nonce Algorithm* [5].

3   Methodology

Our work, in this paper, is based on LISP. However, except in the lisp4.net pilot network[3], LISP is not yet deployed and, there is no clear clue on how to build the mapping between identifiers and locators. In Sec. 3.1, we explain how we construct the mapping as if LISP were deployed today. Next, in Sec. 3.2, we explain our delay measurement campaign.

3.1   Building a LISP Mapping

As LISP is not yet deployed, we have to build, on our own, the mapping between EIDs and RLOCs. In this section, we envision two ways for creating the mapping: the first one is based on BGP information and active measurements, while the second one only takes into account BGP.

For the first technique, we rely on three assumptions to determine how the mappings could look like: (*i*) as contiguous addresses tend to be used similarly [14], EID prefixes follow the current BGP prefixes decomposition; (*ii*) EIDs are used only at the stubs, not in the Internet core; (*iii*) ITRs and ETRs are deployed at the edge between the stubs and the providers, while locator addresses are allocated in a *Provider Aggregetable (PA)* mode.

We base our evaluation on several datasets. First, the BGP prefixes and their stub nature are extracted from the Route Views Project dataset [15] (labeled as *Route Views* in the remainder of this paper), we use BGP dump from the Oregon-IX collected on August, 12th 2010. Second, the mapping between BGP prefixes and locators is done with the Archipelago dataset [11] (labeled as *Ark* in the following) collected on July, 24th 2010.

---

[3] See http://www.lisp4.net

Following the first two assumptions (see above), we filter the prefixes from Route Views to extract the stub prefixes using the AS ranking provided by CAIDA [16]. We further take only into account the most specific prefixes. Less specific prefixes, typically used for resiliency, are thus filtered. By doing so, we have an upper bound on the number of EID prefixes to be supported by the mapping system. Indeed, filtering the less specific prefixes does not filter the prefixes deaggregated for load balancing reasons. After filtering out the Route Views dataset, 41.21% of the prefixes (i.e., 138,123 prefixes) can be considered as EID prefixes.

To determine the active locators for each EID prefix, we rely on the Ark dataset and the third assumption. In this paper, we consider the edge from a IP routing point of view: the router at the edge of a stub is the first router whose address it not inside the EID prefix. For each destination address in the Ark dataset, we determine its most specific BGP prefix, i.e., its EID prefix, and backtrack the traceroute until a hop address that does not belong to this BGP prefix. The hop address is then considered as one RLOC of the EID prefix. Computing this on all the completed traceroutes from the Ark dataset gives the list of RLOCs associated to a given EID prefix. As EID prefixes determined with Route Views are not always traceroutable, we were only able to extract 15,337 EIDs.

The second mapping construction technique is based only on BGP. We simply count the number of different neighbor ASes for each stub prefix. This gives us an approximation of the stub prefixes multi-homing degree. Neighboring ASes can be seen as the prefix locator from an AS-level point of view.

Fig. 2 provides the distribution, as a cumulative mass, of the number of locators (horizontal axis, in log-scale) associated to each EID for both mapping construction techniques described above in this section. The plain line labeled "BGP" has been obtained with the second technique, while the dashed line labeled "traceroute" exhibits results for the first one.

We first observe that, whatever the mapping construction technique considered, we face a large proportion of EID having a single locator: 61% for BGP and 75% for traceroute. This difference can be explained by the fact that some routers do not reply to traceroute probes while some others are not on the path to the EID during the traceroutes as they can be backup routers. However, the maximum number of locators associated to an EID is more important when traceroutes are considered: 13 for BGP and 194 for traceroute with an outlier network offering satellite broadcasting measured with 1,465 locators. This can be explain by the fact that some sites have many points of presence but these points of presence are all connected to the same providers.

In this paper, we are interested to see the impact of the locator selection on the performance and availability. We thus focus our study on prefixes that appear to have at least two locators in the traceroute evaluation. It is therefore important to observe that 68% of multihomed EIDs have either 2 or 3 locators. This small number of locators limits the choices of locators for the EIDs. In the remainder of this paper, we will only consider mapping obtained with the first mapping construction technique (i.e., traceroute).

3.2   Delay Data Collection

In this paper, the RLOC selection is evaluated based on the *round-trip time* (RTT). We built our delay dataset by pinging each locator identified as explained above every five minutes from a vantage point located in Louvain-la-Neuve University campus in Belgium. The campus is single connected to the Internet and the provider is the national research network. It means that without LISP, the campus cannot do traffic engineering (neither incoming nor outgoing). If LISP was deployed, the campus could do performance based outgoing traffic engineering by selecting the preferred destination locators. The measurement campaign started on September, 3rd 2010 and lasted until September, 24th 2010.[4] Measurements have been done in cycles, i.e., all locators are measured during a cycle and a new cycle started every five minutes leading to a total of 6,048 measurement cycles. As the measurements towards the whole set of locators lasted, at worst, five minutes, cycles never overlapped. We consider a locator as not reachable if its RTT is higher than one second. The decision of a five minutes sampling period is a tradeoff between the scalability of our measurements and operational reality where a five minutes granularity is often observed for billing and accounting purposes [17].

4   Impact of RLOC Selection

In this section, we investigate how the RLOC selection can impact the performance of the communications between two distant EIDs.

Fig. 3 shows the average distance between the locators and the best locator (i.e., the locator with the lowest delay), for each of the 1419 EID-RLOC mapping. This average distance has been computed as follows: for each EID prefix, at time t, we take the set of associated RLOCs, measure the distance (i.e., the delay) to each locator and, finally, consider the locator with the lowest delay as being the best locator at time t for the current mapping. The distance is then computed between the best locator and the remainder locators of the given EID as $delay_r - delay_b$, where $delay_r$ is the delay of the considered RLOC and $delay_b$ the delay to the best locator. We next compute the average distance and the 95% confidence interval for all distances over the whole measurement campaign, for all EIDs. The EID prefixes are ordered by average distance in Fig. 3.

We observe two trends in Fig. 3. First, for 90% of the EIDs, the average distance to the best locator is lower than 20ms and 77% are not distant by more than 10ms. This is an important result because, on average, for most of the EID prefixes, the traffic can be balanced between all the locators without the risk of dramatically degrading traffic performance. Second, for 7% of the prefixes, the average distance is higher than 50ms and can reach more than 100ms. A so high average distance implies that performance can be degraded considerably if the locator has been badly selected.

---

[4] Our dataset is freely available at
   http://inl.info.ucl.ac.be/content/locator-reachability-dataset

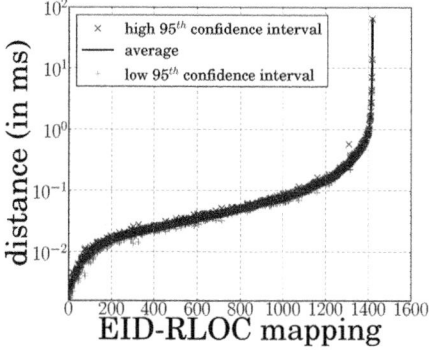
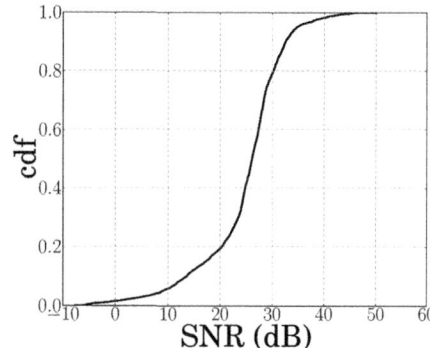

Fig. 3. Average distance between each locator of a mapping and the locator presenting the lowest delay in the mapping

Fig. 4. Stability of the round-trip delay measured for each locator during 21 days represented as the cumulative distribution of the signal-to-noise ratio (i.e., average delay divided by standard deviation, in dB) of the round-trip delay

Fig. 3 only gives an overview of the locator dynamics. The confidence intervals are close around the mean, suggesting that the delay is rather stable. This stability is confirmed by Fig. 4. It shows the cumulative distribution of the signal-to-noise (SNR) ratio (in dB) for the delay measured for each locator during the measurement campaign. The SNR is computed as $snr = \frac{\mu}{\sigma}$, where μ is the average delay and σ the standard deviation. Looking at Fig. 4, we see that a very few locators (i.e., 1.5%) present a negative SNR, meaning that they are completely unstable. On the contrary, 80% have an SNR above 20dB and are thus considered as stable. Said differently, those 80% of locators have a 10:1 ratio between the average and the standard deviation. Finally, 12% of the locators are very stable as they show an SNR over 32.04dB (i.e., a 40:1 ratio).

To refine our observations of the locators dynamics, we compute the *preferred sets* for each EID prefix at every measurement cycle, i.e., the set of locators for which the distance is at most τ percent higher than the locator with the lowest delay. We vary τ between 0%, 5%, 10%, and 100%. A value $\tau = 0\%$ would mean that only the best locator is considered while a value $\tau = 100\%$ would mean that locators with a delay up to twice the lowest one are considered as providing equivalent performance.

Fig. 5 shows the preferred set dynamics. The horizontal axis gives the number of times the preferred set has changed during the measurement campaign and the vertical axis gives the cumulative distribution of this number of changes. A first interesting observation is that, even for a high tolerance of 100%, the preferred set changes over time. This implies that, even if delays are stable and locators relatively close to each others regarding delay, it is likely to find locators for an EID that are more than twice farer (in term of delay) than the best locator. However, for 75% of the EIDs, the number of changes is between 162 and 2,714, depending on the tolerance. It thus means that when only considering the best locator, the preferred set does not change 55% of the time. For a reasonable tolerance of 10%, the preferred set was unchanged 81% of the time. Therefore, when a network decides to potentially use several locators simultaneously, a tradeoff has to be achieved between the delay difference acceptable for the preferred locators and the frequency of the changes in the preferred set. This tradeoff can be observed with the tolerance.

Fig. 6 refines the understanding of the preferred set dynamics. Fig. 6 provides indeed the cumulative distribution of the *dissimilarity*. The dissimilarity, or the *Jaccard distance*, is a metric of the similarity of two sets. The dissimilarity of sets A and B is given by

$$dissimilarity(A, B) = 1 - \frac{|A \cap B|}{|A \cup B|}. \tag{1}$$

A dissimilarity of zero means that the two set are similar, i.e., both sets are made of the same locators. On the contrary, a value of one indicates that no element in the set is present in the other set, i.e., both sets do not share any locators. Fig. 6 shows the distribution of the dissimilarity between a preferred set at time t and the preferred set at time $t - 1$, for all the EID prefixes and all the cycles, for a tolerance of 10%.

We see that, in 85% of the cases, the preferred set does not change between a cycle and the next cycle. In addition, the preferred set was completely different from the previous preferred set in less than 1% of the observations. In other words, the preferred set tend to remain identical from one cycle to another and even if there is a change, this change is only partial. It is worth noticing a small peak for the 0.5 dissimilarity. This peak can be explained by the fact that the majority of the EIDs in our dataset has two locators.

The median value for the time a preferred set is kept is 22 hours and the 95th percentile is 5 days and a half. A low preferred set change rate means that the load of the mapping system because of preferred set remains limited.

5   Availability Issues

Dealing with router failures in LISP is fundamentally different than dealing with router failures in the current Internet. Currently, the path followed by a packet to reach an end-host only depends on the destination IP address. On the contrary, in LISP, the source network determines the ETR through which the packet must transit by selecting a destination locator. In other words, the source network (via the ITR) decides by which border router a packet enters the destination network. Without LISP, when a border router breaks down, the prefix of the network remains reachable as the network prefix is not bound to any specific router. In addition, as every router announces the same prefix, failure recovery is performed locally. Recovery can however not be performed locally with LISP. Indeed, the EIDs are hidden from the network and packets are sent to a particular router. As a consequence, failure recovery is performed by the ITR in LISP. First, the ITR must detect that it cannot reach anymore the desired

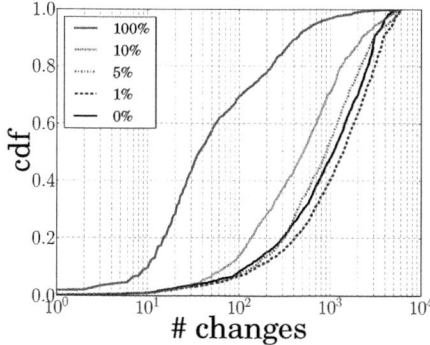

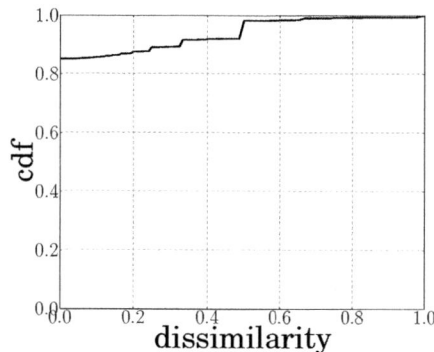

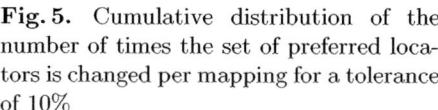

Fig. 5. Cumulative distribution of the number of times the set of preferred locators is changed per mapping for a tolerance of 10%

Fig. 6. Percentage of difference (i.e., dissimilarity) between the preferred set of a mapping at a given time with the preferred set of the same mapping at the previous cycle, expressed as the cumulative distribution of the dissimilarity for all the measured 10% tolerance preferred sets

locator. Once the failure has been detected, the ITR can bypass the failure by selecting another locator (when available). Failures are detected by the mean of active probing. For scalability issue, active probing can only be performed at a low frequency (i.e., a few times per minute). In this section, we evaluate the duration and frequency of locator failures to determine if end-to-end failure detection is acceptable in general.

Fig. 7 gives an insight of failures impacts. For each EID prefixes, we count the number of locator failures as well as the number of preferred set changes. Fig. 7 shows the cumulative distribution of the ratio of the number of locator failures to the number of preferred set changes. A value lower than 100% indicates that the preferred set changed more often than the number of observed failures. On the contrary, a value higher than 100% indicates that the preferred set changed less frequently than the failure occurrence. The plain line is obtained by only considering the failures of a locator that was in the preferred set at the time of its failure. On the contrary, the dashed line curve is obtained by only considering the failure of locators that were not in the preferred set at the time of the failure.

Looking at Fig. 7, it is interesting to see that, in nearly 80% of the cases, no more than 10% of the changes in the preferred set are due to the failure of a locator in the preferred set. This thus means that, most of the time, the preferred set changes because of performance reasons and not because of a locator failure. On the contrary, the dashed line curve shows that failures that do not impact the preferred set are more frequent than failures impacting the locator set. In 10% of the cases, it is even observed that the number of failures that does not impact the preferred set was higher than the number of preferred set changes, potentially by several orders of magnitude. In other words, a potentially important number of failure can have no impact on the forwarding of packets.

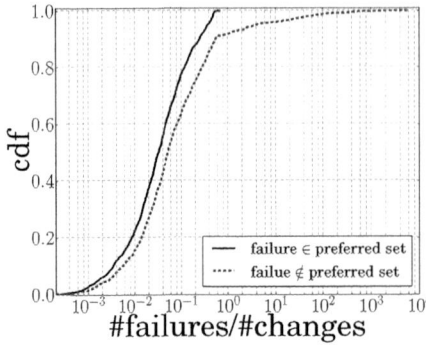
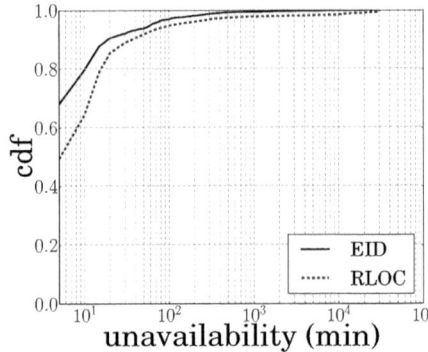

Fig. 7. Ratio between the number of failures and the number of preferred set changes per mapping for a 10% tolerance

Fig. 8. Cumulative distribution of the duration of the failures for every measured failure for both identifiers (e.g., EID) and locators (e.g., RLOC)

Fig. 8 gives the distribution of failures duration (in minutes) observed for the EIDs (curve labeled *EID*) and RLOCs (curve labeled *RLOC*). As we are measuring the availability once every five minutes (see Sec. 3.2), the minimum unavailability of a given EID (or RLOC) is five minutes. In addition, it is worth to notice that Fig. 8 considers in row unavailability.

Fig. 8 shows that in 68% of the cases, the EID has never been considered unavailable by more than one measurement cycle. This unavailability can come from network failure of less than five minutes but also because of the loss of the probe. It also means that 32% of the EIDs have been unavailable for ten minutes or more. Beside, 5% of the EIDs have been unavailable for at least one hour. If we compare this EID maximum consecutive unavailabilities with those from RLOCs, we can see that multiconnected EID sites reduce the unavailability duration.

In this section, we have seen that the failure of a locator is seldom and rarely lasts more that 10 minutes. This result is promising as it suggests that doing end-to-end failure detection with low frequency probes will be effective. However, as we measured the network at low frequency, we might have miss short living failures. Understanding these short lived failures is left for future work.

6   Conclusion

The Internet is facing scalability issues. For that reason, LISP has been proposed. LISP improves the Internet scalability by separating the identifier and the locator role of IP addresses. In LISP, the core Internet is addressed with routing locators bound to the topology. On the contrary, end-hosts receive stable identifiers, independent of the topology. While the locators are globally routable, the identifiers are not. LISP then provides an encapsulation mechanism to transmit packets between identifiers. The association between the identifiers and their locators is provided by a mapping. A key feature of this separation is that several locators can be attached to a given identifier. This means that an end-site has more control on the path selection to reach a given destination by controlling the locators. However, paths are not of equal quality and the selected locator can have an impact on the traffic performance.

In this paper, we performed a three weeks delay measurement campaign to determine the impact of the locator choice on the traffic. Our first result is that for 75% percent of the stub networks, the traffic is not impacted by the locator selection as the measured delay does not vary by more than 10%. Hence, the traffic can be load balanced between the locators without severely impacting the communication performance. However, we also observed that for 25% of the network, a bad locator choice can severely impact the delay. The delay cannot thus be ignored when a locator must be chosen. Our measurement campaign also shows that locators presenting a short delay tend to keep this short delay for long periods. We are currently working on techniques to allow a network to efficiently determine if it can perform load balancing among the locators or if some locators must be avoided.

The ability for a network to control the paths it is using comes at the cost of a robustness loss. Indeed, LISP relies on encapsulation. Hence if a locator is not available, the network cannot recover the failure until the network that encapsulates the packets detects that failure. In LISP, failures are discovered with low frequency active probing. With our dataset, we showed that failures are seldom meaning that probing at low frequency can be effective. However, active probing will never be able to protect against short lived failures. For that reason, we are currently working on a local failure recovery techniques that do not require active probing but that can be applied directly to LISP.

References

1. Huston, G.: BGP routing table analysis reports (2004), http://bgp.potaroo.net
2. Meng, X., Xu, Z., Zhang, B., Huston, G., Lu, S., Zhang, L.: IPv4 address allocation and the BGP routing table evolution. ACM SIGCOMM Computer Communcation Review 35(1), 71–80 (2005)
3. Meyer, D., Zhang, L., Fall, K.: Report from the IAB workshop on routing and addressing. RFC 4984, Internet Engineering Task Force (September 2007)
4. Savola, P.: A survey of IPv6 site multihoming proposals. In: Proc. IEEE Internet Conference of Telecommunications (ConTel 2005) (June 2005)
5. Farinacci, D., Fuller, V., Meyer, D., Lewis, D.: Locator/ID separation protocol (LISP). Internet Draft (Work in Progress) draft-ietf-lisp-22, Internet Engineering Task Force (February 2012)
6. Meyer, D.: The locator identifier separation protocol (LISP). Internet Protocol Journal 11(1), 23–36 (2008)
7. Akella, A., A., S., Sitaraman, R.: A measurement-based analysis of multihoming. In: Proc. ACM SIGCOMM (August 2003)
8. Quoitin, B., Iannone, L., de Launois, C., Bonaventure, O.: Evaluating the benefits of the locator/identifier separation. In: Proc. ACM SIGCOMM MobiArch Workshop (August 2007)

9. de Launois, C., Quoitin, B., Bonaventure, O.: Leveraging networking performance with IPv6 multihoming and multiple provider-dependent aggregatable prefixes. Computer Networks 50(8), 1145–1157 (2006)
10. Zhou, X., Jacobsson, M., Uijterwaal, H., Van Mieghem, P.: IPv6 delay and loss performance evolution. International Journal of Communication Systems 21(6) (June 2007)
11. claffy, k., Hyun, Y., Keys, K., Fomenkov, M.: Internet mapping: from art to science. In: Proc. IEEE Cybersecurity Applications and Technologies Conference for Homeland Security (CATCH) (March 2009)
12. Jakab, L., Cabellos-Aparicio, A., Coras, F., Saucez, D., Bonaventure, O.: Lisp-tree: a dns hierarchy to support the lisp mapping system. IEEE J. Sel. A. Commun. 28, 1332–1343 (2010)
13. Fuller, V., Farinacci, D., Meyer, D., Lewis, D.: LISP alternative topology (LISP+ALT). Internet Draft (Work in Progress) draft-ietf-lisp-alt-10, Internet Engineering Task Force (December 2011)
14. Cai, X., Heidemann, J.: Understanding block-level address usage in the visible Internet. In: Proc. ACM SIGCOMM (August 2010)
15. University of Oregon: Route views, University of Oregon Route Views project, http://www.routeviews.org/
16. CAIDA: AS-Rank (2010), http://as-rank.caida.org/
17. Dimitropoulos, X., Hurley, P., Kind, A., Stoecklin, M.P.: On the 95-Percentile Billing Method. In: Moon, S.B., Teixeira, R., Uhlig, S. (eds.) PAM 2009. LNCS, vol. 5448, pp. 207–216. Springer, Heidelberg (2009)

A Local Approach to Fast Failure Recovery of LISP Ingress Tunnel Routers

Damien Saucez[1], Juhoon Kim[2], Luigi Iannone[2],
Olivier Bonaventure[3], and Clarence Filsfils[4]

[1] INRIA Sophia Antipolis, France
damien.saucez@inria.fr
[2] Telekom Innovation Laboratories – Technische Universität Berlin, Germany
{jkim,luigi}@net.t-labs.tu-berlin.de
[3] Université catholique de Louvain, Belgium
olivier.bonaventure@uclouvain.be
[4] Cisco Systems
cfilsfil@cisco.com

Abstract. LISP (Locator/ID Separation Protocol) has been proposed as a future Internet architecture in order to solve the scalability issues the current architecture is facing. LISP tunnels packets between border routers, which are the locators of the non-globally routable identifiers associated to end-hosts. In this context, the encapsulating routers, which are called *Ingress Tunnel Routers* (ITR) and learn dynamically identifier-to-locators mappings needed for the encapsulation, can cause severe and long lasting traffic disruption upon failure. In this paper, thanks to real traffic traces, we first explore the impact of ITR failures on ongoing traffic. Our measurements confirm that the failure of an ITR can have severe impact on traffic. We then propose and evaluate an ITR synchronization mechanism to locally protect ITRs, achieving disruptionless traffic redirection. We finally explore how to minimize the number of ITRs to synchronize in large networks.

Keywords: Locator/ID Separation, Next Generation Internet, Addressing and Routing Architectures, Measurements, Emulations.

1 Introduction

It is widely recognized that the current Internet architecture is a victim of its own success and is facing unforeseen scalability issues, in terms of increasing routing tables, multi-homing, and inter-domain traffic engineering ([1], [2], [3]). Both academic researchers and industrial companies concur that the locator/ID separation paradigm, *i.e.*, separating the semantic of end-systems' identifier and location (currently merged in the IP address), will provide the needed scalability improvement. Among the several recently proposed protocols leveraging on this principle, the *Locator/ID Separation Protocol* (LISP [4]) has encountered the wider success and is being developed in the IETF (Internet Engineering Task Force [5]). LISP assigns Endpoint IDentifiers (EIDs) to end-hosts and Routing

R. Bestak et al. (Eds.): NETWORKING 2012, Part I, LNCS 7289, pp. 397–408, 2012.

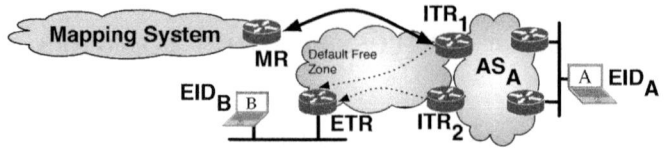

Fig. 1. LISP deployment example

LOCators (RLOCs) to tunnel routers. In the core Internet only RLOCs are globally routable and used to tunnel packets between end-systems, which use EIDs that are globally unique but not globally routable (hence the use of tunnels).

Like the current Internet, LISP is affected by temporary link or node failures that may disrupt end-to-end reachability. More specifically, when an encapsulating router – the so called Ingress Tunnel Router (ITR) – fails, alternate/backup ITRs are not readily able to take over traffic encapsulation because they do not have the necessary identifier-to-locator mappings, resulting in prolonged packet drops and traffic disruption. Some aspects of the performance of LISP have been previously explored in other works, either focusing on the scalability of specific LISP elements, like the LISP cache ([6], [7], [8]) or on the mapping system used to distribute the mappings binding the locators with the identifiers ([9], [10], [11]). However, how LISP reacts to link and node failures has been largely neglected. The only document briefly discussing reachability issues is the draft by Meyer et al. [12], which focuses on the general implications of having an Internet architecture based on the locator/ID separation paradigm. To the best of our knowledge, the present paper is the first analyzing failures in LISP, and, from a more general perspective, the failure of encapsulating routers in the context of a locator/ID separated architecture.

In this paper, we first provide all the relevant information for a complete understanding of the problem, overviewing LISP in Sec. 2. In Sec. 3 we formalize the failure problem, followed by our proposal (Sec. 4) and its evaluation (Sec. 5). Then in Sec. 6 we tackle the recovery problem, showing how it can be solved as well with our approach. Sec. 7 and Sec. 8 explore the synchronization issue in large networks. Finally, Sec. 9 concludes the paper.

2   LISP Basics

LISP is typically implemented on customer-edge border routers, whose upstream IP address (*i.e.*, the one facing the Internet), which is a part of the Border Gateway Protocol (BGP) routing space, is the *Routing LOCator (RLOC)* used to locate end-systems of the local domain. The end-systems in the local domain use globally unique but not globally routable (hence not injected in the BGP routing infrastructure) IP addresses (or *End-point IDentifiers – EID*). Then, as shown in Fig. 1, LISP tunnels packets in the core Internet from one RLOC of the source EID to one RLOC of the destination EID. The *Ingress Tunnel Router (ITR)* prepends a LISP header to each packet using the ITR's RLOC as a source address and the *Egress Tunnel Router's (ETR's)* RLOC as a destination address, while the ETR strips this header and forwards the packet to its final destination.

The use of different addressing spaces, namely the EIDs and RLOCs, implies that there is the need of bindings between EIDs and RLOCs. These bindings specify which RLOC to use when encapsulating packets towards a given EID. These bindings are called *mappings*, hence the reason why LISP is called a *map-and-encap* approach. For scalability reasons mappings are obtained by ITRs on-demand and kept in a dedicated cache [8]. When, no suitable entry is present, a cache-miss is triggered, causing a query to a Map-Resolver (MR [13]), which is the entry point of the Mapping System. The query consists in a Map-Request message to which eventually the MR (or directly the destination ETR) will return a Map-Reply message with the requested mapping.[1] Mapping retrieval is a key point performance wise since a cache-miss causes signaling traffic, increases communication setup latency, and decreases the throughput. Such an impact is due to the fact that current implementations drop packets causing a cache-miss.[2]

In a normal IP network, such as the enterprise network shown in the right part of Fig. 1, the Interior Gateway Protocol (IGP) handles link and node failures. Consider for example that ITR_1 and ITR_2 advertise a default route in the IGP, a very common deployment. In this case, if (the link attached to) ITR_1 fails, then the other local routers will detect the failure and update their routing tables to forward the packets to ITR_2 so that they can reach the Internet through it. During the last years, various techniques have been developed to enable routers to quickly react upon such failures [14]. With LISP, the situation is different. If ITR_1 fails, the IGP will quickly redirect the packets to ITR_2. However, ITR_2 will be able to forward these packets only if it already knows the corresponding mappings. Otherwise, ITR_2 will have to drop packets while querying the mapping system, an operation that can take tens of milliseconds or more per mapping [11].

3   The ITR Failure Problem

In order to estimate how important the ITRs' failure problem is, the first step is to evaluate the level of redundancy present in enterprise and campus networks. Indeed, in today's Internet, these networks often contain several redundant border routers to preserve connectivity in case of failures.

To evaluate such an aspect, we analyze Internet topology information from the Archipelago project [15] and BGP tables of the Routeviews project [16]. We first extract the stub prefixes from the BGP table of Oregon-IX collected on August 12th 2010 [16]. The BGP curve in Fig. 2 shows the number of neighbor ASes for each prefix. This is an approximation of the multi-homing degree. We obtain the traceroute curve in Fig. 2 by filtering an Archipelago trace captured

---

[1] The proposed solution does not rely on any mapping system, hence their description is out of scope. Jakab et al. [11] propose a comparison of different mapping systems.

[2] LISP does not specify any action for packets causing misses, which cannot be encapsulated due to the missing mappings. Current implementations, similarly to the case of missing MAC address in the ARP table, for scalability reasons drop the packet.

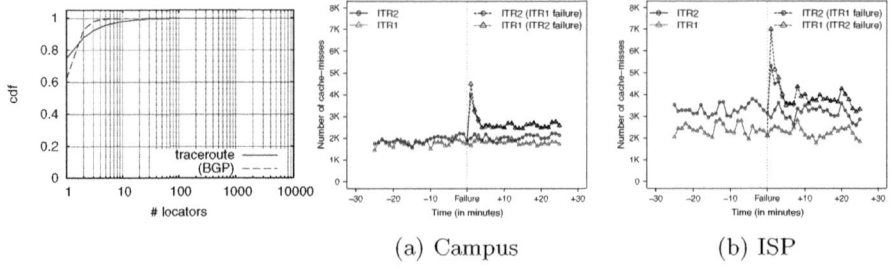

(a) Campus   (b) ISP

Fig. 2. Distribution of the   **Fig. 3.** Cache-misses per-minute due to ITRs failures
number of border routers per
BGP prefix

July 24th 2010 [15]. Archipelago traces are a collection of traceroutes performed
from several vantage points to any possible /24 prefix. We consider that the
last router that does not belong to the stub BGP prefix is a border router for
that prefix, providing an approximation of the number of active routers of the
multi-homed stub prefixes.

The curves in Fig. 2 show that when a stub prefix is multi-homed, most of
the time it has only two border routers. Nevertheless, the long tail of the distribution indicates that for some prefixes, the number of ITRs can be potentially
high. For this reason, to assess to what extent ITRs' failures are a problem, we
can analyze the impact on traffic for the simple scenario of 2 ITRs, by using
trace-driven emulations and the topology presented in Fig. 1. Our emulations
use the software and the methodology used in previous works ([6], [8]), assuming a mapping granularity equivalent to the current BGP table [8]. With this
granularity, previous works ([6], [8]), have shown that cache size is pretty small,
allowing to neglect cache overflow problems, which are unlikely to happen.

We use two different 24h-long traces collected in 2010. A first trace, collected
in September 1st with a NetFlow collector without sampling, is from a middle
sized Campus network (∼9,000 active users), connected via a 1 Gbps link to its
ISP and 122.35 Mbps average traffic. The second trace has been collected within
a large European ISP (∼20,000 active DSL customers). We split the network of
the captured traces into two subnetworks (served by ITR_1 and ITR_2) in order
to implement the multi-homing scenario as in Fig. 1. Since ISP's topology was
unknown, we attach half of the address space to ITR_1 and the other half to ITR_2.
On the contrary, the Campus network has two border routers, which we assigned
the role of ITR_1 and ITR_2, hence attaching the traffic to the ITRs according to
the real topology. From the traces, we selected the busiest hour (*i.e.*, the worst
case) and emulated the failure of one ITR in the middle of the selected hour.

Fig. 3 shows the impact of the failure on the number of cache-misses on the
alternate ITR, for the ISP and the Campus networks, which have a similar
behavior. The figure shows the normal behavior without any failure (solid lines),
as well as the two cases where one of the ITRs fails (dashed lines). During the
first minute after the failure there are up to 5,000 additional cache-misses per

minute (more than three times the normal rate) due to the traffic redirection on the alternate ITR. This failure can affect established TCP flows and cause packet drops for seconds or more. Fig. 3 shows as well that the transient state lasts around 5 minutes.

It has to be pointed out that we are underestimating the impact since we do not consider mapping delay, *i.e.*, we assume that the mapping is retrieved immediately after the cache-miss. In reality, failure-induced cache-misses have much more severe impact because they affect established high traffic volume flows. While a normal cache-miss causes the loss of a few packets at flow setup time [17], a failure-induced cache-miss could cause more losses because disrupting high throughput flows on high bandwidth links. Moreover, the peak of cache-miss causes a load peak on the mapping system to retrieve the missing mappings.

4   Local ITR Failure Protection

To solve the problem presented in the previous section, we propose to synchronize the caches of a group of ITRs of the same site. The set of ITRs belonging to the same group is called the *Synchronization Set*. Replicating the LISP cache on the ITRs ensures that in case of failure and traffic re-routing, the packets of an existing flow will never be dropped because of a cache-miss, whatever the alternate ITR of the set is used.

There are two ways to synchronize ITR caches: either the mappings are *pushed* to the ITRs of the Synchronization Set, or the latter are *notified* that a new mapping has to be retrieved. We discuss both approaches in further details in Sec. 8. Here we just assume that the mapping requested by one ITR of a Synchronization Set is immediately replicated on all ITRs of the set. To ensure that mapping caches remain synchronized, the ITRs should keep it for the same amount of time. This implies that synchronized ITRs cannot anymore use a simple inactivity timeout [6] to purge unused entries from their cache. Indeed, doing so would lead to a loss of synchronization, since the same mapping can expire on one ITR, because it has not been used, while it is kept on the other ITR that forwarded packets towards this EID prefix.

To avoid such loss of synchronization, we propose to use a different policy, namely to keep each mapping in the cache during the TTL (Time-To-Live) that LISP associates to this mapping. When the mapping TTL expires the ITR must check the entry usage during the last minute. If the entry has not been used, it is purged. Otherwise, the mapping entry is renewed by sending a Map-Request. This last action triggers the synchronization mechanism, again replicating the mapping to all ITRs of the Synchronization Set. Such approach guarantees that, if no ITR has used the mapping during the last minute before TTL expiration, all replicas on the different ITRs will be purged. Otherwise, if at least one ITR used the mapping, the mapping will be replicated on all ITRs, renewing it for another TTL time. In both cases there is a consistent state on all ITRs of the Synchronization Set.

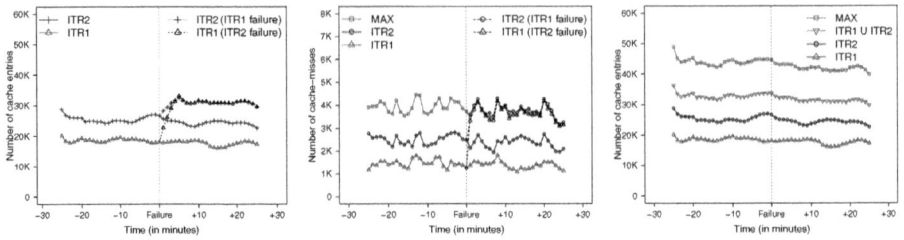

Fig. 4. Cache size without synchronization (ISP)

Fig. 5. Cache-misses with synchronization (ISP)

Fig. 6. Cache size with synchronization (ISP)

5   ITR Failure Protection Evaluation

To evaluate the cache synchronization technique we follow the same methodology and the same traces used in Sec. 3. On the one hand, we emulate the asynchronous cache strategy, *i.e.*, where each ITR manages its own cache independently. On the other hand, we emulate the synchronous cache strategy, *i.e.*, where ITR_1 and ITR_2 belong to the same Synchronization Set. In all of our simulations, we set the TTL to 5 minutes, which we consider as a reasonable worst case. Indeed, a lower value would generate too much overhead [6]. In practice, it is likely that the TTL will be longer than 5 minutes (the default in current LISP implementations being 24h), reducing the number of cache-misses, but increasing the number of entries that are stored in the cache [8]. However, it is worth noticing that the cost of slightly increasing the cache is very small compared to the cost of adding a redundant link. The figures in the remainder of this section show the cache behavior of ITRs in the normal case (*i.e.*, without failures) using solid lines, while dashed lines correspond to scenarios with failures.

In Fig. 3, the observed peak of cache-misses indicates that the traffic behind the two ITRs is not identical and not balanced; hence some mappings are in one ITR but not in the other. This is confirmed by Fig. 4, showing the evolution of the cache size for the ISP network (Campus network showing the same behavior), reinforcing the motivation for cache synchronization. Indeed, it can be observed that the ITRs have in general different cache sizes and when one of the ITRs fails the diverted traffic makes the size of the remaining ITR to grow.

Fig. 5 and Fig. 6 show the evolution of both cache size and cache-misses when our synchronization approach is used. The curves labeled ITR1 and ITR2 show the evolution on the ITRs when they are not synchronized and none fails. The curve labeled MAX shows the maximum obtained when ITRs are synchronized but the content of their cache is assumed to be completely disjoint (*i.e.*, the worst case). Finally, the curve labeled ITR1 (ITR2 failure) (resp. ITR2 (ITR1 failure)) corresponds to the actual values obtained with our synchronization approach if ITR_2 (resp. ITR_1) fails. Fig. 5 is interesting for two reasons. First, no peak is observed when the ITR fails, rather, the miss rate corresponds to the miss rate that would have been observed if the network only had one ITR. As one could expect, the miss rate measured in steady state after the failure is identical to the miss rate observed in Fig. 3 once the steady state has been reached. Second, comparing Fig. 3 and Fig. 5, it turns out that the miss rate when no failure occurs is lower when the ITRs are synchronized than when they are not. This difference can be explained by the fact that some form of locality occurs between the traffic of the two ITRs.

In summary, our emulations on the 2 ITRs scenario clearly show that ITRs' cache synchronization brings two main advantages: (i) it avoids a miss storm (hence induced packet drops) upon ITR failure; (ii) reduces the number of misses (hence packet drops) in the normal case. These benefits come at a small cost of increased cache size. Indeed, Fig. 6 confirms that the ITRs have naturally some entries in common, which makes the burden of synchronization acceptable.

6   ITR Recovery: Problem, Protection, and Evaluation

The cache-miss storm in case of failure, as investigated so far, can also be observed when an ITR boots (or comes back online after failure). Indeed, in this case, its cache is empty, hence the traffic attracted for encapsulation will cause misses and will be dropped. Our synchronization mechanism can be used also in this scenario. In case of synchronization, the time needed for an ITR to be synchronized with the other ITRs of the Synchronization Set is at most equal to the TTL. In this situation, when the ITR starts, it could begin the synchronization process and wait TTL time before announcing itself as an ITR able to encapsulate packets. With this approach the miss rate is not different than if the ITR had always encapsulated traffic.

This naive approach is very simple but has a major drawback. The TTL can be set to a high value, refraining the ITR from being used for a long period of time. To overcome this issue, we suggest allowing the ITR to receive a copy of the cache from another ITRs in the Synchronization Set. The transfer of the cache's information must be done reliably, $e.g.$, using TCP. In this way the ITR can announce itself right after the cache transfer, shortening the start up time to a minimum, while preserving the benefits of the synchronization approach.

To better understand the impact on the traffic in the recovery case, we perform emulations in the two ITRs scenario, with ITR_2 recovering after ITR_1 runs alone for the last 30 minutes (using the ISP trace). Once back online, ITR_2 starts synchronizing with ITR_1, $i.e.$, it receives mapping information, but the traffic is still all routed toward ITR_1. After 5 minutes ($i.e.$, the TTL value we are using throughout all emulations), when the cache is synchronized with ITR_1, the original setup is restored, sending traffic again to ITR_2.

Fig. 7 shows the miss rate observed at the ITRs for both the synchronized and the non-synchronized cases. The curve ITR1 (ITR2 failure) gives the miss rate observed on ITR_1 before the recovery of ITR_2, while ITR1 (sync) (resp. ITR2 (sync)) shows the miss rate as observed on ITR_1 (resp. ITR_2) after ITR_2 has recovered when synchronization is used. For comparison, the curve ITR1

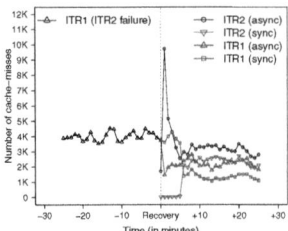

Fig. 7. Cache-miss on ITR$_2$ after boot

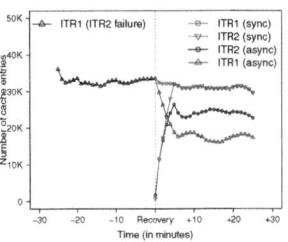

Fig. 8. Cache size on ITR$_2$ after boot

Table 1. Increase of cache size and misses due to ITR failure

	Failure	Increase	Increase*	Network
Entries (avg.)	ITR1	55.54%	46.59%	Campus
	ITR1	20.57%	20.30%	ISP
	ITR2	68.54%	56.98%	Campus
	ITR2	63.93%	56.54%	ISP
Misses (avg.)	ITR1	40.32%	55.25%	Campus
	ITR1	17.02%	28.22%	ISP
	ITR2	52.46%	72.95%	Campus
	ITR2	67.14%	96.34%	ISP

* Increase without counting the 5 min. transient period right after failure

(async) (resp. ITR2 (async)) shows the miss rate when no synchronization is used, with a peak of 10,000 cache-misses confirming that waiting TTL instants to lazily synchronize the cache avoids cache-miss storm. For completeness, Fig. 8 shows the evolution of the cache size before and after recovery. In all cases, the cache smoothly moves towards its new steady state TTL time after the recovery. This smooth convergence motivates the use a fast cache synchronization method (*i.e.*, explicit cache copy request) to speed-up the recovering of an ITR.

7   Synchronization Set in Large-Scale Networks

As shown in Sec. 3, in case of failure, when no synchronization is used, both the cache size and the cache-misses increase, as summarized in Tab. 1. While relatively similar increases can be observed comparing the results of the Campus and the ISP traces, the difference of the increases between ITR$_1$ and ITR$_2$ may be significant. Indeed, the average miss rate can be as low as 28% when ITR$_1$ fails, but as high as 96%, for the same trace, when ITR$_2$ fails. This result suggests that the Synchronization Set is a key point and should be carefully computed. The issue is exacerbated in large networks containing potentially many ITRs, which cannot all be synchronized.

For the above-mentioned reasons, it is important to have an idea of how large the Synchronization Set can be. To this end, we exhaustively calculated the Synchronization Set of 8 different real large-scale network topologies. Each of these topologies belong to one of the three following categories: (i) *TIER1* grouping

Table 2. Topologies characteristics

Name	Routers	Links
TIER1a	500^+	$2,000^+$
TIER1b	200^+	800^+
ISPa	50^+	200^+
ISPb	20^+	60^+
UCL	11	41
NREN	30^+	70^+
Internet2	11	30
Géant	22	72

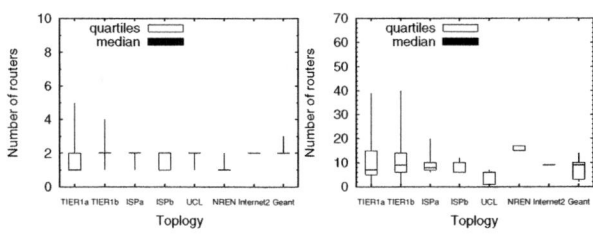

Fig. 9. Number of used ITRs per EID routers

Fig. 10. Number of routers behind an ITR

Tier 1 ISPs; (ii) *ISP* grouping national ISPs; (iii) *Research* grouping university campuses or research networks. A summary of these topologies is shown in Tab. 2. We model the networks as a directed acyclic graph where the nodes are the routers and the edges are the links. Then we differentiate the routers in three different types: the *ITRs*, the *EID routers*, and the *backbone routers*. For the topologies where we had enough information, we considered the routers connected to another ISP as ITRs; the routers connected to the customers or to LANs with end-hosts are considered the EID routers; all the other routers are classified as backbone routers. For the topologies where no enough information was available, we classified the routers based on their connectivity degree. All the routers with a connectivity degree higher or equal to the 90th percentile of the connectivity degree are considered ITRs. The routers with a connectivity degree lower or equal to the 80th percentile of the connectivity degree are EID routers. All the other routers are classified as backbone routers. Applying this heuristic provides results similar to the topologies where information was available. To construct the Synchronization Set, we exhaustively enumerate the topology changes for any possible single failure (*i.e.*, router or link).

Fig. 9 gives the quartiles (including min., max., and median) computed for the number of ITRs potentially used by each EID router. For this evaluation, we consider that only one EID router serves an EID prefix. This is a reasonable assumption since for the transit networks we evaluated (*i.e.*, TIER1a, TIER1b, NREN, Internet2, and Géant) the EID prefixes belong to the customers, which have only one router toward their ISP (but are multi-homed with several ISPs). The figure shows that the number of ITRs potentially used by the EID routers is relatively independent of the type of topology and is between 1 and 2. Meaning that only a small portion of the ITRs are serving the same EIDs, due to the fact that large networks are segmented in relatively independent points of presence. This result clearly shows that synchronizing all the ITRs of a large network would be inefficient, while in our case the burden will be reasonable as an ITR will synchronize only with few other ITRs.

Fig. 10 goes the other way around, showing the number of routers that are potentially behind a single ITR, using again quartiles. The figure shows that in general an ITR has about half a dozen of routers behind it, going up to few tens for large networks. Therefore, the failure of an ITR can potentially impact on an important portion of traffic, which has to be shifted to the other ITRs. If the caches are not synchronized, the ITRs to which the traffic is diverted might experience a very important miss rate, with high packet drops. Synchronizing the ITRs' caches avoids such drops, as we demonstrate in Sec. 5.

In summary, protecting the ITRs is important as the failure of one of them can impact a large portion of the network. Fortunately, topologies of real networks seem to be segmented so to facilitate grouping the ITRs in small Synchronization Sets.

8   Synchronization Techniques

Even if, as discussed above, the Synchronization Set is in practice small also for large-scale networks, it is important to implement a mechanism to achieve such synchronization without introducing excessive overhead or management complexity. There are two possible ways to synchronize ITRs: either, the mappings are *pushed* to the ITRs, or the ITRs are *notified* of the presence of a new mapping and they retrieve the mapping on their own.

In both cases, the synchronization can be implemented either by leveraging on a routing protocol extension, by using a specific existing protocol (*e.g.*, [18], [19], [20], [21]), or by creating a brand new protocol. Extending a routing protocol to synchronize ITRs implies that they must be in the same routing protocol instance (*e.g.*, the same IGP). However, this assumption is too restrictive as it might exist cases where the ITRs are in different networks (*e.g.*, the ITRs are operated by the ISPs and the LISP site is multi-homed). Hence, we do not consider the extension of the routing protocol as an acceptable solution. Fortunately, LISP already proposes features that are handy to implement a synchronization mechanism. Indeed, LISP specifies the *Included Map-Reply* [4] feature to push mappings in ITR caches. An included Map-Reply is a special Map-Request, which piggybacks a mapping. LISP also specifies the *Solicit Map-Request* (SMR) bit in Map-Requests to force ITRs to refresh their cache [4]. When an ITR receives a Map-Request with the SMR bit set, it sends a Map-Request to retrieve a mapping for the EID indicated in the SMRed Map-Request.

Since both ITRs and MRs are involved in the mapping resolution, these are good candidates to trigger cache synchronization. However, ITRs are data-plane devices that need to forward packets at line rate. Therefore, imposing them to actively manage the synchronization protocol might cause excessive overhead with consequences on the data-plane performance. On the contrary, Map Resolvers are purely control-plane devices that are not intended to forward packet at line rate. Hence, MRs look like the best candidates to manage the synchronization protocol. To implement the cache synchronization based on notification messages, the MR only has to send an SMRed Map-Request to all the ITRs, when it receives a Map-Request. However, if the mappings are pushed to the caches, then the MR has to proxy the Map-Requests. The MR performs two operation when it receives a Map-Reply. First, it forwards the Map-Reply to the ITR that requested the mapping. Second, it sends the mapping to the other ITRs by using an included Map-Reply.

As discussed above, the Synchronization Set is in practice small and can be statically configured in the MR. However, for large or very dynamic networks, the configuration burden might still become cumbersome and a dynamic ITR discovery protocol coupled with an automatic Synchronization Set computation algorithm should be considered. It is out of the scope of this paper to propose any specific solution for these two mechanisms. Furthermore, there is still the open question of what is the best trade-off between tight synchronization and signaling overhead. Depending on the importance accorded by the operator to the accuracy of the cache synchronization, the mapping distribution between to the ITRs can be performed with a reliable protocol (*i.e.*, TCP) or not. In addition, batching of synchronization messages can be used to reduce the number of exchanged synchronization messages. Moreover, when the network allows it, the mappings can be distributed by using IP multicast.

9 Conclusion and Further Work

In this paper, we propose a thorough study of failure protection and recovery in the context of the Locator/Identifier Separation Protocol (LISP) and propose a local failure protection mechanism for Ingress Tunnel Routers (ITRs). We first showed that ITR failures can indeed have large disruptive impact on ongoing traffic. Then, we explored how to minimize packet losses due to cache-misses caused by the redirected traffic on the alternate ITRs. Our proposal synchronizes the cache of set of ITRs in order to avoid such a phenomena. We thoroughly evaluate our proposal, showing that the load increase due to the synchronization is acceptable and suppresses the loss of packets upon ITRs failure/recovery.

Our ongoing work is aiming at extending the synchronization mechanism to ETRs and developing detailed specifications for implementation and experimentation in the *lisp4.net* testbed.

Acknowledgements. This work was partially supported by a Cisco URP grant. The research leading to these results has received funding from the French National Research Agency under grant agreement n° ANR-11-EITS-007-01. We would like to thank Simon van der Linden for his help.

References

1. BGP Routing Table Analysis Report, http://bgp.potaroo.net
2. Quoitin, B., Iannone, L., de Launois, C., Bonaventure, O.: Evaluating the benefits of the locator/identifier separation. In: The 2nd ACM/IEEE Workshop on Mobility in the Evolving Internet Architecture (MobiArch 2007) (August 2007)
3. Meyer, D., Zhang, L., Fall, K.: Report from the IAB Workshop on Routing and Addressing. RFC 4984, Internet Engineering Task Force (September 2007)

4. Farinacci, D., Fuller, V., Meyer, D., Lewis, D.: Locator/ID separation protocol (LISP). IETF, Internet Draft draft-ietf-lisp-22.txt (February 2012)
5. Locator/ID Separation Protocol (LISP) Working Goup, http://datatracker.ietf.org/wg/lisp/charter/
6. Kim, J., Iannone, L., Feldmann, A.: A Deep Dive into the LISP Cache and What ISPs Should Know about It. In: Domingo-Pascual, J., Manzoni, P., Palazzo, S., Pont, A., Scoglio, C. (eds.) NETWORKING 2011, Part I. LNCS, vol. 6640, pp. 367–378. Springer, Heidelberg (2011)
7. Zhang, H., Chen, M., Zhu, Y.: Evaluating the Performance on ID/Loc Mapping. In: The Global Communications Conference (Globecom 2008) (November 2008)
8. Iannone, L., Bonaventure, O.: On the cost of caching locator/id mappings. In: The 3rd ACM Annual CoNEXT Conference (CoNEXT 2007) (December 2007)
9. Sriram, K., Gleichmann, P., Kim, Y.-T., Montgomery, D.: Enhanced Efficiency of Mapping Distribution Protocols in Scalable Routing and Addressing Architectures (August 2010)
10. Choi, N., You, T., Park, J., Kwon, T., Choi, Y.: ID/LOC Separation Network Architecture for Mobility Support in Future Internet (February 2009)
11. Jakab, L., Cabellos-Aparicio, A., Coras, F., Saucez, D., Bonaventure, O.: LISP-TREE: A DNS Hierarchy to Support the LISP Mapping System. IEEE Journal on Selected Areas in Communications (October 2010)
12. Meyer, D., Lewis, D.: Architectural Implications of Locator/ID Separation. IETF, Internet Draft draft-meyer-loc-id-implications-01 (January 2009)
13. Fuller, V., Farinacci, D.: LISP Map Server. IETF, Internet Draft draft-ietf-lisp-ms-15.txt (January 2012)
14. Vasseur, J., Demeester, P., Pickavet, M.: Network Recovery: Protection and Restoration of Optical, SONET-SDH, IP, and MPLS, Elsevier, Ed. (2004)
15. Claffy, K., Hyun, Y., Keys, K., Fomenkov, M., Krioukov, D.: Internet mapping: From art to science. In: Proceedings of the 2009 Cybersecurity Applications & Technology Conference for Homeland Security, pp. 205–211. IEEE Computer Society, Washington, DC (2009)
16. University of Oregon, Route views, University of Oregon Route Views project, http://www.routeviews.org/
17. Ohmori, M., Okamura, K., Tanizaki, F.: Analyses on first packet drops of lisp in end-to-end bidirectional communications. In: Internet Conference, Fukuoka, Japan (October 2011)
18. Luciani, J., Armitage, G., Halpern, J., Doraswamy, N.: Server Cache Synchronization Protocol (SCSP). RFC 2334 (Proposed Standard), Internet Engineering Task Force (April 1998), http://www.ietf.org/rfc/rfc2334.txt
19. Saito, Y., Shapiro, M.: Optimistic replication. ACM Comput. Surv. 37, 42–81 (2005)
20. Obraczka, K., Dazing, P.: Evaluating the performance of flood-d: A tool for efficiently replicating internet information services. IEEE Journal on Selected Areas in Communications 16, 369–382 (1998)
21. Petersen, K., Spreitzer, M., Terry, D., Theimer, M., Demers, A.: Flexible update propagation for weakly consistent replication. In: Proceedings of the 16th ACM Symposium on Operating Systems Principles (SOSP), pp. 288–301 (October 1997)

An Analytical Model for the LISP Cache Size

Florin Coras, Albert Cabellos-Aparicio, and Jordi Domingo-Pascual

Universitat Politècnica de Catalunya,
Barcelona, Spain
{fcoras,acabello,jordi.domingo}@ac.upc.edu

Abstract. Concerns regarding the scalability of the inter-domain routing have encouraged researchers to start elaborating a more robust Internet architecture. While consensus on the exact form of the solution is yet to be found, the need for a semantic decoupling of a node's *location* and *identity* is generally accepted as the only way forward. One of the most successful proposals to follow this guideline is LISP (Loc/ID Separation Protocol). Design wise, its aim is to insulate the Internet's core routing state from the dynamics of edge networks. However, this requires the introduction of a mapping system, a distributed database, that should provide the binding of the two resulting namespaces. In order to avoid frequent lookups and not to penalize the speed of packet forwarding, map-caches that store temporal bindings are provisioned in routers. In this paper, we rely on the working-set theory to build a model that accurately predicts a map-cache's performance for traffic with time translation invariance of the working-set size. We validate our model empirically using four different packet traces collected in two different campus networks.

Keywords: LISP, Internet architecture, cache modeling, working set.

1 Introduction

The fast paced growth [8] of the global inter-domain routing table and infrastructure has recently raised a series of concerns and spurred debate regarding the Internet's architectural inability to scale. Reasons for the growth are partly organic in nature, as new domains are continuously added to the topology, but also related to operational practices that defeat provider-based aggregation of prefixes. In this sense, multihoming, traffic engineering and allocations of provider-independent prefixes are the main drivers behind prefix de-aggregation [16].

In a recent Internet Advisory Board workshop [16] participants deemed the routing table growth as one of the Internet's most important problems and tracked down its origins to the semantic overloading of the IP addresses with both *location* and *identity* information. As a result, a set of architectures inspired by the loc/id split paradigm have been proposed [14] to overcome the limitation. Most notably, LISP [5] aims to alleviate the routing state pressure exerted by edge networks on core networks by means of an *identity-location* separation at

R. Bestak et al. (Eds.): NETWORKING 2012, Part I, LNCS 7289, pp. 409–420, 2012.
© IFIP International Federation for Information Processing 2012

network level. It is designed for an incremental deployment and it relies on a pilot network [1] for experimentation.

For the binding of identifiers and locators, LISP introduces a distributed database called the *mapping-system*. As LISP routers are typically domain border routers, the identifier resolution delays through the mapping system (hundreds of ms [11]) are several orders of magnitude higher than packet inter-arrival times (ns). Therefore, to speed up intra-router packet forwarding and to protect the mapping system from floods of resolution requests, the LISP routers are provisioned with mapping caches (*map-caches*) that temporarily store in use mappings. Given the map-cache's primordial role in assuring LISP's data-plane (packet forwarding) and control-plane (mapping-system) scalability, a better understanding of its behavior is crucial.

In this paper we aim to improve the understanding of the map-cache's performance by developing an analytical model that links miss rate with cache size. Specifically, we rely on the working-set theory [3] to build a model that accurately predicts the performance based on simple and measurable parameters of packet traces that possess time translation invariance of the working-set curves. We validate our model empirically using four different packet traces collected in two different campus networks. Our findings show that the miss rate decreases at an accelerated pace with the cache's size, and that even for a relative small size (5% of the Internet's aggregated routing table) the performance is acceptable (below 0.6% miss-rate).

2   LISP Overview

LISP [5] is one [14] of the recently emerged architectural solutions to the Internet's scalability problem [16]. Its main goal is that of splitting the semantics of IP addresses with the aim of forming two namespaces that unambiguously identify core (locators) and edge (identifiers) network objects. To facilitate transition from the current Internet infrastructure, both of the resulting namespaces use the existing IP addressing scheme. As a result, the split does not affect routing within existing stub or transit networks. Nevertheless, as identifiers and locators bear relevance only within their respective namespaces, a form of conversion, from one to the other, has to be performed at border points between core and edge networks. Traditionally, the two techniques that may be employed for a fast translation are address rewriting [17] and map-and-encap [7]. Unfortunately, besides their need for data plane modifications, both also require the introduction of a new control-plane *mapping function* able to provide bindings that link identifiers to locators. Out of the two, LISP enabled border routers make use of the latter. Therefore, prior to forwarding a host generated packet, a LISP router maps the destination address, or Endpoint IDentifier (EID), to a corresponding destination Routing LOCator (RLOC) by means of a LISP specific distributed database, called the *mapping system* [2,19,11,15]. Once a mapping is obtained, the border router tunnels the packet from source edge to corresponding destination edge network by means of an encapsulation with a

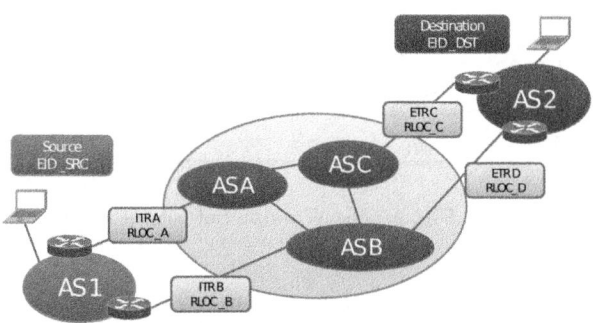

Fig. 1. Example LISP Architecture

LISP-UDP-IP header. The outer IP header addresses are the RLOCs pertaining to the corresponding border routers (see Fig. 1). At the receiving router, the packet is decapsulated and forwarded to its intended destination. In LISP parlance, the source router, that performs the encapsulation, is called an Ingress Tunnel Router (ITR) whereas the one performing the decapsulation is named the Egress Tunnel Router (ETR). One that performs both functions is referred to as an xTR.

Because all packets need to be encapsulated, the packet throughput of an ITR is highly dependent on the time needed to obtain a mapping. In consequence, all ITRs are provisioned with a cache, called map-cache, that stores the recently used EID-prefix-to-RLOC mappings to avoid frequent EID resolutions through the mapping system. Stale entries are avoided with the help of validity periods (timeouts) that mappings carry as attributes. One can easily observe that the map-cache is most efficient in situations when destination EIDs present high temporal and/or spatial locality and that its size depends on the diversity of the visited destinations. As a result, a cache's performance depends entirely on its provisioned size and on traffic characteristics. Given the map-cache's critical role in the LISP architecture, a good understanding of the properties linking its size and performance with the traffic's characteristics is fundamental.

3   LISP Cache Modeling

This section presents the theoretical background and methodology used to model the LISP map-cache's performance. After a brief presentation of the working-set model we perform an analysis of the traffic pertaining to several network traffic traces. The observed traffic properties enable us to analytically link cache size and miss rate.

3.1   Working-Set Theory

As our goal is to determine suitable caching strategies and to better model their performance for LISP routers, we leverage previous resource allocation and cache

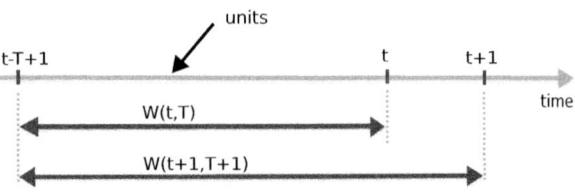

Fig. 2. The Working Set Model

performance research carried out in the field of operating systems. Consequently, in order to unify language, we are forced to draw parallels between concepts pertaining to our two disciplines. We associate to what systems understand by a *program* a network *traffic trace*, and in a similar vein, to a *page* an *address prefix*. Although seemingly unrelated, from an abstract point of view, the properties of the above pairs of objects influence in similar ways a cache's performance. We depict this point further in what follows.

For operating systems, a general resource allocation treatment was possible after it was observed that programs more or less obey the so called *principle of locality*. In such situations, a program's past referenced pages may be used as a good predictor for near future, to be re-referenced pages. We speculate that network traffic traces might roughly obey the same principle mainly because of flow burstiness in time but also due to traffic aggregation. Consequently, we try to evaluate the feasibility of in-router *prefix caching* by first building and then using a working-set model for network traffic.

In what follows, in order to avoid repetitive text, we shall be using the term *unit* of reference as a substitute for both pages and prefixes; and *reference set* to represent the set of all referenced pages or prefixes. Further, given a *reference set* N, we define a *reference string* as the sequence $\rho = r_1 r_2 \ldots r_i \ldots$ where each unit r_i belongs to N.

As explained by Denning in [4], a working-set analysis in this context may be performed only if based on three constraints that provide for a more rigorous definition for *locality of reference*:

1. Reference strings are unending
2. The stochastic mechanism underlying the generation of a reference string is stationary, i.e. independent of absolute time origin.
3. For all $t > 0$, r_t and r_{t+x} become uncorrelated as $x \to \infty$

The first of the constraints, though obviously not fulfillable, introduces an insignificant error because the reference strings generated by practical programs or traces are long from a statistical standpoint. The third of the assumptions requires that the references become uncorrelated as the distance (in units) goes to infinity. This can usually be asserted as being true in practice. The most restrictive of the three is the second assumption which limits the analysis to a locality where all three constraints, including stationarity are satisfied.

If all of the above requirements are satisfied, and t is a measure of time in *units*, then we can formally define the working-set $W(t, T)$, as: the set of distinct *units* that have been referenced among the T most recent references, or in the time interval $[t - T + 1, t]$ (see Fig. 2). In accordance with [4] we shall refer to T as the *window size*. Also, we will be using the term *working-set curve* when referring to $W(t, T)$ as a function of T, when t is constant.

We exploit the above definition to present some of the working-set's properties of further use in our analysis. For a broader scope discussion of the subject and complete proofs, the interested reader is referred to [4]. The first of the properties we are interested in is the *average working-set size, s(T)*. It measures the growth of the working-set with respect to the size of the window T, extending in the past, but independent of absolute time t. If $w(t, T)$ measures the number of units in $W(t, T)$ then, considering the first locality property, one can compute $s(T)$ as the averaged working-set size over an infinite number of windows:

$$s\left(T\right) = \lim_{k \to \infty} \frac{1}{k} \sum_{t=1}^{k} w(t, T) \tag{1}$$

It is also shown in [4] that the the *miss rate, m(T)*, which measures the number of units of reference per unit time returning to the working-set, is the derivative of the previous function:

$$m(T) = s(T + 1) - s(T) \tag{2}$$

In other words, if both $s(T)$ and $m(T)$ are viewed as functions of T, the miss rate can be regarded as the slope of the average working-set size. Finally, the sign inverted slope of the miss rate function, the second slope of $s(T)$, represents the *average interreference density* function.

$$-f(T + 1) = m(T + 1) - m(T) \tag{3}$$

3.2   Traffic Properties

In order to ascertain the practical feasibility of using the working-set model to evaluate a LISP cache's performance we applied the theory presented in Section 3.1 to several network traffic traces. Details regarding the network captures can be found in Section 4.1.

Foremost, we were interested in discovering if our traces satisfy the three constraints introduced in Section 3.1. As previously explained, the first and third introduce negligible errors and thus present no practical limitations. In order to verify the second, and therefore determine the trace segments generated by distinct stationary processes, we devised a simple experiment. For each trace we computed multiple empirical destination prefix working-set curves with start times spanning one day and spaced by half an hour. Intuitively, we were expecting to discover that curves with close start times have a similar growth shape (cluster), whereas those separated by larger time gaps behave differently.

By necessity [4], if working-set curves are generated by the same stationary process then, they will tend to cluster. For a certain window size the working-set

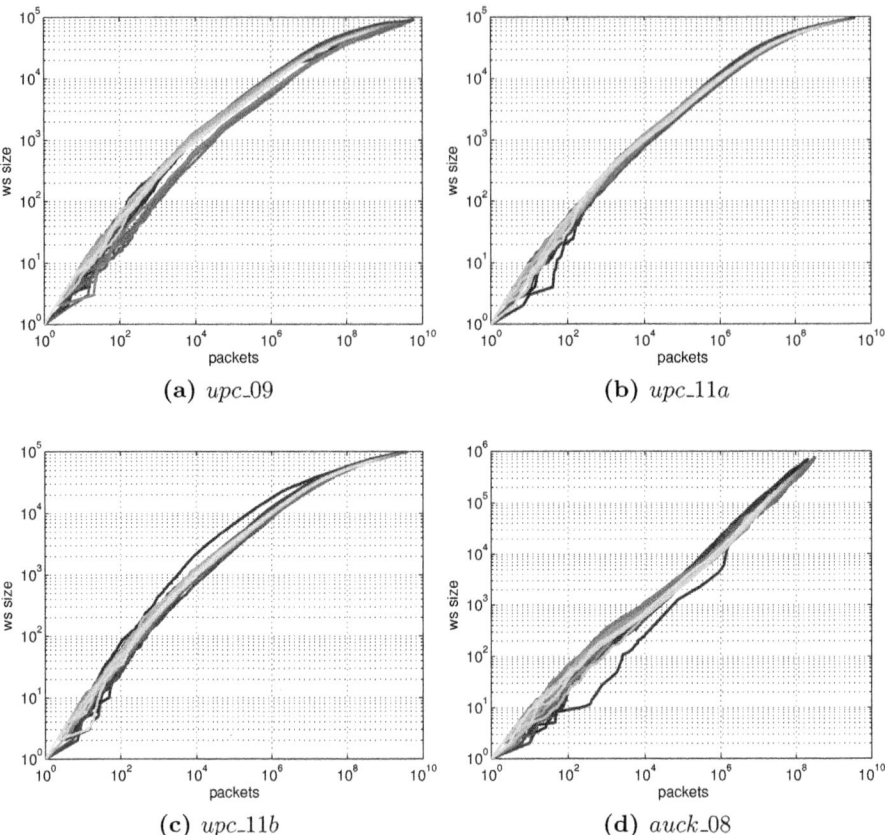

(a) *upc_09*

(b) *upc_11a*

(c) *upc_11b*

(d) *auck_08*

Fig. 3. Empirical working-set curves. Closeness of color nuance reflects closeness of start time.

size should be normally distributed. In other words, one may use the clusters of the working-set curves to discriminate between trace segments with different underlying generating processes. Figure 3 presents the results obtained in log-log plots. Surprisingly, all traces exhibit a strong clustering and sublinear growth, due to locality, of the working-set curves. We also tested and confirmed (not shown here) that the interreference interval and the relative frequency of reference distributions for our traces are approximately time translation invariant. Only *upc_09* presents a subtle time-of-day behavior with two perceptible clusters while *upc_11b* and *auck_08* each have one discordant curve that seems to be due to localized scanning attacks. Nevertheless, the results suggest that one could approximate the traces as being generated by a stationary process and as an implication, that any conclusions drawn after analyzing the average working set size function would characterize the whole trace. Similarly, Kim et al. observed in [12] that the working-set size for prefixes tends to be highly stable with time for traffic pertaining to a large ISP. It's worth noting that we do not assume that all Internet traffic possesses this approximate time translation invariance of the working-set curves. And in fact, we require that the model we develop further be applied only to traces that have it.

3.3   Analytical Model

Considering that the working-set, $W(t,T)$, always contains the $w(t,T)$ most recently referenced units then, the time translation invariance of $w(t,T)$ for a constant T implies that the working-set actually models a cache of fixed size. Similarly, a cache of size c that uses the Least Recently Used (LRU) eviction policy will always store the c most recently referenced units. Therefore, if $s(T) = c$, the working-set simulates a LRU eviction policy. In other words, due to the time translation invariance of the working-set curves the working-set can be seen as modeling a LRU cache.

With hindsight, one can recognize that the empirical working-set curves from Fig. 3 are piecewise linear when depicted in log-log scale. The observation enabled us to compute for each trace the average working-set size function by means of a piecewise linear fit of the log-log scale plot. As a result, we obtained estimates of both the slope α and the y-intercept β for all of a working-set's segments. We limited the number of knots to just three, however if better fits are desired more knots can be used. Figure 4 presents the goodness of fit for two of our traces. Next, through conversion to linear scale the equations become piecewise power law of the type:

$$s_{fit}(u) = \exp(\beta(u))u^{\alpha(u)} \qquad (4)$$

Where u represents the number of referenced destination prefixes, $s_{fit}(u)$ is the fitted working-set size function and $\alpha(u)$ and $\beta(u)$ are piecewise constant functions obtained through fitting. With the help of (2) one can estimate the miss rate for a trace by computing the derivative of s_{fit} like:

$$m_{est}(u) = \exp(\beta(u))\,\alpha(u)\,u^{\alpha(u)-1} \qquad (5)$$

Combing the last two equations one obtains an analytical relation that links the cache size and the estimated miss rate.

$$m_{est}(c) = \exp\left(\frac{\beta^*(c)}{\alpha^*(c)}\right)\alpha^*(c)\,c^{1-\frac{1}{\alpha^*(c)}} \qquad (6)$$

Where, c represents the cache size in number of entries and $\alpha^*(c)$ and $\beta^*(c)$ are piecewise constant functions with knees dependent on c. We put to test the accuracy of the above equation in Section 4.3.

4   Evaluation

In this section, we present the four data sets used to build and evaluate the map-cache model. We then describe the simulator employed in the empirical evaluation of the cache and we finish by comparing the theoretical and empirical cache performance results.

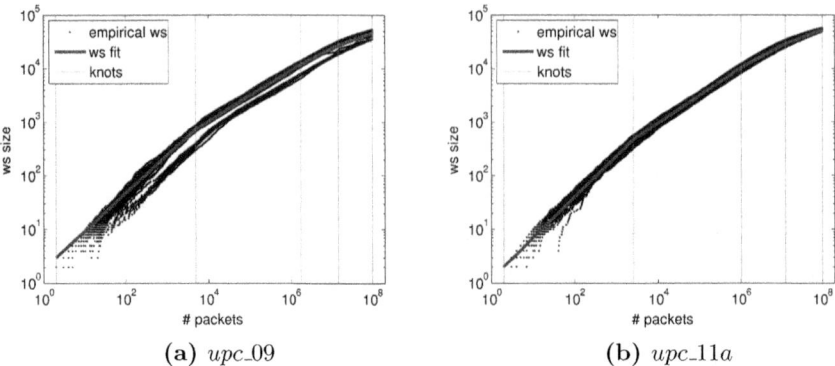

(a) *upc_09*   (b) *upc_11a*

Fig. 4. Fit of empirical working-set curves

4.1   Traces and IP Aggregation

Four one-day packet traces were used for the purpose of our experiments. Three of them have been captured at the 2Gbps link connecting several of our University's campus networks to the Catalan Research Network (CESCA) and consist only of egress traffic. UPC Campus has more than 36k users. One dates back to May 2009 while the other two are from October 2011. The fourth trace belongs to the University of Auckland and has been downloaded from a public trace repository [20]. Its capture date is March 2008 and it consists of University egressing traffic. Further, the addresses have been anonymized and one hour's worth of midday traffic, from 12 to 13, is missing. We shall further refer to these four traces as *upc_09*, *upc_11a*, *upc_11b* and *auck_08* respectively.

Both for the evaluation of the working-set in Section 3.3 and for the cache performance evaluation to be presented in Section 4.3, IP addresses had to be mapped to their corresponding prefixes. We considered EID-prefixes to be of BGP-prefix granularity. In the case of UPC traces, we used BGP routing tables downloaded from the RouteViews archive [18] that matched the exact capture date of the traffic traces. The only preprocessing we performed was to filter out more specific prefixes. Generally, such prefixes are used for traffic engineering purposes however, the LISP protocol provides mechanisms for a more efficient management of these functions that do not require EID-prefix de-aggregation. In the case of *auck_08*, the anonymized traffic is prefix preserving but the lack of an anonymized routing table made it impossible for us to determine the exact clustering of anonymized IPs into prefixes. Consequently, we constructed a virtual routing table made up of all prefixes with a 24 bit netmask. We chose the 24 bit limit as it is the longest netmask (corresponding to the most specific prefix) currently accepted for BGP announcements in the Internet's DFZ.

4.2   Simulator

We indirectly assess the effectiveness of the working-set model as a tool for LISP cache performance evaluation by practically testing the soundness of equation (6). For this we implemented a packet trace based simulator that mimics a subset of a LISP ITR's functionality. In short, for each processed packet, the simulator maps the destination IP address to a prefix in accordance with the methodology presented in Section 4.1. If this prefix is already stored in the ITR's cache, no further processing is done and the simulation continues with the next packet. Should the prefix not yet be stored in the cache, two possibilities arise. First, if the cache is not full, the destination prefix is stored in and the processing proceeds to the next packet. Second, if the cache is full, an entry is evicted, the new prefix is stored in and then the simulator moves to the next packet. The entry to be evicted is chosen according to the LRU eviction policy. We use LRU because, as mentioned in Section 3.3, its performance should be appropriately described by equation (6) for our traffic.

4.3   Results

We ran simulations with several cache sizes for all traces. Figure 5 presents the results for just two, upc_09 and upc_11a, of our traffic captures as they are representative for all four. Cache size is normalized with routing table size. It can be noticed that the absolute error of the estimation is negligible when compared to simulation results. Even in the case of upc_09, the only trace to present a more pronounced time-of-day behavior, the absolute error is of small proportions. Further, the model appropriately predicts that even for small LISP cache sizes the performance still stays acceptable, fact also observed by [9,13,11]. This is even more remarkable as we do not consider TTL, a timeout after which stale entries are removed, in our simulation.

Because of its proven good performance, we can now recommend the use of the LRU eviction policy in LISP caches, at least for traces that present an approximate time translation invariance of the working-set curves. Still, the diversity of our data sets and previous results from Kim [12] suggest that this property is not due to an isolated event. In this situation, equation (6) may be used to dimension the cache size according to the desired miss rate. Further, the prediction of its mathematical expression, due to discontinuity at the knots, is that miss rate decreases at an accelerated pace with cache size and finally settles to a power law decrease. However, the speed of the decrease depends on the degree of locality present in the trace. This leads us to conclude that cache sizes need not be very large for obtaining good performance. For instance, in the case of upc_11a, for a cache having 8.4k entries (0.05% of the Internet's aggregated DFZ routing table) the miss rate is around 0.6%.

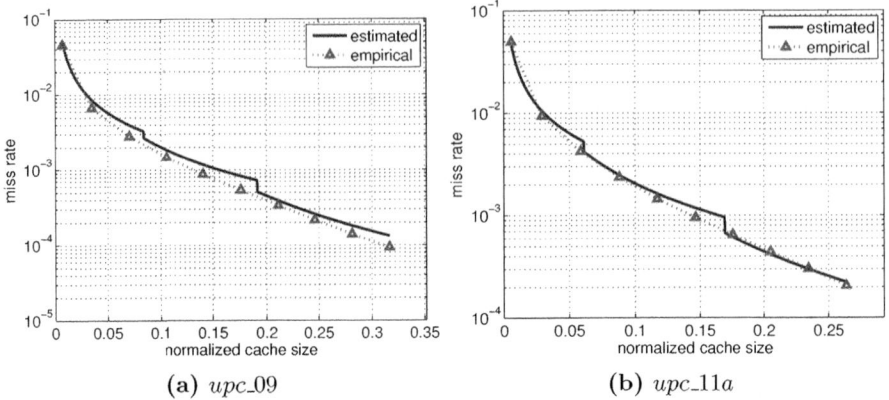

Fig. 5. Comparison of empirical and estimated miss rates

5 Related Work

Feldmeier [6] and Jain [10] were among the first to evaluate the possibility of performing destination address caching by exploiting the locality of the reference strings in network environments. Feldmeier analyzed traffic on a gateway router and showed that caching of both flat and prefix addresses can significantly reduce routing table lookup times. Jain however performed his analysis in an Ethernet environment. He was forced to concede that despite the locality observed in the traffic, small cache performance was struggling due to traffic generated by protocols with deterministic behavior. Both works were fundamental to the field however their results were empirical and date back to the beginning of the 1990s, years when the Internet was still in its infancy. Recently, Kim et al. [12] showed the feasibility of route caching after performing a measurement study within the operational confinement of an ISP's network. They show by means of an experimental evaluation that LRU performs close to an optimal eviction policy and that working-set size is generally stable with time. We also observe the stability of the working-set for our data sets but we further leverage it to build a LRU model instead of just empirically evaluating its performance.

Iannone et al. [9] performed an initial trace driven study of the LISP map-cache performance. Instead of limiting the cache size by using an eviction policy, their cache implementation evicted stale entries after a configurable timeout value. Further, Kim et al. [13] have both extended and confirmed the previous results with the help of a larger, ISP trace and by considering LISP security aspects in their evaluation. Ignoring security concerns, which we did not consider, and despite differences regarding the cache eviction policy, the results of these last two works seem to be in agreement with ours. Zhang et al. [21] performed a trace based Loc/ID mapping cache performance analysis assuming a LRU eviction policy. They used two 24-hour traffic traces captured at two egressing links of the China Education and Research Network backbone network.

They concluded that a small cache can offer good results. Finally, using trace *upc_09*, Jakab et al. [11] analyzed the performance of several LISP mapping systems and, without focusing on a cache analysis, also observed very low miss rates for a cache model similar to that used in [9]. Our work confirms previous LISP cache analysis results however, it also tries to provide a better understanding of the reasons behind the relatively good performance of the LISP cache. In this sense it introduces a cache model that could be used to theoretically evaluate or dimension for operational needs the caching performance.

6   Conclusions

The location/identity split paradigm has been recently recommended as the solution to Internet's scalability problems. Out of the proposals to follow its guidelines, LISP seems to be the one to have drawn the largest amounts of attention and support. In the set of newly introduced architectural components, the map-cache is one of the most important to LISP's own scalability and also the subject of our analysis. Given that caches have their performance driven by the characteristics of the traffic they serve, we proposed an evaluation that links the LISP map-cache performance with measurable, intrinsic properties of the traffic processed by a LISP router. In this sense, we advanced the use of the working-set model [3] as a tool to capture the mentioned traffic properties but also as a performance predictor. Accordingly, we also noted that the time-translation invariance of the working-set curves is a constraint that traffic traces need to observe for the theory to be applicable. After we established that our data sets are amenable to working-set modeling we deduced an analytical link between the cache size and miss rate. Finally, we validated our theoretical result by means of simulation and with the help of traffic traces collected in two different campus networks.

We note that the predictions of our model fit with previous observations however, we also remark the versatility of our result. Besides the possibility of using it as a cache dimensioning tool in operational environments it may also be used in more complex models that evaluate the scalability of LISP or other loc/id architectures. Another interesting application could be to evaluate the effects of trashing attacks on the LISP map-cache.

Acknowledgments. This work has been partially supported by the Spanish Ministry of Education under scholarship number AP2009-3790, research project TEC2011-29700-C02, Catalan Government under project 2009SGR-1140 and a Cisco Unsolicited Research Proposal Grant.

References

1. LISP Testbed, http://www.lisp4.net/
2. Farinacci, D., Fuller, V., Meyer, D., Lewis, D.: LISP Alternative Topology (LISP+ALT). draft-ietf-lisp-alt-10 (December 2011)(work in progress)

3. Denning, P.J.: The working set model for program behavior. Commun. ACM 11(5), 323–333 (1968)
4. Denning, P.J., Schwartz, S.C.: Properties of the working-set model. Commun. ACM 15(3), 191–198 (1972)
5. Farinacci, D., Fuller, V., Meyer, D., Lewis, D.: Locator/ID Separation Protocol (LISP). draft-ietf-lisp-17 (December 2011) (work in progress)
6. Feldmeier, D.: Improving gateway performance with a routing-table cache. In: INFOCOM 1988. Networks: Evolution or Revolution, Proceedings. Seventh Annual Joint Conference of the IEEE Computer and Communcations Societies, pp. 298–307. IEEE (1988)
7. Hinden, R.: New Scheme for Internet Routing and Addressing (ENCAPS) for IPNG. RFC 1955 (Informational) (June 1996)
8. Huston, G.: BGP Report, http://bgp.potaroo.net/
9. Iannone, L., Bonaventure, O.: On the Cost of Caching Locator/ID Mappings. In: Proceedings of the 3rd International Conference on emerging Networking EXperiments and Technologies (CoNEXT 2007), pp. 1–12. ACM (December 2007)
10. Jain, R.: Characteristics of destination address locality in computer networks: A comparison of caching schemes. Computer Networks and ISDN Systems 18(4), 243–254 (1990)
11. Jakab, L., Cabellos-Aparicio, A., Coras, F., Saucez, D., Bonaventure, O.: LISP-TREE: A DNS Hierarchy to Support the LISP Mapping System. IEEE Journal on Selected Areas in Communications 28(8), 1332–1343 (2010)
12. Kim, C., Caesar, M., Gerber, A., Rexford, J.: Revisiting Route Caching: The World Should Be Flat. In: Moon, S.B., Teixeira, R., Uhlig, S. (eds.) PAM 2009. LNCS, vol. 5448, pp. 3–12. Springer, Heidelberg (2009)
13. Kim, J., Iannone, L., Feldmann, A.: A Deep Dive into the LISP Cache and What ISPs Should Know about It. In: Domingo-Pascual, J., Manzoni, P., Palazzo, S., Pont, A., Scoglio, C. (eds.) NETWORKING 2011, Part I. LNCS, vol. 6640, pp. 367–378. Springer, Heidelberg (2011)
14. Li, T.: Recommendation for a Routing Architecture. RFC 6115 (Informational) (February 2011)
15. Menth, M., Hartmann, M., Hofling, M.: FIRMS: a future Internet mapping system. IEEE Journal on Selected Areas in Communications 28(8), 1326–1331 (2010)
16. Meyer, D., Zhang, L., Fall, K.: Report from the IAB Workshop on Routing and Addressing. RFC 4984 (Informational) (September 2007)
17. O'Dell, M.: GSE - An Alternate Addressing Architecture for IPv6. draft-ietf-ipngwg-gseaddr-00.txt (1997)
18. University of Oregon: RouteViews Project, http://www.routeviews.org
19. Fuller, V., Farinacci, D., Lewis, D.: LISP Delegated Database Tree (LISP-DDT) draft-ietf-lisp-ddt-00 (November 2011) (work in progress)
20. WITS: Waikato Internet Traffic Storage, http://www.wand.net.nz/wits/catalogue.php
21. Zhang, H., Chen, M., Zhu, Y.: Evaluating the performance on ID/Loc mapping. In: Global Telecommunications Conference (GLOBECOM 2008), pp. 1–5 (2008)

Path Computation in Multi-layer Multi-domain Networks[*]

Mohamed Lamine Lamali[1], Hélia Pouyllau[1], and Dominique Barth[2]

[1] Alcatel-Lucent Bell Labs France, Route de Villejust, 91620 Nozay, France
{mohamed_lamine.lamali,helia.pouyllau}@alcatel-lucent.com
[2] Lab. PRiSM, UMR8144, Université de Versailles,
45, av. des Etas-Unis, 78035 Versailles Cedex, France
dominique.barth@prism.uvsq.fr

Abstract. Carrier-grade networks have often multiple layers of protocols. To tackle this heterogeneity, the Pseudo-Wire architecture provides encapsulation and decapsulation functions of protocols over Packet-Switched Networks. At the scale of multi-domain networks, computing a path to support an end-to-end service requires various encapsulations and decapsulations that can be nested but for which manual configurations are needed. Graph models are not expressive enough for this problem. In this paper, we propose an approach using graphs and Push-Down Automata (PDA) to capture the compatibility among encapsulations and decapsulations along an inter-domain path. They are respectively modeled as pushes and pops on a PDA's stacks. We provide polynomial algorithms that compute either the shortest path in hops, or in the number of encapsulations and decapsulations to optimize interfaces' configuration

Keywords: Multi-layer networks, Pseudo-Wire, Push-Down Automata.

1 Introduction

Most carrier-grade networks are composed of multiple layers of protocols (e.g. Ethernet, IP, etc.). Such layers are administrated by different control and management plane instances. The Pseudo-Wire (PWE3) architecture [2] is a standard which aims to unify control plane functions in such heterogeneous environments allowing multi-layer services (e.g. Layer 2 VPN). It describes encapsulation (a protocol is encapsulated in another one) and decapsulation functions, also called "adaptation functions", emulating services (e.g. Frame Relay, SDH, Ethernet, etc.) over Packet-Switched Networks (PSN, e.g. IP or MPLS).

The management of these functions is achieved within each carrier's network domain: when an encapsulation function is used, its corresponding decapsulation function is applied within the domain boundaries. In large-scale carrier-grade

---

[*] This work is partially supported by the ETICS-project, funded by the European Commission. Grant agreement no.: FP7-248567 Contract Number: INFSO-ICT-248567.

R. Bestak et al. (Eds.): NETWORKING 2012, Part I, LNCS 7289, pp. 421–433, 2012.

networks or in multi-carrier networks, restricting the management of adaptation functions to network boundaries might lead to ignore feasible end-to-end paths and thus to refuse service demands. Hence, the path computation process that precludes the signaling phase must encompass the notion of *encapsulation/decapsulation compatibility*: when a protocol is encapsulated into another at one node, it must be decapsulated at another, and the possibility to nest such functions (e.g. Ethernet over MPLS over SDH). Furthermore, as such function are manually configured on routers' interfaces, minimizing their number would simplify the signaling phase.

The authors of [8,4] focused on the problem of computing a path in a multi-layer network under bandwidth constraints. In [10], we demonstrated that the problem under multiple Quality of Service constraints is NP-Complete. As a first step in our research, we focus on the problem of finding a path across multiple domains involving compatible - possibly nested - adaptation functions.

In this paper, we demonstrate that this reduced problem can be solved by a polynomial-time algorithm. We consider as an objective function, either the number of adaptation functions to simplify the signaling or the number of nodes to minimize the cost of a path. The proposed approach combines both graph and automata theory: the encapsulation and decapsulation functions are designed as pushes and pops in a Push-Down Automaton (PDA) respectively. To determine the shortest path in adaptation functions, such a PDA is transformed to bypass path segments without such functions. The PDA or transformed PDA is then converted into a Context-Free Grammar (CFG) using the method of [7]. A shortest word, either corresponding to the shortest path in nodes or in adaptation functions, is generated from this CFG.

This paper is organized as follows: sec. 2 recalls the context of multi-layer multi-domain networks and the related work on path computation; sec. 3 provides a formal definition of the problem; sec 4 explains the transformation from a network to a PDA; finally, sec. 5 details the different algorithms computing the shortest path in nodes or in encapsulations/decapsulations.

2   Path Computation in Pseudo-Wire Networks

In order to mitigate multi-layer issues, some standards define the emulation of lower layer connection-oriented protocols over a PSN (e.g. Ethernet over MPLS, [12]). For instance, the layer 2 frames are encapsulated in layer 3 packets at one network node and decapsulated at another, bursting the OSI model.

The PWE3 architecture [2] assumes an exhaustive knowledge of the network states. This assumption is also used in the multi-layer networking description of [13] and is not valid in a multi-domain context. This issue has been identified by the IETF PWE3 working group. The authors of [1] defined the multi-segment Pseudo-Wire architecture for multi-domain networks. The authors of [3] mention the problem of path determination and suggest the use of the Path Computation Element architecture (PCE) [5], which is adapted to the multi-domain context, to figure it out. It could be a control plane container for the approach described in this paper, requiring some protocol extensions to add encapsulation/decapsulation capabilities in the coding of the data model used by PCEs.

Related Work on Path Computation. In [4], the authors addressed the problem of computing the shortest path in the context of the ITU-T G.805 recommendations on adaptation functions. They stress the lack of solutions on path selection and the limitations of graph theory to handle this problem. In [8], the authors addressed the same problem in a multi-layer network qualifying it as an NP-Complete problem. The NP-Completeness comes from the problem definition as they allow loops across layers but under a limited bandwidth. They aim to select the shortest path in nodes and provide an exponential-time algorithm accordingly. The model used in [8] is close to a PDA.

In the problem we consider, we exclude bandwidth constraints as the PCE architecture handles them already and propose a solution for minimizing the number of encapsulations and decapsulations. Our algorithm does not allow loops without adaptation functions, the only loops that may exist involve encapsulations or decapsulations. Thus, minimizing the number of adaptation functions in the path also leads to minimize the number of loops - and to avoid them if a loop-free feasible path with less encapsulations exists. Our contribution is a generalization based on graph and automata theory providing further theoretical assets and a different modeling leading to a polynomial-time algorithm.

Proposed Approach. To the best of our knowledge no previous work has considered this problem at the multi-domain scale. It induces to go further domain boundaries allowing *multi-domain compatibility* to determine a *feasible inter-domain path*: when an encapsulation for a given protocol is realized in one domain its corresponding decapsulation must be done in another. Furthermore, we consider two kind of objectives: either the well-known objective of minimizing the number of nodes or the objective of minimizing the number of adaptation functions. This latter is motivated by the fact that it is equivalent to minimize the number of configuration operations, which are often done manually and can be quite complex. To express encapsulations and decapsulations, the network model is converted into a PDA as the stack allows memorizing encapsulations. Hence, our approach encompasses the two shortest path problems either in nodes or in adaptation functions:

1. Convert a multi-domain Pseudo-Wire network into a PDA,
   (a) If the goal is to minimize the number of adaptation functions, transform the PDA to bypass the "passive" functions (i.e. no protocol adaptation),
   (b) else let the PDA as is,
2. Derive a CFG from the PDA or the transformed PDA,
3. Determine the "shortest" word generated by the CFG and
4. Identify the shortest path from the shortest word.

3   Heterogeneous Multi-domain Network Model

A multi-domain network having routers with encapsulation/decapsulation capabilities can be defined as a 3-tuple: a directed graph $G = (\mathcal{V}, E)$ modeling the routers of a multi-domain network, we consider a pair of vertices (S, D) in G corresponding to the source and the destination of the path we focus on; a finite alphabet $\mathcal{A} = \{a, b, c, \dots\}$ in which each letter is a protocol; an encapsulation or a decapsulation function is a pair of different letters in the alphabet $\mathcal{A}$:

- Figure 1(a) illustrates the encapsulation of the protocol x by the node U in the protocol y;
- Figure 1(b) illustrates that the protocol x is unwrapped by the node U from the protocol y;
- Figure 1(c) illustrates that the protocol x transparently crosses the node U (no encapsulation or decapsulation function is applied). Such pairs are referred as *passive* further in this paper.

We denote by $\mathcal{ED}$ and by $\overline{\mathcal{ED}}$ the set of all possible encapsulation functions and decapsulation functions respectively, (i.e., $\mathcal{ED} \subseteq \mathcal{A}^2$). A subset $P(U)$ of $\mathcal{ED} \cup \overline{\mathcal{ED}}$ indicates the set of encapsulation, passive and decapsulation functions supported by vertex $U \in \mathcal{V}$. We define $In(U) = \{a \in \mathcal{A} \text{ s.t. } \exists b \in \mathcal{A} \text{ s.t. } (a, b) \text{ or } \overline{(b, a)} \in P(U)\}$ and $Out(U) = \{b \in \mathcal{A} \text{ s.t. } \exists a \in \mathcal{A} \text{ s.t. } (a, b) \text{ or } \overline{(b, a)} \in P(U)\}$. The set $\mathcal{A}(U) = \{(a, a) \in P(U)\}$ is the set of protocols that can passively cross node U.

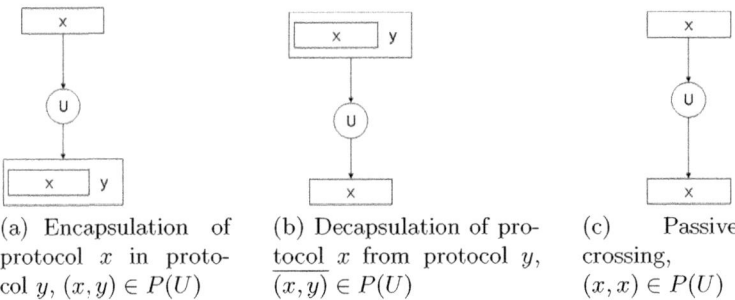

(a) Encapsulation of protocol x in protocol y, $(x, y) \in P(U)$

(b) Decapsulation of protocol x from protocol y, $\overline{(x, y)} \in P(U)$

(c)   Passive crossing, $(x, x) \in P(U)$

Fig. 1. Different transitions when a protocol crosses a node U

Considering a network $N = (G = (\mathcal{V}, E), \mathcal{A}, P)$, a source $S \in \mathcal{V}$, a destination $D \in \mathcal{V}$ and a path $C = S, x^1, U_1, x^2, \dots, U_{n-1}, x^n, D$ where each U_i is a vertex in $\mathcal{V}$ and each $x^i \in \mathcal{A} \cup \overline{\mathcal{A}}$ (where $\overline{\mathcal{A}} = \{\overline{a} : a \in \mathcal{A}\}$).

- $T_C = x^1 \dots x^n$ represents the sequence of protocols which is used over the path C. It is called the *trace* of C. For each x^i:
  - $x^i = a$ and $x^{i+1} = b$ or $\overline{b}$, means that U_i encapsulates protocol a in b ($a, b, \overline{b} \in \mathcal{A} \cup \overline{\mathcal{A}}$)
  - $x^i = \overline{a}$ and $x^{i+1} = b$ or $\overline{b}$ means that U_i unwraps protocol b from a
  - $x^i = a$ and $x^{i+1} = a$ or $\overline{a}$ means that U_i passively transports a
- We denote H_C the sequence $\beta_1, \dots, \beta_n$ obtained from C s.t. for $i = 1..n$:
  - if $x^i = a \in \mathcal{A}$ and $x^{i+1} = b \in \mathcal{A}$ or $x^{i+1} = \overline{b} \in \overline{\mathcal{A}}$ then $\beta_i = (a, b)$
  - if $x^i = \overline{b} \in \overline{\mathcal{A}}$ and $x^{i+1} = a \in \mathcal{A} \backslash \{b\}$ or $x^{i+1} = \overline{a} \in \overline{\mathcal{A}} \backslash \{\overline{b}\}$ then $\beta_i = \overline{(a, b)}$

- We also denote $M_C = \beta'_1, \ldots, \beta'_m$ obtained from H_C by deleting each passive transition β_i s.t. $\beta_i = (a, a)$ and $a \in \mathcal{A}$

Example. Consider the path $C = S, a, U, b, V, \bar{b}, W, a, D$ in the network illustrated by Fig. 2(a). The sequence corresponding to C is $H_C = (a, b), (b, b), \overline{(a, b)}$ and its trace is $T_C = ab\bar{b}a$. Finally, the well parenthesized sequence from C is $M_C = (a, b), \overline{(a, b)}$.

Let ϵ denotes the empty word, and H_C a sequence obtained from a path C as explained above. The following definitions give a formal characterization of the feasible paths.

Definition 1. *A sequence M_C from H_C is valid if and only if $M_C \in \mathcal{L}$ where $\mathcal{L}$ is the formal language recursively defined by: $\mathcal{L} = \epsilon \cup ((a, b).\mathcal{L}.\overline{(a, b)}).\mathcal{L}$ for each $(a, b) \in \mathcal{A}^2$.*

Definition 2. *A path C is a* feasible *path in N from S to D if:*

- *$U_1, \ldots, U_{n-1}$ is a path in G*
- *the sequence M_C from C is valid*
- *for each $i = 1..n$:*
  - *if $x^i = a$ and $x^{i+1} = b$ or $\bar{b}$ then $\underline{(a, b)} \in P(U_i)$*
  - *if $x^i = \bar{a}$ and $x^{i+1} = b$ or $\bar{b}$ then $\overline{(a, b)} \in P(U_i)$*
  - *if $x^i = a$ and $x^{i+1} = a$ or $\bar{a}$ then $a \in \mathcal{A}(U_i)$*

The language $\mathcal{L}$ of valid sequences is known as the *generalized Dyck language* [11]. It is well-known that this language is context free but not regular. Thus, push-down automata are naturally adapted to model this problem.

Example. The multi-domain network illustrated by Fig. 2(a) has 6 routers and two protocols labeled a and b. Adaptation function capabilities are indicated below each node. In this multi-domain network, the only feasible path between S and D is $S, a, U, b, V, \bar{b}, W, a, D$ and involves functions (a, b) on U, (b, b) on V, and $\overline{(a, b)}$ on node W.

Problem definition. As explained above, our goal is to find a feasible multi-domain path. Furthermore, we set as an objective function either the size of the sequence of adaptation functions or the size of the path in number of nodes. Hence, the problem we aim to solve can be defined as follows,

$$\min_C |H_C| \text{ or } |M_C| \text{ s.t. } C \text{ is a feasible path}$$

4   From the Network Model to a PDA

In this section, we address the conversion from a network to a PDA. The **algorithm 1** takes as input a network $N = (G = (\mathcal{V}, E), \mathcal{A}, P)$ and converts it to a PDA $A_N = (Q, \Sigma, \Gamma, \delta, S, \emptyset, \{D\})$, where Q is the set of states of the PDA, Σ the input alphabet, Γ the stack alphabet, δ the transition relation, S the start state, Z_0 the initial stack symbol and D the accepting state, ϵ is the empty string. The automaton A_N from N is obtained as follows:

- Create a state U_x of the automaton for each $U \in V$ and each $x \in In(U)$, except for S for which create only one state,
- An encapsulation of a protocol x in a protocol y by a node U_x is modeled as a push of the character x in the stack between the state U_x and its successor V_y. It is denoted $(U_x, < x, \alpha, x\alpha >, V_y)^1$,
- A decapsulation of y from x by the node U_x is modeled as a pop of the protocol y from the stack. it is denoted $(U_x, < \overline{x}, y, \emptyset >, V_y)$,
- The top of the stack is the last encapsulated protocol,
- If the current state is U_x then the current protocol is x.

Example. Figure 2(b) is an example of output of the algorithm 1. The algorithm transforms the network illustrated by Fig. 2(a) into a PDA of Fig 2(b). For instance, the link (U, V) is transformed into the transitions $(U_a, < a, Z_0, aZ_0 >, V_b)$ and $(U_a, < a, b, ab >, V_b)$ w.r.t. adaptation capabilities of U and V.

Algorithm 1. Convert a network to a PDA

Input: a network $N = (G = (V, E), \mathcal{A}, P)$, a source S and a destination D
Output: push-down automaton $A_N = (Q, \Sigma, \Gamma, \delta, S, \emptyset, \{D\})$
(1) $\Sigma \leftarrow \mathcal{A} \cup \overline{\mathcal{A}}$; $\Gamma \leftarrow \mathcal{A} \cup \{Z_0\}$
(2) Create a single state for the node S
(3) For each node $U \neq S$ and each protocol $x \in IN(U)$, create a state U_x
(4) For each state U_x s.t. $(S, U) \in E$ and $x \in Out(S)$
        Create a transition $(S, < \epsilon, Z_0, Z_0 >, U_x)$
(5) For each link $(U, V) \in E$ s.t. $U \neq S$ and for each $(x, y) \in P(U)$ and each $\alpha \in \Gamma \backslash \{x\}$
    (5.1) If $x \in \mathcal{A}(U) \cap In(V)$ Create a transition $(U_x, < x, \alpha, \alpha >, V_x)$ {passive trans.}

    (5.2) If $x \neq y$ and $y \in In(V)$ Create a transition $(U_x, < x, \alpha, x\alpha >, V_x)$ {encap.}
(6) For each link $(U, V) \in E$ s.t. $U \neq S$ and for each $\overline{(x, y)} \in P(U)$
    (6.1) If $x \in In(V)$ Create a transition $(U_y, < \overline{y}, x, \emptyset >, V_x)$ {decap.}
(7) Create a fictitious final state D.
(8) For each $x \in In(D)$ and each $\alpha \in \Gamma \backslash \{x\}$ Create a transition $(D_x, < x, Z_0, \emptyset >, D)$

Complexity. Each node U from the graph generates $|In(U)|$ states, except the source node S. A fictitious final state is added. Thus, the number of states is at worst $2 + (|V| - 1) \times |\mathcal{A}|$ so in $O(|V| \times |\mathcal{A}|)$. The worst case complexity of **algorithm 1** is in $O(\max((|V| \times |\mathcal{A}|), (|E| \times ((|\mathcal{A}| \times |\mathcal{ED}|) + |\overline{\mathcal{ED}}|)))))$. We assume that the network is connected, so $|E| \geq |V| - 1$. Since $\mathcal{ED}$ is a subset of $\mathcal{A}^2$, then $|\mathcal{ED}| < |\mathcal{A}|^2$ and $|\overline{\mathcal{ED}}| < |\mathcal{A}|^2$. Thus, the upper bound complexity is in $O(|E| \times |\mathcal{A}|^3)$, which is also an upper bound for the number of transitions.

Proposition 1. *Considering a network $N = (G = (V, E), \mathcal{A}, P)$, a source $S \in V$ and a destination $D \in V$, the language recognized by A_N is the set of traces of the feasible paths from S to D in N.*

---

[1] Note that, even if $x = \overline{a} \in \overline{\mathcal{A}}$, the transition has the form $(U_a, < \overline{a}, \alpha, a\alpha >, V_y)$. Characters in $\overline{\mathcal{A}}$ are only used as input characters. Characters indexing states and pushed characters in the stack are their equivalent in $\mathcal{A}$.

Sketch of Proof. The main idea is to show that a path C is feasible if and only if its trace T_C is accepted by the PDA. Thus, for each feasible path there is a sequence of transitions in the PDA that accepts its trace. And for each accepted word, the sequence of transitions, which accepts this word, corresponds to a feasible path. The complete proof is available in the Appendix online [9].

5   The Shortest Feasible Path

In sec. 4, we provided the method to build a PDA allowing to determine the feasible paths. The next step is to minimize either the number of nodes or the number of adaptation functions. The method to minimize the number of nodes uses directly the PDA as described in sec. 5.1. But, to minimize the number of adaptation functions, as detailed in sec. 5.2, the PDA is transformed in order to bypass the sub-paths without any adaptation function. Then, a CFG derived from the PDA (or the transformed PDA) generates words whose length is equivalent to the number of nodes (or to the number of adaptations). An algorithm browses the CFG to determine the shortest word. Finally, another algorithm identifies the multi-domain path corresponding to this shortest word.

5.1   Minimizing the Number of Nodes

The number of characters in a word accepted by the automaton A_N is the number of links in the corresponding feasible path (each character is a protocol used on a link). Thus the step of automaton transformation (sec. 5.2) should be skipped. The automaton A_N is directly transformed into a CFG, then the shortest word is generated as described in sec. 5.3. The corresponding feasible path is computed by the algorithm 4 described in sec. 5.3.

5.2   Minimizing the Number of Adaptation Functions

To enumerate only encapsulations and decapsulations in the length of each word (and thus minimize adaptation functions by finding the shortest word accepted), a transformed automaton A'_N in which all sequences involving passive transitions are bypassed must be determined. The length of the shortest word accepted by A'_N is the number of adaptation functions plus a fixed constant.

Let define $\mathcal{Q}_a$ ($a \in \mathcal{A}$) as $\mathcal{Q}_a = \{V_x \in \mathcal{Q} | x = a\}$, and let A_N^a be the *sub-automaton* induced by $\mathcal{Q}_a$. By analogy with an induced subgraph, an induced sub-automaton is a multigraph with labeled edges such that the set of vertex is $\mathcal{Q}_a$ and the set of edges is the set of transitions between elements of $\mathcal{Q}_a$. Since there are only passive transitions between two states in $\mathcal{Q}_a$, all paths in the sub-automaton are passive. Let define $P(U_x, V_x)$ as the shortest path length between U_x and V_x. This length can be computed by any well-known Shortest Path Algorithm. Let $Succ(V_x)$ be the set of successors of V_x in the original automaton A_N, i.e., $Succ(V_x) = \{W_y \in \mathcal{Q} | \exists (V_x, < x, \alpha, \beta >, W_y) \in \delta\}$.

The algorithm 2 takes as input A_N and computes the transformed automaton $A'_N = (Q', \Sigma', \Gamma', \delta', S', \emptyset, \{D'\})$. A'_N is initialized with the values of A_N. Then, the algorithm computes the sub-automaton for each character $x \in \mathcal{A}$ (step (1)) and the length values $P(U_x, V_x)$ for each pair of states in the sub-automaton (step (2.1)). Each path between a pair of states is a sequence of passive transitions. If such a path exists (step (2.2)), the algorithm adds transitions to δ' from U_x to each state in $Succ(V_x)$ (steps (2.2.2) and (2.2.3)). These added transitions are the same that those which connect V_x to its successors W_y, but with an input character indexed by the number of passive transitions between U_x and V_x, (i.e., $P(U_x, V_x)$) plus one (indicating that there is a transition sequence which matches with a sequence of protocols $xx \ldots x$ of length $P(U_x, V_x) + 1$). The indexed character is added to the input alphabet Σ' (step (2.2.1)).

Example. The algorithm 2 transforms the PDA in Fig. 2(b) into the PDA in Fig. 2(c). the transition $(V_b, < \overline{b_2}, a, \emptyset >, D_a)$ is added to bypass the sequence of transitions $(V_b, < b, a, a >, W_b)$ $(W_b, < \overline{b}, a, \emptyset >, D_a)$.

Algorithm 2. Transform automaton A_N

Input: push-down automaton $A_N = (Q, \Sigma, \Gamma, \delta, S, \emptyset, \{D\})$
Output: transformed push-down automaton $A'_N = (Q', \Sigma', \Gamma', \delta', S', \emptyset, \{D'\})$
$Q' \leftarrow Q, \Sigma' \leftarrow \Sigma, \Gamma' \leftarrow \Gamma, \delta' \leftarrow \delta, S' \leftarrow S, D' \leftarrow D$
For each $x \in \mathcal{A}$
  (1) Compute A_N^x
  (2) For each $U_x \in Q_x$ and each $V_x \in Q_x$ s.t. $U_x \neq V_x$
    (2.1) Compute $P(U_x, V_x)$
    (2.2) If $P(U_x, V_x) < \infty$ {there is a path between U_x and V_x}
      (2.2.1) Add $x_{P(U_x,V_x)+1}$ and $\overline{x}_{P(U_x,V_x)+1}$ to Σ
      (2.2.2) For each $W_y \in Succ(V_x) \backslash \{U_x\}$ and each $(V_x, < x, \alpha, \beta >, W_y) \in \delta$
        Add the transition $(U_x, < x_{P(U_x,V_x)+1}, \alpha, \beta >, W_y)$ to δ'
      (2.2.3) For each $W_y \in Succ(V_x) \backslash \{U_x\}$ and each $(V_x, < \overline{x}, \alpha, \beta >, W_y) \in \delta$
        Add the transition $(U_x, < \overline{x}_{P(U_x,V_x)+1}, \alpha, \beta >, W_y)$ to δ'

Complexity. Steps (2.2.2) and (2.2.3) are bounded by $O(|Q| \times |\delta|)$. Step (2.1) (computing the shortest path) is bounded by $O(|Q_x|^2)$. Since the automaton is connected, $|Q_x| \leq |Q| \leq |\delta| + 1$. Thus, the complexity of each iteration in the loop *For* (step (2)) is bounded by $O(|Q| \times |\delta|)$. The number of iterations of step (2) is in $O(|Q_x|^2)$. However, a state cannot belong to two different sub-automata. The complexity of the algorithm 2 is in $O(\sum_{x \in \mathcal{A}} |Q_x|^2 \times |Q| \times |\delta|)$, which is maximized when $\exists x \in \mathcal{A}$ s.t. $|Q_x| = |Q|$ and $|Q_{x'}| = 0$ for each $x' \neq x$. Thus, the complexity of algorithm 2 is in $O(|Q|^3 \times |\delta|)$. In the network model, this is equivalent to $O(|\mathcal{A}|^6 \times |\mathcal{V}|^3 \times |E|)$.

Let $L(A_N)$ be the set of words accepted by A_N, and let $L(A'_N)$ be the set of words accepted by A'_N. Let $f : \Sigma' \to \Sigma^*$ be a function s.t.:

- if $x_i = a_i \in \mathcal{A}'$ then $f(x_i) = \underbrace{aa \ldots aa}_{i \text{ occurrences}}$

– if $x_i = \overline{a_i} \in \overline{\mathcal{A}'}$ then $f(x_i) = \underbrace{aa \ldots a\overline{a}}_{i \text{ occurrences}}$

The domain of f is extended to $(\Sigma')^*$:

$$f : (\Sigma')^* \to \Sigma^* \text{ s.t. } w' = x_i^1 x_j^2 \ldots x_k^n \to f(w') = f(x_i^1)f(x_j^2) \ldots f(x_k^n)$$

For simplicity, we consider that x and x_1 are the same character. $f(L(A_N'))$ denotes the set of words accepted by A_N' transformed by f (i.e. $f(L(A_N')) = \{f(w') \text{ s.t. } w' \in L(A_N')\}$).

It is clear that f is not a bijection ($f(x_i x_j) = f(x_{i+j})$). So to operate the transformation between $L(A_N)$ and $L(A_N')$, we define $g : \Sigma^* \to (\Sigma')^*$ s.t. : for each $w = \underbrace{xx \ldots x}_{i \text{ occurrences}} \underbrace{yy \ldots y}_{j \text{ occurrences}} \ldots \underbrace{zz \ldots z}_{k \text{ occurrences}} \in \Sigma^*$, $g(w) = x_i y_j \ldots z_k$.

In other words, $w' = g(w)$ is the shortest word in $(\Sigma')^*$ s.t. $f(w') = w$.

The following proposition shows the relation between the number of encapsulations in A_N and the size of a word accepted by A_N'.

Proposition 2. *The word accepted by A_N which minimizes the number of pushes is $f(w')$, where w' is the shortest word (i.e., with minimal number of characters) accepted by A_N'.*

Sketch of Proof. There is a linear relation between the length of a word accepted by $L(A_N')$ and the number of pushes and pops involved to accept it, and the same number of pushes and pops is involved to accept w' (in A_N') and $f(w')$ (in A_N). The complete proof is available in [9].

5.3 The Shortest Path as a Shortest Word

In order to find the shortest word accepted by A_N (resp. A_N'), the CFG G_N such that $L(G_N) = L(A_N)$ (resp. $L(A_N')$) is computed. Backtracking from terminals to the start symbol, the shortest sequence of derivations is then computed. **From the PDA to the CFG.** The transformation of a PDA into a CFG is well-known. We adapted a general method described in [7] to transform A_N(resp. A_N') into a CFG $G_N = (\mathcal{N}, \Sigma, S_G, \mathcal{P})$ (resp.$(\mathcal{N}, \Sigma', S_G, \mathcal{P})$) where $\mathcal{N}$ is the set of nonterminals (variables), Σ (resp. Σ') is the input alphabet, S_G is the initial symbol (initial nonterminal) and $\mathcal{P}$ is the set of production rules. Nonterminals are in the form $[UXV]$ where $U, V \in \mathcal{Q}$ (resp. $\mathcal{Q}'$) and $X \in \Gamma$ (resp. Γ').The worst case complexity of this algorithm is in $O(|\delta| \times |\mathcal{Q}|^2)$ (resp. $O(|\delta'| \times |\mathcal{Q}'|^2)$). W.r.t. the definition of the network, the upper bound is in $O(|\mathcal{A}|^5 \times |\mathcal{V}|^2 \times |E|)$ and the number of production rules in the wost case is $1 + |\mathcal{Q}| + (|\delta| \times |\mathcal{Q}|^2)$ (resp. $O(|\mathcal{A}|^8 \times |\mathcal{V}|^5 \times |E|^2)$ and $1 + |\mathcal{Q}'| + (|\delta'| \times |\mathcal{Q}'|^2)$). The number of nonterminals is bounded by $O(|\mathcal{Q}|^2 \times |\mathcal{A}|) = O(|\mathcal{A}|^3 \times |V|^2)$ (resp. $O(|\mathcal{Q}'|^2 \times |\mathcal{A}|) = O(|\mathcal{A}|^3 \times |V|^2)$). The interested reader can refer to [9] for the detailed algorithm.

Example. This method transforms the PDA in Fig. 2(c) into a CFG. Figure 2(d) is a subset of production rules of the obtained CFG. This subset allows generating the shortest trace of a feasible path in the network in Fig. 2(a).

The Shortest Word Generated by a CFG. To find the shortest word generated by G_N, a function ℓ associates to each nonterminal the length of the shortest word that it generates.

More formally, $\ell : \{\mathcal{N} \cup \Sigma \cup \{\epsilon\}\}^*$ or $\{\mathcal{N} \cup \Sigma' \cup \{\epsilon\}\}^* \rightarrow \mathbb{N} \cup \infty$ s.t.:
- if $w = \epsilon$ then $\ell(w) = 0$,
- if $w \in \Sigma$ or Σ' then $\ell(w) = 1$,
- if $w = \alpha_1 \ldots \alpha_n$ (with $\alpha_i \in \{\mathcal{N} \cup \Sigma \cup \{\epsilon\}\}$ or $\{\mathcal{N} \cup \Sigma' \cup \{\epsilon\}\}$) then $\ell(w) = \sum_{i=1}^{n} \ell(\alpha_i)$.

The algorithm 3 computes the value of $\ell([x])$ for each $[x] \in \mathcal{N}$.

Algorithm 3. Compute the values $\ell([x])$ for each nonterminal $[x] \in \mathcal{N}$

Input: $G_N = (\mathcal{N}, \Sigma, S_G, \mathcal{P})$ or $(\mathcal{N}, \Sigma', S_G, \mathcal{P})$
Output: $\ell([x])$ for each nonterminal $[x]$
(1) Initialize each $\ell([x])$ to ∞
(2) While there is at least one $\ell([x])$ updated at the previous iteration do
      (2.1) For each production rule $[x] \rightarrow \alpha_1 \ldots \alpha_n$ in $\mathcal{P}$
           (2.1.1) $\ell([x]) \leftarrow \min\{\ell([x]), \sum_{i=1}^{n} \ell(\alpha_i)\}$

Proposition 3. *The algorithm 3 terminates at worst after $|\mathcal{N}|$ iterations, and each $\ell([x])$ ($[x] \in \mathcal{N}$) obtained is the length of the shortest word produced by $[x]$.*

Sketch of Proof. During each iteration, at least one $\ell([x])$ is updated to its correct value, until all values are correct. The complete proof is available in [9].

Complexity. The complexity of the algorithm 3 is in $O(|\mathcal{N}| \times |\mathcal{P}|)$ which is $O(|\mathcal{A}|^8 \times |V|^4 \times |E|)$ (resp. $O(|\mathcal{A}|^{11} \times |V|^7 \times |E|^2)$) in the network model.

There are several algorithms which allow generating a random word of some length from a CFG. The *boustrophedonic* and the *sequential* algorithms described in [6] generate a random labeled combinatorial object of some length from any decomposable structure (including CFGs). The boustrophedonic algorithm is in $O(n \log n)$ (where n is the length of the object) and the sequential algorithm is in $O(n^2)$ but may have a better average complexity. Both algorithms use a precomputed table of linear size. This table can be computed in $O(n^2)$. These algorithms require an unambiguous CFG, but this requirement is only for the randomness of the generation. Recall that our goal is to generate the trace of the shortest feasible path. Thus, we do not take into consideration the randomness and the distribution over the set of shortest traces.

In order to generate the shortest word in $L(G_N)$, the boustrophedonic algorithm uses G_N and $\ell(S_G)$ as input ($\ell(S_G)$ is the length of the shortest word generated by G_N). Thus, the generation of the shortest word w (resp. w') is in $O(|w|^2)$ (resp. $O(|w'|^2)$) including the precomputation of the table.

Example. The algorithm 3 gives $\ell(S_G) = 3$. The boustrophedonic algorithm computes the shortest word with the production rules in Fig. 2(d). The derivation is:
$$S_G \overset{(1)}{\vdash} [SZ_0D] \overset{(2)}{\vdash} [U_a Z_0 D] \overset{(3)}{\vdash} a[V_b Z_0 D_a][D_a Z_0 D] \overset{(4)}{\vdash} a\bar{b}_2[D_a Z_0 D] \overset{(5)}{\vdash} a\bar{b}_2 a$$

Thus, the shortest word accepted by the transformed PDA is $a\bar{b}_2a$. And the shortest trace of a feasible path is $f(a\bar{b}_2a) = ab\bar{b}a$.

From the Shortest Word to the Path. If the goal is to minimize the number of nodes in the path, the algorithm 4 takes as input the shortest word w accepted by A_N. Otherwise, as w' is the shortest word accepted by A'_N and generated by G_N, by prop. 2, $f(w')$ is the word which minimizes the number of pops and pushes in A_N. In such a case it is the trace T_C of the shortest feasible path C in the network N. It is possible that several paths match with the trace $T_C = w$ (resp. $f(w')$). In such a case, a load-balancing policy can choose a path.

The algorithm 4 is a dynamic programming algorithm that computes C. It starts at the node S and takes at each step all the links in E which match with the current character in T_C. Let $T_C = x^1x^2 \ldots x^n$ ($x^i \in \mathcal{A} \cup \overline{\mathcal{A}}$). At each step i, the algorithm starts from each node U in $Nodes[i]$ and adds to $Links[i]$ all links (U, V) which match with x^i. It also adds each V in $Nodes[i+1]$. When reaching D, it backtracks to S and selects the links from D to S.

Example. From the shortest trace $ab\bar{b}a$, the algorithm 4 computes the only feasible path in the network on Fig. 2(a), which is $S, a, U, b, V, \bar{b}, W, a, D$.

Algorithm 4. Find the shortest path

Input: Network N and T_C
Output: Shortest path C
(1) $Nodes[1] \leftarrow S$; $i \leftarrow 1$
(2) While D is not reached do
        (2.1) for each $U \in Nodes[i]$ and each $V \in \mathcal{V}$ s.t. $(U, V) \subset E$ do
                (2.1.1) If $x^i \in \mathcal{A}$, $x^i \in Out(U)$, $x^i \in In(V)$ and $(x^{i-1}, x^i) \in P(U)$
                        (2.1.1.1) Add (U, V) to $Links[i]$ and V to $Nodes[i+1]$
                (2.1.2) If $x^i \in \overline{\mathcal{A}}$, $x^i \in Out(U)$, $x^i \in In(V)$ and $\overline{(x^i, x^{i-1})} \in P(U)$
                        (2.1.2.1) Add (U, V) to $Links[i]$ and V to $Nodes[i+1]$
        (2.2) $i{+}{+}$
(3) Backtrack from D to S by adding each covered link in the backtracking to C.

Complexity. The *while* loop stops exactly after T_C steps, because it is sure that there is a feasible path of length $|T_C|$ if T_C is accepted by the automaton

Table 1. Algorithms and their complexities

Step	Algorithm	Upper-Bound Complexity	
		Minimizing hops	Minimizing enc./dec.
Network to PDA	Algorithm 1	$O(|E| \times |\mathcal{A}|^3)$	$O(|E| \times |\mathcal{A}|^3)$
PDA to trans. PDA	Algorithm 2	/	$O(|\mathcal{A}|^6 \times |\mathcal{V}|^3 \times |E|)$
Trans. PDA to CFG	Algorithm 5[9]	$O(|\mathcal{A}|^5 \times |\mathcal{V}|^2 \times |E|)$	$O(|\mathcal{A}|^8 \times |\mathcal{V}|^5 \times |E|)$
Shortest word length	Algorithm 3	$O(|\mathcal{A}|^8 \times |\mathcal{V}|^4 \times |E|)$	$O(|\mathcal{A}|^{11} \times |\mathcal{V}|^7 \times |E|^2)$
Shortest word	Boustrophedonic[6]	$O(|w|^2)$	$O(|w'|^2)$
Trace to shortest path	Algorithm 4	$O(|T_C| \times |\mathcal{V}| \times |E|)$	$O(|T_C| \times |\mathcal{V}| \times |E|)$

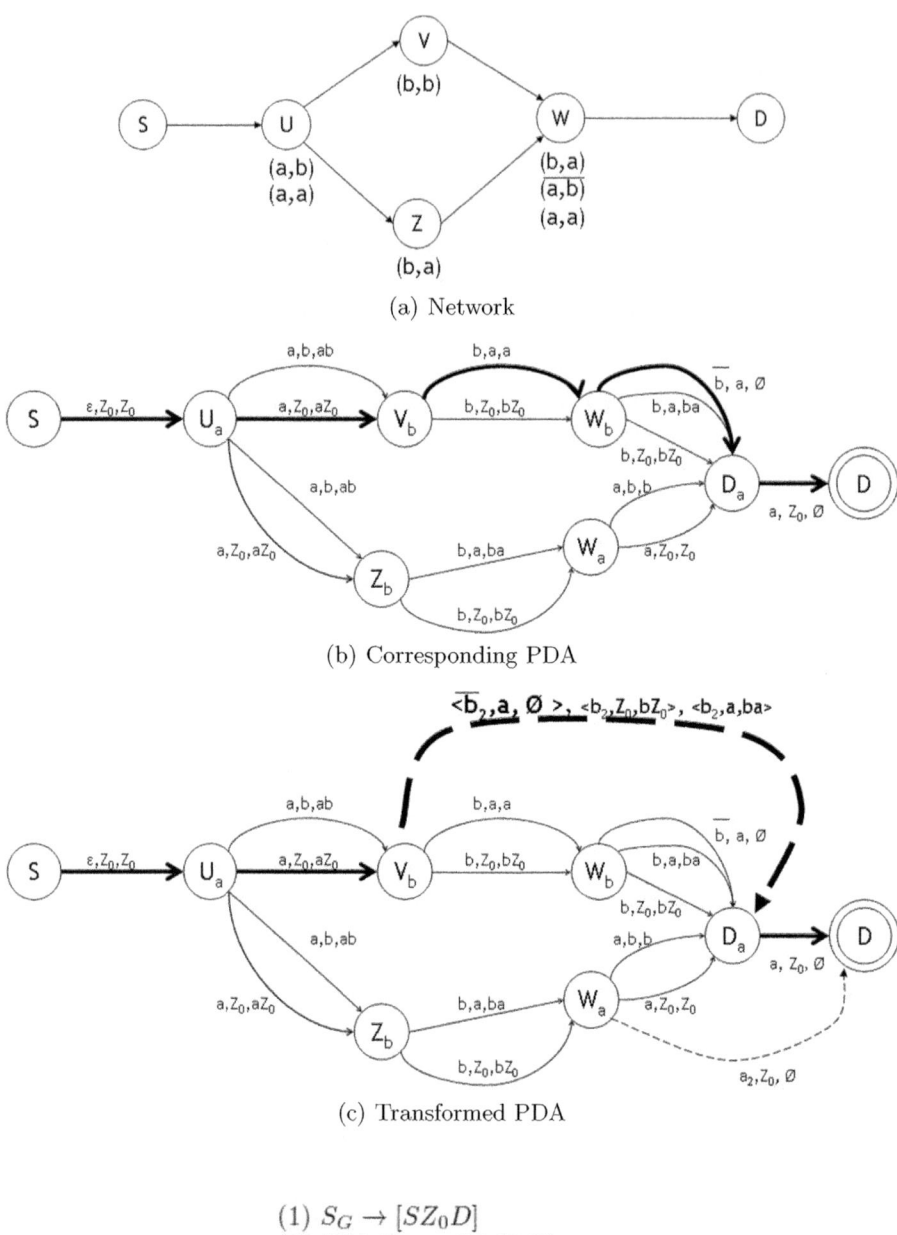

(a) Network

(b) Corresponding PDA

(c) Transformed PDA

(1) $S_G \rightarrow [SZ_0D]$
(2) $[SZ_0D] \rightarrow \epsilon[U_aZ_0D]$
(3) $[U_aZ_0D] \rightarrow a[V_bZ_0D_a][D_aZ_0D]$
(4) $[V_bZ_0D_a] \rightarrow \overline{b_2}$
(5) $[D_aZ_0D] \rightarrow a$

(d) Subset of G_N which generates T_C

Fig. 2. Example

A_N. At each step, all links and nodes are checked in the worst case. Thus, the algorithm 4 is in $O(|T_C| \times |\mathcal{V}| \times |E|)$ in the worst case.

6   Conclusion

In this paper, we presented a new model for heterogeneous networks involving automata theory to provide the shortest path in number in nodes or in encapsulations/decapsulations. The proposed solution can be applied over the Pseudo-Wire architecture [2], but also over other network architectures which involve adaptation functions and protocol changes. The different algorithms of our methodology have polynomial upper-bounds as summarized by Table 1. These worst-case upper-bounds can be quite high but correspond to unrealistic situations. In future work, we aim to extend our model to support QoS constraints and to define a distributed solution.

References

1. Bocci, M., Bryant, S.: RFC5659 - An Architecture for Multi-Segment Pseudowire Emulation Edge-to-Edge (2009)
2. Bryant, S., Pate, P.: RFC3985 - Pseudo Wire Emulation Edge-to-Edge (PWE3) Architecture (2005)
3. Cho, H., Ryoo, J., King, D.: Stitching dynamically and statically provisioned segments to construct end-to-end multi-segment pseudowires (2011), http://www.ietf.org/id/draft-cho-pwe3-mpls-tp-mixed-ms-pw-setup-01.txt
4. Dijkstra, F., Andree, B., Koymans, K., van der Ham, J., Grosso, P., de Laat, C.: A multi-layer network model based on ITU-T G.805. Comput. Netw. (2008)
5. Farrel, A., Vasseur, J.P., Ash, J.: RFC4655 - A Path Computation Element (PCE)-Based Architecture (2006)
6. Flajolet, P., Zimmermann, P., Van Cutsem, B.: A calculus for the random generation of labelled combinatorial structures. Theoretical Computer Science (1994)
7. Hopcroft, J.E., Motwani, R., Ullman, J.D.: Introduction to Automata Theory, Languages, and Computation. Addison-Wesley Longman Publishing (2006)
8. Kuipers, F.A., Dijkstra, F.: Path selection in multi-layer networks. Computer Communications (2009)
9. Lamali, M.L., Pouyllau, H., Barth, D.: Appendices to Path computation in multi-Layer multi-domain Networks (2011), https://www.ict-etics.eu/fileadmin/documents/publications/scientific_papers/report.pdf
10. Lamali, M.L., Pouyllau, H., Barth, D.: End-to-End Quality of Service in Pseudo-Wire Networks. In: ACM CoNEXT Student Workshop (2011)
11. Liebehenschel, J.: Lexicographical Generation of a Generalized Dyck Language. SIAM J. Comput. (2003)
12. Martini, L., Rosen, E., El-Aawar, N., Heron, G.: RFC4448 - Encapsulation Methods for Transport of Ethernet over MPLS Networks (2008)
13. Shiomoto, K., Papadimitriou, D., Le Roux, J., Vigoureux, M., Brungard, D.: RFC5212 - Requirements for GMPLS-based multi-region and multi-layer networks, MRN/MLN (2008)

Author Index

Abboud, Osama   II-1
Akshay Uttama Nambi, S.N.   II-291
Ali, Arshad   II-277
Almeida, Jussara M.   II-84
Altman, Eitan   II-211, II-277, II-343, II-392
Amigo, Isabel   II-252
An, Xueli   I-175
Apolónia, Nuno   I-82
Applegate, D.   II-44
Archer, A.   II-44
Arya, Vijay   I-289

Barla, Isil Burcu   I-161
Barlet-Ros, Pere   II-111
Barth, Dominique   I-421
Belzarena, Pablo   II-252
Ben-Porat, Udi   I-135
Bernardi, Giacomo   I-344
Bhattacharya, Parantapa   I-190
Blenn, Norbert   I-56, I-97
Bocek, Thomas   II-15
Bonaventure, Olivier   I-397
Boreli, Roksana   II-198
Bremler-Barr, Anat   I-135

Cabellos-Aparicio, Albert   I-409
Calinescu, Gruia   II-366
Carle, Georg   I-161
Casetti, Claudio E.   I-68
Chahed, Tijani   II-277
Chai, Wei Koong   I-27
Chang, Rocky K.C.   II-124
Chang, Yi-Yin   II-304
Charalampidou, Kassandra   I-97
Chen, Ling-Jyh   II-304
Cherukuru, Himalatha   II-136
Chiluka, Nitin   II-238
Choi, Yanghee   II-97
Chu, Xiaowen   II-316
Chung, Taejoong   II-97
Claffy, Kc   II-136
Clegg, Richard G.   II-223
Coras, Florin   I-409

Cunche, Mathieu   II-198
Cunha, Ítalo   II-84

Daenen, Koen   I-175
DaSilva, Luiz A.   II-331
Dhamdhere, Amogh   II-111, II-136
Djatmiko, Mentari   II-198
Doerr, Christian   I-56, I-97
Domingo-Pascual, Jordi   I-409
Donnet, Benoit   I-385
Dovrolis, Constantine   II-111, II-136
Dutta, Partha   I-190

El-Ansary, Sameh   II-29
Elayoubi, Salah Eddine   II-343
El-Azouzi, Rachid   II-343
Engel, Thomas   I-370, II-161
Epema, Dick   II-238
Evans, Nathan   I-304

Fang, Binxing   II-71
Fenacci, Damon   I-344
Ferreira, Paulo   I-82
Fiems, Dieter   II-277, II-392
Filsfils, Clarence   I-397
Forlivesi, Claudio   I-175
Fotiou, Nikos   I-13
François, Jérôme   I-370

Gaillard, Julien   II-392
Gaito, Sabrina   II-265
Garcia-Jimenez, Santiago   I-358
Genge, Béla   I-121
Giroire, Frédéric   I-247
Gkorou, Dimitra   II-238
Gonçalves, Glauber D.   II-84
Gopalakrishnan, V.   II-44
Gorinsky, Sergey   II-149
Griffin, David   II-223
Grothoff, Christian   I-304
Guan, Hongtao   I-205
Guimarães, Anna   II-84
Guo, Li   II-71
Guo, Shuo   I-41
Gupta, Akanksha   I-190

Haddad, Majed   II-343, II-392
Han, Jinyoung   II-97
Haridi, Seif   II-29
Hasan, Syed   II-149
Hausheer, David   II-1
Hayel, Yezekael   II-211
He, Cheng   I-318
He, Diliang   I-27
He, Ting   II-71
He, Xiuqiang   I-318
Hecht, Fabio V.   II-15
Hommes, Stefan   II-161
Hwang, K.-W.   II-44

Iannone, Luigi   I-397
Irwin, Ryan E.   II-331
Iyer, Aakash   I-190
Izal, Mikel   I-358

Janssens, Nico   I-175
Jerschow, Yves Igor   II-184
Jiang, Jin   I-68
Jiménez, Tania   II-211

Katsaros, Konstantinos V.   I-13
Kim, Hyun-chul   II-97
Kim, Juhoon   I-397
Kim, Seungbae   II-97
Kuipers, Fernando   II-404
Kvalbein, Amund   I-217
Kwon, Ted "Taekyoung"   II-97

Lamali, Mohamed Lamine   I-421
Landa, Raul   II-223
Lee, HyunYong   I-1
Lee, Patrick P.C.   I-318
Lee, S.   II-44
Leung, Yiu-Wing   II-316
Levy, Hanoch   I-135
Li, Cong   I-149
Lin, Cheng-Yu   II-304
Lin, Qunyang   I-331
Lin, Zhiyong   II-316
Liu, Alex X.   II-71
Liu, Hai   II-316
Liu, Hongying   I-232
Liu, Jiangchuan   II-59
Liu, Junjie   I-205
Liu, Yujing   II-124
Lui, John C.S.   I-318, II-172

Luo, Layong   I-205
Luo, Xiapu   II-124
Lysne, Olav   I-217

MacKenzie, Allen B.   II-331
Magaña, Eduardo   I-358
Maggiorini, Dario   II-265
Mahmud, Sabbir   II-378
Maillé, Patrick   II-355
Mann, Vijay   I-190
Marchal, Samuel   I-370
Marina, Mahesh K.   I-344
Mauve, Martin   II-184
Mikians, Jakub   II-111
Misra, V.   II-44
Morató, Daniel   I-358
Moulierac, Joanna   I-247
Mykoniati, Eleni   II-223

Nakao, Akihiro   I-1
Niemegeers, Ignas   II-291
Ning, Qingsong   I-205

Pan, Lujia   I-318
Panda, Manoj   II-277
Pavlou, George   I-27
Pezaros, Dimitrios P.   I-344
Phan, Truong Khoa   I-247
Plattner, Bernhard   I-135
Poddar, Rishabh   I-190
Polot, Bartlomiej   I-304
Polyzos, George C.   I-13
Pouwelse, Johan   II-238
Pouyllau, Hélia   I-421
Prabhakar, T.V.   II-291
Prakruthi, K.   II-291
Psaras, Ioannis   I-27

Quadri, Christian   II-265

Ramakrishnan, K.K.   II-44
Rio, Miguel   II-223
Rossi, Gian Paolo   II-265
Roudaut, Frédéric   I-247
Roverso, Roberto   II-29
Rückert, Julius   II-1

Salles, Ronaldo M.   I-109
Santos, Flávio Roberto   II-15
Sarac, Kamil   I-274

Sassatelli, Lucile   II-277
Saucez, Damien   I-385, I-397
Schupke, Dominic A.   I-161
Seneviratne, Aruna   II-198
Shen, Meng   I-232
Shen, Zhiyong   I-331
Shi, Guangyu   I-41
Shi, Xingang   I-259
Shilpa, S.   II-291
Siaterlis, Christos   I-121
Solé-Pareta, Josep   II-111
State, Radu   II-161
Steinmetz, Ralf   II-1
Stiller, Burkhard   II-15
Stojmenovic, Ivan   II-316
Su, Jinshu   II-124
Swayne, D.F.   II-44

Tang, Linpeng   I-331
Tozal, M. Engin   I-274
Tsilopoulos, Christos   I-13
Tuffin, Bruno   II-355

van de Bovenkamp, Ruud   II-404
Van Kester, Bas   I-56
Van Mieghem, Piet   I-56, I-149, II-404
Vasconcelos, Marcelo F.   I-109
Vasilakos, Xenofon   I-13
Vaton, Sandrine   II-252
Veiga, Luís   I-82
Veitch, Darryl   I-289
Venkatesha Prasad, R.   II-291
Ververidis, Christopher N.   I-13

Vieira, Alex Borges   II-84
Vigne, Jean-Marc   II-355
Vinkó, Tamás   II-238
Vishnoi, Anilkumar   I-190
Vo, Hung Quoc   I-217

Wagner, Cynthia   I-370
Wang, Feng   II-59
Wang, Haiyang   II-59
Wang, Huijuan   I-149
Wang, Ning   I-232
Wang, Zhiliang   I-259
Wu, Hui   II-378
Wu, Jianping   I-259
Wu, Xiaolong   I-205

Xiang, Yang   I-259
Xie, Gaogang   I-205
Xie, Haiyong   I-41
Xie, Junqing   I-331
Xie, Yingke   I-205
Xu, Ke   I-232, II-59
Xu, Yuedong   II-343
Xylomenos, George   I-13

Yang, Zichao   II-172
Yin, Xia   I-259

Zhang, Fuxing   I-205
Zhang, Zhibin   II-71
Zhao, Yong   II-71
Zhong, Yifeng   I-232
Zinner, Thomas   II-1